Neurogenetics

METHODS IN MOLECULAR BIOLOGY™

John M. Walker, SERIES EDITOR

224. **Functional Genomics:** *Methods and Protocols,* edited by *Michael J. Brownstein and Arkady Khodursky, 2003*
223. **Tumor Suppressor Genes:** *Volume 2: Regulation, Function, and Medicinal Applications,* edited by *Wafik S. El-Deiry, 2003*
222. **Tumor Suppressor Genes:** *Volume 1: Pathways and Isolation Strategies,* edited by *Wafik S. El-Deiry, 2003*
221. **Generation of cDNA Libraries:** *Methods and Protocols,* edited by *Shao-Yao Ying, 2003*
220. **Cancer Cytogenetics:** *Methods and Protocols,* edited by *John Swansbury, 2003*
219. **Cardiac Cell and Gene Transfer:** *Principles, Protocols, and Applications,* edited by *Joseph M. Metzger, 2003*
218. **Cancer Cell Signaling:** *Methods and Protocols,* edited by *David M. Terrian, 2003*
217. **Neurogenetics:** *Methods and Protocols,* edited by *Nicholas T. Potter, 2003*
216. **PCR Detection of Microbial Pathogens:** *Methods and Protocols,* edited by *Konrad Sachse and Joachim Frey, 2003*
215. **Cytokines and Colony Stimulating Factors:** *Methods and Protocols,* edited by *Dieter Körholz and Wieland Kiess, 2003*
214. **Superantigen Protocols,** edited by *Teresa Krakauer, 2003*
213. **Capillary Electrophoresis of Carbohydrates,** edited by *Pierre Thibault and Susumu Honda, 2003*
212. **Single Nucleotide Polymorphisms:** *Methods and Protocols,* edited by *Piu-Yan Kwok, 2003*
211. **Protein Sequencing Protocols, 2nd ed.,** edited by *Bryan John Smith, 2003*
210. **MHC Protocols,** edited by *Stephen H. Powis and Robert W. Vaughan, 2003*
209. **Transgenic Mouse Methods and Protocols,** edited by *Marten Hofker and Jan van Deursen, 2002*
208. **Peptide Nucleic Acids:** *Methods and Protocols,* edited by *Peter E. Nielsen, 2002*
207. **Recombinant Antibodies for Cancer Therapy:** *Methods and Protocols.* edited by *Martin Welschof and Jürgen Krauss, 2002*
206. **Endothelin Protocols,** edited by *Janet J. Maguire and Anthony P. Davenport, 2002*
205. **E. coli Gene Expression Protocols,** edited by *Peter E. Vaillancourt, 2002*
204. **Molecular Cytogenetics:** *Protocols and Applications,* edited by *Yao-Shan Fan, 2002*
203. **In Situ Detection of DNA Damage:** *Methods and Protocols,* edited by *Vladimir V. Didenko, 2002*
202. **Thyroid Hormone Receptors:** *Methods and Protocols,* edited by *Aria Baniahmad, 2002*
201. **Combinatorial Library Methods and Protocols,** edited by *Lisa B. English, 2002*
200. **DNA Methylation Protocols,** edited by *Ken I. Mills and Bernie H, Ramsahoye, 2002*
199. **Liposome Methods and Protocols,** edited by *Subhash C. Basu and Manju Basu, 2002*
198. **Neural Stem Cells:** *Methods and Protocols,* edited by *Tanja Zigova, Juan R. Sanchez-Ramos, and Paul R. Sanberg, 2002*
197. **Mitochondrial DNA:** *Methods and Protocols,* edited by *William C. Copeland, 2002*
196. **Oxidants and Antioxidants:** *Ultrastructure and Molecular Biology Protocols,* edited by *Donald Armstrong, 2002*
195. **Quantitative Trait Loci:** *Methods and Protocols*, edited by *Nicola J. Camp and Angela Cox, 2002*
194. **Posttranslational Modifications of Proteins:** *Tools for Functional Proteomics,* edited by *Christoph Kannicht, 2002*
193. **RT-PCR Protocols,** edited by *Joe O'Connell, 2002*
192. **PCR Cloning Protocols, 2nd ed.,** edited by *Bing-Yuan Chen and Harry W. Janes, 2002*
191. **Telomeres and Telomerase:** *Methods and Protocols,* edited by *John A. Double and Michael J. Thompson, 2002*
190. **High Throughput Screening:** *Methods and Protocols,* edited by *William P. Janzen, 2002*
189. **GTPase Protocols:** *The RAS Superfamily,* edited by *Edward J. Manser and Thomas Leung, 2002*
188. **Epithelial Cell Culture Protocols,** edited by *Clare Wise, 2002*
187. **PCR Mutation Detection Protocols,** edited by *Bimal D. M. Theophilus and Ralph Rapley, 2002*
186. **Oxidative Stress Biomarkers and Antioxidant Protocols,** edited by *Donald Armstrong, 2002*
185. **Embryonic Stem Cells:** *Methods and Protocols,* edited by *Kursad Turksen, 2002*
184. **Biostatistical Methods,** edited by *Stephen W. Looney, 2002*
183. **Green Fluorescent Protein:** *Applications and Protocols,* edited by *Barry W. Hicks, 2002*
182. **In Vitro Mutagenesis Protocols, 2nd ed.**, edited by *Jeff Braman, 2002*
181. **Genomic Imprinting:** *Methods and Protocols,* edited by *Andrew Ward, 2002*
180. **Transgenesis Techniques, 2nd ed.:** *Principles and Protocols,* edited by *Alan R. Clarke, 2002*
179. **Gene Probes:** *Principles and Protocols,* edited by *Marilena Aquino de Muro and Ralph Rapley, 2002*
178. **Antibody Phage Display:** *Methods and Protocols,* edited by *Philippa M. O'Brien and Robert Aitken, 2001*
177. **Two-Hybrid Systems:** *Methods and Protocols,* edited by *Paul N. MacDonald, 2001*
176. **Steroid Receptor Methods:** *Protocols and Assays,* edited by *Benjamin A. Lieberman, 2001*
175. **Genomics Protocols,** edited by *Michael P. Starkey and Ramnath Elaswarapu, 2001*
174. **Epstein-Barr Virus Protocols,** edited by *Joanna B. Wilson and Gerhard H. W. May, 2001*
173. **Calcium-Binding Protein Protocols, Volume 2:** *Methods and Techniques,* edited by *Hans J. Vogel, 2001*
172. **Calcium-Binding Protein Protocols, Volume 1:** *Reviews and Case Histories,* edited by *Hans J. Vogel, 2001*
171. **Proteoglycan Protocols,** edited by *Renato V. Iozzo, 2001*
170. **DNA Arrays:** *Methods and Protocols,* edited by *Jang B. Rampal, 2001*
169. **Neurotrophin Protocols,** edited by *Robert A. Rush, 2001*
168. **Protein Structure, Stability, and Folding,** edited by *Kenneth P. Murphy, 2001*
167. **DNA Sequencing Protocols,** *Second Edition,* edited by *Colin A. Graham and Alison J. M. Hill, 2001*
166. **Immunotoxin Methods and Protocols,** edited by *Walter A. Hall, 2001*
165. **SV40 Protocols**, edited by *Leda Raptis, 2001*
164. **Kinesin Protocols,** edited by *Isabelle Vernos, 2001*
163. **Capillary Electrophoresis of Nucleic Acids, Volume 2:** *Practical Applications of Capillary Electrophoresis,* edited by *Keith R. Mitchelson and Jing Cheng, 2001*
162. **Capillary Electrophoresis of Nucleic Acids, Volume 1:** *Introduction to the Capillary Electrophoresis of Nucleic Acids,* edited by *Keith R. Mitchelson and Jing Cheng, 2001*
161. **Cytoskeleton Methods and Protocols,** edited by *Ray H. Gavin, 2001*
160. **Nuclease Methods and Protocols,** edited by *Catherine H. Schein, 2001*
159. **Amino Acid Analysis Protocols,** edited by *Catherine Cooper, Nicole Packer, and Keith Williams, 2001*
158. **Gene Knockoout Protocols,** edited by *Martin J. Tymms and Ismail Kola, 2001*
157. **Mycotoxin Protocols,** edited by *Mary W. Trucksess and Albert E. Pohland, 2001*
156. **Antigen Processing and Presentation Protocols,** edited by *Joyce C. Solheim, 2001*
155. **Adipose Tissue Protocols,** edited by *Gérard Ailhaud, 2000*
154. **Connexin Methods and Protocols,** edited by *Roberto Bruzzone and Christian Giaume, 2001*

METHODS IN MOLECULAR BIOLOGY™

Neurogenetics

Methods and Protocols

Edited by
Nicholas T. Potter

*University of Tennessee Medical Center,
Knoxville, Tennessee*

Humana Press ✻ Totowa, New Jersey

©2003 Humana Press Inc.
999 Riverview Drive, Suite 208
Totowa, New Jersey 07512

www.humanapress.com

All rights reserved. No part of this book may be reproduced, stored in a retrieval system, or transmitted in any form or by any means, electronic, mechanical, photocopying, microfilming, recording, or otherwise without written permission from the Publisher. Methods in Molecular Biology™ is a trademark of The Humana Press Inc.

All authored papers, comments, opinions, conclusions, or recommendations are those of the author(s), and do not necessarily reflect the views of the publisher.

This publication is printed on acid-free paper. ∞
ANSI Z39.48-1984 (American Standards Institute)
Permanence of Paper for Printed Library Materials.

Cover design by Patricia F. Cleary.

Production Editor: Mark J. Breaugh.

For additional copies, pricing for bulk purchases, and/or information about other Humana titles, contact Humana at the above address or at any of the following numbers: Tel.: 973-256-1699; Fax: 973-256-8341; E-mail: humana@humanapr.com or visit our Website: http://humanapress.com

Photocopy Authorization Policy:
Authorization to photocopy items for internal or personal use, or the internal or personal use of specific clients, is granted by Humana Press Inc., provided that the base fee of US $10.00 per copy, plus US $00.25 per page, is paid directly to the Copyright Clearance Center at 222 Rosewood Drive, Danvers, MA 01923. For those organizations that have been granted a photocopy license from the CCC, a separate system of payment has been arranged and is acceptable to Humana Press Inc. The fee code for users of the Transactional Reporting Service is: [0-89603-990-0/03 $10.00 + $00.25].

Printed in the United States of America. 10 9 8 7 6 5 4 3 2 1

Library of Congress Cataloging-in-Publication Data

Neurogenetics / edited by Nicholas T. Potter
    p. cm. -- (Methods in molecular biology ; v. 217)
  Includes bibliographical references and index.
  ISBN 0-89603-990-0 (alk. paper)
  1. Neurogenetics--Laboratory manuals. 2. Genetic disorders--Diagnosis--Laboratory manuals. I. Potter, Nicholas T. II. Series.

QP356.22 .N485 2003
616'.042-dc21

                                                       2002068584

Preface

The rapid identification and characterization of genes of neurological relevance holds great potential for offering insight into the diagnosis, management, and understanding of the pathophysiologic mechanisms of neurological diseases. This volume in the *Methods in Molecular Biology*™ series was conceived to highlight many of the contemporary methodological approaches utilized for the characterization of neurologically relevant gene mutations and their protein products. Although an emphasis has been placed upon descriptions of methodologies with a defined clinical utility, it is hoped that *Neurogenetics: Methods and Protocols* will appeal not only to clinical laboratory diagnosticians, but also to clinicians, and to biomedical researchers with an interest in advances in disease diagnosis and the functional consequences of neurologically relevant gene mutations.

To meet this challenge, more than 60 authors graciously accepted my invitation to contribute to the 32 chapters of this book. Through their collective commitment and diligence, what has emerged is a comprehensive and timely treatise that covers many methodological aspects of mutation detection and screening, including discussions on quantitative PCR, trinucleotide repeat detection, sequence-based mutation detection, molecular detection of imprinted genes, fluorescence *in situ* hybridization (FISH), in vitro protein expression systems, and studies of protein expression and function. I would like to take this opportunity to formally thank my colleagues for their effort and dedication to this work.

This book would not have been possible without the guidance and wisdom of the Series Editor, Professor John M. Walker, whose intimate knowledge of the nuances of the editorial process made my job somewhat less intimidating. I would also like to thank Thomas Lanigan, President of Humana Press, who enthusiastically embraced the book concept and my original prospectus from the very beginning, and Craig Adams, also at Humana Press, for transforming the individual chapters into their final form.

Nicholas T. Potter

Contents

Preface v
Color Plates x
Contributors xi

PART I. QUANTITATIVE PCR

1 Determination of Gene Dosage:
   Utilization of Endogenous and Exogenous Internal Standards
   Thomas W. Prior 3
2 Semiquantitative PCR for the Detection of Exon Rearrangements
   in the Parkin Gene
   Christoph B. Lücking and Alexis Brice 13

PART II. TRINUCLEOTIDE REPEAT DETECTION

3 Detection of FMR1 Trinucleotide Repeat Expansion Mutations
   Using Southern Blot and PCR Methodologies
   Jack Tarleton 29
4 Extreme Expansion Detection in Spinocerebellar Ataxia Type 2 and Type 7
   Karen Snow and Rong Mao 41
5 Repeat Expansion Detection (RED) and the RED Cloning Strategy
   Qiu-Ping Yuan, Kerstin Lindblad-Toh, and Martin Schalling 51
6 Repeat Analysis Pooled Isolation and Detection (RAPID) Cloning
   of Microsatellite Expansions
   Laura P. W. Ranum 61
7 DIRECT Technologies for Molecular Cloning of Genes Containing
   Expanded CAG Repeats
   Kazuhiro Sanpei, Takeshi Ikeuchi, and Shoji Tsuji 73
8 Antibody-Based Detection of CAG Repeat Expansion Containing Genes
   Yvon Trottier 83
9 Detection of Trinucleotide Repeat Containing Genes
   by Matrix-Assisted Laser Desorption/Ionization (MALDI)
   Mass Spectrometry
   Chung-Hsuan Chen, Nicholas T. Potter, and Nelly T. Taranenko 91
10 Fluorescence PCR and GeneScan® Analysis for the Detection
   of CAG Repeat Expansions Associated with Huntington's Disease
   Cindy L. Vnencak-Jones 101

PART III. SEQUENCE-BASED MUTATION DETECTION

11 Molecular Detection of Galactosemia Mutations by PCR-ELISA
   Kasinathan Muralidharan and Wei Zhang 111
12 Denaturing High-Performance Liquid Chromatography
   and Sequence Analyses for *MECP2* Mutations in Rett Syndrome
   Inge M. Buyse and Benjamin B. Roa 119

13 Multiplexed Fluorescence Analysis for Mutations Causing
Tay-Sachs Disease
Tracy L. Stockley and Peter N. Ray *131*

14 Single-Strand Conformational Polymorphism Analysis (SSCP)
and Sequencing for Ion Channel Gene Mutations
Kylie A. Scoggan and Dennis E. Bulman *143*

15 Pulse Field Gel Electrophoresis for the Detection of Facioscapulohumeral
Muscular Dystrophy Gene Rearrangements
Luciano Felicetti and Giuliana Galluzzi *153*

16 Denaturing Gradient Gel Electrophoresis (DGGE) for Mutation Detection
in Duchenne Muscular Dystrophy (DMD)
Luciana C. B. Dolinsky *165*

17 Genetic Diagnosis of Charcot-Marie-Tooth Disease
Frank Baas *177*

18 Analysis of Human Mitochondrial DNA Mutations
Antonio L. Andreu, Ramon Martí, and Michio Hirano *185*

19 Detection of Mitochondrial DNA Mutations Associated
with Leber Hereditary Optic Neuropathy
Kasinathan Muralidharan *199*

PART IV. MOLECULAR DETECTION OF IMPRINTED GENES

20 PCR-Based Strategies for the Diagnosis of Prader-Willi/Angelman
Syndromes
Milen Velinov and Edmund C. Jenkins *209*

PART V. FLUORESCENCE *IN SITU* HYBRIDIZATION (FISH)

21 Fluorescence *In Situ* Hybridization (FISH) for Identifying the Genomic
Rearrangements Associated with Three Myelinopathies:
*Charcot-Marie-Tooth Disease, Hereditary Neuropathy with Liability
to Pressure Palsies, and Pelizaeus-Merzbacher Disease*
Mansoor S. Mohammed and Lisa G. Shaffer *219*

PART VI. IN VITRO EXPRESSION SYSTEMS AND STUDIES
OF PROTEIN EXPRESSION AND FUNCTION

22 *Drosophila* Models of Polyglutamine Diseases
H. Y. Edwin Chan and Nancy M. Bonini *241*

23 A Comparative Gene Expression Analysis
of Emery-Dreifuss Muscular Dystrophy Using a cDNA Microarray
Toshifumi Tsukahara and Kiichi Arahata *253*

24 The COS-7 Cell In Vitro Paradigm to Study Myelin Proteolipid Protein 1
Gene Mutations
Alexander Gow *263*

25 In Vitro Expression Systems for the Huntington Protein
Shi-Hua Li and Xiao-Jiang Li *277*

26 Heterologous Expression of Ion Channels
Andrew R. Tapper and Alfred L. George, Jr. *285*

27 An Assay for Characterizing In Vitro the Kinetics
of Polyglutamine Aggregation
Valerie Berthelier and Ronald Wetzel 295
28 Characterization of Prion Proteins
**Wenquan Zou, Monica Colucci, Pierluigi Gambetti,
and Shu G. Chen** 305
29 Detection of *NF1* Mutations Utilizing the Protein Truncation Test (PTT)
Meena Upadhyaya, Michael Osborn, and David N. Cooper 315
30 Application of the Protein Truncation Test (PTT) for the Detection
of Tuberosis Sclerosis Complex Type 1 and 2 (TSC1 and TSC2) Mutations
Karin Mayer 329
31 Development and Characterization of Antibodies
that Immunoprecipitate the FMR1 Protein
**Stephanie Ceman, Fuping Zhang, Tamika Johnson,
and Stephen T. Warren** 345
32 Immunological Methods for the Analysis of Protein Expression
in Neuromuscular Diseases
Mariz Vainzof, Maria Rita Passos-Bueno, and Mayana Zatz 355
Index 379

Color Plates

Color Plates 1–11 appear as an insert following p. 82.

PLATE 1 Representative gel image for HD analysis using fluorescence PCR. (See full caption on p. 106, Chapter 10.)

PLATE 2 Illustration of representative DHPLC and direct sequencing data for an *MECP2* missense and insertion mutation. (See full caption on p. 127, Chapter 12.)

PLATE 3 Results of the Tay-Sachs ASA assay for a normal sample (no mutations). (See full caption on p. 138, Chapter 13.)

PLATE 4 Example of Tay-Sachs ASA assay results for a carrier of the +TATC$_{1278}$ mutation. (See full caption on p. 139, Chapter 13.)

PLATE 5 Schematic representations to depict the typical hybridization patterns of test and control probes in the detection of microdeletions and cryptic translations. (See full caption on p. 220, Chapter 21.)

PLATE 6 Typical microarray image. (See full caption on p. 259, Chapter 23.)

PLATE 7 Schematic representation of CMT1A duplication, HNPP deletion; and PMD duplication with accompanying FISH images from representative patient test samples. (See full caption on pp. 224–225, Chapter 21.)

PLATE 8 Neurodegeneration induced by expression of expanded Machado-Joseph Disease (MJD) protein in the *Drosophila* retina, and its suppression by the molecular chaperone Hsp70. (See full caption on p. 245, Chapter 22.)

PLATE 9 Neurodegenerative phenotype caused by expression of pathogenic Machado-Joseph Disease (MJD) protein in the *Drosophila* retina and the modulatory effects of the molecular chaperone Hsp70. (See full caption on p. 245, Chapter 22.)

PLATE 10 (See discussion and captions on pp. 356–366, Chapter 32.)

PLATE 11 (See discussion and captions on pp. 367–370, Chapter 32.)

Contributors

ANTONIO L. ANDREU • *Centre d'Investigacions en Bioquímica i Biologia Molecular, Hospital Vall d'Hebron, Barcelona, Spain*
KIICHI ARAHATA • *Deceased, Formerly of Department of Neuromuscular Research, National Institute of Neuroscience, NCNP, Kodaira, Japan*
FRANK BAAS • *Neurogenetics Laboratory, Academic Medical Center, Amsterdam, The Netherlands*
VALERIE BERTHELIER • *Graduate School of Medicine, University of Tennessee Medical Center, Knoxville, TN*
NANCY M. BONINI • *Department of Biology, Howard Hughes Medical Institute, University of Pennsylvania, Philadelphia, PA*
ALEXIS BRICE • *INSERM U289, Hôpital de la Salpetríere, Paris, France*
DENNIS E. BULMAN • *Ottawa Health Research Institute, Ottawa, ON, Canada*
INGE M. BUYSE • *Department of Molecular and Human Genetics, Baylor College of Medicine, Houston, TX*
STEPHANIE CEMAN • *Howard Hughes Medical Institute, Emory School of Medicine, Atlanta, GA*
H. Y. EDWIN CHAN • *Department of Biology, University of Pennsylvania, Philadelphia, PA*
CHUNG-HSUAN CHEN • *Oak Ridge National Laboratory, Oak Ridge, TN*
SHU G. CHEN • *Institute of Pathology, Case Western Reserve University and National Prion Disease Pathology Surveillance Center, Cleveland, OH*
MONICA COLUCCI • *Institute of Pathology, Case Western Reserve University and National Prion Disease Pathology Surveillance Center, Cleveland, OH*
DAVID N. COOPER • *Institute of Medical Genetics, University of Wales College of Medicine, Cardiff, UK*
LUCIANA C. B. DOLINSKY • *Departamento de Genética, Instituo de Biologia, Universidade Federal do Rio de Janeiro (UFRJ), Rio de Janeiro, Brazil*
LUCIANO FELICETTI • *Institute of Neurology, Faculty of Medicine, Catholic University, Rome, Italy*
GIULIANA GALLUZZI • *Unione Italiana Distrofia Muscolare (UILDM), Rome, Italy*
PIERLUIGI GAMBETTI • *Institute of Pathology, Case Western Reserve University and National Prion Disease Pathology Surveillance Center, Cleveland, OH*
ALFRED L. GEORGE, JR. • *Departments of Medicine and Pharmacology, Vanderbilt University, Nashville, TN*
ALEXANDER GOW • *Center for Molecular Medicine and Genetics, Departments of Pediatrics and Neurology, Wayne State University School of Medicine, Detroit, MI*

MICHIO HIRANO • *Department of Neurology, Columbia University College of Physicians and Surgeons, New York, NY*

TAKESHI IKEUCHI • *Department of Neurology, Brain Research Institute, Niigata University, Niigata, Japan*

EDMUND C. JENKINS • *New York State Institute for Basic Research in Developmental Disabilities, Staten Island, NY*

TAMIKA JOHNSON • *Howard Hughes Medical Institute, Emory School of Medicine, Atlanta, GA*

SHI-HUA LI • *Department of Genetics, Emory University School of Medicine, Atlanta, GA*

XIAO-JIANG LI • *Department of Genetics, Emory University School of Medicine, Atlanta, GA*

KERSTIN LINDBLAD-TOH • *Department of Molecular Medicine, Karolinska Institute, Stockholm, Sweden*

CHRISTOPH B. LÜCKING • *Neurologische Klinik der Ludwig-Maximilians-Universität München (LMU), München, Germany*

RONG MAO • *DNA Diagnostic Laboratory, University of Utah School of Medicine, Salt Lake City, UT*

RAMON MARTÍ • *Department of Neurology, Columbia University College of Physicians and Surgeons, New York, NY*

KARIN MAYER • *Laboratory for Medical Genetics, Martinsried, Germany*

MANSOOR S. MOHAMMED • *Department of Molecular and Human Genetics, Baylor College of Medicine, Houston, TX*

KASINATHAN MURALIDHARAN • *Department of Pediatrics, Division of Genetics, Emory University School of Medicine, Atlanta, GA*

MICHAEL OSBORN • *Department of Molecular Immunology, The Babraham Institute, Cambridge, UK*

MARIA RITA PASSOS-BUENO • *Departamento de Biologia, Centro de Estudos do Genoma Humano, Instituto de Biociências, Universidade de Sao Paulo, Sao Paulo, Brazil*

NICHOLAS T. POTTER • *Department of Medical Genetics, University of Tennessee Medical Center, Knoxville, TN*

THOMAS W. PRIOR • *Department of Pathology, Ohio State University, Columbus, OH*

LAURA P. W. RANUM • *Department of Genetics, Cell Biology, and Development, Institute of Human Genetics, University of Minnesota, Minneapolis, MN*

PETER N. RAY • *Division of Molecular Genetics, Hospital for Sick Children, Toronto, ON, Canada*

BENJAMIN B. ROA • *Department of Molecular and Human Genetics, Baylor College of Medicine, Houston, TX*

KAZUHIRO SANPEI • *Department of Neurology, Brain Research Institute, Niigata University, Niigata, Japan*

MARTIN SCHALLING • *Department of Molecular Medicine, Karolinska Institute, Stockholm, Sweden*

Contributors

KYLIE A. SCOGGAN • *Ottawa Health Research Institute, Ottawa, ON, Canada*

LISA G. SHAFFER • *Department of Molecular and Human Genetics, Baylor College of Medicine, Houston, TX*

KAREN SNOW • *Molecular Genetics Laboratory, Mayo Clinic, Rochester, MN*

TRACY L. STOCKLEY • *Division of Molecular Genetics, Hospital for Sick Children, Toronto, ON, Canada*

ANDREW R. TAPPER • *Department of Pharmacology, Vanderbilt University, Nashville, TN*

NELLY T. TARANENKO • *Oak Ridge National Laboratory, Oak Ridge, TN*

JACK TARLETON • *Fullerton Genetics Center, Asheville, NC*

YVON TROTTIER • *IGBMC, CNRS/INSERM/ULP, Strasbourg, France*

SHOJI TSUJI • *Department of Neurology, Brain Research Institute, Niigata University, Niigata, Japan*

TOSHIFUMI TSUKAHARA • *Department of Neuromuscular Research, National Institute of Neuroscience, NCNP, Kodaira, Japan*

MEENA UPADHYAYA • *Institute of Medical Genetics, University of Wales College of Medicine, Cardiff, UK*

MARIZ VAINZOF • *Departamento de Biologia, Centro de Estudos do Genoma Humano, Instituto de Biociências, Universidade de Sao Paulo, Sao Paulo, Brazil*

MILEN VELINOV • *New York State Institute for Basic Research in Developmental Disabilities, Staten Island, NY*

CINDY L. VNENCAK-JONES • *Department of Pathology, Vanderbilt University Medical Center, Nashville, TN*

STEPHEN T. WARREN • *Howard Hughes Medical Institute, Emory School of Medicine, Atlanta, GA*

RONALD WETZEL • *Graduate School of Medicine, University of Tennessee Medical Center, Knoxville, TN*

QIU-PING YUAN • *Department of Molecular Medicine, Karolinska Institute, Stockholm, Sweden*

MAYANA ZATZ • *Departamento de Biologia, Centro de Estudos do Genoma Humano, Instituto de Biociências, Universidade de Sao Paulo, Sao Paulo, Brazil*

FUPING ZHANG • *Howard Hughes Medical Institute, Emory School of Medicine, Atlanta, GA*

WEI ZHANG • *Department of Pediatrics, Division of Genetics, Emory University School of Medicine, Atlanta, GA*

WENQUAN ZOU • *Institute of Pathology, Case Western Reserve University and National Prion Disease Pathology Surveillance Center, Cleveland, OH*

I

QUANTITATIVE PCR

1

Determination of Gene Dosage

Utilization of Endogenous and Exogenous Internal Standards

Thomas W. Prior

1. Introduction

There are currently several screening methods for the detection of point mutations, such as single-stranded conformation polymorphism, heteroduplex analysis, denaturing gradient gel electrophoresis, and chemical cleavage. These are powerful tools for the identification of small sequence changes, but fail to detect heterozygous deletions or duplications of exons, genes or chromosomes. There are many genetic disorders where the primary defect is either owing to allelic deletions (Duchenne muscular dystrophy, spinal muscular atrophy, alpha thalassemia, growth hormone deficiency, familial hypercholesterolemia, and so on) or duplications (Charcot-Marie-Tooth, Klinefelter syndrome, Down syndrome, and so on). Furthermore, for the determination of the carrier state, for disorders such as Duchenne muscular dystrophy and spinal muscular atrophy, the accurate determination of heterozygous deletions is essential. This chapter will describe two methods for the determination of gene dosage, using Duchenne muscular dystrophy and spinal muscular atrophy as examples.

1.1. Duchenne Muscular Dystrophy Dosage Testing Utilization of an Endogenous Internal Standard

Duchenne muscular dystrophy is an X-linked neuromuscular disease characterized by progressive muscular weakness and degeneration of skeletal muscle. Approximately 60% of the DMD and BMD patients have deletions of the dystrophin gene *(1–3)*. Originally, in order to identify female carriers in Duchenne muscular dystrophy, one performed gene dosage using quantitative Southern blot analysis, whereby one determines whether the female at risk exhibits no reduction (noncarrier status) or 50% reduction (carrier status) in hybridization intensity in those bands that are deleted in the affected male *(4,5)*. The dosage determinations permit direct carrier analysis and eliminates the inherent problems of the restriction fragment length polymorphism (RFLP) technique (recombinations, noninformative meioses, unavailability of family members, and sporadic mutations). To further increase the accuracy of the dosage analysis, the autoradiographic bands can be scanned with a densitometer *(6)*.

Although dosage analysis has significantly improved carrier studies, particularly in the isolated cases of the disease, there are technical limitations. Dosage analysis of

Southern blots requires optimal conditions; very good quality blots are necessary, with even transfer and hybridization, and low background. In order to obtain this high quality we have found that approx 20% of the Southerns have to be repeated, resulting in increased time and labor. Rather than directly comparing single bands, band ratios are calculated as a means of decreasing the error caused by differences in the amount of DNA in each lane. The normal control ratio is established by comparing a band lacking against a band present in the patient (which serves as an internal control) in an unaffected female. When this ratio in a female (at risk) is approx half the control ratio, this indicates that she has a single copy of the band deleted in the patient and therefore is a carrier. Depending on the extent of the deletion, the restriction fragments involved in the deletion, and the specific cDNA probe that identifies the deletion, one may be extremely limited as to what bands are used in the control ratio. We have found that bands greater than 10 kb and less than 0.5 kb typically result in weaker intensities and are not always adequate for scanning purposes. Mao et al. stated that for deletions in the center of the gene (cDNA 8 hybridizations), they prefer to make a statement of the carrier status only if at least one of the strong hybridizing fragments (7, 3.8, 3.7, or 3.1 kb) is deleted in the patient *(7)*. Furthermore the difference between one or two copies is relatively easy to detect but differences between two and three copies, or sometimes three or four copies, in the case of a duplication or comigrating bands can be very difficult. Lastly, due to the extent of a deletion in an affected individual, no hybridizing bands may be detected with a cDNA probe and comparison of hybridization bands within a lane is not possible in these cases.

The determination of carrier status has significantly improved by using the polymerase chain reaction (PCR). Since the extension product of each primer serves as a template for the other primer, each cycle essentially doubles the amount of the DNA product produced in the previous PCR cycle. This results in the exponential accumulation of the specific fragment, up to several millionfold in a few hours. However, to obtain quantitative results, the PCR products must be estimated during the exponential phase of the amplification process, because it is during the exponential phase where the amount of amplified products is proportional to the abundance of starting DNA *(8)*. This occurs when the primers, nucleotides, and Taq polymerase are in a large excess over that of the template concentration. In our experience, after the completion of an adequate number of cycles (25–30) to visualize the PCR products on an ethidium-bromide-stained gel, the PCR reaction is no longer in the exponential quantitative range. Therefore the gene dosage-PCR is accomplished by amplifying the genomic DNA at lower cycle numbers (before visualization by ethidium bromide), running the products out on an agarose gel, Southern transferring the products, and hybridizing the amplicons with a radiolabeled probe. We have found that linearity is well-maintained within 10–15 cycles and hybridization band intensity is still strong *(9)*.

A case study using quantitative PCR is shown in **Fig. 1**. A DMD patient was found to have a molecular deletion for exons 8–19. This was an isolated case of the disease and the mother and two daughters were tested for carrier status. Therefore exons 19 and 50 in the mother, daughters, proband, and a normal female control were amplified for 12 cycles, hybridized with the corresponding cDNA probes, and the autoradiogram is shown in **Fig. 1**. Exon 50 serves as an endogenous internal control, because this is an exon that is not deleted in the patient. The endogenous internal standard is coamplified

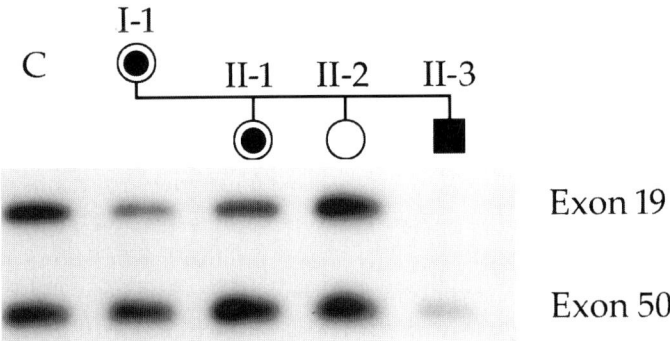

Fig. 1. Duchenne muscular dystrophy carrier determination by gene dosage using an endogenous internal standard. The affected son is deleted for exon 19. Exon 50 is the endogenous internal standard, since the affected son is not deleted for this exon. The mother (I-1) and daughter (II-1) show a 50% reduction in the exon 19/50 ratio compared to the C (noncarrier female control). Daughter II-2 is a noncarrier since her exon 19/50 is equivalent to C.

with the target of interest (deleted exon) and serves as a control for several factors: differences in initial template concentrations between different samples, sample-to-sample variations in the PCR, the extent of any DNA degradation, and differences in the amounts of amplicon loaded onto the gel. Thus, rather than directly comparing single bands, band ratios are calculated. The 19/50 exon ratios in the mother and daughter (II-1) were approx half the normal control ratio, and the ratio in daughter (II-2) was the same as the control. The ratios were confirmed by densitometer. Therefore, the mother, daughter (II-1) are carriers and daughter (II-2) is a noncarrier of the exon 19. Dosage determinations permit direct carrier analysis and eliminates the inherent problems of the RFLP technique (recombinations, noninformative meioses, unavailability of family members, and spontaneous mutations). This is important since unlike the affected males, the heterozygous females are generally asymptomatic and creatine kinase (CK) is only elevated in approx 50–60% of known carriers *(10)*.

1.2. Spinal Muscular Atrophy Dosage Testing: Utilization of an Exogenous Internal Standard

Spinal muscular atrophy (SMA) is an autosomal recessive disorder characterized by degeneration of the motor neurons in the spinal cord, resulting in symmetrical limb and trunk paralysis. With a prevalence of about 1 in 10,000 live births, and a carrier frequency of approx 1 in 50, SMA is the second most common fatal autosomal recessive disorder after cystic fibrosis. The survival motor neuron gene (SMN) has been shown to be deleted in approx 95% of patients with SMA *(11)*. Although direct diagnostics have significantly improved by the identification of homozygous SMN gene deletions, carrier detection for the determination of a single copy of the gene is a technical challenge. This is mainly due to the fact that the SMA region is characterized by the presence of many repeated elements. The SMN gene itself is present in two almost identical copies, a telomeric (SMN1) and a centromeric copy (SMN2). The two genes differ in exons by only two base pairs, one in exon 7 and one in exon 8, and it is SMN1 that is deleted in cases of SMA. Although the centromic and telomeric copies can be readily separated by a restriction enzyme digestion, using the centromeric copy as the internal endogenous standard will not be accurate since it is not constant. In the normal population we have observed the following: approx 10% are homozygously deleted, 40% have one copy, and 50% have two copies of the SMN2 gene, respectively.

Dosage determination of the SMN copy number is performed by a competitive PCR method using an exogenous in vitro synthesized DNA internal standard *(12)*. In the competitive PCR method, a known number of copies of a synthetic mutated internal standard is introduced with the patient sample into the PCR mixture. The major advantage of this technique is that the internal standard is amplified with the same primers that amplify the target sequence. Thus, the efficiency of the amplification of the patient DNA and the internal standard DNA should be very similar and allow one to accurately determine the gene copy number. The internal standard is synthesized using the same forward specific primer. However, the reverse primer has now been moved 50 bases upstream from the original reverse primer and is tagged at it 5' end with the original reverse primer sequence *(13)*. The resulting PCR product will thus be identical to the original specific PCR product, but will lack 50 base pairs and thus be distinguished from the endogenous sequence by size (**Fig. 2A**). The internal standard is quantitated by UV spectrophotometry, and diluted to an appropriate concentration. The quantitative PCR dosage assay then consists of spiking a known amount of the internal standard to the patient sample and amplifying the sample with the original forward and reverse primers that are against common sequences (**Fig. 2B**). One of the primers is ^{32}P-end-labeled. With this approach, two products will be generated: one derived from the patient and a second 50 bp smaller product from the internal standard. The PCR products are then diluted in loading buffer, electrophoresed on a 6% denaturing gel, and autoradiography is performed.

Our dosage assay also uses an exon from the cystic fibrosis transmembrane regulator (CFTR) as an internal standard *(14)*. Thus, multiple ratios can be utilized for the accurate determination of carrier status and, most importantly, changes in the SMN1 dosage as a result of fluctuations in the SMN2 copy number are avoided. Furthermore, the use of two internal standards (SMN-IS and CFTR-IS) allows one to monitor the efficiency of the PCR reaction and ensures that equal amounts of target DNA is added to each tube. Similar quantitative PCR approaches have been used successfully to identify deletions in the insulin receptor gene *(15)*, to detect duplications in Down syndrome patients *(15)*, and to quantitate oncogene amplification *(16)*.

Figure 3 shows several carriers and noncarriers identified using the competitive PCR with the exogenous internal standards. As shown, although the SMN2 copies varies from 0–2 copies, the dosage ratios are maintained. Furthermore, multiple ratios can be used in determining carrier status and thereby improve the overall quality assurance of the assay. Our present protocol utilizes the SMN1/CFTR ratio.

2. Materials

2.1. Genomic Isolation

1. Genomic DNA was extracted from leukocytes harvested from whole blood anticoagulated with EDTA using a salting out procedure *(17)*. DNA concentrations were determined using a spectrophotometer, as well as by monitoring the intensity of ethidium bromide staining on a test gel.

Determination of Gene Dosage

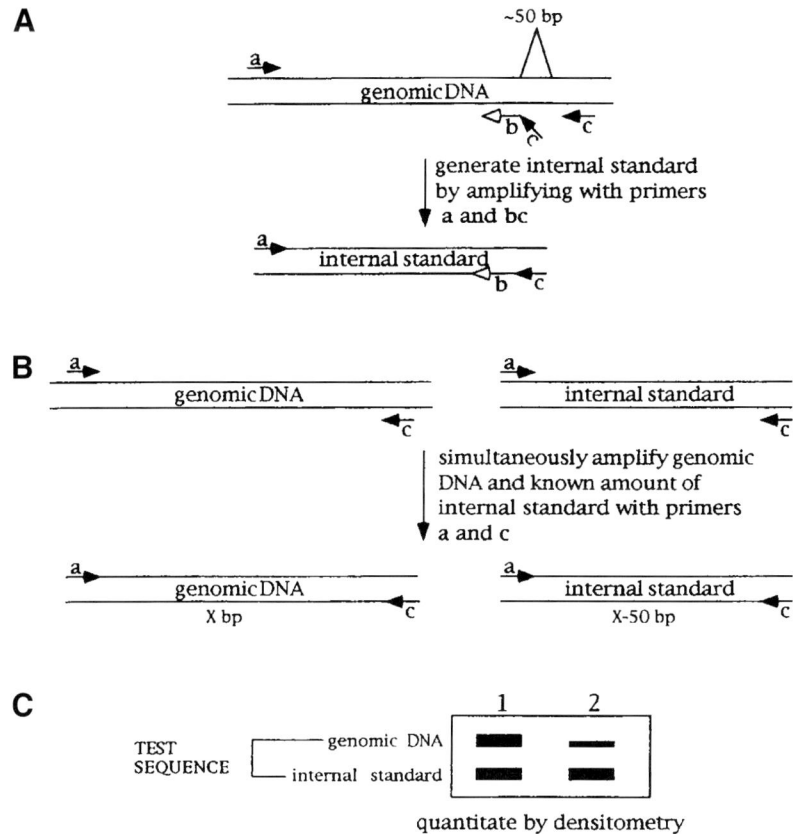

Fig. 2. Schematic representation of gene dosage using an exogenous internal standard. The internal standard is synthesized using primers a and bc (**A**). Forward primer a is a conventional PCR primer. Primer bc is tagged at its 5' end with a 20 bp sequence which lies 50 bp downstream of sequence b. The genomic DNA and the internal standard are then coamplified using primers a and c (**B**). The coamplification results in two products, one derived from the genomic DNA and a second product which is 50 bp smaller derived form the internal standard (**C**). As shown the carrier has a ratio (lane 2), which is half the normal (lane 1).

2.2. Duchenne Muscular Dystrophy Dosage Testing

1. AmpliTaq DNA polymerase, 5 U/μL (Applied Biosystems, Foster City, CA, USA).
2. PCR buffer: 0.5 mmol of deoxynucleotide triphosphates, 3 mmol of $MgCl_2$, 67 mmol of Tris, pH 8.8, 16.6 mmol of ammonium sulfate, 6.7 mol of EDTA, 10 mmol of 2-mercaptoethanol per liter.
3. The DMD primer sequences have been previously described (*18*).
4. Southern blot transfer buffer: 0.5 *M* NaOH, 0.6 *M* NaCl.
5. Prehybridization buffer: 5X SSC (1X SSC = 0.15 *M* NaCl, 150 m*M* Na citrate), sodium phosphate buffer (50 m*M*, pH 6.5), 5X Denhardt's (1.0 g/L ficoll, 1.0 g/L polyvinylpyrrolidone, 1.0 g/L bovine serum albumin [BSA]), and containing 1 g of SDS and 250 mg of yeast RNA per liter

2.3. SMA Dosage Testing

1. AmpliTaq DNA polymerase, 5 U/μL (Applied Biosystems).

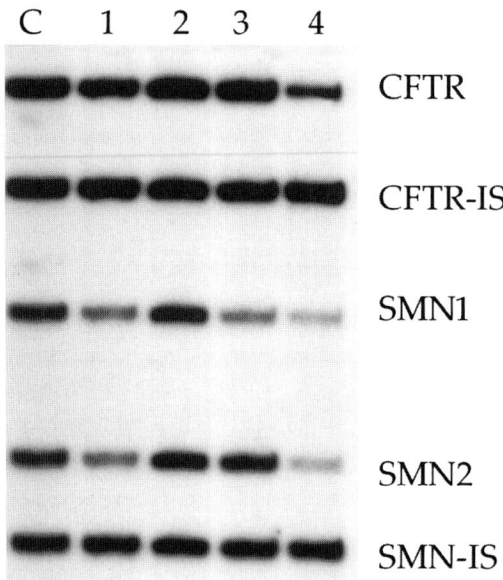

Fig. 3. Spinal muscular atrophy carrier determination by gene dosage using an exogenous internal standard. The SMN1/CFTR ratios in SMA carriers (lanes 1, 3, and 4) are half the ratio of the normal control (C). Lane 2 is a noncarrier. Lanes 1 and 4 have 1 SMN2 copies, whereas C and lanes 2 and 3 have 2 SMN2 copies. As shown, the SMN1/CFTR ratios are not changed as a result of variable SMN2 copy number.

2. Generation of internal standards: The 50 µL PCR reaction contained 200 ng genomic DNA, 3 mM MgCl$_2$, 1X Taq DNA polymerase buffer, 200 µM each dNTP, 30 ng each of F621F(5'-AGTCACCAAAGCAGTACAGC-3') and CFTR-IS(5'GGGCCTGTG CAAGGAAGTGTTAAGCTATTCTCATCTGCATTCCA-3') primers, and 0.5 U DNA polymerase. A 50 µL reaction, containing the same components to those in the CFTR internal standard reaction, was used to generate the SMN internal standard except that primers R111 (5'-AGACTATCAACTTAAATTTCTGATCA-3') and SMN-IS (5'-CCTTCCTTCTTTTGATTTTGTTTATAGCTATATAGACATAGATAGCTA-3') were used.
3. Competitive PCR reaction mix:R111 and CF621F primers (15 ng each) are end-labeled with 0.1 µL [gamma] ATP (10 µCi/µL;Amersham) and T4 DNA kinase (Gibco-BRL) at 37°C for 20 min. The 25 µL PCR reaction contained 200 ng genomic DNA, 3 mM MgCl$_2$, 1X Taq DNA polymerase buffer (USB), 200 µM each dNTP, end-labeled forward primers 15 ng each of CF621R (5'-GGGCCTGTGCAAGGAAGTGTTA-3') and X-7Dra (5'-CCTTCCTTCTTTTGATTTTGTTT-3') reverse primers, 0.5 U Taq DNA polymerase (USB) and 1 µL each of CFTR and SMN internal standards.

3. Methods

3.1. Duchenne Muscular Dystrophy Dosage Testing

1. Amplification was performed at 55°C as the annealing temperature, 72°C as the extension temperature, and 94°C as the denaturation temperature in a thermal cycler (Ericomp, San Diego, CA, USA). For quantitative analysis the samples were subjected

to 14–16 cycles. For visual deletion analysis, via ethidium-bromide-stained gel, amplification was carried out for 30 cycles. Optimal resolution was obtained by electrophoresing 20 µL of the 100-µL reaction mixtures through 25 g/L agarose gels.
2. The PCR products were transferred to a nylon membrane (Zetabind; CUNO, Inc., Meriden, CT, USA) by Southern blotting with 0.5 mol/L NaOH, 0.6 mol/L NaCl. The filter was prehybridized at 37°C in prehybridization buffer.
3. Hybridization was performed in the same solution at 37°C to which random primed cDNA probe was added, complementary to the amplified DNA sequences.
4. The filter was washed first in a 5X SSC, followed by a second wash in 2X SSC, 0.5 % SDS solution at 45°C for 15–30 min.
5. Autoradiography was performed for approx 18 h at –70°C. Multiple exposures of each filter were made to ensure that quantitated films were in the linear range.
6. The degree of hybridization was assessed by using a CS-9000 scanning densitometer (Shimadzu Corp., Kyoto, Japan). The area under the peak represents the intensity of each hybridization band.

3.2. SMA Dosage Testing

1. The reaction conditions consisted of an initial denaturation at 95°C for 5 min, followed by 16 cycles of 95°C 1 min, 55°C 2 min, and 72°C 3 min.
2. The PCR product (8 µL) is digested with 20 U DraI (New England Biolabs) overnight at 37°C. The digestion serves to separate SMN1 from SMN2 *(19)*.
3. The digested samples are run on a 6% denaturing polyacrylamide gel and are quantitated by autoradiography, The gel is exposed for 16–24 and 48–72 h at –70°C.
4. Densitometry of the bands was performed on a Shimadzu CS-9000.

3.3. Conclusions

In this chapter two methods were described that allow for the accurate determination of gene dosage. Both methods are PCR-based, rapid, do not require major equipment purchases, and thus can be performed in a diagnostic laboratory. Both examples, DMD and SMA carrier determinations, are routinely performed in our diagnostic laboratory. Both methods are based on amplifying not only the target of interest but also an internal standard, which is then used in determining the dosage ratio. Dosage ratios are used as a means of correcting any errors due to differences in initial template concentrations and efficiencies of PCR. It is essential that the standard and the target have similar yields per cycle. It has often been observed that when different-sized targets are simultaneously amplified, the yield is often higher for smaller product. We have found that by keeping the internal standard no more or less that 20% the size of the genomic target DNA of interest, the efficiencies are usually quite similar. Furthermore, since PCR is so sensitive, contamination with small amounts of carryover products can lead to inaccurate results. All precautions should be used to eliminate any potential sources of contamination.

The first dosage application described was used for determining DMD carrier status. The internal standard was an endogenous target (nondeleted exon) and the two were co-amplified simultaneously and the dosage ratio was calculated. As long as one is careful in choosing an internal standard that is not a member of a homologous gene family or a repetitive DNA sequence, this is a straightforward technique and can often be used for the determination of the heterozygous state. Due to the complexities of the SMA gene, the more technically complex competitive PCR was used for the identification of SMA carriers. This technique requires the synthesis of an exogenous internal standard, which is then added to the PCR reaction. Since the genomic DNA and the internal standard utilize the same primers, efficiencies of amplification should be equivalent, thus allowing for extremely sensitive dosage measurements. This technique is not only applicable for complex gene loci, but is often utilized for reverse transcriptase and polymerase chain reaction (RT-PCR), when one may be interested in small differences of gene expression *(19)*.

4. Notes
4.1. Duchenne Muscular Dystrophy Dosage Testing

1. The determination of gene dosage via the polymerase chain reaction and using an endogenous internal standard has several advantages. The amplification of only two specific exons reduces the background problems that are often present on Southern blots. We have not repeated any of the hybridizations due to high background or unresolvable autoradiographic bands, unlike our experience with the Southern analysis. Furthermore this assay requires less DNA, can be performed more rapidly than Southern analysis, and is both cost- and labor-effective.
2. In order to reliably quantitate the amount of DNA, the range of concentrations of template and the number of amplification cycles must be determined such that they stay within the exponential phase of the PCR. It is critical that samples are assayed within the exponential phase of the PCR reaction, before the plateau phase when the amplification efficiency begins to decrease and the relative concentration of amplicons begin to vary. Therefore, it is necessary to define the range of concentrations that give an exponential amplification over a defined range of cycle numbers. The simplest way to establish the exponential range is to remove small aliquots from a trial PCR reaction every few cycles during amplification, measure the band intensity (using a densitometer) and determine the product linearity over several cycles. We have found the conditions are relatively liberal, for the PCR reaction can be accurately quantitated over a range of cycle numbers and starting DNA template concentrations.
3. It is important to choose an internal endogenous standard that amplifies equivalently with the target of interest. Ideally the normal dosage control ratio should be approx 1 and the heterozygous state should have a ratio of about 0.5. Occasionally we have found exons that do not amplify well together and as a result the control ratios are much greater or less than one. Distinguishing a 50% reduction in these situations can be more difficult, and therefore we recommend using targets that possess similar PCR efficiencies during the linear range. The internal standard should be different enough in size to be easily resolved from the PCR product of interest, but close enough in size so that there are not major differences in transfer efficiency.
4. An alternative approach to transfer and hybridization is to use PCR primers, end-labeled with polynucleotide kinase. The products are then quantified by direct autoradiography of the gel *(20)*. In our experience this procedure results in higher background due to nonspecific labeling. In our protocol the specificity of the PCR products is guaranteed by specific hybridization, which reduces the level of background bands.

4.2. SMA Dosage Testing

5. Optimization of the assay requires determining the exact amount of internal standard to be added to the patient's sample. An internal standard titration over several concentrations allows one to establish the range. A standard concentration curve of the internal control can be produced by amplifying serial dilutions of the spiked internal standard with a con-

stant amount of target DNA. The goal is to have the normal control ratio approx one, which will allow unequivocal detection of deletions in carriers who demonstrate ratios of approx 0.5.

6. The described competitive PCR assay has demonstrated high precision between different gel runs *(14)*. Furthermore, by preparing a sufficient quantity of the internal standard and storing it frozen, it can be used over multiple runs, which will further minimize interassay variability. It is critical that the dosage ratios be maintained, when the SMN2 copy number changes. We have found, that by using 16 cycles, the SMN1 copy number is unaffected by changes in the SMN2 copy number. Furthermore, the use of 16 cycles not only allows one to quantitate in the linear range of the PCR, but reduces the formation of heterduplexes (between SMN1 and SMN2), which may occur at higher cycle number and contribute to assay imprecision.

7. It is imperative that carrier and noncarrier controls with varying SMN2 copy number be included on every assay run. It is also important to control for the amount of input genomic DNA. We have observed a small degree of variability in the amplification efficiency of SMN-IS when comparing DNA samples extracted by different methods. Therefore, we recommend the use of normal, carrier, and affected controls prepared by the same extraction method as is used for the samples being tested.

8. Autoradiography is an effective and accurate method for the quantitation of PCR amplicons, but standard precautions such as monitoring the exposure times and preflashing the film should be taken in order to ensure the linearity of film response.

References

1. Darras, B. T., Blattner, P., Harper, J. F., Spiro, A. J., Alter, S., and Franke, U. (1988) Intragenic deletions in 21 Duchenne muscular dystrophy (DMD)/ Becker muscular dystrophy (BMD) families studied with the dystrophin cDNA: location of breakpoints on HindIII and BglII exon-containing fragment maps, meiotic and mitotic orgin of mutations. *Am. J. Hum. Genet.* **43,** 620–629.
2. Read, A. P., Mountford, R. C., Forrest, S. M., Kenwrick, S. J., Davies, K. E., and Harris R. (1988) Patterns of exon deletions in Duchenne and Becker muscular dystrophy. *Hum. Genet.* **80,** 152–156.
3. Gillard, E. F., Chamberlain, J. S., Murphy, E. G., Duff, C. L., Smith, B., Burghes, A. H. M., et al. (1989). Molecular and phenotypic analysis of patients with deletions within the deletion-rich region of the Duchenne muscular dystrophy (DMD) gene. *Am. J. Hum. Genet.* **45,** 507–520.
4. Darras, B. T., Koenig, M., Kunkel, L. M., and Francke U. (1988) Direct method for prenatal diagnosis and carrier detection in Duchenne/Becker muscular dystrophy using the entire dystrophin cDNA. *Am. J. Med. Genet.* **29,** 713–726.
5. Prior, T. W., Friedman, K. J., Highsmith, W. E., Perry, T. R., and Silverman L. M. (1990) Molecular probe protocol for determining carrier status in Duchenne and Becker muscular dystrophies. *Clin. Chem.* **36,** 441–445.
6. Prior, T. W., Friedman, K. J., and Silverman, L. M. (1989) Detection of Duchenne/Becker muscular dystrophy carrier by densitometric scanning. *Clin. Chem.* **35,** 1256–1257.
7. Mao, Y. and Cremer M. (1989) Detection of Duchenne muscular dystrophy carriers by dosage analysis using the DMD cDNA clone 8. *Hum. Genet.* **81,** 193–195.
8. Lubin, M., Elashoff, J. D., Wang, S., Rotter, J., and Toyoda, H. (1991) Precise gene dosage determination by polymerase chain reaction: Theory, methodology, and statistical approach. *Mol. Cell. Probe* **5,** 307–317.

9. Prior, T. W., Papp, A. C., Snyder, P. J., Highsmith, W. E., Friedman, K. J., Perry, T. R., et al. (1990) Determination of carrier status in Duchenne and Becker muscular dystrophies by quantitative polymerase chain reaction and allele-specific oligonucleotides. *Clin. Chem.* **36,** 2113–2117.
10. Thompson, M. W., Murphy, E. G., and McAlpine, P. J. (1967) An assessment of the creatine kinase test in the detection of carriers of Duchenne muscular dystrophy. *Pediatrics* **71,** 82–93.
11. Lefebvre, S., Burglen, L., Reboullet, S., Clermont, O., Burlet, P., Viollet, L., et al. (1995) Identification and characterization of a spinal muscular atrophy-determining gene. *Cell* **80,** 155–165.
12. Gilliland, G., Perrin, S., and Bunn, H. F. (1990) Competitive PCR for quantitation of mRNA, in *PCR Protocols: A Guide to Methods and Applications* (Innis, M. A., Gelfand, D. H., Sninsky, J. J., and White, T. J., eds.), Academic Press, San Diego, CA, USA, pp. 60–69.
13. Celi, F. S., Zenilman, M. E., and Shuldiner, A. R. (1993) A rapid and versatile method to synthesize internal standards for competitive PCR. *Nucleic Acids Res.* **21,** 1047.
14. McAndrew, P. E., Parsons, D. W., Simard, L. R., Rochette, C., Ray, P. N., Mendell, J. R., et al. (1997) Identification of proximal spinal muscular atrophy carriers and patients by analysis of SMNT and SMNC gene copy number. *Am. J. Hum. Genet.* **60,** 1411–1422.
15. Celi, F. S., Cohen, M. M., Antonarakis, S. E., Wertheimer, E., Roth J., and Shuldiner, A. R. (1994) Determination of gene dosage by a quantitative adaptation of the polymerase chain reaction (gd-PCR): rapid detection of deletions and duplications of gene sequences. *Genomics* **21,** 351–356.
16. Sestini, R., Orlando, C., Zentilin, L., Lami, D., Gelmini, S., Pinzani, P., and Giacca, M. (1995) Gene amplification for c-erbB-2, c-myc, epidermal growth factor receptor, int-2, and N-myc measured by quantitative PCR with a multiple competitor template. *Clin. Chem.* **41,** 826–832.
17. Miller, S. A., Dykes, D. D., and Polesky, H. F. (1988) A simple salting out procedure for extracting DNA from human nucleated cells. *Nucleic Acids Res.* **264,** 1215.
18. Chamberland, J. S., Gibbs, R. A., Ranier, R. A., Nguyen, P. N., and Caskey, C. T. (1988) Deletion screening of the Duchenne muscular dystrophy locus via multiplex DNA amplification. *Nucleic Acids Res.* **16,** 11,141–11,156.
19. Foley, K. P., Leonard, M. W., and Engel, J. D. (1993) Quantitation of RNA using the polymerase chain reaction. *Trend Genet.* **9,** 380–384.
20. Arigo, S. J., Weitsman, S., Rosenblatt, J. D., and Chen, I. S. Y. (1989) Analysis of *rev* gene function on human immunodeficiency virus type 1 replication in lymphoid cells by using quantitative polymerase chain reaction. *J. Virol.* **63,** 4875–4881.

2

Semiquantitative PCR for the Detection of Exon Rearrangements in the Parkin Gene

Christoph B. Lücking and Alexis Brice

1. Introduction
1.1. Frequency and Type of Parkin Mutations

Mutations in the *parkin* gene have been shown to be responsible for a substantial number of cases of autosomal recessive early onset parkinsonism (AR-JP, PARK2, OMIM 602544) worldwide *(1–4)*. The gene on chromosome 6q25.2-27 consists of 12 coding exons with an open reading frame of 1395 bp. The gene is estimated to cover >1.5 Mb. The gene product, Parkin, functions an an E3-ubiquitin-protein ligase *(5)*. The only known substrate to date is CDCrel-1 *(6)* but the existence of other substrates is likely *(5)*.

The mutations in the *parkin* gene identified so far have been extremely varied in both location and type. About 50% of affected Caucasian sibpairs without affected parents (suggesting autosomal-recessive inheritance) in which onset occured before or at the age of 45 yr in at least one of the affecteds, have mutations in the parkin gene *(3)*. In sporadic cases, 80% had mutations when the age at onset was ≤ 20, 25% when onset was at age 21–30, and only 2% when age of onset was 31–45. About 50% of the mutations in the Caucasian population are exon rearrangements, consisting mainly of deletions but multiplications also occur *(3)*. In nonconsanguineous cases, these rearrangements might be present in the heterozygous state and thus escape detection by nonquantitative polymerase chain reaction (PCR) because there is still a normal copy of the deleted exon. Direct sequencing will not permit detection of such deletions either, since only the undeleted allele, normal in most cases, will be analyzed. Thus, a simple, fast, and accurate assay is needed to quantify the number of exons present in a genomic DNA sample.

1.2. Principle of the Semiquantitative Multiplex PCR Dosage Assay

Quantitative PCR is considered to be superior to Southern blotting for detecting gene dosage *(7)*. Since absolute quantification of the template DNA is not necessary, we have chosen a semi-quantitative PCR assay that compares the relative amounts of template DNA. This is sufficient to answer to the question whether the amount of template DNA corresponding to one or more exons is abnormal. This assay can be performed with any fluorescence-based automated sequencer, now available in many laboratories, unlike real-time PCR analyzers that remain relatively rare.

In order to detect homozygous or heterozygous exon deletions or multiplications in the *parkin* gene, several exons amplified simultaneously in one multiplex PCR are compared, the co-amplified exons serving as internal standards for quantification. Fluorescently labeled PCR primers are used, so that the PCR product can be quantified from the height of the peak detected by the automated sequencer. The amount of PCR product is directly related to the amount of template DNA as long as PCR amplification is exponential. Each multiplex PCR for a given combination of exons results in a typical pattern of peak heights for normal control cases and thus in typical ratios between the peaks. If an exon is heterozygously deleted, for example, its peak is half as high as expected and the ratios between the deleted and the undeleted exons change. The resulting ratios are then compared to the ratios of the peaks obtained with DNA from known normal subjects that are run in parallel. If the ratios for a given exon differ from those of a control, the exon has been rearranged. In order to detect a heterozygous deletion of the entire *parkin* gene, an 328 bp PCR product (C328) of a distant gene (*transthyretin* gene on chromosome 18) is amplified as external standard in one of the multiplex PCR combinations. The fact that only ratios between peak heights are compared, renders the assay relatively independent of the quantity and quality of the template DNA. In addition, small-scale deletions and insertions within the PCR products can be detected by comparing the size of the patient's PCR product to the product of a control. Seventy percent of the parkin mutations identified by our group have been reliably detected with this technique *(3)*.

2. Materials

The products used are listed below. Comparable products from other suppliers should also be effective.

2.1. Multiplex PCR

1. High quality genomic DNA extracted by standard procedures. Prepare a working dilution at 20 ng/µL (diluted in water) just before the experiment. It can be kept at +4°C for short periods of time, or at −20°C for long-term storage.
2. 96-well thin-walled PCR plates with appropriate lid-strips.
3. 8-well PCR r-tube strips with appropriate lid-strips.
4. TaqPolymerase (5 U/µL, Life Technologies, store at −20°C) with the supplied buffers and MgCl$_2$ (kept at 4°C).
5. 100 mM single dNTP solutions (Life Technologies, store at −20°C). Combine them (resulting in 25 mM for each dNTP), and prepare aliquots to be stored at −20°C. The working aliquot is kept at +4°C to avoid freeze-thaw-cycles.
6. HEX-labeled (*see* **Notes 1–3**) forward primers (light sensitive) and unlabeled reverse primers (Life Technologies, sequences in **Table 1**). The primers are re-suspended in water (we did not try dilution in TE) to 200 µM (stock solution) and further diluted to 20 µM (working dilution). Aliquots (50–200 µL) of the working dilution should be kept at −20°C in order to minimize freezing and thawing of the stock solution. Once an aliquot of the working dilution is thawed, it should be kept at 4°C and used within 4 wk.
7. "Hybaid PCR Express" PCR machine.

Table 1
Primers Used for the Multiplex PCR[a]

Primer	Forward (5'- 3')	Reverse (5'- 3')	Product (bp)
Ex 1*	CGCGCATGGGCCTGTTCCTG	GCGGCGCAGAGAGGCTGTAC	94
Ex 2	ATGTTGCTATCACCATTTAAGGG	AGATTGGCAGGCGCAGGCGGCATG	308
Ex 3i*	AATTGTGACCTGGATCAGC	CTGGACTTCCAGCTGGTGGTGAG	243
Ex 3iDos*	as Ex 3i For	AGTCAAGCTCTGGGGCTCC	138
Ex 4o	ACAAGCTTTAAAGAGTTTCTTGT	AGGCAATGTGTTAGTACACA	261
Ex 5	ACATGTCTTAAGGAGTACATTT	TCTCTAATTTCCTGGCAAACAGTG	227
Ex 6	AGAGATTGTTTACTGTGGAAACA	GAGTGATGCTATTTTTAGATCCT	268
Ex 7o	TGCCTTTCCACACTGACAGGTCT	TCTGTTCTTCATTAGCATTAGAGA	239
Ex 8	TGATAGTCATAACTGTGTGTAAG	ACTGTCTCATTAGCGTCTATCTT	206
Ex 9	GGGTGAAATTTGCAGTCAGT	AATATAATCCCAGCCCATGTGCA	278
Ex 10	ATTGCCAAATGCAACCTMTGTC	TTGGAGGAATGAGTAGGGCATT	165
Ex 11	ACAGGGAACATAAACTCTGATCC	CAACACCAGGCACCTTCAGA	303
Ex 12	GTTTGGGAATGCGTGTTT	AGAATTAGAAAATGAAGGTAGACA	255
C328*	ACGTTCCTGATAATGGGATC	CCTCTCTCTACCAAGTGAGG	328

[a]The primers of the exons marked with an * were designed for this assay, the others are as in Kitada et al. (1). Forward primers are 5'-fluorescently labeled (with the same fluorchrome or see **Note 3**).

2.2. Polyacrylamide Gel Electrophoresis

1. Prepare stocks of 5% denaturing polyacrylamide gels using Page-Plus 40% (Amersham) according to the manufacturer's recommendations. Keep 35-mL aliquots at 4°C and use within 4 wk.
2. An automated sequencer (e.g., ABIPRISM 377 upgraded for 96 wells by Applied Biosystems, *see* **Note 3**).
3. Internal size standard (TAMRA-500 and TAMRA-500 XL by Applied Biosystems) and its blue loading buffer.
4. Deionized formamide (Amersham).

2.3. Interpretation of Data

1. Fragment analysis software (e.g., Genescan 3.1 and Genotyper 1.1.1 software by Applied Biosystems).
2. Calculation software (e.g., Excel 97 by Microsoft).

3. Methods

3.1. Multiplex PCR

3.1.1. For a 25 µL Reaction

1. Set up premixes (13 µL per reaction) corresponding to the first 3 exon combinations (for combination 4, *see* **Note 4**) by adding 2.5 µL of 10X buffer (final concentration 1X), 1.5 µL of 50 mM MgCl$_2$ (final concentration 3 mM), the appropriate amount of each primer (between 0.3 and 2.5 µL, final concentration between 0.4 and 2.0 µM, *see* **Note 5**), 0.2 µL of 25 mM dNTP-Mix (final concentration 0.2 mM per dNTP) and complete with H$_2$O up to 13 µl per reaction. Prepare enough premix for duplicate (or more) reactions for each case.
2. Spot 2 µL of the DNA (= 40 ng) on the wall of a 96-well PCR-plate. This is an easy way to control which well contains DNA (*see* **Note 6**).
3. Add 13 µL of the appropriate premix to the well (*see* **Note 7**). Close the wells with lid strips, vortex briefly, and spin down the contents in a centrifuge equipped for PCR plates.
4. Prepare for a semi-hotstart (*see* **Note 8**) by heating the PCR apparatus to 94°C. Open the wells carefully, add 10 µL of Taq solution (2 U/10 µL water kept on ice) using a distribution pipet, close the wells with new strips, vortex briefly, tap the plate on the lab bench to collect all liquid at the bottom of the wells, and place it on the preheated PCR block.
5. Use the following cycling conditions in "tube simulation mode" for combinations 1–3: 94°C for 5 min (initial denaturation), 23 cycles of 94°C for 30 s, 53°C for 45 s and 68°C for 2.5 min, 68°C for 5 min (final extension). For combination 4, *see* **Note 4**.

3.2. Polyacrylamide Gel Electrophoresis

1. Prepare the sequencing gel by adding ammonium persulfate and TEMED, following the manufacturer's recommendations, to an aliquot of the acrylamide.
2. Prepare the loading buffer by pipetting together a premix containing: 0.3 µL of TAMRA-500 XL, 0.6 µL of loading blue, and 3 µL of formamide (for sequencer with 36 lanes; *see* **Note 3**) per sample.
3. Pipet 3.5 µL aliquots of this loading premix into the wells or tubes used to prepare the samples for loading.
4. Add 2–2.5 µL of the PCR products to the premix. Close the tubes, centrifuge briefly, vortex, and centrifuge again (for sequencers with 36 lanes; *see* **Note 3**).

5. Load 1–1.5 µL of this mixture into the wells of the sequencing gel and let run it as for a standard fragment length analysis protocol.
6. Analyze the gel image with appropriate software for calculating the peak heights (e.g., Genescan 3.1 for gel analysis and Genotyper 1.1.1 for peak labeling; *see* **Note 9**). Print the results.

3.3. Interpretation of Data

1. Transfer the peak heights, indicated by Genotyper, to an Excel sheet that has been configured to automatically calculate all the ratios that are needed (*see* **Notes 10–12**).
2. Check the size of the PCR products in order to detect small deletions and insertions. This is done by placing a ruler over the respective peaks on the printed pages. Alternatively, Genotyper can label peak size/length and height.

3.4. Verification that PCR Amplification is in the Exponential Phase

Correct quantification requires that PCR amplification is in the exponential phase for each of the 4 exon combinations:

1. Prepare 50 µL PCR reactions in duplicate for each multiplex combination using DNA from a control subject as template. Ideally, use an 8-well-PCR-strip.
2. Run the aforementioned PCR programs.
3. Take 8 µL aliquots after 20, 21, 22, 23, 24, and 25 cycles by stopping ("pause") the PCR machine a few seconds before the end of the respective extension phase. Take the aliquot with a multipipet to work as quickly as possible. Continue the PCR program afterwards.
4. Collect the aliquots in an appropriately labeled 96-well plate on ice.
5. Transfer 2 µL of each aliquot into loading buffer, load the gel, and perform the fragment analysis.
6. Label and print the electrophoregrams of the exon combinations and the size marker.
7. Prepare an "Excel" sheet where you can enter the crude peak height values. The program can correct for differences in the amount of sample loaded. For this purpose, calculate the ratio of the peak height of a TAMRA-peak (e.g., 200 bp) in the first lane to the TAMRA-200-bp-peaks of the other samples on the same gel. This results in a factor of normalization by that the exon peak heights of the corresponding sample should be multiplied.
8. Finally, let "Excel" plot the corrected peak heights as a function of cycle number on a logarithmic scale. The exponential phase is shown by a straight ascending line. The lines should be parallel for all exons. Ideally, the should coincide.
9. If amplification is not exponential for one or more exons, adjust the primer concentrations (decrease the concentration for exons that amplify too rapidly and increase the concentration for exons that amplify too slowly).

4. Notes

1. During preliminary experiments, it was noted that exons that amplify well might negatively affect the amplification of the other exons. Thus, we partly followed the recommendations of Henegariu et al. *(8)* and grouped together exons that amplify to the same extent. In addition, since smaller PCR products often amplified better than longer products, we grouped together exons of similar size. Thus, the following 3 exon combinations were chosen for multiplex PCR (primers are as specified in **Table 1**):
   comb 1: Ex 4o (261 bp) + 7o (239 bp) + 8 (206 bp) + 11 (303 bp),
   comb 2: Ex 5 (227 bp) + 6 (268 bp) + 8 (206 bp) + 10 (165 bp) and
   comb 3: Ex 2 (308 bp) + 3i (243 bp) + 9 (278 bp) + 12 (255 bp) + C328
   (external control of 328 bp, derived from the transthyretine gene on chromosome 18).

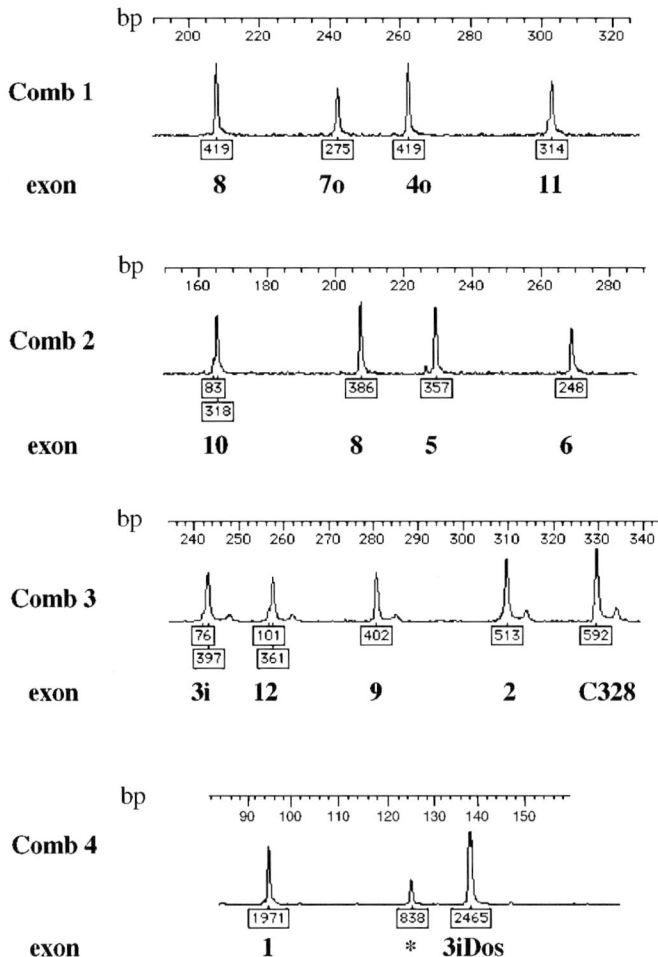

Fig. 1. Four exon combinations used in multiplex PCRs. Representative electrophoregrams obtained after multiplex PCR with HEX-labeled forward primers and separation on 5% denaturing gels on an automated Sequencer (ABI PRISM 377). Peak length (indicated in bp above each electrophoregram) and peak height (indicated below each electrophoregram) are given as calculated by Genotyper 1.1.1 software (Applied Biosystems). Exons are numbered below each peak. Please note that two peak heights are given for double peaks the sizes of which have to be added (*see* **Note 9**). *, unspecific peak.

Representative examples of electrophoregrams are given in **Fig. 1**. However, other peak patterns are also possible, as long as all exons amplify exponentially (*see* **Subheading 3.4.**). Exon 8 was amplified in comb 1 and comb 2, so that the same positive control (heterozygously deleted for exons 8 and 9) could be used in all three combinations.

2. More recently, we added a fourth exon combination: Ex1 (94 bp) + Ex3iDos (138 bp) in order to co-amplify exon 1. This is difficult, since the 5'- untranslated region of the parkin gene (containing the forward primer binding site for exon 1) is very GC-rich. We obtained the best results with the primers mentioned in **Table 1** for exon 1 and exon 3iDos. However, a nonspecific peak is also obtained (*see* **Fig. 1**) that co-amplifies exponentially with the two exons (for PCR conditions; *see* **Note 8**). Since all known exon 3 deletions were

detected with this combination, we assume that exon 1 deletions should also have been detected. However, we have not yet detected an exon 1 deletion in the patients studied.

3. In order to save space on the sequencing gel (e.g., with a 36 lanes sequencer), the exons of the first 3 combinations can be labeled with 3 different fluorchromes (e.g., comb 1 with TET, comb 2 with FAM, and comb 3 with HEX). Aliquots of the PCRs are then pooled in the loading buffer. Under these conditions, we obtained good results with 1 µL of comb 1, 1.5 µL of comb 2 and 3 µL of comb 3, added to 1.5 µL of TAMRA-500, 3 µL of formamide and 0.7 µL of loading blue. Combination 4 is run separately, since no other fluorchrome is available.

4. For multiplex combination 4, the best results were obtained with the following PCR protocol: in a final volume of 25 µL, 40 ng of DNA, 10% DMSO, 1.5 μM of each primer, 0.2 mM of each dNTP and 2 U of Taq polymerase. The cycling protocol consisted of 94°C for 4 min followed by 22 cycles of 94°C for 30 s, 53°C for 30 s and 68°C for 45 s. Final elongation was performed at 68°C for 10 m.

5. Concentrations of the primers varied from 0.40 µM to 2 µM. The exact primer concentrations have to be determined experimentally because they vary among batches.

   In order to obtain approximately equal peak heights for the exons in each multiplex PCR combination and to be in the exponential phase for each exon, we followed in part the recommendations of Henegariou et al. *(8)*, in particular: lowering the annealing temperature to 53°C and extension temperature to 68°C, increasing the MgCl$_2$ concentration to 3 µM and the extension time to 2.5 min (*see* **Note 4** for combination 4). In addition, the primer concentrations have to be adjusted. We recommend starting with 0.8 µM for each primer and adapting the concentration according to the resulting peak height (i.e., primer concentrations have to be decreased for exons that amplify well and increased for exons that amplify less well).

6. In addition to the cases, at least one normal control and a negative control (without template DNA) should be run in parallel. We also included a positive control with known heterozygous deletions of exons 8 and 9 in order to control for false results due to variation in the premix. However, if no positive control with known exon rearrangements is available, the dosage assay can still be considered to be valid as long as PCR amplification is exponential.

7. In order to facilitate mixing of the reagents, pipet the premix onto the drop of DNA (e.g., with a Gilson Distriman with Distritips) but avoid touching the DNA to prevent carryover contamination.

8. If the PCR premix is kept on ice or if a Taq Polymerase with automated hot-start is used, the TaqPolymerase could also probably be added directly to the premix, although we have not done this.

9. Peak labeling programs can determine peak length (size of the PCR product), peak height, and peak area. Peak area would seem to be the most precise reflection of the amount of PCR product, but often the algorithms used only approximate the area. We have had more consistent results using peak height. It is important to sum the peak heights when a PCR product forms a double peak with 1 bp difference, caused by the addition of an A to the PCR products by the polymerase (*see* examples in **Figs. 1–3**). Genotyper does not always distinguish double peaks, so vigilence is necessary.

10. The interpretation of the data is based on the relative peak heights obtained within a given multiplex reaction. The method is therefore only semi-quantitative and does not give absolute values that would be affected by the amount and quality of template DNA as well as slight variations in the premix. In order to establish the peak height patterns for cases and controls, the ratios of the peak heights of each exon to every other exon in a given exon combination are calculated (*see* **Tables 2** and **3**). The average ratio of the duplicate/triplicate/quadruplicate reactions are then calculated and normalized with respect to the

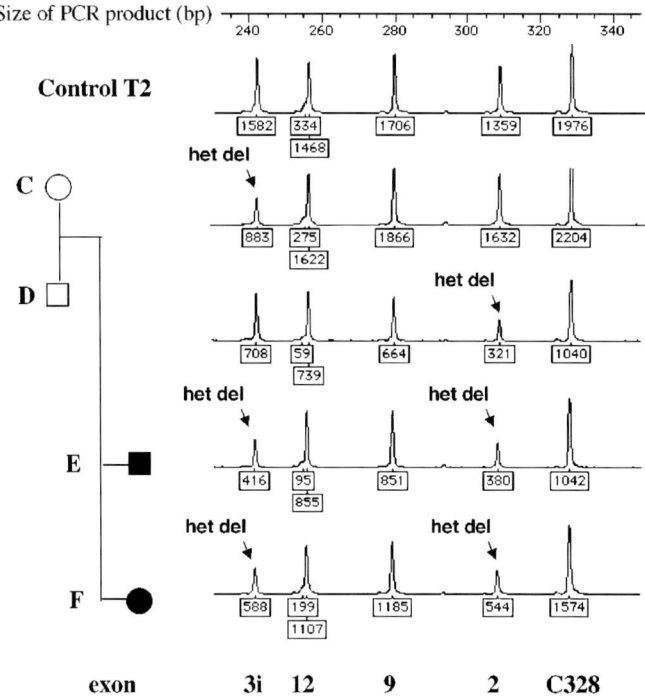

Fig 2. Compound heterozygous deletions of exon 2 and 3. Representative electrophoregrams of the cosegregation of two different exon deletions in a family compared to a control. Peak labeling is as in **Fig. 1**. For each case, one of the triplicate PCRs is shown. However, in **Table 2**, all three PCRs are used for the NR calculation. Circles, women; squares, men; filled symbols, patients; het, heterozygous; del, deletion.

control(s). The normalized ratios (NRs) obtained in different experiments (e.g., if experiments are performed several times with different batches of DNA) are therefore comparable and mean normalized ratios (MNRs) can be calculated.

In practice, a template "Excel" sheet should be established containing all the calculation instructions described below, so that only the crude peak heights need to be entered. Two examples are given in **Tables 2** and **3**, that correspond to the electrophoregrams in **Figs. 2** and **3**. In the left part of the table, the peak heights for each case are entered, e.g., the duplicate PCRs of controls 1 to 3 and the quadruplicate PCRs of the cases A and B (**Table 3**). In the right part of the table, next to the peak heights, the program calculates the peak height ratios as indicated at the top of the tables. Below, the program calculates the mean values (in bold) and the standard deviation of the ratios, respectively. Rearrangements are deduced from the division of the mean peak height ratios of all available controls (i.e., one control in triplicate in **Table 2** and three controls in duplicate in **Table 3**) by the mean peak height ratios of the case. This results in the NRs for each exon peak height ratio (shown in bold italics for normal values or bold underlined for pathological values in the table). The NRs have a value of approx 1 (i.e., 0.8 to 1.2) for nonrearranged exons (2 copies of an exon). Values of 0.6 or less are interpreted as one copy of an exon (i.e., heterozygous deletion), values of 1.3 to 1.7 as 3 copies (i.e., heterozygous duplication), values of 1.8 to 2.3 as 4 copies (i.e., homozygous duplication or heterozygous triplication)

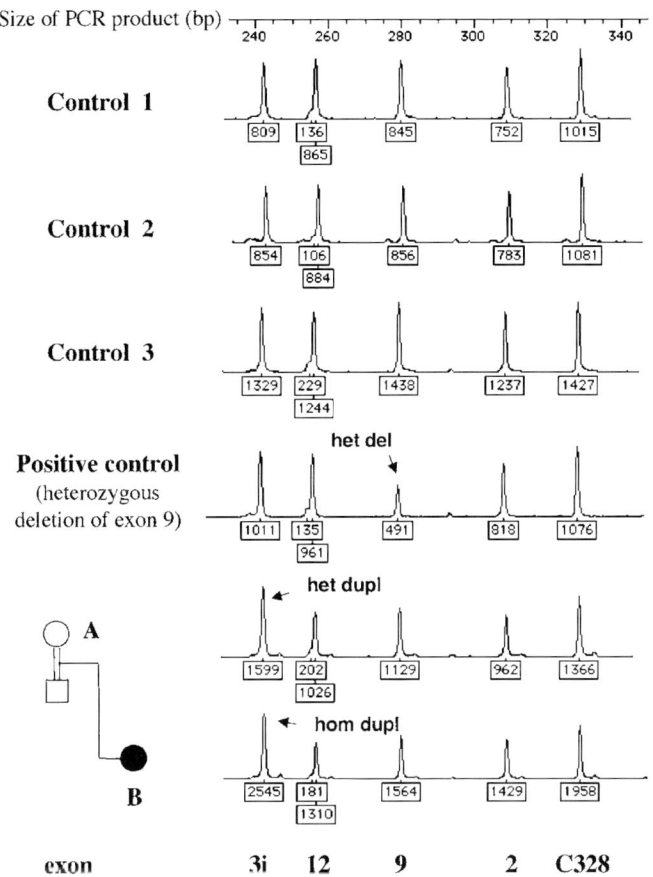

Fig 3. Homozygous duplication of exon 3. Representative electrophoregrams of the cosegregation of an exon duplication in a family compared to 3 controls. In addition, a positive control with known heterozygous deletions of exons 8 and 9 was run in parallel. All values of the duplicate or quadruplicate PCRs are shown in table 3. Peak labeling and symbols are as in **Figs.1** and **2**. Het, heterozygous; hom, homozygous; del, deletion; dupl, duplication.

and values above 2.6 as indication for 6 copies (i.e., homozygous triplication) (*see* **Fig. 4**). Please note that the NR values given above apply only if the exon under investigation is the dividing value in the exon ratio, i.e., C328/3i. If the ratio is Ex3i/12 (e.g., case C in **Table 2**), the NR becomes approx 2 (1/0.5). An exon rearrangement is confirmed only if all of the ratios concerning the exon are abnormal (*see* tables). If the results are ambiguous, the PCR should be repeated, if possible with a new DNA sample and a larger number of replicates (e.g., up to quadruplicate). In addition, several control cases can be included to obtain mean values for the control ratios.

11. Reproducibility. As seen from the tables, standard deviations are usually about 10% of the mean, and rarely exceed 20%, for both the crude exon ratios calculated from the PCR replicates and the mean NRs from several different experiments (data not shown).

12. Search for homozygous exon deletions. In addition to the detection of heterozygous exon deletions, the multiplex PCR assay also detects homozygous exon deletions that are obvious, since one or several peaks are missing.

Table 2
Calculation of Exon Ratios, SD, and Normalized Ratios Corresponding to Fig. 2[a]

Case	3i	12	9	2	C328	C328/3i	C328/12	C328/9	C328/2	Ex3i/12	Ex3i/9	Ex3i/2	Ex12/9	Ex12/2	Ex9/2
Control T2	1491	1550	1338	1143	2031	1.36	1.31	1.52	1.78	0.96	1.11	1.30	1.16	1.36	1.17
	1582	1802	1706	1359	1976	1.25	1.10	1.16	1.45	0.88	0.93	1.16	1.06	1.33	1.26
	1445	1580	1525	1379	1990	1.38	1.26	1.30	1.44	0.91	0.95	1.05	1.04	1.15	1.11
Mean						**1.33**	**1.22**	**1.33**	**1.56**	**0.92**	**1.00**	**1.17**	**1.08**	**1.28**	**1.18**
SD						0.07	0.11	0.18	0.19	0.04	0.10	0.13	0.07	0.11	0.07
Case C	693	1593	1606	1441	1990	2.87	1.25	1.24	1.38	0.44	0.43	0.48	0.99	1.11	1.11
	760	1746	1646	1435	2143	2.82	1.23	1.30	1.49	0.44	0.46	0.53	1.06	1.22	1.15
	883	1897	1866	1632	2204	2.50	1.16	1.18	1.35	0.47	0.47	0.54	1.02	1.16	1.14
Mean						**2.73**	**1.21**	**1.24**	**1.41**	**0.45**	**0.46**	**0.52**	**1.02**	**1.16**	**1.13**
SD						0.20	0.05	0.06	0.08	0.02	0.02	0.03	0.03	0.06	0.02
Mean T2/mean C						*0.47*	*1.01*	*1.07*	*1.11*	*2.15*	*2.28*	*2.37*	*1.06*	*1.10*	*1.04*
Case D	657	762	757	276	862	1.31	1.13	1.14	3.12	0.86	0.87	2.38	1.01	2.76	2.74
	708	798	664	321	1040	1.47	1.30	1.57	3.24	0.89	1.07	2.21	1.20	2.49	2.07
	811	962	894	414	1278	1.58	1.33	1.43	3.09	0.84	0.91	1.96	1.08	2.32	2.16
Mean						**1.45**	**1.25**	**1.38**	**3.15**	**0.86**	**0.95**	**2.18**	**1.09**	**2.52**	**2.32**
SD						0.13	0.11	0.22	0.08	0.02	0.11	0.21	0.10	0.22	0.37
Mean T2/mean D						*0.88*	*0.97*	*0.96*	*0.49*	*1.11*	*1.10*	*0.56*	*0.99*	*0.51*	*0.51*
Case E	430	973	881	428	1333	3.10	1.37	1.51	3.11	0.44	0.49	1.00	1.10	2.27	2.06
	416	950	851	380	1042	2.50	1.10	1.22	2.74	0.44	0.49	1.09	1.12	2.50	2.24
	463	1024	943	427	1248	2.70	1.22	1.32	2.92	0.45	0.49	1.08	1.09	2.40	2.21
Mean						**2.77**	**1.23**	**1.35**	**2.93**	**0.44**	**0.49**	**1.06**	**1.10**	**2.39**	**2.17**
SD						0.30	0.14	0.15	0.19	0.01	0.00	0.05	0.02	0.11	0.10
Mean T2/mean E						*0.46*	*0.99*	*0.98*	*0.53*	*2.16*	*2.12*	*1.16*	*0.98*	*0.53*	*0.54*
Case F	588	1306	1185	544	1574	2.68	1.21	1.33	2.89	0.45	0.50	1.08	1.10	2.40	2.18
	915	2078	1943	853	2413	2.64	1.16	1.24	2.83	0.44	0.47	1.07	1.07	2.44	2.28
	944	1913	1825	844	2428	2.57	1.27	1.33	2.88	0.49	0.52	1.12	1.05	2.27	2.16
Mean						**2.63**	**1.21**	**1.30**	**2.87**	**0.46**	**0.49**	**1.09**	**1.07**	**2.37**	**2.21**
SD						0.05	0.05	0.05	0.03	0.03	0.02	0.02	0.03	0.09	0.06
Mean T2/mean F						*0.49*	*1.01*	*1.02*	*0.54*	*2.08*	*2.10*	*1.12*	*1.01*	*0.54*	*0.53*

[a]The steps of the calculation are explained in **Note 10**. Note that the exon ratios are independent of the absolute peak heights.

Table 3
Calculation of Exon Ratios, SD, and Normalized Ratios Corresponding to Fig. 3[a]

Case	3i	12	9	2	C328	C328/3i	C328/12	C328/9	C328/2	Ex3i/12	Ex3i/9	Ex3i/2	Ex12/9	Ex12/2	Ex9/2
Control 1(C1)	809	1001	845	752	1015	1.25	1.01	1.20	1.35	0.81	0.96	1.08	1.18	1.33	1.12
	953	1157	1000	959	1344	1.41	1.16	1.34	1.40	0.82	0.95	0.99	1.16	1.21	1.04
Mean						**1.33**	**1.09**	**1.27**	**1.38**	**0.82**	**0.96**	**1.03**	**1.17**	**1.27**	**1.08**
SD						0.11	0.10	0.10	0.04	0.01	0.00	0.06	0.02	0.09	0.06
Mean C1-3/mean C1						*0.90*	*0.95*	*0.92*	*0.96*	*1.05*	*1.02*	*1.06*	*0.97*	*1.01*	*1.05*
Control 2(C2)	745	897	764	669	828	1.11	0.92	1.08	1.24	0.83	0.98	1.11	1.17	1.34	1.14
	854	990	856	783	1081	1.27	1.09	1.26	1.38	0.86	0.98	1.11	1.17	1.34	1.14
Mean						**1.19**	**1.01**	**1.17**	**1.31**	**0.85**	**0.99**	**1.10**	**1.17**	**1.30**	**1.12**
SD						0.11	0.12	0.13	0.10	0.02	0.02	0.02	0.01	0.05	0.03
Mean C1-3/mean C2						*1.01*	*1.02*	*1.00*	*1.01*	*1.01*	*0.98*	*1.00*	*0.97*	*0.99*	*1.01*
Control 3(C3)	1329	1473	1438	1237	1427	1.07	0.97	0.99	1.15	0.90	0.92	1.07	1.02	1.19	1.16
	1444	1578	1423	1146	1616	1.12	1.02	1.14	1.41	0.92	1.01	1.26	1.11	1.38	1.24
Mean						**1.10**	**1.00**	**1.06**	**1.28**	**0.91**	**0.97**	**1.17**	**1.07**	**1.28**	**1.20**
SD						0.03	0.04	0.10	0.18	0.01	0.06	0.13	0.06	0.13	0.06
Mean C1-3/mean C13						*1.10*	*1.03*	*1.10*	*1.03*	*0.94*	*1.00*	*0.94*	*1.06*	*1.00*	*0.94*
Mean of C1-3						1.21	1.03	1.17	1.32	0.86	0.97	1.10	1.13	1.29	1.13
SD of C1-3						0.13	0.09	0.13	0.10	0.04	0.03	0.09	0.06	0.08	0.07

(*continued*)

Table 3 (continued)
Calculation of Exon Ratios, SD and Normalized Ratios Corresponding to Fig. 3[a]

Case	3i	12	9	2	C328	C328/3i	C328/12	C328/9	C328/2	Ex3i/12	Ex3i/9	Ex3i/2	Ex12/9	Ex12/2	Ex9/2
Positive Control (PC)	937	1041	494	901	1294	1.38	1.24	2.62	1.44	0.90	1.90	1.04	2.11	1.16	0.55
	1011	1096	491	818	1076	1.06	0.98	2.19	1.32	0.92	2.06	1.24	2.23	1.34	0.60
Mean						**1.22**	**1.11**	**2.41**	**1.38**	**0.91**	**1.98**	**1.14**	**2.17**	**1.25**	**0.57**
SD						0.22	0.18	0.30	0.09	0.02	0.11	0.14	0.09	0.13	0.04
Mean C1-3/mean PC						*0.99*	*0.93*	*0.49*	*0.96*	*0.94*	*0.49*	*0.97*	*0.52*	*1.03*	*1.98*
CaseA	2116	1634	1465	1372	1930	0.91	1.18	1.32	1.14	1.29	1.44	1.54	1.12	1.19	1.07
	1375	1252	1087	963	1259	0.92	1.01	1.16	1.31	1.10	1.26	1.43	1.15	1.30	1.13
	1599	1228	1129	962	1366	0.85	1.11	1.21	1.42	1.30	1.42	1.66	1.09	1.28	1.17
	2569	1700	1586	1377	1965	0.73	1.10	1.18	1.35	1.51	1.62	1.87	1.07	1.23	1.15
Mean						**0.85**	**1.10**	**1.22**	**1.37**	**1.30**	**1.44**	**1.62**	**1.11**	**1.25**	**1.13**
SD						0.09	0.07	0.07	0.05	0.17	0.15	0.19	0.04	0.05	0.05
Mean C1-3/mean A						*1.42*	*0.94*	*0.96*	*0.96*	*0.66*	*0.68*	*0.68*	*1.02*	*1.03*	*1.00*
Case B	2975	1704	1545	1428	1881	0.63	1.10	1.22	1.32	1.75	1.93	2.08	1.10	1.19	1.08
	2452	1273	1144	991	1370	0.56	1.08	1.20	1.38	1.93	2.14	2.47	1.11	1.28	1.15
	2545	1491	1564	1429	1958	0.77	1.31	1.25	1.37	1.71	1.63	1.78	0.95	1.04	1.09
	2776	1523	1435	1107	1495	0.54	0.98	1.04	1.35	1.82	1.93	2.51	1.06	1.38	1.30
Mean						**0.62**	**1.12**	**1.18**	**1.36**	**1.80**	**1.91**	**2.21**	**1.06**	**1.22**	**1.16**
SD						0.10	0.14	0.09	0.03	0.10	0.21	0.35	0.07	0.14	0.10
Mean C1-3/mean B						*1.93*	*0.92*	*0.99*	*0.98*	*0.48*	*0.51*	*0.50*	*1.07*	*1.05*	*0.98*

[a] The steps of the calculation are explained in Note 10. Note that the exon ratios are independent of the absolute peak heights.

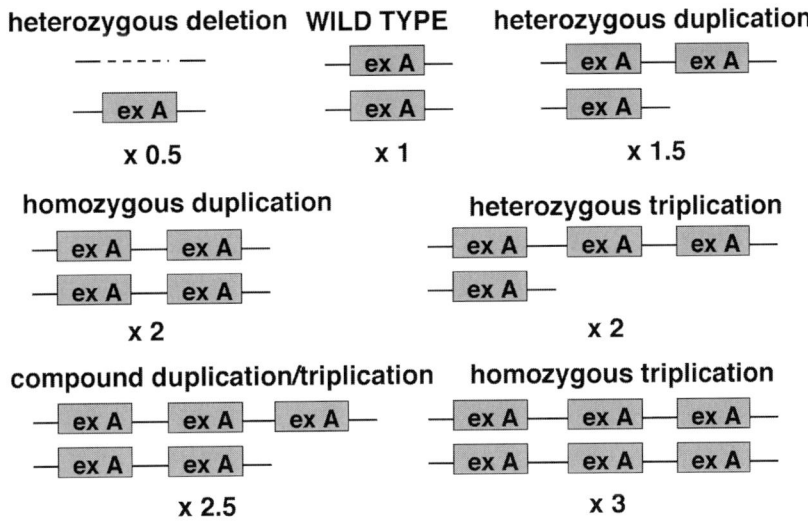

Fig 4. Schematic representation of the exon rearrangements. Below each type of rearrangement, the NR is indicated. Since the presence of two exons results in an NR of 1, the presence of one exon (heterozygous deletion) results in a NR of 0.5, more than 2 exons (as shown for the different multiplications) in NRs >1, i.e., 4 exons resulting in a NR of 2 and 6 exons in a NR of 3. The positions of the multiplied exons are theoretical, because we do not know whether they are arranged in tandem.

Acknowledgments

This work was supported by the AP-HP (Assistance publique - Hôpitaux de Paris), the Association France-Parkinson, the European Community Biomed 2 (BMH4CT960664), and Aventis Pharma France. C.B.L. was supported by the Deutsche Forschungsgemeinschaft (DFG). We thank the family members for their participation, N. Rawal for the establishment of the multiplex combination 4, and Dr. Merle Ruberg for helpful discussions.

References

1. Kitada, T., Asakawa, S., Hattori, N., Matsumine, H., Yamamura,Y., Minoshima, S., et al. (1998) Mutations in the parkin gene cause autosomal recessive juvenile parkinsonism. *Nature* **392,** 605–608.
2. Hattori, N., Kitada, T., Matsumine, H., Asakawa, S., Yamamura, Y., Yoshino, H., et al. (1998) Molecular genetic analysis of a novel parkin gene in Japanese families with autosomal recessive juvenile parkinsonism: evidence for variable homozygous deletions in the parkin gene in affected individuals. *Ann. Neurol.* **44,** 935–941.
3. Lücking, C. B., Dürr , A., Bonifati, V., Vaughan, J., De Michele, G., Gasser, T., et al. (2000) Association between early-onset Parkinson's Disease and mutations in the *parkin* gene. *N. Engl. J. Med.* **342,** 1560–1567.
4. Klein, C., Pramstaller, P. P., Kis, B., Page, C. C., Kann, M., Leung, J., et al. (2000) Parkin deletions in a family with adult-onset, tremor-dominant parkinsonism: expanding the phenotype. *Ann. Neurol.* **48,** 65–71.

5. Shimura, H., Hattori, N., Kubo, S., Mizuno, Y., Asakawa, S., Minoshima, S., et al. (2000) Familial Parkinson disease gene product, parkin, is a ubiquitin-protein ligase. *Nat. Genet.* **25,** 302–305.
6. Zhang, Y., Gao, J., Chung, K. K. K., Huang, H., Dawson, V. L., and Dawson, T. M. (2000) Parkin functions as an E2-dependent ubiquitin-protein ligase and promotes the degradation of the synaptic vesicle-associated protein, CDCrel-1. *Proc. Natl. Acad. Sci. USA* **97,** 13,354–13,359.
7. Prior, T. W. (1998) Determining gene dosage. *Clin. Chem.* **44,** 703–704.
8. Henegariu, O., Heerema, N. A., Dlouhy, S. R., Vance, G. H., and Vogt, P. H. (1997) Multiplex PCR: Critical Parameters and Step-by-Step Protocol. *BioTechniques* **23,** 504–511.

II

TRINUCLEOTIDE REPEAT DETECTION

3

Detection of FMR1 Trinucleotide Repeat Expansion Mutations Using Southern Blot and PCR Methodologies

Jack Tarleton

1. Introduction

Fragile X syndrome, caused by the loss or diminution of the FMR1 (FRAXA - chromosomal locus Xq27.3) encoded protein, FMRP, results in mild to moderate mental retardation as its hallmark. Patients with the syndrome often vary dramatically in presentation with a range of intellectual and behavioral deficits, and provide a diagnostic challenge for clinicians due to the subtle nature of the physical phenotype *(1,2)*. Instability of a CGG repeat segment contained within FMR1 exon 1 is the molecular basis for nearly all mutations (>99%) in the gene and leads to reduced or complete loss of FMRP *(3–8)*. The variable phenotype occurs related to variation in FMR1 expression mediated by the extent of CGG repeat expansion and a secondary epigenetic feature: the aberrant hypermethylation of CpG dinucleotides contained in the CGG repeat segment and surrounding regions of the gene *(9)*. Thus, molecular genetic studies of FMR1 are utilized to confirm a clinical diagnosis of fragile X syndrome, and perhaps just as importantly, to exclude an alteration in FMR1 as an explanation for nonspecific mental retardation in a patient. For clinical molecular diagnosis, the variety of FMR1 alleles and the myriad of possible alterations in the gene present a diagnostic challenge for which no one detection method has proven fully satisfactory. Here, a dual approach to FMR1 repeat expansion mutation detection utilizing Southern blot and polymerase chain reaction (PCR) methodologies is presented *(10,11)*. The reader is referred to published technical standards for fragile X analysis to supplement the interpretation of molecular genetic results for patients *(12)*.

Routine clinical molecular genetic testing for mutations in the FMR1 gene has as its goal the determination of both the trinucleotide repeat number and the FMR1 methylation status. This latter feature of aberrant hypermethylation in FMR1 typically accompanies trinucleotide repeat expansion when more than approx 200 repeats are present. Gross expansion to many hundreds of repeats with complete methylation ("full mutation") is the most common finding in patients with fragile X syndrome. However, there are numerous exceptional patients with unusual FMR1 alterations, mirroring the variable clinical presentations.

Table 1
Risk for Maternal Premutation Expansion (Data from *13,14*)

Number of maternal premutation CGG repeats	Approximate % risk for affected son	Approximate % risk for affected daughter
56–59	7	3.5
60–69	10	5
70–79	29	15
80–89	36	18
90–99	47	24
>100	50	25

In general, the number of CGG repeats ranges from 5–44 in normal (stable) alleles. In a range of 45 to 54 repeats occasional minor instability may be observed leading to a "gray zone" of borderline alleles. Alleles in the gray zone can be considered normal as women with these alleles have not been observed to have children with fragile X syndrome. Repeats from 55–58 have been observed to have more dramatic instability of repeat copy number, yet no direct transmissions to a full mutation have been reported. To date, all known mothers of affected children have had alleles of 59 repeats or higher *(13)*. However, generally alleles with repeats ranging from ~55 to ~200 are considered "premutation" alleles, which have the capacity for instability and expansion (*see* **Notes 1** and **2** regarding technical approaches diagnostic testing).

Premutation alleles are associated with no symptoms of fragile X syndrome, or perhaps only very mild effects, on intellect or behavior. A female with a premutation allele has variable risk for repeat instability upon transmission to her children based on the number of repeats. When transmitted through the mother, premutations may stay essentially unchanged or may expand into full mutations. For example, a female having a premutation allele containing 70 repeats statistically has a risk for producing a child with fragile X syndrome greater than another female having a premutation allele with 69 repeats (**Table 1**)—illustrating the need for precise repeat copy number analysis for genetic counseling purposes *(13,14)*. Interestingly, dramatic repeat instability does not occur when a premutation is transmitted from father to daughter. However, daughters of males with a premutation then are obligate carriers of a premutation themselves, which may expand upon transmission to her children.

Figure 1 demonstrates the complexity of Southern blot patterns associated with FMR1 repeat expansion and instability (*see* **Notes 2** and **3**). The vast majority of patients with >200 repeats will have aberrant methylation of deoxycytosine residues in FMR1 generating a heterogenous "smear" upon Southern blot analysis (**Fig. 1**, lane 5). (This indistinct autoradiographic signal illustrates the repeat instability from cell to cell). Occasionally, a patient with a full mutation will be found to have some cells in which FMR1 methylation has occurred and some cells in which there is no methylation of FMR1 (**Fig. 1**, lane 7). These individuals have been referred to as "methylation mosaics" and their Southern blot pattern represents the presence of cellular mosaicism involving FMR1 methylation events. Less frequently, a patient will be identified who appears to have no cells in which the abnormal methylation events have occurred even

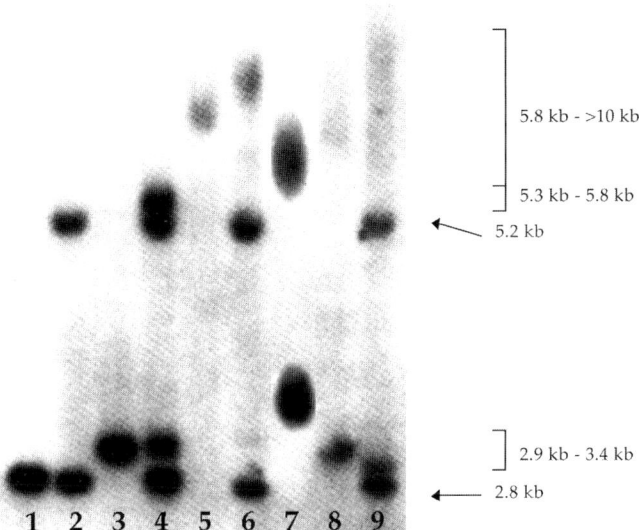

Fig. 1. Southern blot analysis of CGG repeat/methylation in FMR1. Patient DNA was digested with EcoRI and EagI and probed with StB12.3 as described. Lane 1, normal male with 30 repeats; lane 2, normal female with both FMR1 alleles bearing 30 repeats; lane 3, male with premutation containing ~70 repeats; lane 4, female with premutation containing ~70 repeats; lane 5, male with full mutation (smear indicates differences in repeat expansion from cell to cell with a midpoint average of ~800 repeats); lane 6, female with a full mutation (smear midpoint average of ~1000 repeats); lane 7, methylation mosaicism in a male (i.e., partial methylation of a full mutation); lane 8, male with a mixture of premutation and full mutation cell lines ("premutation/full mutation size mosaicism"); lane 9, female with premutation/full mutation size mosaicism.

when >200 CGG repeats are found. This latter group is referred to as having "unmethylated full mutations." An additional common type of mosaicism for molecular events in FMR1 describes patients who have some cells with methylated full mutations and some cells with premutations. These individuals have been termed "size mosaics" or "premutation/full mutation mosaics" (**Fig. 1**, lanes 8 and 9). Collectively, all the mosaicism types comprise perhaps 15% of the observed FMR1 mutations. Partial gene expression may occur in mosaics and lessen the severity of the mental and physical phenotype leading to a continuum of affectation when patients with FMR1 alterations are considered *in toto*. Rarely, a patient DNA sample will not yield an autoradiographic signal due to deletion of the FMR1 region complementary to the Southern blot probe or PCR primers (*see* **Note 4**).

The presence of AGG repeats embedded within the CGG repeat segment appears to mitigate the risk of repeat instability - apparently by disrupting DNA secondary structures which increase the likelihood of DNA strand slippage during replication *(15,16)*. Typically, most individuals have AGG repeats at about repeat 10 and 20. These interrupting repeats appear to "anchor" the segment against expansion. Long stretches of uninterrupted CGG repeats (greater than about 34–38 repeats) beyond the last AGG repeat ("pure repeats") appear to increase the risk for instability of maternal alleles

Fig. 2. PCR Analysis of FMR1. ^{32}P-labeled PCR products were separated on a 6% polyacrylamide DNA sequencing gel for 6 h at 85 W and exposed to X-ray film overnight (approx 18 h). Signals are not distinct but composed of multiple bands due to stuttering of the DNA polymerase during amplification. The most intense band is presumed to represent the actual allele. Lane M, marker; lane 1, female with alleles of 20 and 30 repeats; lanes 2–8, males with 40, 30, 53, 20, 28, 30, and 30 repeats, respectively.

upon transmission to offspring. FMR1 alleles in the premutation range contain long stretches of pure repeats and clearly are at risk for expansion to full mutations. While the number and position of AGG repeats are known to be important in the overall stability of the CGG repeat sequence, sequence analysis is typically available only in research-oriented laboratories.

The consequences of FMR1 repeat expansion are fascinating from the molecular biology perspective, but disconcerting from the patient care perspective as many patients have remarkable variation in FMR1. There is a tremendous responsibility for mastering the information needed for both technical considerations and risk assessment for individuals impacted by the fragile X syndrome. The need to detect both large repeat expansions, FMR1 methylation, and accurately size normal, gray zone, and premutation alleles often leads to the use of multiple technical approaches. Southern blot analysis is not sensitive enough to detect the minor differences between normal, gray zone, and small premutation alleles. Standard PCR specific for FMR1 (**Fig. 2**) often is problematic for amplifying large premutation and full mutation alleles, and does not detect methylation. Presented below are the techniques used in many laboratories for detection of all possible FMR1 mutations involving the trinucleotide repeat expansion and hypermethylation.

2. Materials
2.1. Solutions for FMR1 Southern Blot Analysis

Known reagent suppliers are listed but not recomended or endorsed. Other suppliers may provide the same reagents.

1. Restriction Enzymes: EcoRI (American Allied Biochemical) (HindIII can be substituted) and EagI (New England Biolabs) (BssHII, NruI, or SacI can be substituted).
2. 1.5 mL microcentrifuge tubes (USA Scientific).
3. 10X Bovine serum albumin (BSA): Add 10 mg BSA (Sigma) to 1 mL sterile, deionized H_2O.
4. 10X Spermidine (40 mM): Add 152 mg spermidine (Sigma) to 15 mL sterile, deionized H_2O.
5. 100 mM Magnesium chloride ($MgCl_2$): Add 203 mg $MgCl_2$ (Sigma) to 1 mL sterile, deionized H_2O.
6. 7.5 M Ammonium acetate (NH_4Ac) (1 L): Add 578.1 g NH_4Ac (Sigma) to deionized water. Bring volume to 1 L. Autoclave.
7. 100% ethanol.
8. 2 M Tris-HCl (4 L): Add 968.8 g Trizma Base (Sigma) to sterile, deionized water. Bring volume to 4 L and pH 7.5.
9. 0.5 M EDTA (1 L): Add 186.12 g EDTA (Sigma) to sterile, deionized H_2O. Bring volume to 1 L.
10. TE (low EDTA) buffer (1 L): Add 5 mL 2 M Tris-HCl, pH 7.5, to 0.2 mL 0.5 M EDTA. Bring volume to 1 L and pH to 7.5–8.0.
11. 20X Tris Acetate EDTA (TAE) Electrophoresis Buffer (4 L): Add 387.2 g Trizma Base to sterile, deionized water. Add 91.5 mL glacial Acid and 160 mL 0.5 M EDTA. Bring volume to 4 L and pH 8.1.
12. SeaKem (BioWhittaker) agarose.
13. 10 mg/mL ethidium bromide (Sigma).
14. Bromophenol blue tracking dye (Sigma).
15. 5 M Sodium chloride (NaCl) (4 L): Add 1169 g NaCl (Sigma) to deionized water and bring volume to 4 L. Autoclave.
16. 4 N Sodium hydroxide (NaOH) (4 L): Carefully add 640.6 g of NaOH (Sigma) to deionized water and bring the volume to 4 L. Filter-sterilize and put in plastic bottles.
17. DNA Denaturing solution for agarose gel (1 L): 0.6 M NaCl, 0.2 N NaOH (120 mL 5 M NaCl, 50 mL 4 N NaOH to 1 L with deionized water).
18. Random primer probe labeling kit (Roche).
19. 3 M Sodium acetate (4 L): Add 984.36 g sodium acetate (Sigma) to deionized water and bring volume to 4 L and pH 5.2 with acetic acid. Autoclave.
20. 0.5 M Sodium acetate neutralizing solution for agarose gel (1 L): Add 167 mL 3 M sodium acetate to deionized water and bring to a volume of 1 L. Filter-sterilize.
21. 1 M phosphate solution (4 L): Add 772.04 g $Na_2PO_4 \cdot 7 H_2O$ (dibasic heptahydrate) and 174.71 g $NaHPO_4 \cdot 2 H_2O$ (monobasic dihydrate) (Sigma) to deionized water. Bring volume to 4 L and autoclave.
22. 25 mM phosphate Southern Blot Transfer Solution (1 L): Add 25 mL 1 M phosphate solution to deionized water and bring volume to 1 L.
23. Neutral nylon transfer membrane (Fisher).
24. 3 mm Whatman paper (Fisher).
25. Paper towels.
26. Blot gel transfer basin (e.g., baking sheet or other container with sides greater than about 3 centimeters).
27. Formamide (Sigma).
28. Sodium citrate (Sigma).
29. 20X sodium citrate buffer (SSC) (1 L): add 175.2 g NaCl and 88.0 g sodium citrate to deionized water; bring total volume to 1 L and adjust pH to 7.4 with HCl.

30. Dextran sulfate (Sigma).
31. 50% Dextran Sulfate: add 100 g dextran sulfate to 130 mL sterile, deionized water; mix at 37°C overnight. Store in brown bottles at 4°C.
32. Ficoll 400 (Sigma).
33. Polyvinyl pyrrolidone (Sigma).
34. BSA - fraction V (Sigma).
35. Denhardt's Solution (100 mL): add 2 g Ficoll 400, 2 g polyvinyl pyrrolidone, and 2 g BSA (Fraction V) to sterile, deionized water. Bring total volume to 100 mL. Mix overnight. Cover container to prevent contamination. Store in a plastic bottle at 4°C.
    salmon sperm DNA (Sigma): sonicate
36. Prehybridization Solution (1 L): 500 mL formamide, 250 mL 20X SSC, 100 mL Denhardt's Solution; 50 mL 1 M phosphate; 50 mL 50% dextran sulfate, 50 mL denatured salmon sperm DNA (boil 7 min before adding); Store in brown bottles at 4°C.
37. Hybridization Solution (1 L): 500 mL formamide, 250 mL 20X SSC, 10 mL Denhardt's Solution, 20 mL 1 M phosphate, 200 mL 50% dextran sulfate, 10 mL denatured salmon DNA (boil 7 min before adding). Store in brown bottles at 4°C.
38. 10% sodium lauryl sulfate (Sigma).

2.2. Solutions for Polymerase Chain Reaction Amplification of FMR1

1. Primer C: 5'-AGG CGC TCA GCT CCG TTT CGG TTT CAC TTC-3'
2. Primer F: 5'-AGC CCC GCA CTT CCA CCA CCA GCT CCT CCA-3'
3. 0.2 mL PCR tubes (USA Scientific).
4. 10X PCR buffer (GeneAmp buffer II - Perkin Elmer/Applied Biosystems): 100 mM Tris-HCl, pH 8.3; 500 mM KCl.
5. 25 mM magnesium chloride (Perkin Elmer/Applied Biosystems).
6. 6 M Betaine (Sigma).
7. Deoxynucleotide mix: using 10 mM deoxynucleotide stocks (Perkin Elmer/Applied Biosystems), add 60 μL deoxyadenosine, 60 μL deoxythymidine, 15 μL deoxycytidine, 15 μL deoxyguanosine, and 50 μL sterile, deionized water (final concentrations: 3 mM deoxyadenosine, 3 mM deoxythymidine, 0.75 mM deoxycytidine, and 0.75 mM deoxyguanosine).
8. α^{32}phosphorus-deoxycytidine 5'triphosphate (New England Nuclear).
9. 5 mM 7-deaza-2'-deoxyguanosine-5-triphosphate (Pharmacia Biotech).
10. Dimethylsulfoxide (Fisher).
11. DNA polymerase (AmpliTaq) (Perkin Elmer/Applied Biosystems).
12. 40% (w/v) acrylamide solution (38% acrylamide/2% bis-acrylamide) (Amresco).
13. 10X Tris Borate EDTA (TBE) Electrophoresis Buffer (1 L): Add 108 g Trizma Base to sterile, deionized water. Add 55 g of boric acid (Sigma) and 9.3 g of sodium EDTA (Sigma). Bring volume to 1 L and pH 8.3 with sodium hydroxide (Sigma). Autoclave.
14. Xylene cyanol tracking dye.

3. Methods

3.1. Overview of FMR1 Mutation Detection

Several genomic DNA probes (**Fig. 3**) have been isolated from FMR1 and may be used for molecular diagnosis (3–6). In combination with PCR specific for FMR1, Southern blot analysis using StB12.3, p5.1, or pfxa3 as a probe after methylation-sensitive restriction enzyme digestion covers all contingencies for normal, gray zone, and premutation alleles plus repeat expansion and methylation (*see* **Note 3**). In this analy-

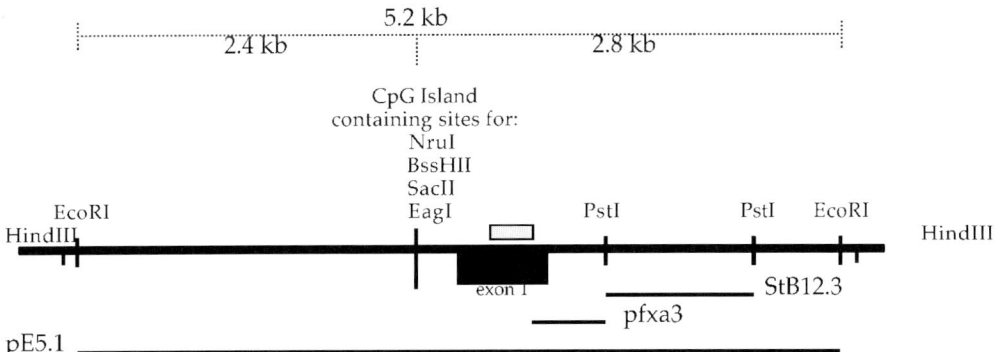

Fig. 3. Restriction map of FMR1 promoter and exon 1 region containing the CGG repeat. DNA probes (StB12.3, pfxa3, and pE5.1) often used in FMR1 analysis are noted. Shaded box represents the relative location of the CGG repeat segment within exon 1.

sis, typically EcoRI or HindIII is used in a double digestion with a methylation-sensitive enzyme having a site present in the CpG island upstream of the repeat segment, e.g., EagI, BssHII, or NruI *(10)*.

Southern blot analysis allows visualization of premutations and the heterogeneous "smears" generated upon autoradiography from full mutations. The Southern blot analysis is complemented by a high resolution PCR assay described by Fu et al *(11)*. This procedure utilizes incorporation of a radioactive isotope, 32-phosphorus, along with DMSO, betaine, and 7-deazaguanosine triphosphate. Analysis is performed on a denaturing 6% polyacrylamide sequencing gel, allowing more precise sizing of gray zone and premutation alleles.

3.2. FMR1 Southern Blot Protocol

The protocol described below uses a sodium phosphate solution for transfer onto a neutral nylon membrane. The use of a neutral membrane minimizes background in our hands on the resulting autoradiograph but other protocols for transfer using different transfer conditions and charged membranes may be utilized. Minimizing autoradiographic background is imperative because the smears generated from FMR1 expansion may be diffuse and difficult to visualize against a high background. DNA probes noted in **Fig. 3** can be used for Southern blot hybridization with genomic DNA digested with both methylation-sensitive and insensitive restriction enzymes. Genomic DNA digestion with EcoRI and EagI is used in many laboratories but other enzyme combinations are possible (*see* **Fig. 3**). Upon probing the blotted digestion products, male and female specific patterns for repeat expansion mutations plus methylation status are generated (*see* **Fig. 1**).

The following is a protocol for FMR1 Southern blot analysis.

1. In a 1.5 mL microcentrifuge tube, digest 5–10 μg of genomic DNA with both EcoRI and EagI using the 10X restriction enzyme buffer for EagI (10 U of enzyme per μg of DNA). Incubate overnight or 12–20 h at 37°C. Following incubation, precipitate the samples using one-half volume of 7.5 *M* ammonium acetate (2.5 *M* final concentration) and two volumes of 100% ethanol. Place at –20°C for 30 min or –70°C for 20 min then spin for 15 min at 12,000*g*. Remove the supernatant and dry the samples on a countertop at room

temperature or in a vacuum dessicator. Resuspend DNA in a small volume of sterile, deionized water or Tris-EDTA (TE) buffer plus electrophoresis tracking dye (bromophenol blue). The resuspension volume should be small enough to be contained in the agarose gel wells. (We use thin-toothed combs that accomodate 15 µL total volume [improves autoradiographic band resolution] when casting agarose gels. The dried digestion products are resuspended in 10 microliters of TE buffer plus 5 µL of tracking dye).

2. Prepare a 0.8% agarose gel (at least 20 centimeters in length) with ethidium bromide (5 µL of a 10 mg/mL solution for every 100 mL of gel solution) and 1X Tris-acetate-EDTA (TAE) buffer for electrophoresis of the digestion products. Use 1X TAE buffer plus ethidium bromide (again using 5 µL of a 10 mg/mL solution for every 100 mL of solution) as electrophoresis running buffer. A 1 kilobase ladder (Sigma) molecular weight marker and at least two control samples of known mutational status for FMR1 are included. Controls should include (at minimum) a normal and abnormal sample, for example, a normal female, producing signals corresponding to 2.8 kb and 5.2 kb, and a full mutation male. Control cell lines (transformed lymphocytes) can be obtained through the Coriell Institute (Camden, NJ, USA). Electrophorese the samples for ~16–18 hours at 30–60 volts to allow for optimal separation of the expected autoradiographic signals. Gel length, electrophoresis time, and electrophoresis voltage may need to be optimized for each laboratory setting.

3. Following electrophoresis, photograph the gel using a fluorescent ruler then denature the gel in DNA Denaturing Solution (0.6 M sodium chloride/0.2 N sodium hydroxide) using a plastic container for 30 min with gentle shaking. Rinse the gel briefly in deionized water then neutralize in 0.5 M sodium acetate solution in a separate plastic container for a minimum of 45 min with gentle shaking. The time for the denaturation is more critical than that for neutralization.

4. Set up a Southern transfer onto a nylon membrane using 0.025 M sodium phosphate, pH 7.4. Allow the DNA to transfer for approx 18 h or overnight. Note: Although we use a neutral membrane and sodium phosphate transfer buffer, we are familiar with protocols utilizing positively charged membranes and sodium citrate or alkaline transfers for FMR1 Southern transfers.

5. Following transfer, fix the genomic DNA by baking the membrane at 80°C for 2–3 h or UV crosslinking according to the manufacturer's specifications.

6. Label 30–50 nanograms of the DNA probe utilizing α^{32}phosphorus-deoxycytidine and a random primer protocol according to the manufacturer's specifications. Removal of the unincorporated radioactive-labeled deoxycytosine is critical to maintain low background signals on the subsequent autoradiograph.

7. Prehybridization and hybridization are performed at 42°C using a 50% formamide hybridization buffer or at 65°C in solutions containing no formamide. Following prehybridization for at least 4 h, the labeled probe is added at $2-3 \times 10^6$ cpm. Hybridization is carried out for 18–24 h. Longer hybridization times tend to increase the background signal on the autoradiograph.

8. Posthybridization washes vary from probe to probe. The StB12.3 probe is used for hybridization followed by two 10-min washes at room temperature in 2X SSC buffer/0.1% sodium lauryl sulfate (SLS), one 15-min wash at room temperature in 0.5X SSC/0.1% SLS, and two 10–15 min washes at 65°C in 0.5X SSC/0.1% SLS.

9. Following the posthybridization washes, the membrane is blotted dry on Whatman 3MM paper and wrapped in Saran wrap (or comparable plastic wrap). Autoradiography is performed using Kodak X-OMAT or BioMax MS film with intensifier screen for 1–3 d exposures at −70°C to −80°C.

3.3. FMR1-Specific Polymerase Chain Reaction Protocol

This assay was the original PCR amplification for FMR1 described by Fu et al. *(11)*, in 1991 shortly after isolation of FMR1. It remains one of the most accurate methods for sizing normal and premutation alleles. A modification of the original protocol that improves amplification of large premutation alleles is the inclusion of 2.5 *M* betaine (final concentration) in the amplification reaction. This amino acid analog destabilizes the FMR1 repeat region by altering the melting point of double stranded template DNA. This modified protocol accurately sizes normal alleles, gray zone alleles, and most premutations.

1. Set up a 25 µL PCR reaction containing 1X PCR buffer II (no magnesium chloride); 2 m*M* magnesium chloride (2.0 µL of 25 m*M* solution); 2.5 *M* betaine (10.4 µL of 6 *M* solution); 2% DMSO (0.5 µL of 100% solution); 1.6 µL of deoxynucleotide mix; 0.15 m*M* 7-deazaguanosine triphosphate (0.75 µL of 5 m*M* solution); 16 picomoles of each primer (primers c and f); 5 microcuries (µCi) of α^{32}phosphorus-deoxycytidine (3000 µCi/mL); 250 nanograms of genomic DNA as template; and 2 U of AmpliTaq polymerase. Bring total volume to 25 µL with sterile, deionized water.
2. The thermocycling profile consists of a 10-min initial denaturation at 95°C followed by 28 cycles of 95°C for 1.5 min, 65°C for 1 min, and 72°C for 3 min. A final extension step at 72°C for 5 min is used to complete the assay.
3. The amplification products are separated on a 6% denaturing polyacrylamide sequencing gel for 4–6 h. Greater resolution of alleles varying by 1–2 repeats and shorter electrophoresis times may be obtained using Long Ranger gels (BioWhittaker). ^{32}Phosphorus-labeled DNA fragments of known size are used for molecular weight standards.
4. After electrophoresis the gel is removed from the glass plates using Whatman 3MM paper, wrapped in Saran wrap, and placed in a film cassette using Kodak X-OMAT AR or BioMax MS film. No intensifier screens are used for an overnight exposure (16–24 h) at –80°C. A sample with 30 repeats yields a PCR product of 311 base pairs.

4. Notes

1. Some clinical laboratories first screen for potential FMR1 alterations by the described PCR approach. (This procedure may easily be modified using automated DNA sequencing instruments and fluorescent detection schemes.) When PCR of the FMR1 repeat segment reveals a normal or premutation allele in male patients, or two alleles within the normal or premutation range in female patients, further testing may not be indicated. However, rare patients who have cellular mosaicism for the FMR1 repeat may demonstrate PCR signals in the normal or premutation range, when in fact a more complex FMR1 mutation is present leading to potential false-negative misdiagnosis *(17,18)*.
2. While much slower than PCR testing, Southern blot analysis of the FMR1 repeat segment using a double restriction enzyme digestion (i.e., using methylation-sensitive and -insensitive restriction endonucleases) can reveal a greater degree of information for genotype-phenotype correlations. Southern blotting has the advantage of detecting methylation status but has the limitations of relatively imprecise repeat sizing, higher costs, and slower turnaround time. On the other hand, PCR can size the repeat copy number with good precision, but amplification may fail with large alleles and is not informative regarding methylation status. Thus, it is recommended that Southern blot and PCR be used as complimentary techniques for diagnosis of fragile X syndrome, not as competing techniques.
3. The StB12.3, p5.1, and pfxa3 probes detect a 2.8 kb EagI (or BssHII)-EcoRI fragment containing the CGG repeat region in individuals with normal alleles (**Figs. 1** and **3**).

In addition, the p5.1 probe will detect a constant 2.4 kb fragment in all patients. Irregardless of the selected probe, blockage by methylation of the EagI or BssHII sites provides a means for determining the methylation status of the FMR1 promoter region. However, the presence of DNA methylation related to normal X-chromosome inactivation is a complicating factor. In females with normal alleles, normal X-inactivation (involving *normal* methylation of FMR1) results in a two fragments being detected: the 2.8 kb fragment arising from active (unmethylated) X chromosomes and a 5.2 kb EcoRI-EcoRI fragment from inactive (methylated) X chromosomes. Premutations (containing less than about 200 repeats) remain unmethylated on active X chromosomes and migrate at approx 2.9–3.4 kb. In females only, premutations contained on the inactive X chromosome migrate at 5.3–5.8 kb. Full mutations containing greater than 200 repeats become hypermethylated and migrate from 5.8 to approx 10 kb depending upon the extent of repeat expansion.

4. As previously mentioned, expansion of the trinucleotide repeat in FMR1 exon 1 accounts for >99% of the mutations. However, occasionally other alterations may be identified. Patients with deletions of all or part of FMR1, or point mutations in FMR1, have been reported but probably account for much less than 1% of patients with fragile X mental retardation *(19–24)*. The deletion patients are detected when an autoradiographic signal is completely absent or in a position lower than or above the expected normal signal and "tight" when a smeary signal is expected. Restriction enzyme mapping using the known restriction map flanking the CGG repeat and probes from the region can help clarify where deletion endpoints are located. Deletion patients have confirmed that fragile X syndrome is due to the lack of FMR1 expression because their phenotype is indistinguishable from patients with repeat expansion.

References

1. Tarleton, J. C. and Saul, R. A. (1993) Molecular genetic advances in fragile X syndrome. *J. Pediatr.* **122,** 169–185.
2. Tarleton, J. and Saul, R. A. (updated June, 2000) Fragile X Syndrome, in: GeneClinics: Medical Genetics Knowledge Base [database online]. Copyright, University of Washington, Seattle, USA. Available online at http://www.geneclinics.org/profiles/fragile X. Accessed 1/8/01.
3. Oberle, I., Rousseau, F., Heitz, D., Kretz, C., Devys, D., Hanauer, A., et al. (1991) Instability of a 550-base pair DNA segment and abnormal methylation in fragile X syndrome. *Science* **252,** 1097–1102.
4. Kremer, E. J., Pritchard, M., Lynch, M., Yu, S., Holman, K., Baker, E., et al. (1991) Mapping of DNA instability at the fragile X to a trinucleotide repeat sequence p(CCG)$_n$. *Science* **252,** 1711–1714.
5. Verkerk, A. J. M. H., Pieretti, M., Sutcliffe, J. S., Fu, Y-H., Kuhl, D. P. A., Pizzuti, A., et al. (1991) Identification of a gene (FMR-1) containing a CGG repeat coincident with a breakpoint cluster region exhibiting length variation in fragile X syndrome. *Cell* **65,** 905–914.
6. Yu, S., Pritchard, M., Kremer, E., Lynch, M., Nancarrow, J., Baker, E., et al. (1991) Fragile X genotype characterized by an unstable region of DNA. *Science* **252,** 1179–1181.
7. Sutcliffe, J. S., Nelson, D. L., Zhang, F., Pieretti, M., Caskey, C. T., Saxe, D., and Warren, S. T. (1992) DNA methylation represses FMR-1 transcription in fragile X syndrome. *Hum. Mol. Genet.* **1,** 397–400.
8. Pieretti, M., Zhang, F., Fu, Y-H., Warren, S. T., Oostra, B. A., Caskey, C. T., and Nelson, D. L. (1991) Absence of expression of the FMR-1 gene in fragile X syndrome. *Cell* **66,** 817–822.

9. McConkie-Rosell, A., Lachiewicz, A. M., Spiridigliozzi, G. A., Tarleton, J., Schoenwald, S., Phelan, M. C., et al. (1993) Evidence that methylation of the FMR-1 locus is responsible for variable phenotypic expression of the fragile X syndrome. *Am. J. Hum. Genet.* **53,** 800–809.
10. Rousseau, F., Heitz, D., Biancalana, V., Blumenfeld, S., Kretz, C., Boue, J., et al. (1991) Direct diagnosis by DNA analysis of the fragile X syndrome of mental retardation. *N. Engl. J. Med.* **325,** 1673–1681.
11. Fu, Y-H., Kuhl, D. P. A., Pizzuti, A., Pieretti, M., Sutcliffe, J. S., Richards, S., et al. (1991) Variation of the CGG repeat at the fragile X site results in genetic instability: resolution of the Sherman paradox. *Cell* **67,** 1047–1058.
12. Maddalena, A., Richards, C.S., McGinniss, M.J., Brothman, A., Desnick, R.J., Grier, R.E., et al. (2001) Technical standards and guidelines for fragile X: the first of a series of disease-specific supplements to the Standards and Guidelines for Clinical Genetics Laboratories of the American College of Medical Genetics. *Genet. Med.* **3,** 200–205.
13. Nolin, S. L., Lewis, F.A., III, Ye, L. L., Houck, G.E., Glicksman, A. E., Limprasert, P., et al. (1996) Familial transmission of the FMR1 repeat. *Am. J. Hum. Genet.* **59,** 1252–1261.
14. Warren, S. T. and Nelson, D. L. (1994) Advances in molecular analysis of fragile X Syndrome. *JAMA* **271,** 536–542.
15. Eichler, E. E., Holden, J. J. A., Popovich, B. W., Reiss, A. L., Snow, K., Thibodeau, S. N., et al. (1994) Length of uninterrupted CGG repeats determines instability in the FMR1 gene. *Nat. Genet.* **8,** 88–94.
16. Kunst, C. B. and Warren, S. T. (1994) Cryptic and polar variation of the fragile X repeat could result in predisposing normal alleles. *Cell* **77,** 853–861.
17. Orrico, A., Galli, L., Dotti, M. T., Plewnia, K., Censini, S., and Federico, A. (1998) Mosaicism for full mutation and normal-sized allele of the FMR1 gene: a new case. *Am. J. Med. Genet.* **78,** 341–344.
18. Schmucker, B. and Seidel, J. (1999) Mosaicism for a full mutation and a normal size allele in two fragile X males. *Am. J. Med. Genet.* **84,** 221–225.
19. Gedeon, A. K., Baker, E., Robinson, H., Partington, M. W., Gross, B., Manca, A., et al. (1992) Fragile X syndrome without CCG amplification has an FMR1 deletion. *Nat. Genet.* **1,** 341–344.
20. Tarleton, J., Richie, R., Schwartz, C., Rao, K., Aylsworth, A. S., and Lachiewicz, A. (1993) An extensive de novo deletion removing FMR1 in a patient with mental retardation and the fragile X syndrome phenotype. *Hum. Mol. Genet.* **2,** 1973–1974.
21. Wohrle, D., Kotzot, D., Hirst, M. C., Manca, A., Korn, B., Schmidt, A., et al. (1992) A microdeletion of less than 250 kb, including the proximal part of the FMR-1 gene and the fragile-X site, in a male with the clinical phenotype of fragile X syndrome. *Am. J. Hum. Genet.* **51,** 299–306.
22. Hammond, L. S , Macias, M. M , Tarleton, J. C., and Pai, G. S. (1997) Fragile X syndrome and deletions in FMR1: new case and review of the Lature. *Am. J. Med. Genet.* **72,** 430–434.
23. De Boulle, K., Verkerk, A. J. M. H., Reyniers, E., Vits, L., Hendrickx, J., Van Roy, B., et al. (1993) A point mutation in the FMR-1 gene associated with fragile X mental retardation. *Nat. Genet.* **3,** 31–35.
24. Lugenbeel, K. A., Peier, A. M., Carson, N. L., Chudley, A. E., and Nelson, D. L. (1995) Intragenic loss of function mutations demonstrate the primary role of FMR1 in fragile X syndrome. *Nat. Genet.* **10,** 483–485.

4

Extreme Expansion Detection in Spinocerebellar Ataxia Type 2 and Type 7

Karen Snow and Rong Mao

1. Introduction

The autosomal dominant cerebellar ataxias are each defined by progressive ataxia and variable association with other clinical findings *(1)*. Numerous spinocerebellar ataxia (SCA) loci have been identified and several of the SCA genes have expansion of a CAG-repeat as the underlying mutation *(2,3)*. Assays that determine the CAG-repeat length in *SCA1*, *SCA2*, *SCA3*, *SCA6*, and *SCA7* are used for both diagnostic and predictive testing purposes. Molecular genetic testing is necessary to establish a diagnosis of the SCAs that do not have unique clinical features. It is also a valuable tool in confirming a clinical diagnosis of SCAs that have characteristic clinical findings (such as retinal dystrophy in SCA7) *(2,3)*. Assays typically utilize polymerase chain reaction (PCR) amplification of the repeat region, separation of PCR products by gel electrophoresis or capillary electrophoresis, and visualization of products by incorporation of radioactivity or dye into PCR products or staining with dye after product separation *(4–6)*.

As for other trinucleotide repeat disorders, the genetic phenomenon of anticipation (i.e., decreasing age of onset, increasing disease severity, and faster rate of progression in subsequent generations) in the SCAs is explained by increasing size of the CAG repeat from generation to generation (reviewed **ref. 3**). The majority of cases of SCA1, SCA2, SCA3, SCA6, and SCA7 have fewer than 70 CAG repeats and the standard techniques described above allow sensitive detection and good size resolution of alleles.

Several cases of infantile- and juvenile-onset SCA have been reported where *SCA2* and *SCA7* alleles have between 130 and 500 CAG repeats *(7–13)*. In SCA7, anticipation is more frequently associated with male transmission and sperm of carriers demonstrate massive increases in CAG repeat size *(14)*. A concern is that extreme expansion alleles in SCA2 and SCA7 are not adequately detected by standard methods *(13)*. Similarly, very large CAG-repeat expansions in the *IT-15* gene in juvenile onset Huntington disease (HD) have been difficult to detect by routine methods. A recommendation is that samples yielding apparently homozygous normal alleles be tested by additional methods to ensure that a second allele of a different size is not present *(15)*. For example, Southern blot and probe hybridization to PCR products have been successfully used to detect HD alleles with more than 100 CAG repeats *(16,17)*.

Table 1
Oligonucleotide Sequences

Name	DNA sequence	Reference
SCA2A	5' - GGG CCC CTC ACC ATG TCG - 3'	*(22)*
SCA2B	5' - CGG GCT TGC GGA CAT TGG - 3'	*(22)*
SCA7B	5' - GTA GGA GCG AAA AGA ATG TC - 3'	*(13)*
SCA7C	5' - CCC CGA CCG TCG CCA TTG - 3'	*(13)*
(CTG)6	5' - CTG CTG CTG CTG CTG CTG - 3'	*(13)*

This chapter describes a PCR-blot-oligo hybridization assay for detection of extreme expansion alleles in the *SCA2* and *SCA7* genes. Utilization of this assay is appropriate for cases of infantile- or juvenile-onset ataxia where routine assays such as PCR-polyacrylamide gel electrophoresis (PAGE) show apparent homozygosity for a normal repeat size in the *SCA2* and *SCA7* genes and where there is a suspected or confirmed diagnosis of SCA2 or SCA7 in a parent.

2. Materials

2.1. DNA Extraction from Whole Blood

1. Puregene™ DNA extraction kit (Gentra, Minneapolis, MN, USA).
2. Isopropyl alcohol, high-performance liquid chromatography (HPLC) grade (2-propanol) (Fisher-Scientific, Hampton, NH, USA)

2.2. PCR

1. GeneAmp® DNA Amplification Reagent Kit or individual components: 5 U/μL Taq Polymerase, 10 mM dATP, 10 mM dCTP, 10 mM dGTP, 10 mM dTTP, 25 mM MgCl$_2$, 10X reaction buffer without Mg (Applied Biosystems, Foster City, CA, USA).
2. Sterile H$_2$O for Molecular Biology, Dnase and Rnase free (Sigma, St. Louis, MO, USA).
3. To make 5X reaction buffer, mix 10X reaction buffer, 10 mM dGTP, 10 mM dATP, 10 mM dTTP, 10 mM dCTP, sterile H$_2$O in ratio 5:1:1:1:1:1 by volume (*see* **Note 1**).
4. Dimethyl sulfoxide (DMSO) (Sigma) (*see* **Note 2**).
5. 50% Glycerol: mix glycerol (Sigma) and sterile H$_2$O in 1:1 ratio by volume (*see* **Note 2**).
6. Oligonucleotide primers, 20 μM (**Table 1**) (*see* **Note 3**).

2.3. Agarose Gel Electrophoresis

1. Molecular biology grade agarose, e.g., Ultra Pure™ agarose (Invitrogen Corporation, Carlsbad, CA, USA)
2. 1X Tris-acetate-EDTA (TAE) buffer: 40 mM Tris-acetate, 1 mM EDTA.
3. 6X gel loading buffer: 0.25% bromophenol blue (Sigma), 15% ficoll type 400 (Sigma).
4. 10 mg/mL ethidium bromide (BioRad Laboratories, Hercules, CA, USA). Store in the dark at 4°C. Caution: ethidium bromide is a mutagen and may be a carcinogen and/or teratogen. Avoid inhalation or skin contact. Wear gloves when handling and use hazardous waste container for disposal of ethidium bromide containing solutions.

2.4. Blotting to Nylon Membrane

1. Gel Denaturation Buffer: 0.4 N NaOH, 0.6 M NaCl.
2. Gel Neutralization Buffer: 1.5 M NaCl, 0.6 M Tris-HCl, pH 7.5.

3. 20X SSC buffer: 3 M NaCl, 0.3 M sodium citrate, pH 7.0 (*see* **Note 4**).
4. Positively charged nylon membrane, e.g., Biodyne® membrane (Invitrogen Corp.).

2.5. Radiolabeling of 1 kb Ladder and Probe

1. 1 kb DNA ladder, 1 μg/μL (Invitrogen Corp.)
2. (CTG)$_6$ oligonucleotide, 2 μM (**Table 1**).
3. End-labeling reagents: T4 polynucleotide kinase and 10X PNK buffer (Promega, Madison, WI, USA)
4. Gamma ^{32}P ATP (Adenosine 5'-[γ^{32}P] triphosphate, triethylammonium salt, >5000 Ci/mmol; Amersham Biosciences, Piscataway, NJ, USA).
5. ProbeQuant™ G-50 Micro Columns (Amersham Biosciences).
6. STE buffer: 100 mM NaCl, 10 mM Tris-HCl, 1 mM EDTA, pH 8.0. Autoclave prior to use.

2.6. Hybridization

1. 20X SSPE buffer: 0.2 M phosphate, 3.0 M NaCl, 0.02 M EDTA. Autoclave prior to use.
2. Hybridization Solution: 4X SSPE, 0.2% SDS, 0.1 mg/mL yeast tRNA (Sigma).

2.7. Post-hybridization Washes

1. 5X SSPE buffer (diluted from 20X SSPE buffer).

2.8. Signal Detection

1. Autoradiography cassette with intensifying screen (Kodak BioMax™ TranScreen HE or similar) and X-Ray film (e.g., Kodak BioMax™ MS),
   OR
2. Phosphorimager.

3. Methods

3.1. Importance of Family History and Results from Standard SCA2 or SCA7 Assays

1. In most of the previous reports of infantile- or juvenile-onset SCA2 or SCA7, a parent of the affected child was diagnosed as having SCA prior to evaluation of the child. Thus, family history is important in suggesting a potential diagnosis of hereditary ataxia in an infant or child who has findings similar to those previously reported (infantile hypotonia, seizures, dysphagia, developmental delay, cerebellar atrophy, visual impairment). However, it should be noted that a carrier parent may be asymptomatic or undiagnosed when the affected child is first evaluated *(12,13)* (*see* **Note 5**).
2. Side-by-side analysis of parental and child DNA samples using the standard assays may reveal apparent noninheritance of an allele from the carrier or affected parent *(13)*. This is explained by failure to detect the very large expansion in DNA from the child.
3. Extreme expansions may give rise to very faint smudges or faint bands in standard assays as shown in **Fig. 1**. These may be in regions of the gel or scan not typically scrutinized in scoring of results.

3.2. DNA Extraction

1. Use Puregene™ reagents and manufacturer's instructions to extract genomic DNA from 1–3 mL whole blood (*see* **Notes 6–8**).
2. Determine the DNA concentration by measuring the A260 on a spectrophotometer. Adjust all DNA samples to the same concentration (between 0.1 and 0.5 μg/μL) (*see* **Notes 9** and **10**).

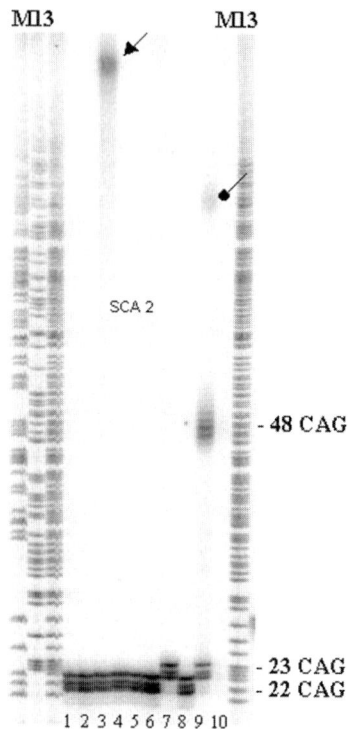

Fig. 1. Analysis of CAG repeats in *SCA2* by a PCR-PAGE assay. Lanes 1, 2, 4, 5, 6, and 8 show results for unaffected individuals who are each homozygous for 22 CAG repeats. This repeat length is the most prevalent in the general population. Lane 3 shows results from a case of infantile onset SCA2 in which the expanded allele has approx 230 CAG, as determined by the PCR-blot-oligo hybridization assay. As indicated by the arrow, the expanded allele appears as a smudge in the upper part of the gel. Lane 7 shows a normal repeat of 23 CAG. Lane 9 shows results from an affected patient who has 23 and 48 CAG repeats. The circle indicates artifact that is probably heteroduplex. The 'no template' control shows no bands (Lane 10). M13 sequence is used to determine the size of alleles.

3.3. PCR

1. Prepare the SCA2 reaction mix by adding together the following volumes for each sample: 12.0 µL sterile H_2O, 5 µL 5X buffer, 1 µL $MgCl_2$, 1 µL primer SCA2A, 1 µL primer SCA2B, 1.25 µL DMSO, 2.5 µL 50% glycerol. Multiply volumes to prepare enough mix for all DNA samples to be tested plus a 'no template' control. Vortex the mix. Add the appropriate multiple of 0.25 µL Taq polymerase. Vortex gently or use pipet to mix. Aliquot 24 µL of mix into PCR tubes. Add 1 µL DNA to each tube (*see* **Notes 11–15**).

2. Prepare the SCA7 reaction mix by adding together the following volumes for each sample: 9.75 µL sterile H_2O, 5 µL 5X buffer, 1 µL $MgCl_2$, 1 µL primer SCA7B, 1 µL primer SCA7C, 1.25 µL DMSO, 5.0 µL 50% glycerol. Multiply volumes to prepare enough mix for all DNA samples to be tested plus a 'no template' control. Vortex the mix. Add the

appropriate multiple of 0.25 µL Taq polymerase. Vortex gently or use pipet to mix. Aliquot 24 µL of mix into PCR tubes. Add 1 µL DNA to each tube (*see* **Notes 11–15**).
3. Perform PCR using the following conditions on a GeneAmp® PCR System 9600 (Applied Biosystems). 35 cycles of 94°C, 30 s; 60°C, 30 s; 72°C, 30 s; 10 min hold at 72°C; hold at 5°C until samples removed from the instrument (*see* **Note 16**).

3.4. Preparation of Labeled 1 kb Ladder

1. Mix reagents in the order listed: 13.5 µL sterile H_2O, 5 µL 1 kb ladder, 2.5 µL 10X PNK buffer, 3.0 µL γ^{32}P-ATP, 1.0 µL T4 polynucleotide kinase (*see* **Note 17**).
2. Vortex reaction mix and spin briefly.
3. Incubate reaction mix at 37°C for 30 min.
4. Add 25 µL STE.
5. Inactive kinase by incubation at 90°C for 2 min.
6. Prepare ProbeQuant™ spin column as follows: resuspend the resin in the column by vortexing, loosen the cap one-fourth turn and snap off the bottom closure, place the column in a 2-mL conical tube, pre-spin the column for 1 min at 750g (*see* **Note 18**).
7. Place the column in a new 2-mL conical tube support tube.
8. Slowly apply the entire sample to the top center of the resin, being careful not to disturb the resin bed.
9. Spin the column at 750g for 1 min to retrieve labeled ladder into the collection tube. Unincorporated ^{32}P-dATP remains in the column.
10. Dilute purified product to 100 µL by adding 50 µL of STE buffer.
11. Take an aliquot (1 or 2 µL) of the purified probe and determine the CPMs. Counts should be approx 4×10^3 CPM/µL (*see* **Note 19**).

3.5. Agarose Gel Electrophoresis

1. Prepare a 2% agarose gel by adding 2.0 g agarose to 100 mL of 1X TAE buffer. Heat in a microwave oven to dissolve. Cool to a temperature comfortable to handle (approx 50°C). Add 5 µL ethidium bromide (10 mg/mL) per 100 mL of gel. Pour into gel tray and allow to set for approx 2 h (*see* **Note 20**).
2. Add 5 µL 6X gel loading buffer to each PCR product and load 25 µL of each sample on the gel (*see* **Note 21**).
3. Include size standards on the gel. One lane should include unlabeled 1 kb ladder. Another lane should contain $5-10 \times 10^3$ CPM of radiolabeled 1 kb ladder.
4. Perform electrophoresis at 100V for approx 5–6 h (*see* **Note 22**).
5. Visualize PCR products on a UV transilluminator and photograph for record. Take care to not damage the gel.

3.6. Blotting to Nylon Membrane

1. Incubate gel in gel denaturing solution for 30 min at room temperature with gentle agitation.
2. Transfer gel to gel neutralizing solution and incubate for 30 min at room temperature with gentle agitation.
3. Cut the nylon membrane slightly larger than the size of the gel (*see* **Note 23**).
4. Lay nylon membrane into a tray of fresh 10X SSC solution and allow the membrane completely to wet by capillary action (*see* **Note 24**).
5. Assemble the blot on top of a gel holder that is in a tray filled with 10X SSC. The order of blot components is: filter paper wick, gel with wells facing downwards, nylon membrane, filter paper, stack of paper towels, weight (*see* **Notes 25** and **26**).

6. Allow the transfer to continue for 6–16 h.
7. After the blotting is complete, carefully remove the membrane and immerse in an excess of 2X SSC for 1–2 min with gentle agitation. The membrane can be rubbed very gently with gloved fingers to remove any remaining fragments of agarose.
8. Place the membrane with transferred DNA face up on a piece of filter paper and bake for 2 h at 70°C.

3.7. Probe Preparation

1. Mix reagents in the order listed: 5.5 µL sterile H_2O, 2 µL $(CTG)_6$ oligonucleotide (2 µLM), 1.0 µL 10X polynucleotide kinase buffer, 1.0 µL γ-^{32}P ATP, 0.5 µL T4 polynucleotide kinase (*see* **Note 17**).
2. Vortex reaction mix and spin briefly.
3. Incubate reaction mix at 37°C for 30 min.
4. Add 90 µL STE buffer.
5. Inactive kinase by incubation at 90°C for 2 min.
6. Purify probe using a ProbeQuant™ column as described in (**Subheading 3.4., steps 6–9**).
7. Count a 1 or 2 µL aliquot of purified labeled probe to determine CPMs incorporated. The expected range is 0.2 to 1.0×10^5 cpm/µL (*see* **Note 19**).

3.8. Hybridization

1. Wet membrane in 5X SSPE buffer.
2. Add 20 mL hybridization solution to hybridization tube, roll to wet sides of the tube (*see* **Note 27**).
3. Drain excess fluid off membrane and place in the hybridization tube with the DNA side facing inwards.
4. Incubate at least 2 h at 55°C (*see* **Note 28**).
5. Discard hybridization solution. Add 20 mL of warm, fresh hybridization solution. Add labeled $(CTG)_6$ oligo to hybridization tube, seal, swirl to mix, and incubate overnight at 55°C. The hybridization solution should contain $1–5 \times 10^5$ cpm/mL.

3.9. Post-hybridization Washes

1. Pour out hybridization solution into radioactive waste container.
2. Add 100 mL of 5X SSPE buffer, seal, and swirl. Pour into radioactive waste container.
3. Transfer the membrane to a plastic container, add 1 L of 5X SSPE buffer, cover, and agitate for 15 min at room temperature.
4. Add 1 L 5X SSPE buffer to a different container, warm to 60°C.
5. Transfer membrane to the warm container and agitate for 10 min at 60°C.
6. Air-dry membrane and wrap in plastic wrap (e.g., Saran wrap or Glad wrap).

3.10. Signal Detection

1. X-Ray Film:
    a. Place wrapped membrane in an autoradiography cassette with an intensifying screen.
    b. In dark room, place X-ray film on top of membrane and close cassette.
    c. Develop film after 4–6 h exposure.
    d. Obtain a longer or shorter exposure as determined by the original exposure.
2. Phosphorimager:
    a. Load wrapped membrane into a phosphorimager cassette.
    b. Scan the cassette after 4–6 h of exposure.
    c. Obtain a longer exposure scan if needed.

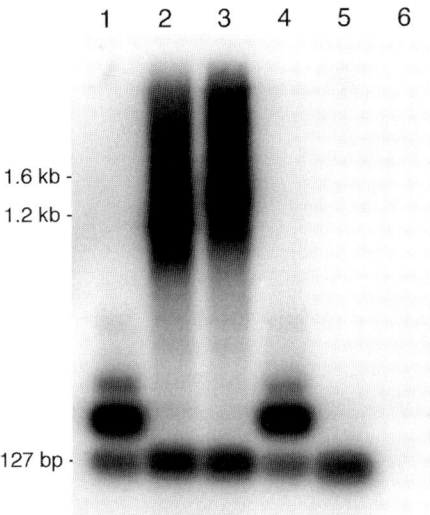

Fig. 2. Analysis of CAG repeats in *SCA2* by a PCR-blot-oligo hybridization assay. Lane 1, 22 and 45 CAG; Lane 2, 23 and 350 CAG; Lane 3, 23 and 400 CAG; Lane 4, 23 and 48 CAG; Lane 5, homozygous for 22 CAG; Lane 6, 'no template' control.

3.11. Interpretation of Results (see Fig. 2)

1. Normal *SCA2* and *SCA7* alleles demonstrate bands at approx 130 bp and 255 bp, respectively.
2. Typical expansions seen in adult-onset SCA2 and SCA7 patients appear as bands at approx 210 bp and 375 bp, respectively.
3. Extreme expansions seen in infantile- or juvenile onset SCA2 and SCA7 appear as disperse 'smears' of hybridization between approx 0.7 and 4 kb.

4. Notes

1. Prepare working aliquots of PCR buffers and reagents and avoid unnecessary repeat freeze-thawing of stocks. Keep stock reagents on ice when thawed. Nucleotides in particular are labile and subject to hydrolysis.
2. Use molecular biology grade reagents for all parts of the procedure.
3. Oligonucleotides may be stored in sterile TE buffer or in sterile H_2O.
4. Molecular biology buffers and solutions such as SSC, TE, and SDS solution can be made from component reagents or can be purchased pre-made from vendors of molecular biology products.
5. See **refs.** *(18)* and *(19)* for reviews of the clinical features, molecular genetics and counseling issues for SCA2 and SCA7, respectively.
6. Whole blood should be collected in EDTA or ACD anticoagulant. DNA isolated from heparinized blood may fail in PCR due to the inhibitory effect of heparin. In such cases, dilution of the DNA (e.g., 1/10 in TE) may remove some of the inhibition.
7. Apply the following tips to ensure a good yield of high-quality DNA. If any cell clumps are visible at the cell lysis step, try to break-up the clumps by pulling the cell clump up into and out of a disposable transfer pipet vigorously several times. Incubate at the cell lysis step for a prolonged period if necessary until all cells are lysed. It is important that the cell lysate be no warmer than 20°C when protein precipitation solution is added. Effective precipitation of protein at this step requires very vigorous vortexing.

8. Alternative DNA purification procedures include the standard proteinase K/phenol/chloroform method and commercially available column extraction methods.
9. Adjusting the DNA concentration of each sample to a standard value typically yields greater uniformity of results between specimens but is not a necessary step.
10. Concentration of DNA may also be determined by fluorometer or by commercially available detection kits.
11. Follow precautions to minimize the risk for PCR contamination. Set up all PCR reactions in a designated 'clean' area. Wear clean gloves when setting up a PCR and use dedicated pipets and barrier tips for all pipetting. Wipe down the setup bench with water both before and after setting up a PCR. Reagents used in PCR reactions must stay in the PCR set up area. At least weekly, the PCR setup area should be cleaned with a 1/10 dilution of bleach. However, bleach is an inhibitor of PCR so the cleaned area should be thoroughly wiped clean with water following the bleach treatment. If available, turn on UV lights after use of the area. Amplified products must not be brought into the PCR set-up area. Store PCR products away from DNA samples and PCR reagents. Do not handle blood samples or genomic DNA in the post-PCR area of the laboratory. Reagents suspected of being contaminated must be discarded (e.g., if the 'no template' control shows product).
12. Keep all reagents on ice during set up of PCR reactions. The Taq polymerase should be kept at $-20°C$ until ready to use and then returned to $-20°C$ as soon as possible.
13. It is essential to mix the reaction mix thoroughly before addition of the Taq polymerase to avoid denaturation of the enzyme by DMSO.
14. To avoid false patient results and allow detection of PCR contamination, PCR reactions should be set up in the order: patient samples, negative control, positive control, 'no template' control.
15. To minimize formation of nonspecific PCR products, keep reactions on ice until they are placed in the thermocycler. Place reaction tubes in the thermocycler after the block temperature reaches approx 80°C. It may be helpful to use the pause feature of the thermocycler while reaction tubes are being placed in the heating block.
16. Cycling parameters on other thermocyclers may need to be adjusted slightly to obtain comparable results. This is most likely because of some differences in ramping characteristics between different thermocyclers.
17. T4 polynucleotide kinase is added last to maximize incorporation of radiolabeled nucleotide. If kinase is added before the $\gamma^{32}P$ ATP, DNA synthesis would incorporate unlabeled nucleotides that are present in the PNK buffer.
18. ProbeQuant™ preparation and purification steps should be performed as quickly as possible to prevent drying out of the resin within the column.
19. The incorporation values listed are for Cerenkov counting. Cerenkov counting is performed in the absence of scintillation fluid and measures radioactivity from ^{32}P in the ^{3}H channel of a liquid scintillation counter.
20. Never microwave an agarose solution containing ethidium bromide. Ethidium bromide is a mutagen and may be a carcinogen and/or teratogen. Avoid inhalation or skin contact. Wear gloves when handling and use hazardous waste container for disposal of ethidium bromide containing solutions.
21. Volumes of PCR products and standards to load are for 1 mm × 8 mm well sizes. Load smaller amounts for smaller wells. Samples should be pipetted into the wells slowly to minimize mixing of sample and running buffer.
22. DNA migrates toward the anode (+).
23. Wear gloves when handling the nylon membrane or filter paper that may come in contact with the nylon.

24. To avoid trapping of air in the membrane, wet the membrane by placing a corner into the buffer then gently lower the remainder of the membrane into the buffer. Trapped air, which is visible as a whitening of the membrane, will interfere with transfer of DNA onto the membrane.
25. The procedure provided is for blotting under neutral conditions and requires subsequent baking of the membrane to covalently attach DNA to the membrane. In contrast, blotting to nylon using alkaline transfer does not need subsequent baking or UV crosslinking. However, alkaline transfer may result in higher levels of background hybridization *(20)*
26. Alternative DNA transfer methods are downward transfer or electrotransfer *(21)*.
27. The volume of hybridization solution used may be less than 20 mL, depending on if a hybridization tube or hybridization bag is used. The volume of 20 mL is for a 20 × 20 cm membrane in a hybridization tube of 10 cm diameter. There should be enough solution to allow complete coverage of the membrane as it is being rotated or rocked.
28. This prehybridization step allows blocking of non-specific binding sites on the membrane and reduces background signal.

References

1. Harding, A. E. (1982) Clinical features and classification of the late onset autosomal dominant cerebellar ataxias. *Brain* **105,** 1–28.
2. Evidente, V. G., Gwinn-Hardy, K. A., Caviness, J. N., and Gilman, S. (2000) Hereditary ataxias. *Mayo Clin. Proc.* **75,** 475–490.
3. Bird, T. D. Hereditary Ataxia Overview, in *GeneClinics: Medical Genetics Knowledge Base*. University of Washington, Seattle, WA, USA. Available online at http://www.geneclinics.org/profiles/ataxias. Accessed 4 February 2002. (Updated 14 June 2000)
4. Moseley, M.L., Benzow, K.A., Schut, L.J., Bird, T.D., Gomez, C.M., Barkhaus, P.E., et al. (1998) Incidence of dominant spinocerebellar and Friedreich triplet repeats among 361 ataxia families. *Neurology* **51,** 1666–1671.
5. Vuillaume, I., Schraen, S., Rousseaux, J., and Sablonniere, B. (1998) Simple nonisotopic assays for detection of (CAG)n repeats expansions associated with seven neurodegenerative disorders. *Diagn. Mol. Pathol.* **7,** 174–179.
6. Potter, N.T. and Nance, M.A. (2000) Genetic testing for ataxia in North America. *Mol. Diagn.* **5,** 91–99.
7. Babovic-Vuksanovic, D., Snow, K., Patterson, M. C., and Michels, V. V. (1998) Spinocerebellar ataxia type 2 (SCA 2) in an infant with extreme CAG repeat expansion. *Am. J. Med. Genet.* **79,** 383–387.
8. Benton, C., de Silva, R., Rutledge, S., Bohlega, S., Ashizawa, T., and Zoghbi, H. (1998) Molecular and clinical studies in SCA-7 define a broad clinical spectrum and the infantile phenotype. *Neurology* **51,** 1081–1086.
9. David, G., Durr, A., Stevanin, G., Cancel, G., Abbas, N., Benomar, A., et al. (1998) Molecular and clinical correlations in autosomal dominant cerebellar ataxia with progressive macular dystrophy (SCA7). *Hum. Mol. Genet.* **7,** 165–170.
10. Johansson, J., Forsgren, L., Sandgren, O., Brice, A., Holmgren, G., and Holmberg, M. (1998) Expanded CAG repeats in Swedish spinocerebellar ataxia type 7 (SCA7) patients: effect of CAG repeat length on the clinical manifestation. *Hum. Mol. Genet.* **7,** 171–176.
11. Giunti, P., Stevanin, G., Worth, P., David, G., Brice, A., and Wood, N. (1999) Molecular and clinical study of 18 families with ADCA type II: evidence for genetic heterogeneity and de novo mutation. *Am. J. Hum. Genet.* **64,** 1594–1603.

12. Grattan-Smith, P. J., Healey, S., Grigg, J. R., and Christodoulou, J. (2001) Spinocerebellar ataxia type 7: a distinctive form of autosomal dominant cerebellar ataxia with retinopathy and marked genetic anticipation. *J. Paediatr. Child. Health* **37,** 81–84.
13. Mao, R., Aylsworth, A. S., Potter, N., Wilson, W.G., Breningstall, G., Wick, M. J., et al. Childhood-onset ataxia: testing for large CAG-repeats in SCA2 and SCA7. *Am. J. Med. Genet.* **110,** 338–345.
14. Monckton, D. G., Cayuela, M. L., Gould, F. K., Brock, G. J., Silva, R., and Ashizawa, T. (1999) Very large (CAG)(n) DNA repeat expansions in the sperm of two spinocerebellar ataxia type 7 males. *Hum. Mol. Genet.* **8,** 2473–2478.
15. The American College of Medical Genetics/American Society of Human Genetics Huntington Disease Genetic Testing Working Group. (1998) ACMG/ASHG statement. Laboratory guidelines for Huntington disease genetic testing. *Am. J. Hum. Genet.* **62,** 1243–1247.
16. Guida, M., Fenwick, R. G., Papp, A. C., Snyder, P. J., Sedra, M., and Prior, T. W. (1996) Southern transfer protocol for confirmation of Huntington disease. *Clin. Chem.* **42,** 1711–1712.
17. Nance, M., Mathias-Hagen, V., Breningstall, G., Wick, M., and McGlennen, R. (1999) Analysis of a very large trinucleotide repeat in a patient with juvenile Huntington's disease. *Neurology* **52,** 392–394.
18. Pulst, S-M. Spinocerebellar Ataxia Type 2, in *GeneClinics: Medical Genetics Knowledge Base*. University of Washington, Seattle, WA, USA. Available online at http://www.geneclinics.org/profiles/sca2. Accessed 4 February 2002. (Updated 9 January 2001)
19. Gouw, L. G. and Ptacek, L. J. Spinocerebellar Ataxia Type 7, in *GeneClinics: Medical Genetics Knowledge Base*. University of Washington, Seattle, WA, USA. Available online at http://www.geneclinics.org/profiles/sca7. Accessed 4 February 2002. (Updated 22 May 2001)
20. Sambrook, J., Fritsch, E. F., and Maniatis, T., eds. (1989) *Molecular Cloning. A Laboratory Manual*. Cold Spring Harbor Laboratory Press, Cold Springs Harbor, NY, USA.
21. Moore, D. D. Preparation and analysis of DNA, in: *Current Protocols in Molecular Biology* (Ausubel I. and Frederick M., eds.), Unit 2.9, John Wiley & Sons, Inc., New York, NY, USA.
22. Pulst, S., Nechiporuk, A., Nechiporuk, T., Gispert, S., Chen, X., Lopes-Cendes, I., et al. (1996) Moderate expansion of a normally biallelic trinucleotide repeat in spinocerebellar ataxia type 2. *Nat. Genet.* **14,** 269–276.

5

Repeat Expansion Detection (RED) and the RED Cloning Strategy

Qiu-Ping Yuan, Kerstin Lindblad-Toh, and Martin Schalling

1. Introduction

1.1. Background

Trinucleotide repeat sequences are present at approx 30,000–40,000 loci in the human genome *(1)*. The majority of these repeats are below 35 copies and are stably transmitted. However, unstable trinucleotide repeat expansions at some loci have been found to be the causal mutation for nearly 20 genetic neurodegenerative disorders in human *(2–4)*. Most of these disorders occur at repeat lengths above 35 copies, with a tendency towards further expansion upon successive transmissions. An inverse correlation between the repeat length and disease severity/earlier age of onset, known as anticipation, has been observed in most of the families transmitting such types of diseases, suggesting that the length change of the repeats may play a role in the manifestation of anticipation. Only three motifs, CAG/CTG, CGG/CCG, and GAA/TTC, of the 10 possible trinucleotide repeat permutations have so far been associated with human disease. It remains possible that other disease phenotypes are caused by expansions of any repeat motif at any repeat containing locus. We have established a repeat detection and gene-isolation system, which allows identification of a repeat-containing gene within a couple of months.

1.2. Repeat Expansion Detection

The repeat expansion detection (RED) method *(5,6)* has been used for the detection of a trinucleotide repeat expansion without the need of prior knowledge of its chromosomal location *(7–10)*. In RED, genomic DNA serves as a template for repeat specific oligonucleotides after DNA denaturation. Oligonucleotides that have annealed at adjacent bases of a repeat sequence in genomic DNA are ligated by a thermostable ligase, generating multimers through multiple rounds of cycling (**Fig. 1**). This is a linear amplification process requiring several hundred cycles of ligation/denaturation. The products are size-separated by gel electrophoresis, blotted onto a membrane, and hybridized with a ^{32}P-labeled repeat probe complementary to the multimer. The maximum product size observed corresponds to the longest repeat sequence existing in the genome tested. The method consists of the following steps: 1) amplification of long repeat

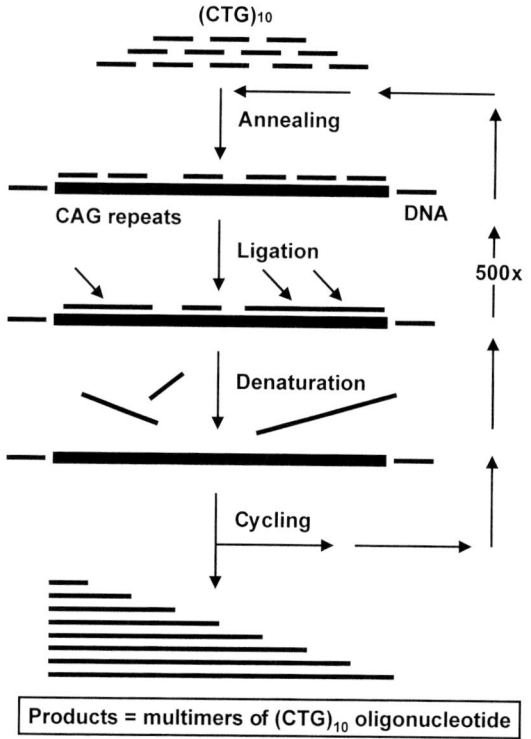

Fig. 1. The principle of RED. DNA is heat-denatured and $(CTG)_{10}$ repeat oligonucleotides are allowed to anneal to the template. Ligation occurs when oligonucleotides anneal at adjacent positions. These reaction steps are repeated several hundred times on a thermocycler yielding multimers with different sizes, where the longest product corresponds to the longest repeat sequence in the tested DNA.

sequences by ligation-denaturation cycling; 2) size separation of the RED products by electrophoresis on a polyacrylamide gel; 3) transfering products onto Hybond-N+ membrane by blotting; 4) hybridization of RED products with a repeat probe labeled with isotope; and 5) autoradiography.

1.3. The RED-Based Cloning Strategy

We have developed a cloning method to facilitate the isolation of disease genes containing trinucleotide repeat expansions. The method is based on size separation of genomic fragments, followed by subcloning and library hybridization with an oligonucleotide probe (**Fig. 2**). Fractions and clones containing expanded repeats are identified by the RED method throughout the cloning procedure. DNA from multiple family members are not required and as little as 10 µg genomic DNA from a single individual is sufficient for this method. A major obstacle of cloning long repeat sequences has been the tendency of repeats to expand or delete during the cloing procedure (*11,12*). The strategy described here appears to avoid this problem, as cloning steps and propagation in bacteria are reduced to a minimum through the use of direct hybridization to identify repeat containing plaques. Using this strategy, we have cloned two DNA frag-

RED and RED Cloning Strategy

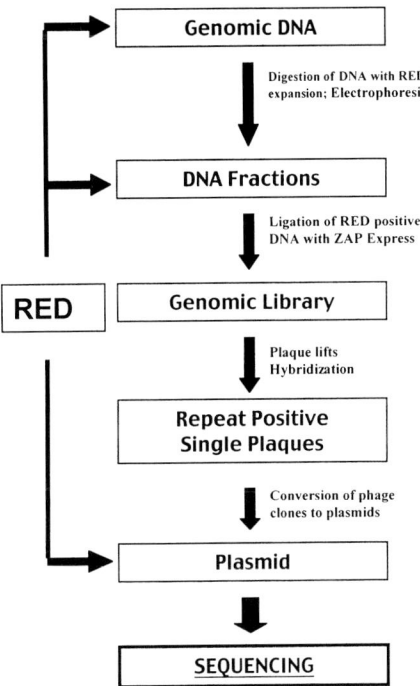

Fig. 2. Diagram illustrating the major steps of the RED-cloning procedure. The RED method was used at three different stages to identify the target repeat expansion.

ments containing expanded repeats from two unrelated patients with a clinical diagnosis of cerebellar ataxia *(13)*. Sequencing of the two fragments showed sequences identical with two disease loci, the Huntington gene and the ataxin 3 gene, respectively. The sequenced repeats were very similar in size to that obtained in the initial RED reaction, providing evidence that repeats essentially remain stable throughout the procedure. The method should be adaptable to the cloning of any long repeat motif in any species. Furthermore the experimental steps can be performed within two months, making it a time-efficient method for disease gene identification. Major steps of this method are illustrated in **Fig. 2**.

2. Materials

2.1. The RED Method

2.1.1. DNA Template

A standard phenol/chloroform extraction method or QIAmp Blood Kit (Qiagen Inc., Chatsworth, CA) can be used for DNA preparation. Dissolve DNA in 5 m*M* Tris buffer, pH 9.0, with final concentration of 0.2 µg/µL.

2.1.2. Oligonucleotides

Oligonucleotides must be 5'phosphorylated. The denaturation/ligation cycling conditions described in this chapter are optimized for use with a $(CTG)_{10}$ oligonucleotide. Please *see* **Note 1** if other oligonucleotides are used.

2.1.3. Reagents

1. Ampligase (100 U/µL) (Epicentre Technologies, Madison, WI, USA) with supplied buffer.
2. Terminal deoxynucleotide transferase (TdT) (Amersham, Little Chalfont, UK) with supplied buffer for end-labeling of hybridization probes.
3. Isotope α^{32}P-dATPs (6000Ci/mmol) (NEG 012Z, NEN DuPont Medical, Wilmington, DE, USA).
4. Rapid Hybe buffer (Amersham).
5. 6% denaturing polyacrylamide/6M Urea gel solution and supplied buffer (National Diagnostics, Atlanta, GA, USA).
6. 10X TBE: 0.89 M Tris-boric acid, 20 mM Na$_2$EDTA.
7. TE-4.
8. 20X SSC.
9. 3 M NaCl, 0.3 M Na$_3$citrate.
10. 20% sodium dodecyl sulfate (SDS).
11. Formamide gel loading dye: 100% formamide, 0.1% xylene cyanol, 0.1% bromophenol blue.

2.1.4. Apparatus and Supplies

We have been using the GeneAmp PCR System 9600 (Perkin Elmer Cetus, Norwalk, CT, USA), and the PTC-225 Peltier Thermal Cycler (MJ Research, Watertown, MA, USA), but any thermocycler with a heated lid should work. In addition, Whatman 3-mm filter paper, Hybond N+ membrane (Amersham) or any similar membrane, DuPont reflection, NEF 495 X-ray film, and intensifying screen or a similar product is needed.

0.2-mL microtube strips and caps are used for the RED reactions (CLP products, San Diego, CA, USA).

2.2. The RED-Cloning Strategy

2.2.1. RED Method (see **Subheading 2.1.**)

2.2.2. Cloning Procedure

1. Phage, and competent cells (XL-1 Blue cells & XLOLR cells).
2. Maltose medium and XLOLR in NZY-broth.
3. LB medium: 10 g of Tryptone, 5 g of Yeast extract, 10 g of NaCl, per liter; pH 7.5.
4. LB-maltose media: LB media, 0.2% maltose, 10 mM MgSO$_4$.
5. NZY-broth: 5 g NaCl, 2 g MgSO$_4$, 5 g Yeast extract, 10 g NZ Amine, per liter; pH 7.5.
6. NZY-top agar (NZY broth + 0.7% (w/v) agarose).
7. NZY-agar dishes (NZY broth + 15 g agar, per liter).
8. LB-tetracycline dishes (LB medium + 12.5 mg tetracycline per liter).
9. LB-kanamycine dishes (LB medium + 50 mg of Kanamycine).
10. SM buffer: 5.8 g NaCl, 2 g MgSO$_4$, 7 H$_2$O, 50 mL 1 M Tris-HCl, 5 mL 2% gelatin; pH 7.5.
11. 0.5 M IPTG; Xgal (250 mg/mL).

2.2.3. Other Reagents (Enzymes, Buffers, etc.)

1. Enzymes: T4 Ligase (10 U/µL), Restriction enzyme (MboI) and supplied buffer, Agarose-Digesting Enzyme AgarACE, (Promega, Madison, WI, USA).
2. SeaPlaque CTG low-melting temperature agarose (FMC BioProducts, Rockland, ME, USA).
3. 6X gel loading dye: 4.0 mL 0.5 M EDTA, 2 mL of 10 M Tris-acetate, 50 mg bromphenol blue, 2.0 g Ficoll.

4. Size standard: 1 kb ladder and 100 bp ladder (GibcoBRL, Life Technologies, Täby, Sweden).
5. Reagents for DNA precipitation: 99.5% and 95% ethanol; 3 M NaOAc pH 5.2; TE Buffer.
6. Denaturation solution: 67.6 g NaCl, 20 g NaOH, per liter.
7. Renaturation solution: 67.6 g of NaCl, 121.1 g Tris, per liter; pH7.0.
8. SNAP kit for phagemid DNA preparation (Invitrogen, Carlsbad, CA, USA).

2.2.4. DNA Sequencing Kit:

1. BigDye™ Terminator Cycle Sequencing Ready Reaction (ABI PRISM, Warrington, UK).

3. Methods

3.1. The RED Method

The principle of RED is presented in **Fig 1**.

3.1.1. RED Reaction

1. RED reaction mixture contains the following: Genomic DNA 1 (0.5–2) µg, $(CTG)_{10}$ Oligonucleotide (50 ng/µL)1.0 µL, TE-4 1.0 µL, Ampligase (100 U/µL) 0.15 µL, Co-Buffer 0.5–1.0 µL and add H_2O to a final volume of 10 µL. Make sure tubes are tightly capped.
2. Amplification in a thermal cycling machine with the following conditions: a primary denaturation at 95°C for 5 min, followed by 500 cycles of annealing/ligation at 80°C for 20 seconds and denaturation at 95°C for 10 s. This step needs approx 10 h.

3.1.2. Electrophoresis and Blotting

1. Electrophoresis: Use a 6% polyacrylamide/6 M urea gel with a wide tooth comb. Pre-electrophorese at 90W for 20 min. Heat denature RED products in 0.5X formamide gel loading dye for 5 min at 95°C and load all content onto gel. Electrophorese at 90W until xylene cyanol (xc) has migrated 12 cm into gel. Separate plates and discard gel below 16 cm to avoid probe hybridization to excess oligonucleotides.
2. Blotting: Place a wet (with 1X TBE) sheet of Hybond N+ membrane on the gel, and overlay with three dry 3mm Whatman papers cut to fit, the top glass plate, and a weight. Let gels sit for 2 h to permit capillary blotting. Thereafter, immobilize the DNA on the membrane by cross-linking in a UV-light box.

3.1.3. Hybridization

1. Labeling probe: 3' end labeling using TdT is prefered because it effectively permits addition of multiple ^{32}P-dATPs to each molecule, yielding a high specific activity. Mix 8.7 µL H_2O, 5 µL 5x Co-buffer and 2.5 µL $(CAG)_{10}$ oligonucleotide (50 ng/µL) on ice. Add 7 µL α^{32}P-dATP and 1.8 µL TdT enzyme. Incubate for 1 h at 37°C. Add 500 µL 0.1 M Tris pH 8.0 to stop the reaction. Probe should be labeled to a specific activity of $2–9 \times 10^9$ cpm/µg.
2. Hybridization: Prehybridize membrane in Rapid Hybe (Amersham) solution for 20 min; add labeled probe and hybridize for 1 h at 60°C. Wash the membrane for 20 min at room temperature and 30 min at 60°C in 1X SSC + 0.1 % SDS with at least one change of wash solution.

3.1.4. Autoradiography

Expose membranes overnight (or up to 7 d) to X-ray film at –70°C using intensifying screens.

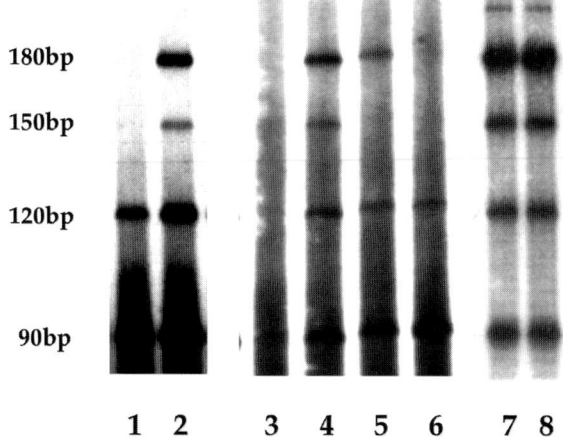

Fig. 3. Autoradiographs showing a series of RED products detected from DNA samples from patient A at different cloning stages. A consistent 180bp RED product was seen in genomic DNA (lane 2), two MboI-DNA fractions (lane 4 and 5) but not in the flanking fractions (lane 3 and 6), as well as two representative cloned DNA fragments (lane 7 and 8). RED was performed using a $(CTG)_{10}$, generating multimers at 30 nt intervals represented as bands after size separation, blotting, hybridization, and autoradiography. The bottom band corresponds to 90 bp.

3.1.5. Results

Products are revealed as a ladder of bands (*see* **Fig. 3**) with 30 nucleotide interval between neighboring bands. The band with the highest molecular weight represents the largest repeat expansion in that particular genome. A "base line" ligation product formed by trimers of the oligonucleotide used should be seen in all lanes as the human genome contains many short repeat sequences. Absence of such a product should be regarded as a reaction failure in need of troubleshooting. A sample with a known repeat expansion should be included as a positive control to deduce RED product expansion sizes in the samples analyzed. See also **Notes 1–5** for trouble shooting guide.

3.2. The RED-Cloning Strategy

An overview of the cloning procedure is presented in **Fig. 2**, detailing the major steps involved.

3.2.1. Physical Enrichment of DNA Fragments Containing Long Trinucleotide Repeats

1. Digestion of genomic DNA: Select genomic DNA from an individual with an expanded trinucleotide repeat that you wish to clone. Set up digestion reaction containing 10 µg of DNA, 20–40 U Restriction enzyme (MboI), 6 µL supplied buffer for the restriction enzyme and add water to a total volume of 60 µL. After incubation in a 37°C water bath for 4–5 h, DNA should be fully digested.
2. Precipitation of digested DNA: Add 1/10 volume of 3 M NaOAc and 2X volume of 99.5% Ethanol. Mix thoroughly by inverting the tube. Place the tube on ice for half an hour or at −20°C for 15 min. Centrifuge at 8,000g at 4°C for 15 min. Discard supernatant. Dry DNA pellet in a vacuum oven.

RED and RED Cloning Strategy

3. Electrophoresis: Resuspend DNA in 20 μL TE buffer and mix with 4 μL 6X loading dye. Load all content to a 0.8% low-melting temperature agarose gel. Load 1 kb and 100 bp ladder as size standards in separate lanes. Electrophorese at 20V for 17–18 h. The 100 bp DNA fragments should then have migrated around 11 cm into the gel.
4. Dissection of agarose lane containing DNA separated by size: Cut the agarose lane and dissect it into uniform 2 mm pieces and place each piece into clean eppendorf tubes.
5. Digestion of agarose using AgarACE: Melt agarose pieces at 70°C for 10 min and then transfer tubes to a water bath or heating block at 42°C. Add 2–3 U AgarACE to each tube and mix by vortexing. Incubate at 42°C for 15 min. Inactivate AgarACE at 70°C for 10 min.
6. Precipitation of DNA: Add 1/10 volume of 3 M NaOAc and 2X volume of 99.5% Ethanol. Mix thoroughly by inverting the tube. Leave the tubes at room temperature for 1.5–2.0 h. Centrifuge at 8,000g at room temperature for 15 min and dry pellet as in step 2. Resuspend DNA pellet in 10 μL H_2O.
7. RED analysis: Perform the RED analysis as described in **Subheading 3.1.** on each DNA fraction to identify the fractions enriched for DNA fragments containing the expanded repeat. Use 3 μL DNA to set up RED reaction (*see* **Note 6**).
8. Select the 1–2 fractions containing the expanded repeat of interest.

3.2.2. Cloning of the DNA Fragments Containing an Expanded Repeat

Every step in the cloning part is performed following the instructions supplied together with the vector kit, with some modifications. Please read the original protocol for details if necessary.

1. Generation of genomic library using the RED positive DNA fraction: Set up ligation reaction mixture containing the following: 1.0 μL of the ZAP Express Vector (1 μg/μL), 1.0 μL of the repeat containing DNA fraction, 0.5 μL of 10X Ligase buffer, 0.5 μL of 10 mM ATP, 0.2 μL of T4 Ligase (10U/μL) and 1.8 μL of H_2O. Incubate at 4°C for 48 h.
2. Packaging Ligation: Take one tube of GigaPack golden III from –70°C. Thaw the content by holding the tube. Add 5 μL ligation into packaging content. Mix by moving tip gently. Incubate at room temperature for 1.5–2 h. Add 500 μL SM buffer and 20 μL chloroform. Invert tube and spin briefly.

3.2.3. Titering

1. Dilute packaged ligation 1:10.
2. Prepare XL-1 Blue cell at OD_{600} = 0.5 in 10 mM $MgSO_4$.
3. Mix 200 μL XL-1 Blue cell with 1 μL of original ligation-package and 1:10 diluted ligation.
4. Incubate at 37°C for 15 min.
5. Melt Top-agar and cool down to 48°C.
6. Mix 3 mL Top-agar with the content of the tube and 15 μL 0.5 M IPTG, 40 μL Xgal.
7. Pour quickly onto big NZY plates. Incubate at 37°C over night.
8. Count white plaques (recombinants).
   The titer of the white plaques should reach 1000/μL of the packaged ligation.. Titers lower than this number can lead to a failure in cloning the long repeat containing fragment (*see* **Note 7**).

3.2.4. Plating Library

1. 600 μL XL-1 Blue cells prepared at OD_{600} = 0.5 in 10 mM $MgSO_4$.
2. Mix with the calculated amount of packaged ligation, so that each plate contains 400,000–500,000 recombinants.
3. Incubate at 37°C for 10–11 h. Wrap plates with plastic film and store them at 4°C for two h. (Plates are now ready for plaque lifts.)

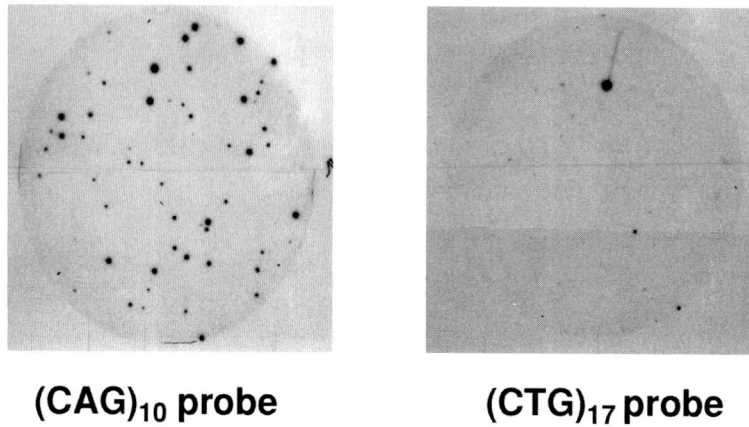

(CAG)₁₀ probe **(CTG)₁₇ probe**

Fig. 4. Autoradiograph of library screening of repeat containing plaques using a $(CAG)_{10}$ repeat probe or a $(CTG)_{17}$ probe. The number of positive spots was reduced when the $(CTG)_{17}$ probe is used. Hybridization stringency used for $(CAG)_{10}$: hybridization at 60°C and washed at 65°C with 1X SSC + 0.1% SDS, $(CTG)_{17}$: 65°C and washed at 72°C with 0.2X SSC + 0.1% SDS.

3.2.5. Plaque Lifts

1. Make duplicate Hybond-+ filters for each plate. Place filter A on the plate for 2 min, remove it and place filter B on the plate for 4 min before removing it.
2. Directly after removing each filter from the plate, slowly place the filter on the surface of denaturation solution, with the plaque-containing side upward for 5 min.
3. Transfer filters to renaturation solution for 5 min.
4. Dip filters in 2X SSC solution for 2 min.
5. Immobilize DNA to the filter by UV light autocrosslinking.
6. Prewash filters at 65°C in wash solution (2X SSC + 0.1%SDS) for 2 h.

3.2.6. Hybridization

A $(CTG)_{17}$ repeat-oligonucleotide-probe is used for the selection of long-repeat containing plaques from the library. Conditions for probe-labeling and library hybridization are the same as the RED hybridization step (*see* **Subheading 3.1.3.**), except for the temperature (65°C for hybridization and 72°C for wash with wash solution [0.2X SSC + 0.1% SDS]). **Figure 4** shows the positive spots after autoradiography, using probes with different length.

3.2.7. Picking Positive Plaques

Pick positive plaques according to their position on the film and place them separately into 1 mL SM buffer. Add 20 µL chloroform. Vortex and store them at 4°C for at least overnight.

3.2.8. Isolation of Single Positive Plaques

Repeat steps from **Subheadings 3.2.4.–3.2.7.** using positive phage-stock in different dilutions (10 µL; 1 µL; 0.1 µL, and 0.01 µL) until single positive plaques are obtained. Pick these single plaques and place them in 200 µL SM buffer and 5 µL chloroform. Incubate at 4°C over night.

3.2.9. Conversion of Phage DNA to Phagemid DNA

1. Prepare Xl-1 Blue cells at $OD_{600} = 1.0$ in 10 mM $MgSO_4$.
2. Mix 100 μL XL-1 Blue cells with 50 μL of concentrated phage and 0.5 μL assistant helper phage (Stratagene ZAP cloning kit).
3. Incubate at 37°C for 15 min.
4. Add 1.5 mL NZY-broth and incubate with shaking at 225 rpm at 37°C for 2.5 h.
5. Heat tubes at 70°C for 20 min. Spin down cells at 1500 rpm for 15 min.
6. Take supernatant 100 μL and mix with 200 μL XLOLR cells at $OD_{600} = 1.0$ in 10 mM $MgSO_4$.
7. Incubate at 37°C for 15 min.
8. Add NZY-broth 300 μL and incubate at 37°C for 45 min.
9. Plate out 200 μL onto kanamysin LB plates. Incubate overnight at 37°C.

3.2.10. Preparation of Phagemid DNA

Collect bacteria from each plate and prepare DNA using the SNAP-kit (Invitrogen). Finally, dissolve the DNA in 60 μL TE buffer.

3.2.11. Sequencing Using 377 ABI DNA Sequencer

1. Sequencing primers: T3 and T7 promotor sequences located on the vector flanking the insert DNA are used to sequence the insert from both ends.
2. Set up the sequencing reaction containing: 2 μL phagemid DNA (0.3 μg/μL), 3.2 pmol of T3 or T7 primer and 2 μL BigDye™ Terminator Cycle Sequencing Ready Reaction. Conditions for cycling: 35 cycles of 96°C for 10 s, 50°C for 5 s and 60°C for 4 min.
3. Sequence on a denaturing 6% gel on ABI 377.

4. Notes

1. The RED method can be used to detect expansions with different trinucleotide repeats. However, the oligonucleotide length may need to be increased if the GC content of the repeat motif to be analyzed is low, and an optimization of the cycling conditions is needed. Oligonucleotides of different sizes may be used when screening for several motifs simultaneously (5). It is important that oligonucleotides are phosphorylated and purified for the RED reaction to work. Any residue of shorter molecules from the synthesis will reduce the signal markedly.
2. Empty tubes due to evaporation following the long RED cycling procedure can be a problem. Select PCR tubes carefully. We have found that Continental LP tubes work well.
3. Light areas on autorad can be caused by poor contact during blotting. Too much buffer may obscure part of the ladder present in a given lane.
4. Dark autorad without visible lanes can sometimes be produced. The probe has randomly hybridized to the whole filter. This can be caused by a spill of RED reaction sample with a large excess of oligonucleotide into the upper buffer tank during gel loading.
5. Dark sample lanes and lack of products may be the case of reaction failure. This could be due to the quality of the DNA. Check the degree of degradation and the actual concentration of high molecular weight DNA on an agarose gel. High salt concentration may reduce the efficiency of the Ampligase. Try to lower the Ampligase buffer concentration to 0.5X. A low pH also lowers the reaction yield and may cause reaction failure. Include 1 μL buffered TE-4, pH 8.0 in the RED reaction if the DNA is in an unbuffered solution. Weak bands could also be due to poor handling of the reagents. We have noticed that the ampligase buffer is sensitive to repeated thawing.
6. Pure DNA is essential for the RED reaction. Therefore the fractionation and extraction steps of the cloning procedure has to work well. Too much agarose or agarose residues may inhibit the RED reaction. This problem can be avoided by repeating the DNA extraction step using a standard DNA extraction protocol using phenol/chloroform/isoamyl alchohol.

7. A titer of recombinants lower than 1,000/µL may lead to a failure of cloning the long repeat containing fragments due to preferential selection against DNA fragments containing long repeat during the cloning procedure. Low ligation efficiency may again be caused by poor DNA quality.

References

1. Lander, E. S., Linton, L. M., Birren, B., Nusbaum, C., Zody, M. C., Baldwin, J., et al. (2001) Initial sequencing and analysis of the human genome. *Nature* **409,** 860–921.
2. Warren, S. T. and Wells, R. D. (1998) *Genetic Instabilities and Hereditary Neurological Diseases.* Academic Press, San Diego, CA, USA.
3. Bowater, R. P. and Wells, R. D. (2000) The intrinsically unstable life of DNA triplet repeats associated with human hereditary disorders. *Prog. Nucleic. Acid. Res. Mol. Biol.* **66,** 159–202.
4. Zoghbi, H. Y. and Orr, H. T. (2000) Glutamine repeats and neurodegeneration. *Annu. Rev. Neurosci.* **23,** 217–247.
5. Schalling, M., Hudson, T. J., Buetow, K. H., and Housman, D. E. (1993) Direct detection of novel expanded trinucleotide repeats in the human genome. *Nat. Genet.* **4,** 135–139.
6. Zander, C., Thelaus, J., Lindblad, K., Karlsson, M., Sjoberg, K., and Schalling, M. (1998) Multivariate analysis of factors influencing repeat expansion detection. *Genome Res.* **8,** 1085–1094.
7. Lindblad, K., Nylander, P. O., De bruyn, A., Sourey, D., Zander, C., Engstrom, C., et al. (1995) Detection of expanded CAG repeats in bipolar affective disorder using the repeat expansion detection (RED) method. *Neurobiol. Dis.* **2,** 55–62.
8. Lindblad, K., Savontaus, M. L., Stevanin, G., Holmberg, M., Digre, K., Zander, C., et al. (1996) An expanded CAG repeat sequence in spinocerebellar ataxia type 7. *Genome Res.* **6,** 965–971.
9. Lindblad, K., Lunkes, A., Maciel, P., Stevanin, G., Zander, C., Klockgether, T., et al. (1996) Mutation detection in Machado-Joseph disease using repeat expansion detection. *Mol. Med.* **2,** 77–85.
10. Holmes, S. E., O'Hearn, E. E., McInnis, M.G., Gorelick-Feldman, D. A., Kleiderlein, J. J., Callahan, C., et al. (1999) Expansion of a novel CAG trinucleotide repeat in the 5' region of PPP2R2B is associated with SCA12. *Nat. Genet.* **23,** 391–392.
11. Wells, R. D. (1996) Molecular basis of genetic instability of triplet repeats. *J. Biol. Chem.* **271,** 2875–2878.
12. Iyer, R. R. and Wells, R. D. (1999) Expansion and deletion of triplet repeat sequences in *Escherichia coli* occur on the leading strand of DNA. *J. Biol. Chem.* **274,** 3865–3877.
13. Yuan, Q. P., Lindblad, K., Zander, C., Burgess, C., Durr, A., and Schalling, M. (2001) A cloning strategy for the identification of novel genes containing trinucleotide repeat expansion. *Int. J. Med. Genet.* **8,** 427–431

6

Repeat Analysis Pooled Isolation and Detection (RAPID) Cloning of Microsatellite Expansions

Laura P. W. Ranum

1. Introduction

Microsatellite repeat expansions have been shown to cause a number of neurodegenerative diseases *(1)*. Most of the disease genes identified to date involve the expansion of a trinucleotide repeat motif, but recently tetra- and pentanucleotide repeat expansions have been shown to cause myotonic dystrophy type 2 (DM2) and spinocerebellar ataxia type 10 (SCA10), respectively *(2,3)*. Most microsatellite diseases are characterized by the presence of anticipation, or a decrease in the age of onset in consecutive generations due to the tendency of the unstable repeat tract to lengthen when passed from one generation to the next *(1,4,5)*. In addition, the involvement of trinucleotide repeat expansions in a number of other diseases including schizophrenia *(6)* and bipolar affective disorder *(7,8)* has been suggested both by the presence of anticipation and by Repeat Expansion Detection (RED) analysis *(9,10)*. The involvement of trinucleotide expansions in these diseases, however, can only be conclusively confirmed by the isolation of the expansions present in these populations and detailed analysis to assess each expansion as a possible pathogenic mutation. We previously described a novel procedure to quickly isolate expanded trinucleotide repeats and the corresponding flanking nucleotide sequence directly from small amounts of genomic DNA using a process of Repeat Analysis, Pooled Isolation, and Detection of individual clones containing expanded trinucleotide repeats (RAPID cloning) *(11)*. We used this technology to clone the pathogenic SCA7 and SCA8 CAG/CTG repeat expansions from banked DNA samples from single individuals affected with ataxia *(11–13)*. In addition, Holmes et al. used RAPID cloning to identify the CAG/CTG expansion responsible for SCA12 *(14)* and a novel CAG/CTG expansion that causes Huntington disease-like 2 *(15)*.

The RAPID cloning procedure, outlined schematically in **Fig. 1**, uses an optimized RED protocol to follow an expanded trinucleotide repeat through a series of enrichment steps until a single, isolated clone is obtained *(11)* The initial step in RAPID cloning from genomic DNA is two-dimensional RED analysis (2D-RED). In the 2D-RED protocol, genomic DNA is digested with a restriction enzyme and run out on an agar-

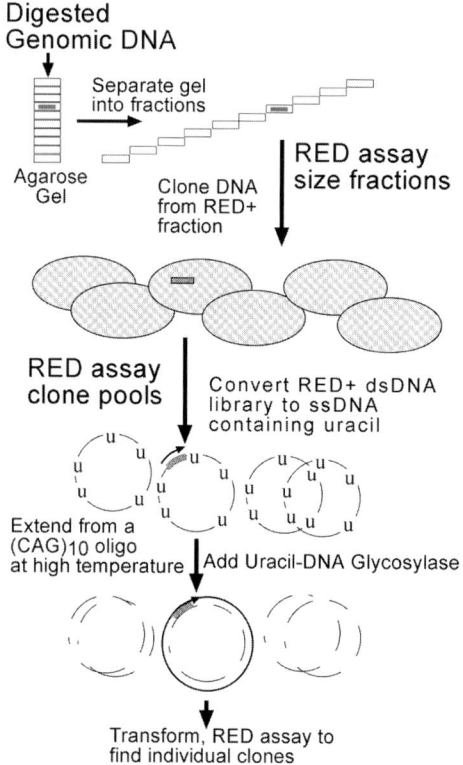

Fig. 1. Overview of RAPID cloning strategy. In general, the Repeat Expansion Detection (RED) assay *(9,10)* is used to follow an expanded trinucleotide repeat present in genomic DNA through a series of enrichment steps until a single, purified clone is obtained. Reprinted with permission from Nature Genetics, **ref.** *(11)*.

ose gel to separate the DNA into discrete size fractions. RED analysis is then performed on each size fraction. In addition to identifying an enriched genomic size fraction for use in the subsequent cloning and enrichment protocol, the 2D-RED assay measures both the number and size of the expansions present in an individual's genome. After 2D RED separation the purified DNA from the fraction enriched for the repeat expansion is cloned and enriched for the expansion *(11)*.

2. Materials

2.1. Genomic DNA Isolation

1. Pure Gene Genomic DNA Isolation Kit #D-5000 (Gentra Systems, Plymouth, MN, USA).

2.2. Optimized RED Assay (see Note 1)

1. (5'P-CTG)$_{10}$ oligonucleotide primer, 5' phosphorylated, gel purified (aliquot, dry and store at $-80°C$, for RED reactions).
2. Hybond N+ (Amersham, Piscataway, NJ, USA).
3. (CAG)$_{10}$ primer for probe.
4. α–P^{32} (6000 Ci/mmole) (NEN, Boston, MA, USA).

5. TdT and 10X 1-phor-all Buffer (Pharmacia Corporation, Peapack, NJ, USA).
6. Ampligase and Buffer (Epicentre, Madison, WI, USA).
7. Formamide.
8. Control genomic DNA from patient samples containing repeat expansions of various sizes (e.g., SCA1, HD, etc., controls).

2.3. Two-Dimensional RED Analysis

1. Genomic DNA.
2. *Eco*RI or another restriction enzyme.
3. SeaPlaque GTG low melting-point agarose (BioWhittaker Molecular Applications, Rockland, ME, USA).
4. Medium Size Gel Box (gel size = 24 × 13 cm).
5. TAE buffer: 0.4 M Tris base, 0.2 M Acetic Acid, 0.01 M EDTA, pH 8.0.
6. Gel-slicing apparatus with microscope coverslips as disposable dissecting blades for even 2-mm slices.
7. AgarACE (Promega, Madison, WI, USA).
8. Glycogen 5 mg/mL (Ambion, Austin, TX, USA).

2.4. Cloning of Genomic Fragments

1. Lambda ZapII Cloning and Packaging Kit (Stratagene, La Jolla, CA, USA): RecA⁻ *Escherichia coli* host strain XL1-Blue MRF'; ExAssist interference-resistant helper phage; SOLR strain; Packaging extracts.
2. 10X Ligase Buffer: 500 mM Tris-HCl, pH 7.5, 70 mM MgCl$_2$, 10 mM dithiothreitol (DTT) plus.
3. 10 mM rATP, pH 7.5.
4. T4 DNA Ligase (Epicentre).
5. SM Buffer (1 L): 5.8 g NaCl, 2.0 g MgSO$_4$ · 7H$_2$O, 50 mL 1 M Tris-HCl, pH 7.5, 2 mL 5% (w/v) gelatin, add H$_2$O to 1 L and autoclave.
6. Chloroform.
7. LB Plates (For 1 L): 10 g NaCl, 10 g tryptone, 5 g yeast extract, 20 g agar adjust to pH 7.0.
8. LB Media (For 1 L): 10g NaCl, 10 g tryptone, 5 g yeast extract, adjust to pH 7.0.
9. LB-Tetracycline Plates: LB plates as above with tetracycline 12.5 µg/mL.
10. LB-MgSO$_4$-Maltose Media: LB media as above with 10 mM MgSO$_4$ and 0.2% maltose.
11. Top Agar: Prepare 1 L of LB broth and add 0.7% [w/v] agarose, autoclave.
12. 10 mM MgSO$_4$.

2.5. Post Cloning Enrichment

1. *Escherichia coli* strain CJ236 (*dut-*, *ung-*; Bio Rad, Hercules, CA, USA).
2. M13K07 helper phage (1 × 10^{10} pfu, Promega, Madison, WI, USA).
3. T4 DNA polymerase (New England Biolabs, Beverly, MA, USA).
4. Amplitaq (Perkin Elmer, Wellesley, MA, USA).
5. (CAG)$_{10}$ oligonucleotide primer (IDT, Hillsboro, OR, USA).

3. Methods

3.1. Optimized 1-D RED Analysis

3.1.1. Optimized RED Reactions

Reactions are performed according to Schalling et al. (*9*) with slight modification:

1. Repeat Expansion Detection Reaction:
   10X Ampligase Buffer.
   5' P (CTG)$_{10}$ oligonucleotide primer (50 ng).
   4% Formamide (final concentration).
   5 U Ampligase (Epicentre).
   2 μg genomic DNA.
   H$_2$O.
   ──────────────────────────
   10 μL = total volume.
2. Reactions are performed in 0.5 mL Eppendorf tubes. To prevent evaporation, each reaction is internally capped with the bottom portion (~2 mm) of another trimmed off 0.5 mL Eppendorf tube. Seal caps by pushing down with a micropipet tip.
3. Sealed tubes are then heated to 94°C for 4 min and 50 s and then cycled 495 times at 94°C for 10 s, 78°C for 40 s with a 2 s/degree ramp when cooling from 94°C to 78°C.

3.1.2. Gel Electrophoresis and Blotting

1. Un-cap the samples by flaming a micropipet tip to melt the end and seal it to the top of the internal cap. Let cool, then slightly heat tubes to break the seal and remove. Repeat if necessary.
2. To increase the amount of RED product that can be loaded onto a gel, dry the samples in a speed vac.
3. Resuspend the RED products with gel loading dye (10% formamide, 2 mM EDTA, 0.005% Bromphenol Blue, and 0.005% Xylene Cyanol).
4. Denature at 94°C for 5 min, cool on ice and load onto a denaturing sequencing gel (6% acrylamide gel in TBE buffer, 6 M urea, TBE buffer).
5. Run gel at 70 watts or until Xylene Cyanol band has run into gel for ~12 cm.
6. Pry gel plates apart leaving the gel in place on one of the glass sequencing plates.
7. Blot gel with Hybond membrane (cut memberane so that it covers the top 15 cm of the gel) that has been pre-wet in 1X TBE buffer. Trim away excess acrylamide. Place 3 sheets of 3 mm blotting paper on top of membrane and then cover with the second glass plate and a pyrex glass tray filled with water as a weight. Allow the DNA to transfer to the membrane for 2 h.
8. Remove membrane from gel and UV crosslink the DNA onto the membrane.

3.1.3. Hybridization of CAG Oligonucleotide to RED Products

1. Label Probe (CAG)$_{10}$:

(CAG)10 oligonucleotide	125 ng
10X 1-phor-all Buffer	2.5 μL
TdT Enzyme	1.5 μL
α-^{32}P dATP (6000 Ci/mM)	3.5 μL
H$_2$O up to 10 μL total volume	
Total Volume	10.0 μL

2. Incubate labeling reaction at 37°C for 2 h.
3. Prehybridize membrane with 25 mL of Rapid Hyb Buffer (Amersham Pharmacia Biotech, Piscataway, NJ, USA) at 60°C for 15 min.
4. Add probe and hybridize at 60°C for 1 h.
5. Wash 2 × 15 min at 45°C in 2X SSC, 0.1% SDS, and expose to X-ray film.

3.2. Two-Dimensional RED Analysis (see Note 2)

1. Ten μg of genomic DNA is digested with a restriction enzyme.
2. Size separate digested DNA on a 0.7% SeaPlaque GTG (FMC, Rockland, ME, USA) low melting-point agarose gel in 1X TAE buffer.

3. Excise the lane containing the DNA with a razor blade and dissect it into uniform 2-mm slices using a gel-slicing device in which microscope coverslips are used as disposable dissecting blades.
4. Place individual gel slices in separate 0.5 mL PCR tubes.
5. Remove the agarose by digestion with AgarACE (0.2 U, Promega).
6. After digestion, spin and cool tubes and the transfer liquid to a new tube leaving behind any remaining undigested agarose.
7. Precipitate DNA by adding 1/10th volume of 3 M NaOAc, 2 volumes of 100% Ethanol and as a carrier 10 µL of a 5 mg/mL glycogen stock solution, incubate O/N at room temperature.
8. On the next day spin samples 30 min at RT in microcentrifuge.
9. Rinse pellets containing the DNA with 70% EtOH to remove salt.
10. Dry pellets in speed vac and resuspend in 7.5 µL of 10 mM Tris-HCl, 1 mM EDTA, pH 7.5 buffer.
11. RED analysis is performed on 2.5 µL of DNA from each fraction to determine which size fraction is most highly enriched for the RED-positive genomic fragments.
12. The RED reactions are as described earlier but the genomic DNA is replaced with the size-separated DNA and 1 µg of genomic *E. coli* DNA is added to each reaction to improve the reliability of the RED reactions on the fractionated DNA.

3.3. Cloning of Genomic Fragments (see Note 3)

1. *Eco*RI digested genomic DNA recovered from the RED-positive gel fraction is then cloned using the predigested Lamdba ZapII cloning and packaging kit (Stratagene, La Jolla, CA, USA).
2. Plates containing 5 × 10^4 primary clones/plate are amplified and mass excised separately as described in the protocols provided by the manufacturer and summarized below (Stratagene).

3.3.1. Ligation Reaction

Stratagene Lambda Zap II/*Eco*RI/ CIAP cloning kit ligation 12–14°C overnight.

1.0 µL Lambda Zap II prepared arms (1 µL) (–70°C aliquoted)
2.8 µL positive, purified 2DRed product
0.5 µL 10X ligase buffer (aliquoted out and stored at 4°C) (Epicentre)
0.5 µL 10 mM rATP, pH 7.5
2 U   of T4 DNA ligase (Epicentre)

H$_2$O up to a final volume of 5 µL

3.3.2. Packaging into Phage

Statagene Gigapack III Gold

1. Remove the appropriate number of packaging extracts from –80°C freezer and place on dry ice.
2. Quick thaw between fingers until it just begins to thaw.
3. Add 5 µL (total volume) of the ligation to thawing extract.
4. Stir the contents with the pipet tip to mix well. Do not introduce air bubbles. Quick spin.
5. Incubate tube at room temperature 90 min to 2 h (do not exceed 2 h).
6. Add 500 µL of SM Buffer:
7. Add 20 µL of chloroform and mix the contents of the tube gently to lyse cells of extracts.
8. Spin the tube briefly to sediment the debris. Store at 4°C for up to 1 mo.
9. Titer supernatant.

3.3.3. Titering Procedure

1. Streak XL-1 Blue onto tetracycline plate
2. Inoculate 1 colony into 50 mL of LB with 10 mM MgSO$_4$ and 0.2% maltose Grow O/N at 30°C with shaking
3. Microwave top agar to melt and place in water bath and equilibrate to 48°C.
4. Equilibrate a water bath to 37°C.
5. Place desired number of small, plain LB plates in 37°C incubator to temperature equilibrate.
6. Spin overnight culture at 1000 g for 10 min.
7. Carefully decant off media and resuspend cells to OD$_{600}$ = 0.5 in 10 mM MgSO$_4$. (Do not vortex.)
8. Dilute out phage at 1:10, 1:100 and 1:1000 in SM buffer.
9. Add 200 µL of above resuspended XL-1 Blue cells to 15 mL snap top Falcon tubes.
10. Label and add 1 µL of each phage dilution to corresponding tube.
11. Incubate tubes at 37°C for 15 min.
12. Add 3mL of 48°C top agar to a tube at 37°C and quickly pour on top of pre-warmed LB plates.
13. Incubate upside down at 37°C overnight.
14. Count the number of plaques and determine the plaque forming units (pfu) per mL concentration of the library based on the dilution.

3.3.4. Amplification of Library

1. Start a 50 mL overnight culture of XL1-Blue in LB with 0.2% maltose.
2. Warm large LB plates to 37°C.
3. Equilibrate water baths to 37°C and 48°C.
4. Melt top agar and equilibrate to 48°C.
5. Centrifuge overnight culture for 10 min at 1000g.
6. Carefully decant media off the cell pellet and gently resuspend pellet to OD ~0.5 in 10 mM MgSO$_4$.
7. Aliquot 600 µL of resuspended cells into 20 Falcon 2059 polypropylene tubes and then add ~50,000 pfu to each aliquot. Use all the mixture.
8. Incubate the tubes containing the phage and host cells for 15 min at 37°C
9. Mix 6.5 mL of top agar, melted and cooled to ~48°C, with each aliquot of infected bacteria and spread evenly onto a 150 mm agar plate.
10. Incubate the plates at 37°C for 6–8 h. The plaques should be ~1–2 mm and should be touching at the end of the incubation.
11. Overlay the plates with 8 mL of SM buffer. Store the plates at 4°C overnight. (on rocker in cold room). This allows the phage to diffuse into the buffer.
12. Transfer SM liquid and phage from plate to individual 15-mL conical tubes. Place plate on a slant and pipet off into corresponding labeled tubes. Rinse with another 2 mL of SM buffer.
13. Add chloroform to a 5% (v/v) final concentration and incubate at RT for 15 min.
14. Centrifuge at RT, 10 min, 2000 g to remove cell debris.
15. Transfer to a new tube and titer using host cells and serial dilutions of the library (Assume ~10^9–10^{12} pfu/mL).

3.3.5. Mass Excision of Phagemid from Lambda Zap II Vector
3.3.5.1. Day 1

1. Grow separate overnight cultures of XL1-Blue MRF' and SOLR cells in LB supplemented with 0.2% (w/v) maltose and 10 mM MgSO$_4$, at 30°C.

3.3.5.2. Day 2

2. Gently spin down the XL1-Blue MRF' and SOLR cells (1000 g). Resuspend the XL1-Blue MRF' and SOLR cells in 10 mM MgSO$_4$, to an OD$_{600}$ of 1.0 (8×10^8 cells/mL).
3. In separate 1.5 mL eppendorf tubes, combine a portion of each aliquot of the separately amplified lambda bacteriophage library with XL1-Blue MRF' cells at a 1:10 lambda phage:cell ratio. Excise 10- to 100-fold more lambda phage than primary library to ensure a statistical representation of the excised clones. To the same tube also add ExAssist helper phage at a 1:1 helper phage-to-cells ratio to ensure that every cell is co-infected with lambda phage and helper phage. Incubate 37°C for 15 min to allow absorption.

    For example: 10^7 pfu lambda phage (10- to 100-fold the primary library)
    10^8 XL1-Blue MRF' cells
    10^8 pfu of ExAssist helper phage

4. Add each to corresponding 125 mL flask containing 20 mL of LB media and incubate at 37°C for 2.5 h.
5. Transfer to 50 mL conical tubes and heat shock at 70°C for 20 minutes. Centrifuge at 1000g for 10 min and decant supernatant into a sterile conical tube.

3.3.6. Rescue

1. Overnight prepare a 50 mL SOLR culture in LB.
2. Next day, start a fresh SOLR culture from O.N. culture- 5 mL into 45 mL of LB.
3. Add 1 mL of each of the excised phagemid sublibraries (supernatant from **Subheading 3.3.5., step 5**) to a 50 mL conical tube.
4. To each tube add 2 mL of 1 h SOLR cell culture and incubate at 37°C for 15 min.
5. To each tube then add 10µL of LB with 100 µg/mL ampicillian and incubate at 37°C with shaking overnight
6. Pellet cells and miniprep the resulting plasmid DNA.
7. From the 20 separately excised and rescued libraries identify the library containing the expanded repeat tract by performing RED analysis on the isolated plasmid DNA representing ~50,000 clones/pool (**Fig. 2B**).

3.4. Postcloning Enrichment of DNA from RED Positive Sublibrary

1. The enrichment of CAG containing clones is an adaptation of the general approach described by Duyk et al. *(16)*, which is based on the selection method described by Kunkel et al. *(17)*.
2. Plasmid DNA from a RED positive pool of clones is electroporated into *E. coli* strain CJ236 (*dut-*, *ung-*, BioRad, Hercules, CA) and M13K07 helper phage (1×10^{10} pfu) is added to generate uracil-substituted ssDNA *(17)*.
3. The purified ssDNA is incubated with T4 DNA polymerase (NEB) overnight without dNTP to eliminate contaminating DNA that could act as primers.
4. The CTG repeat containing ssDNA was then converted to dsDNA by primer extension using ampliTaq (Perkin Elmer) and a (CAG)$_{10}$ primer at 72°C in Promega PCR buffer containing 1.5 mM MgCl$_2$, 200 µM dNTPs, and 4% formamide.
5. One µL of uracil DNA glycosylase (UDG, Gibco-BRL) is added after extension to degrade the remaining ssDNA.
6. After extraction and precipitation, the DNA is electroporated into the SURE strain of *E. coli* (Stratagene).
7. Individual colonies that survived the selection process are picked in duplicate onto LB/Amp plates.
8. Clones from one of the replica plates are then pooled and the plasmid DNA is miniprepped.

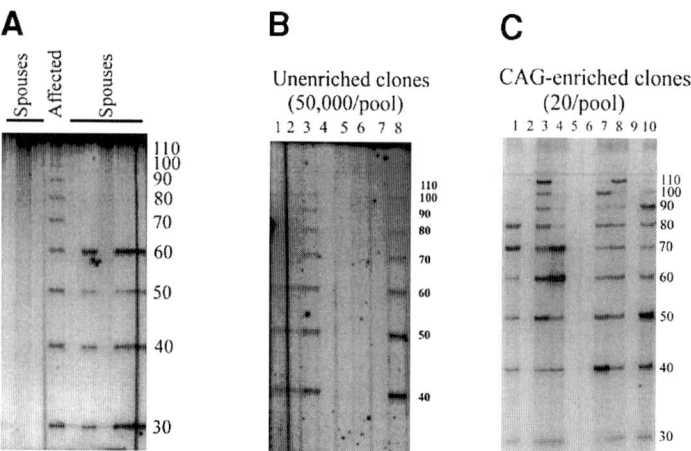

Fig. 2. Cloning and post-cloning enrichment of an expanded CAG repeat from the MN1 kindred. (**A**) RED analysis of genomic DNA samples from a kindred with a novel form of myotonic dystrophy, the MN1 kindred. RED products from an affected individual and from eight unaffected spouses are shown. The sizes of the RED products are indicated as the number of CAG/CTG repeats. (**B**) RED analysis of plasmid DNA isolated from unenriched clone pools of approximately 5×10^4 clones. (**C**) RED analysis of CAG/CTG-enriched clone pools derived from pools 3 and 8 from B. Each enriched clone pool contains DNA from 20 individual clones. Reprinted with permission from Nature Genetics, **ref.** *(11)*.

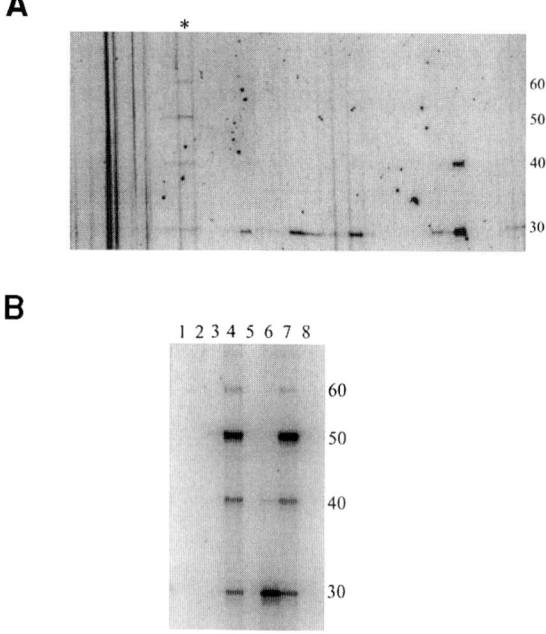

Fig. 3. RAPID cloning of the SCA7 expanded CAG repeat. (**A**), 2D-RED analysis of *Eco*RI-digested genomic DNA isolated from an individual with an autosomal dominant ataxia with rentinopathy (starred individual kindred A, **Fig. 5A**). The genomic DNA size-fraction containing the CAG expansion (indicated by *) was cloned into a lambda vector. The resulting library was amplified in pools that were then converted into plasmid library pools (*see* Methods) (*continued*).

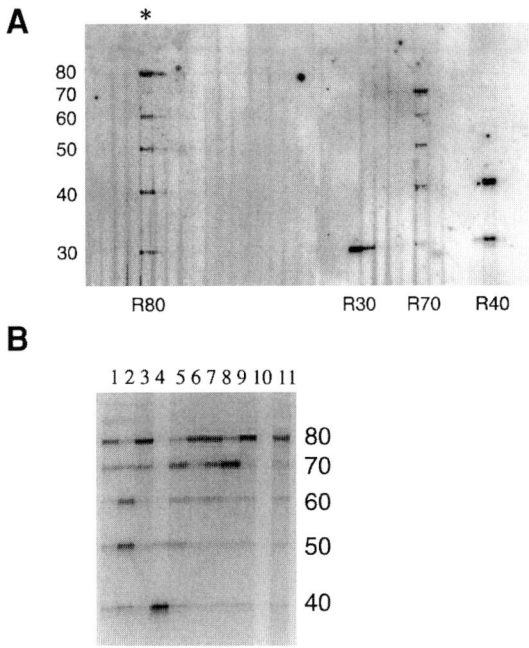

Fig. 4. RAPID cloning of the *SCA8* expanded CTG repeat. (**A**), 2D-RED analysis of EcoRI-digested genomic DNA isolated from an individual with a dominantly inherited ataxia. The number of CAG/CTG repeats for the RED products generated are indicated at the side of the panel. Four separate fractions that generated RED products are indicated below the panel. The size fraction containing the RED80 CTG expansion (indicated by an asterisk) was unique to this ataxia patient and was cloned. (**B**), RED analysis of CAG/CTG-enriched clone pools derived from a RED-positive primary clone pool. Each pool contained DNA from 36 individual clones. Two clones containing the expanded CAG/CTG repeat were isolated from pool 9 and shown to have identical expanded CAG/CTG tracts with 80 uninterrupted repeats. Reprinted with permission from Nature Genetics, **ref. *(11)*.**

9. Individual clones from a RED positive pool (20–36 clones/pool) are picked, separately grown up, mini-prepped, and assayed by RED to identify clones containing both the repeat expansion and the unique flanking DNA sequence.
10. Sequence analysis is performed on individual RED positive clones and unique primers are designed which flank the repeat expansion.
11. PCR is then used to directly test whether or not the candidate repeat expansion co-segregates with a given disease. Examples of repeat tracts that were detected by RED analysis and subsequently cloned by RAPID cloning are shown in **Figs. 2**, **3**, and **4**. The expansion isolated from the MN1 family shown in **Fig. 2** did not co-segregate with the myotonic dystrophy phenotype in the MN1 family *(11)*. The expansions isolated from a family with ataxia

Fig. 3 *(continued)* (**B**), RED analysis of CTG-enriched clone pools derived from a RED-positive primary clone pool. Each enriched clone pool contains DNA from 36 individual clones. RED analysis of plasmid DNA from individual clones in pool 4 identified two clones containing the expanded CAG repeat. Sequence analysis of these clones revealed an expanded CAG tract with 53 uninterrupted repeats. Reprinted with permission from Nature Genetics, **ref. *(11)*.**

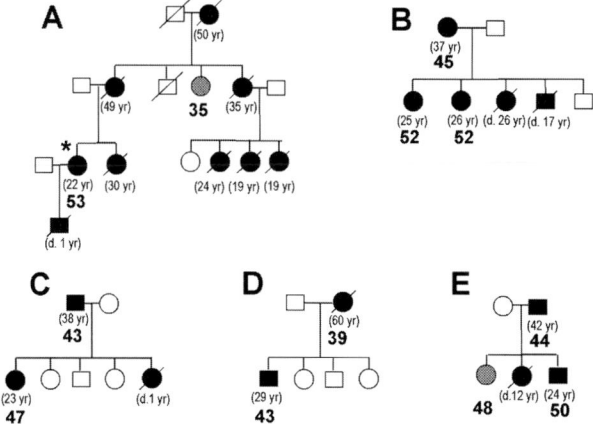

Fig. 5. PCR analysis of the SCA7 CAG alleles. Kindreds diagnosed with autosomal dominant ataxia with retinopathy are shown. Filled symbols represent individuals that have been diagnosed with ataxia and partially filled symbols represent individuals that have symptoms consistent with the early stages of ataxia (the current age of the individual in kindred **[A]** with early signs of ataxia is 64 and in kindred **[E]** is 30) *(11)*. The estimated age of onset or age at death (indicated by **[D]**) is given in parentheses and the number of CAG repeats in the SCA7 expansion is indicated below the symbols. The individual from whom the expanded CAG was isolated is starred in kindred **(A)**. An expanded allele was present only in affected or at-risk individuals. Reprinted with permission from Nature Genetics **ref.** *(11)*.

and retinal degeneration co-segregated with disease (**Fig. 5**) and was shown to be the CAG expansion responsible for SCA7 *(11)*. The expansion shown in **Fig. 4** was isolated and shown to be a novel CTG expansion responsible for SCA8 *(13)*.

4. Notes

1. It is important to carefully optimize the RED reaction with control samples containing repeats of various sizes. For genomic DNA with known repeat expansion sizes (such as SCA1 or HD controls) the RED ladder should stop at the predicted size based on the repeat length as a strong band (e.g., RED60 products due to ERDA background repeats in control samples in **Fig. 2A**). If the RED ladder continues beyond what is predicted by the known repeat sizes for genomic DNA samples then the RED reactions are likely not stringent enough and the primer may be binding non-specifically to CAG/CTG like repeat tracts which contain sequence interruptions. When using DNA isolated from clones, sometimes the RED reaction will overshoot the repeat size (i.e., when the RED assay is being used to identify clone pools containing an expanded repeat).
2. Prescreening for background, "nonpathogenic" CAG/CTG expansions that occur at several different loci simplifies the interpretation of the RED and 2-D RED analysis. These loci include ERDA1 *(18)*, SEF2-1B (also known as MN1) *(11,19)*. The most common background expansion is at the ERDA1 locus *(18)*. An example of an ERDA1 background expansion on *Eco*RI 2-D RED analysis that gives a RED 70 product is shown in **Fig. 4**.
3. When preparing to clone the RED positive size separated genomic fragments for further purification streak out XL1-Blue MRF on LB tetracyline plate 2 d prior to starting and 1 d prior to beginning the cloning procedure start an overnight 10 mL culture of XL-1 Blue in LB with 0.2% maltose.

References

1. Brice, A. (1998) Unstable mutations and neurodegenerative disorders. *J. Neurol.* **245(8)**, 505–510.
2. Liquori, C., Ricker, K., Moseley, M. L., Jacobsen, J. F., Kress, W., Naylor, S., Day, J. W., and Ranum, L. P. W. (2001) Myotonic dystrophy type 2 caused by a CCTG expansion in intron 1 of ZNF9. *Science* **293**, 864–867.
3. Matsuura, T., Yamagata, T., Burgess, D. L., Rasmussen, A., Grewal, R. P., Watase, K. (2000) Large expansion of the ATTCT pentanucleotide repeat in spinocerebellar ataxia type 10. *Nat. Genet.* **26(2)**, 191–194.
4. Warren, S.T. (1996) The expanding world of trinucleotide repeats. *Science* **271**, 1374–1375.
5. Klockgether, T. and Evert, B. (1998) Genes involved in hereditary ataxias. *Trends Neurosci.* **21(9)**, 413–418.
6. O'Donovan, M., Guy, C., Craddock, N., Murphy, K., Cardno, A., Jones, L., et al. (1995) Expanded CAG repeats in schizophrenia and bipolar disorder. *Nat. Genet.* **10**, 380–381.
7. Oruc, L., Lindblad, K., Verheyen, G., Ahlberg, S., Jakovljevic, M., Ivezic, S., et al. (1997) CAG repeat expansions in bipolar and unipolar disorders. *Am. J. Hum. Genet.* **60**, 732–735.
8. Vincent, J. B., Paterson, A. D., Strong, E., Petronis, A., and Kennedy, J. L. (2000) The unstable trinucleotide repeat story of major psychosis. *Am. J. Med. Genet.* **97(1)**, 77–97.
9. Schalling, M., Hudson, T., Buetow, K., and Housman, D. (1993) Direct detection of novel expanded trinucleotide repeats in the human genome. *Nat. Genet.* **4**, 135–139.
10. Lindblad, K., Zander, C., Schalling, M., and Hudson, T. (1994) Growing triplet repeats. *Nat. Genet.* **7**, 124.
11. Koob, M. D., Benzow, K. A., Bird, T. D., Day, J. W., Moseley, M. L., and Ranum, L. P. W. (1998) Rapid cloning of expanded trinucleotide repeat sequences from genomic DNA. *Nat. Genet.* **18**, 72–75.
12. Moseley, M. L., Benzow, K. A., Schut, L. J., Bird, T. D., Gomez, C. M., Barkhaus, P. E. (1998) Incidence of dominant spinocerebellar and Friedreich triplet repeats among 361 ataxia families. *Neurology* **51(6)**, 1666–1671.
13. Koob, M.D., Moseley, M. L., Schut, L. J., Benzow, K. A., Bird, T. D., Day, J. W., and Ranum, L. P. W. (1999) An untranslated CTG expansion causes a novel form of spinocerebellar ataxia (SCA8). *Nat. Genet.* **21(4)**, 379–384.
14. Holmes, S. E., O'Hearn, E. E., McInnis, M. G., Gorelick-Feldman, D. A., Kleiderlein, J. J., Callahan, C. (1999) Expansion of a novel CAG trinucleotide repeat in the 5' region of PPP2R2B is associated with SCA12. *Nat. Genet.* **23(4)**, 391–392.
15. Holmes, S. E., O'Hearn, E. E., Rosenblatt, A., Callahan, C., Hwang, H., Ingersoll-Ashworth, R. G., et al. (2001) A repeat expansion in the gene encoding juctophillin-3 is associated with Huntington disease-like 2. *Nat. Genet.* **29**, 377–378.
16. Ostrander, E. O., Jong, P. M., Rine J., and Duyk, G. (1992) Construction of small-insert genomic DNA libraries highly enriched for microsatellite repeat sequences. *Proc. Natl. Acad. Sci. USA* **89**, 3419–3423.
17. Kunkel, T. A., Roberts, J. D., and Zakour, R. A. (1987) Rapid and efficient site-specific mutagenesis without phenotypic selection. *Methods Enzymol.* **154**, 367–382.
18. Nakamoto, M., Takebayashi, H., Kawaguchi, Y., Narumiya, S., Taniwaki, M., Nakamura, Y., et al. (1997) A CAG/CTG expansion in the normal population. *Nat. Genet.* **17(4)**, 385–386.
19. Breschel, T. S., McInnis, M. G., Margolis, R. L., Sirugo, G., Corneliussen, B., Simpson, S. G., et al. (1997) A novel, heritable, expanding CTG repeat in an intron of the SEF2-1 gene on chromosome 18q21.1. *Hum. Mol. Genet.* **6(11)**, 1855–1863.

7

DIRECT Technologies for Molecular Cloning of Genes Containing Expanded CAG Repeats

Kazuhiro Sanpei, Takeshi Ikeuchi, and Shoji Tsuji

1. Introduction

Expansions of unstable CAG/CTG trinucleotide repeats have been identified as a common pathogenic mechanism in a growing number of hereditary neurodegenerative diseases, including myotonic dystrophy, spinal and bulbar muscular atrophy, Huntington's disease (HD), spinocerebellar ataxia type 1 (SCA1), dentatorubral-pallidoluysian atrophy (DRPLA), Machado-Joseph disease (MJD), SCA2, SCA6, SCA7, SCA8, SCA12, and SCA17 (reviewed in **refs. 1–7**). From the viewpoint of clinical genetics, these diseases are characterized by anticipation, i.e., accelerating age at onset and increasing disease severity in successive generations. It has been discovered that anticipation is a result of intergenerational increase in the size of expanded CAG repeats. These observations suggest that many hereditary neurodegenerative diseases characterized by anticipation and the broad spectrum of the clinical presentations are likely to be caused by the unstable expansion of CAG repeats.

The CAG repeats in the genes of normal individuals are, in general, highly polymorphic, not exceeding 35 repeat units in size, whereas the pathologically expanded CAG repeats range in size from 35 to greater than several 100 repeats (e.g., infantile/juvenile onset SCA2 and SCA7). SCA6 is the only exception, in which the expanded CAG repeats range from 21–26 (8). These results imply that a method that allows selective and sensitive detection of CAG repeats consisting of more than 35 repeats would facilitate the search for the causative genes for a number of neurodegenerative diseases.

To develop a novel and robust technology to identify expanded CAG repeats, we took advantage of hybridization kinetics. We considered that the Tm (melting temperature) of a hybrid between DNA molecules containing CAG/CTG trinucleotide repeats would be determined as a function of the number of CAG repeat units engaged in the hybrid formation. If we can prepare a DNA probe that is long enough to make a hybrid with expanded CAG repeats, the Tm of such a hybrid would be substantially increased compared to that between the probe and CAG/CTG repeats of normal lengths. If this is the case, below a certain temperature, only hybrid formation of the probe with a largely expanded CAG repeat would occur (**Fig. 1**, *see* **Note 1**).

From: *Methods in Molecular Biology, vol. 217: Neurogenetics: Methods and Protocols*
Edited by: N. T. Potter © Humana Press Inc., Totowa, NJ

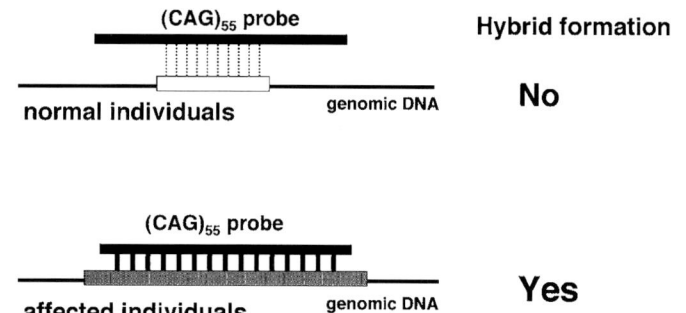

Fig 1. Schematic illustration of the principle of selective hybridization. The melting temperature of hybrids (Tm) between the $(CAG)_{55}$ probe and the genomic DNA segment containing a CAG repeat varies depending on the number of bases engaged in the hybrid formation. Therefore, under stringent hybridization conditions, it is possible to have hybridization conditions at which only the hybrid between the $(CAG)_{55}$ probe and the genomic DNA segment containing expanded CAG repeats (>35 repeats) is formed, while the hybrid between the $(CAG)_{55}$ probe and the genomic DNA segment containing CAG repeats of normal length (< 35 repeats) is not formed.

With this background we have devised a novel and robust technique, the Direct Identification of Repeat Expansion and Cloning Technique (DIRECT), which allows for selective detection of the expanded CAG repeats by genomic Southern blot and subsequent cloning of the genomic segments containing the expanded CAG repeats. Because spinocerebellar ataxia type 2 (SCA2) is one of the autosomal dominant ataxias characterized by genetic anticipation *(9)*, we applied DIRECT to cloning the gene responsible for spinocerebellar ataxia type 2 (SCA2) *(10)*.

The major advantages of DIRECT are as follows. 1) DIRECT allows identification of largely expanded CAG repeats without prior knowledge about the causative gene, meaning that the positional cloning approach including linkage analysis is not required. 2) By adjusting the stringency of hybridization, the "cutoff" values of expanded CAG repeats can be varied. This is particularly important given that some of the hereditary neurodegenerative diseases are caused by mildly expanded CAG repeats. For example, SCA6 has been shown to be caused by expanded CAG repeats ranging from 21 up to 30. Under less stringent hybridization conditions, however, a number of genomic segments are also detected by the $(CAG)_{55}$ probe (Ikeuchi, unpublished), indicating that there are a number of genomic segments containing CAG repeats ranging from 21–30. To overcome this problem, we are currently developing 2D-DIRECT to accomplish much better resolution of genomic segments on agarose gel electrophoresis.

We have been further applying the DIRECT technology to identify additional expanded CAG repeats in human genome. Recently, we have cloned an interesting genomic segment containing a novel long CAG/CTG trinucleotide repeat on chromosome 17 *(12)*. The CAG/CTG repeat is highly polymorphic and exhibits a bimodal distribution. Although the size of the CAG/CTG repeat is within the range of the expanded CAG repeat of disease-causing genes, we did not detect any association with neurodegenerative diseases *(12)*. Nonetheless, the results further confirmed the advantage of DIRECT.

To date, various technologies have been devised to identify largely expanded CAG repeats. Among these, Repeat Expansion Detection (RED) is an alternative approach to identify largely expanded CAG repeats *(13)*. In RED, human genomic DNA is used as a template for a two-step cycling process that generates ligation products of oligonucleotides. Although RED is a simple and robust technique to detect the presence of expanded trinucleotide repeats, molecular cloning of the genomic segment flanking the trinucleotide repeats has been difficult. Recent improvement of the RED technologies, however, made it possible to clone the genomic segment containing largely expanded CAG repeats *(14)*.

Thus, DIRECT and RED are both robust methods to identify largely expanded CAG repeats in the genome, and it is hoped that a much larger number of disease genes caused by expanded CAG repeats will be identified by applying these robust technologies.

2. Materials

2.1. Preparation of Genomic Blots

1. High molecular-weight genomic DNA (*see* **Notes 2**, **3**).
2. Nitrocellulose membrane (BA-S85 Ref. No. 439196, Schleicher & Scheull).

2.2. Preparation of (CAG)$_{55}$ Probe with a High Specific Radioactivity

1. 9.25 MBq of [α-^{32}P]dATP (222 TBq/mmol).
2. Plasmid p-2093 containing a genomic fragment of the CAG repeat (55 repeats) and the flanking sequences of DRPLA gene *(10)*.
3. Primers (primer 1: 5'-CAC CAC CAG CAA CAG CAA CA-3', and primer 2: 5'- biotin-GGC CCA GAG TTT CCG TGA TG-3') designed based on the flanking the CAG repeat of the DRPLA gene.
4. N, N, N-trimethylglycine
5. 10X reaction buffer: 100 mM Tris-HCl, pH 8.3, 500 mM KCl, 15 mM MgCl$_2$.
6. Taq DNA polymerase (5.0 U/mL).
7. Dynabeads M-280 Streptavidin (Dynal A. S, Oslo, Norway).
8. Dynal Magnetic Particle Concentrator (MPC).
9. PBS, pH 7.4, containing 0.1% BSA (bovine serum albumin).
10. 2X concentrated binding and washing buffer (B&W buffer): 10 mM Tris-HCl, pH 7.5, 1 mM EDTA, 2.0 M NaCl.
11. 1X B&W buffer.
12. 0.1 M NaOH.

2.3. Prehybridization

1. 2.75X SSPE (1X SSPE = 150 mM NaCl, 10 mM NaH$_2$PO$_4$, 1 mM EDTA).
2. Formamide.
3. Denhardt's solution.
4. Sheared salmon sperm DNA.
5. Heat-sealable hybridization bag.

2.4. Hybridization

1. 2.75X SSPE (1X SSPE = 150 mM NaCl, 10 mM NaH$_2$PO$_4$, 1 mM EDTA).
2. Formamide.

3. Denhardt's solution.
4. Sheared salmon sperm DNA.

2.5. Washes

1. 1X SSC: 150 mM NaCl, 15 mM sodium citrate.
2. 10% sodium dodecyl sulfate (SDS).

2.6. Exposure to Films

1. Kodak Bio Max MS films.
2. MS intensifying screen (Kodak).

2.7. Cloning of the Genomic DNA Segment Containing Pathologically Expanded CAG Repeats

1. Two-well comb (Bio-Rad 170-4345).
2. Dialysis bag (Spectra/Por 6 Moleculaporus Dialysis Membrane, MWCO: 15000, Spectrum Houston, TX, USA).
3. 1X TAE buffer.
4. Centricon-100 (Amicon, Inc., Beverly, MA, USA).
5. λZAPII vector (Stratagene).

3. Methods

3.1. Preparation of Genomic Blots

1. Digest high-molecular-weight genomic DNA (10–15 µg) with various restriction enzymes (100 U) (*see* **Note 3**) in 200 µL of appropriate buffers. Aliquots of the digest should be run through agarose gels to make sure that the digestion is complete, as well as to estimate the concentration of genomic DNA in each solution.
2. Ethanol precipitate the digested genomic DNA and redissolve in 30 µL of TE. Load the solution onto agarose gels and run through 0.8% agarose gels overnight. For better resolution, electrophoresis at 2 V/cm gel length is recommended. It is important that each lane contains equal amounts of genomic DNA for the comparison of signal intensities between lanes.
3. Transfer to a nitrocellulose membrane by a standard procedure of Southern blotting.
4. Bake the nitrocellulose membrane at 80°C in a vacuum over for 2 h.

3.2. Preparation of (CAG)$_{55}$ Probe with a High Specific Radioactivity (Fig. 2)

The genomic segment that contained 55 CAG repeats from a DRPLA patient was first cloned into a plasmid vector. Using primers (primer 1: 5'-CAC CAC CAG CAA CAG CAA CA-3', and primer 2: 5'- biotin-GGC CCA GAG TTT CCG TGA TG-3') based on the sequences flanking the CAG repeat of the DRPLA gene, polymerase chain reaction (PCR) was performed in the presence of [α-^{32}P]dATP (*see* **Fig. 2**). The method allowed incorporation of 59 molecules of [α-^{32}P] dATP per molecule of the DNA probe, exclusively into the sense strand.

1. Vacuum dry 12.5 µL of 9.25 MBq of [α-^{32}P] dATP (222 TBq/mmol) in a 0.5-mL Eppendorf tube.

DIRECT Technologies

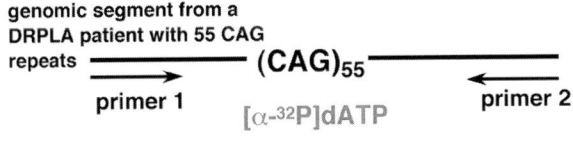

Fig. 2. Preparation of a $(CAG)_{55}$ probe with a high specific radioactivity. The $(CAG)_{55}$ probe was generated by internal labeling with $[\alpha-^{32}P]$ dATP.

2. Add the following "Solution for PCR" to the tube containing the dried $[\alpha-^{32}P]$ dATP, and then add Taq DNA polymerase.

   Solution for PCR
Template (p-2093 [0.73 ng/mL])	0.4 μL
Primer 1 (20 μM)	0.4 μL
Primer 2 (20 μM)	0.4 μL
dCTP (4 mM)	0.4 μL
dGTP (4 mM)	0.4 μL
TTP (4 mM)	0.4 μL
4.0 M N, N, N-trimethylglycine	8.0 μL
10X reaction buffer	1.6 μL
Distilled water	3.6 μL
Taq DNA polymerase (5.0 U/mL)	0.4 μL

3. PCR Cycle
   After an initial 2-min denaturation at 94°C, PCR was performed for 30 cycles with each cycle consisting of denaturation at 94°C for 1 min, annealing at 54°C for 1 min, and extension at 72°C for 3 min, followed by a final extension at 72°C for 10 min.

4. Strand Separation Using Magnetic Beads
   a. Resuspend the Dynabeads M-280 Streptavidin by gently shaking the vial to obtain a homogeneous suspension.
   b. Add 100 μL of Dynabeads M-280 Streptavidin to a 1.5-mL Eppendorf tube and place the tube in the Dynal MPC for at least 30 s.
   c. Remove the supernatant by aspiration with a pipet while the tube remains in the Dynal MPC.
   d. Take the tube from the Dynal MPC. Gently resuspend the Dynabeads M-280 Streptavidin in 100 μL of PBS, pH 7.4, containing 0.1% bovine serum albumin (BSA) along the internal surface of the tube.
   e. Place the tube in the Dynal MPC for at least 30 s and repeat **step 4**.
   f. Remove the tube from the Dynal MPC and add 100 μL of PBS, pH 7.4, containing 0.1% BSA.
   g. Add 20 μL of the washed Dynabeads M-280 Streptavidin into a new 1.5-mL Eppendorf tube and place the tube in the Dynal MPC. Remove the supernatant with a pipet while keeping the tube in the Dynal MPC.

h. Resuspend the Dynabeads in 20 µL of 1X binding and washing buffer (B&W buffer) and mix gently. Remove the supernatant with a pipette while keeping the tube in the Dynal MPC. Resuspend the Dynabeads in 40 µL of 2X B&W buffer.
i. Add 40 µL of the prewashed beads to 16 µL of solution containing the PCR products.
j. Incubate the PCR products with the Dynabeads for 15 min keeping the beads suspended by gentle rotation of the tube. Place the tube containing the immobilized product in a Dynal MPC and remove the supernatant with a pipet.
k. Wash the Dynabeads with 40 µL of 1X B&W buffer.
l. Place the tube in a Dynal MPC and remove the supernatant. Resuspend the Dynabeads in 16 µL of a freshly prepared 0.1 M NaOH solution.
m. Incubate the solution for 10 min at room temperature.
n. Using the Dynal MPC, collect the Dynabeads on the side of the tube and transfer the supernatant solution containing the labeled CAG strand to a new tube (tube A).
o. Wash the Dynabeads with 34 µL of 0.1 M NaOH.
p. Take the supernatant solution and transfer to the tube A to combine the solution containing the $(CAG)_{55}$ probe.

3.3. Prehybridization

Prehybridization solution contains 2.75X SSPE, 50% formamide, 5X Denhardt's solution, and 100 ng/mL sheared salmon sperm DNA. Incubate the nitrocellulose membranes in an appropriate amount of prehybridization solution in a heat-sealable hybridization bag for 2 h at 62°C.

3.4. Hybridization (see Note 4)

Hybridization solution contains 2.75X SSPE, 50% formamide, 5X Denhardt's solution, 100 ng/mL sheared salmon sperm DNA, and $(CAG)_{55}$ probe (6×10^6 cpm/mL). The $(CAG)_{55}$ probe should be prepared freshly for each experiment.

Replace the prehybridization solution with the hybridization solution containing the $(CAG)_{55}$ probe and incubate for 18 h at 62°C with gentle shaking.

3.5. Washes

Wash the membranes twice in 1X SSC containing 0.5% SDS with gentle shaking for 1 h each at room temperature, and finally in 1X SSC; 0.5% SDS for 30 min at 65°C.

3.6. Exposure to Films

The filters are autoradiographed to Kodak Bio Max MS films for 16 h at –70°C using an MS intensifying screen. An exposure time of 16 h is generally sufficient to obtain strong signals. Note that the stringency of the hybridization conditions (the temperature is particularly important) should be monitored using a membrane containing cloned DNAs (50 pg) carrying various lengths of CAG repeats (9, 23, 43, and 51 repeat units) in each experiment. In general, the cutoff points can be adjusted by changing the temperature for hybridization.

Strong signals should be obtained for the 50 pg of cloned DNA with 2hr exposure to Kodak Bio Max MS films. If the signals are weak under these conditions, it is difficult to obtain good signals for genomic DNAs.

3.7. Cloning of the Genomic DNA Segment Containing Pathologically Expanded CAG Repeats (Fig. 3)

Since it was previously reported that the CAG repeats in plasmid vectors can be unstable during propagation through *Escherichia coli*, *(11)* we selected a λ phage vector with consideration that the CAG repeats might be more stable in λ phage vectors than in plasmid vectors.

1. Digest 270 µg of genomic DNA from an affected individual in a 3600 µL volume with 1800 U of an appropriate enzyme.
2. Ethanol precipitate and resuspend the DNA in 360 µL of TE.
3. Prepare an agarose gel (the porosity of agarose gel varies with the size of target DNA) using a two-well comb and apply 180 µL into each well.
4. Electrophoresis should be performed at approx 2 V/cm gel length for a minimum of 10 h. The optimum time required for good resolution should be determined for each experiment.
5. Determine the expected position for the genomic fragment of interest based on the mobility of size markers, and then cut out the gel segments. (Make an effort to avoid damage to the DNA by minimizing the duration of UV exposure of the gel.)
6. Place gel pieces in a dialysis bag and add 20 mL of 1X TAE buffer.
7. Electroelute the DNA fragments at 40 V for 2 h.
8. Reverse the polarity and electrophorese for an additional minute.
9. Carefully take the solution out of the bag and transfer to a 50-mL conical tube. Wash the bag with an additional 10 mL of TE and combine with solution in the 50-mL conical tube.
10. Spin the tube to remove gel pieces. Transfer the supernatant to a new tube.
11. Concentrate the solution to approx 1 mL using a Centricon-100 concentrator.
12. Extract with an equal volume of phenol-chloroform solution.
13. Extract with an equal volume of chloroform.
14. Add 2 µg of λZAPII vector arm digested with an appropriate enzyme and treat with alkaline phosphatase.
15. Ligation and in vitro packaging are performed by standard procedures.
16. Screen the phage genomic library by plaque hybridization using the $(CAG)_{55}$ probe under hybridization conditions identical to those used for the genomic Southern blot hybridization.

4. Notes

1. Principle of direct identification of repeat expansion and cloning technique (DIRECT). As shown in **Fig. 1**, if we use a DNA probe containing a long trinucleotide repeat, $(CAG)_{55}$ for example, such a probe would hybridize to an expanded CAG repeat (>35 repeats) but not to a CAG repeat containing region with normal repeat numbers (<35 repeats) under stringent hybridization conditions. The Tm (melting temperature) of the hybrids between the probe and the genomic DNA segment would be varied depending on the number of bases engaged in the hybrid formation. Once a genomic fragment containing an expanded trinucleotide repeat is detected using Southern blotting hybridization analysis employing this probe under stringent hybridization conditions, the genomic segment containing the expanded trinucleotide repeat can then easily be cloned from the agarose gel using standard techniques (**Fig. 3**). In developing DIRECT, therefore, the following two factors were crucial. First, the hybridization conditions should be sufficiently stringent such that only pathologically expanded CAG repeats are detected. Second, the probe should be sufficiently sensitive to allow the detection of a single-copy genomic segment containing pathologically expanded CAG repeats by genomic Southern blotting hybridization analysis.

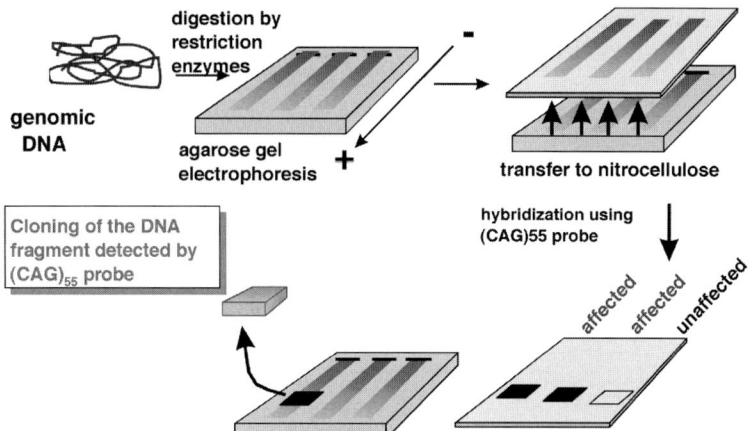

Fig. 3. A schematic illustration of the direct identification of repeat expansion and cloning technique (DIRECT). Genomic DNAs are digested with appropriate restriction enzymes, subjected to agarose gel electrophoresis, and blotted to a nitrocellulose membrane. The membrane is hybridized to the $(CAG)_{55}$ probe with a high specific radioactivity. Under a stringent hybridization condition, only largely expanded CAG repeats are detected. Such genomic segments are easily cloned using a phage vector.

2. Choice of Pedigree. DIRECT allows the detection of pathologically expanded trinucleotide repeats in the genomic DNA of affected individuals. Therefore, prior knowledge about the chromosomal localization of the disease genes is not required. In other words, DIRECT can be applied to small pedigrees where linkage analysis is not feasible. In principle, expanded trinucleotide repeats can be identified even in one affected individual. However, DIRECT can detect multiple bands even in unaffected individuals. This means that even normal individuals may contain CAG repeats larger than 35 repeats. In fact, we have recently identified such a CAG repeat ranging from 10–90 repeats in normal individuals *(12)*. These bands detected in normal individuals are occasionally polymorphic. Based on our experience, we recommend the analysis of multiple affected and unaffected members to confirm that the band detected by DIRECT exhibits perfect cosegregation with the disease.
3. Choice of Restriction Enzymes. There are no *a prori* suggestions for the choice of restriction enzymes. For cloning purposes, it is desirable that the size of the restriction fragments containing the expanded trinucleotide repeat does not exceed several Kb. It is also desired that the restriction fragments have cohesive ends, which can easily be cloned into cloning vectors. Therefore, we strongly recommend trying as many restriction enzymes as possible to find the enzymes which generate restriction fragments of appropriate sizes. Restriction enzymes which have hexametric recognition sites are good candidates since these enzymes produce only a few bands that are detected by $(CAG)_{55}$ probe in normal individuals.
4. Quality Control of Hybridization. To apply DIRECT successfully, quality control of hybridization is also crucial. Particularly, the specific radioactivity must be high enough to detect a single copy genomic DNA fragment, and the "cutoff" point should be strictly controlled by adjusting the hybridization conditions. For this purpose, we recommend a quality-control hybridization using a membrane containing plasmid DNA (50 pg) carrying various lengths of CAG repeats (9, 23, 43, and 51 repeats). (We used a cloned 1.3 Kb androgen receptor gene fragment containing the CAG repeat). If the specific radioactivity of the probe is suffi-

ciently high, strong signals should be obtained with 2 h exposure to Kodak Bio Max MS films. For adjusting the "cutoff" points, the hybridization temperature should be adjusted so that strong signals are obtained for 43 and 51 repeats but not for 9 or 23 repeats. Alternatively the concentration of the ionic strength (the concentration of SSPE) can be adjusted.

References

1. David, G., Abbas, N., Stevanin, G., Durr, A., Yvert, G., Cancel, G., et al. (1997) Cloning of the SCA7 gene reveals a highly unstable CAG repeat expantion. *Nature Genet.* **17,** 65–70.
2. Mandel, J. L. (1997) Breaking the rule of three. *Nature* **386,** 767–769.
3. Paulson, H. L. and Fischbeck, K. H. (1996) Trinucleotide repeats in neurogenic disorders. *Annu. Rev. Neurosci.* **19,** 79–107.
4. Tsuji, S. (1996) Unstable expansion of triplet repeats as a new disease mechanism for neurodgenerative diseases. *Jpn. J. Hum. Genet.* **41,** 279–290.
5. Koob, M. D., Moseley, M. L., Schut, L. J., Benzow, K. A., Bird, T. D., Day, J. W., and Ranum, L. P. W. (1999) An untranslated CTG expansion causes a nove form of spinocerebellar ataxia (SCA8). *Nature Genet.* **21,** 379–384.
6. Holmes, S. E., O' Hearn, E. E., McInnis, M. G., Gorelick-Feldman, D. A., Kleiderlein, J. J., Callahan, C., et al. (1999) Expansion of a novel CAG trinucleotide repeat in the 5' region of *PPP2R2B* is associated with SCA12. *Nature Genet.* **23,** 391–392.
7. Nakamura, K., Jeong, S-Y, Uchihara, T., Anno, M., Nagashima, K., Ikeda, S., Tsuji, S., and Kanazawa, I. (2001) SCA17, a novel autosomal dominant cerebellar ataxia caused by an expanded polyglutamine in TATA-binding protein. *Hum. Mol. Genet.* **14,** 1441–1448.
8. Zhuchenko, O., Bailey, J., Bonnen, P., Ashizawa, T., Stockton, D., Amos, W. et al. (1997) Autosomal dominant cerebellar ataxia (SCA6) associated with small polyglutamine expansions in the alpha 1a-voltage-dependent calcium channel. *Nature Genet.* **15,** 62–69.
9. Gispert, S., Twells, R., Oronzco, G., Brice, A., Weber, J., Heredero, L., et al. (1993) Chromosomal assignment of the second locus for autosomal dominant cerebellar ataxia (SCA2) to chromosome 12q23-24.1. *Nature Genet.* **4,** 295–299.
10. Sanpei, K., Takano, H., Igarashi, S., Sato, T., Oyake, M., Sasaki, H., et al. (1996) Identification of the spinocerebellar ataxia type 2 gene using a direct identifiction of repeat expansion and cloning technique (DIRECT). *Nature Genet.* **14,** 277–284.
11. Kang, S., Jaworski, A., Ohshima, K., and Wells, R. D. (1995) Expansion and detection of CTG repeats from human disease gene are determined by the direction of replication in *E. coli. Nature Genet.* **10,** 213–218.
12. Ikeuchi, T., Sanpei, K., Takano, H., Sasaki, H., Tashiro, K., Cancel, G., et al. (1998) A novel long unstable CAG/CTG trinucleotide repeat on chromosome 17q. *Genomics* **49,** 321–326.
13. Schalling, M. (1993) Direct detection of novel expanded trinucleotide repeats in the human genome. *Nature Genet.* **4,** 135–139.
14. Koob, M. D., Moseley, M. L., Schut, L. J., Benzow, K. A., Bird, T. D., Day, J. W, and Ranum L. P. (1999) An untranslated CTG expansion causes a novel form of spinocerebellar ataxia (SCA8) *Nature Genet.* **21,** 379–384.

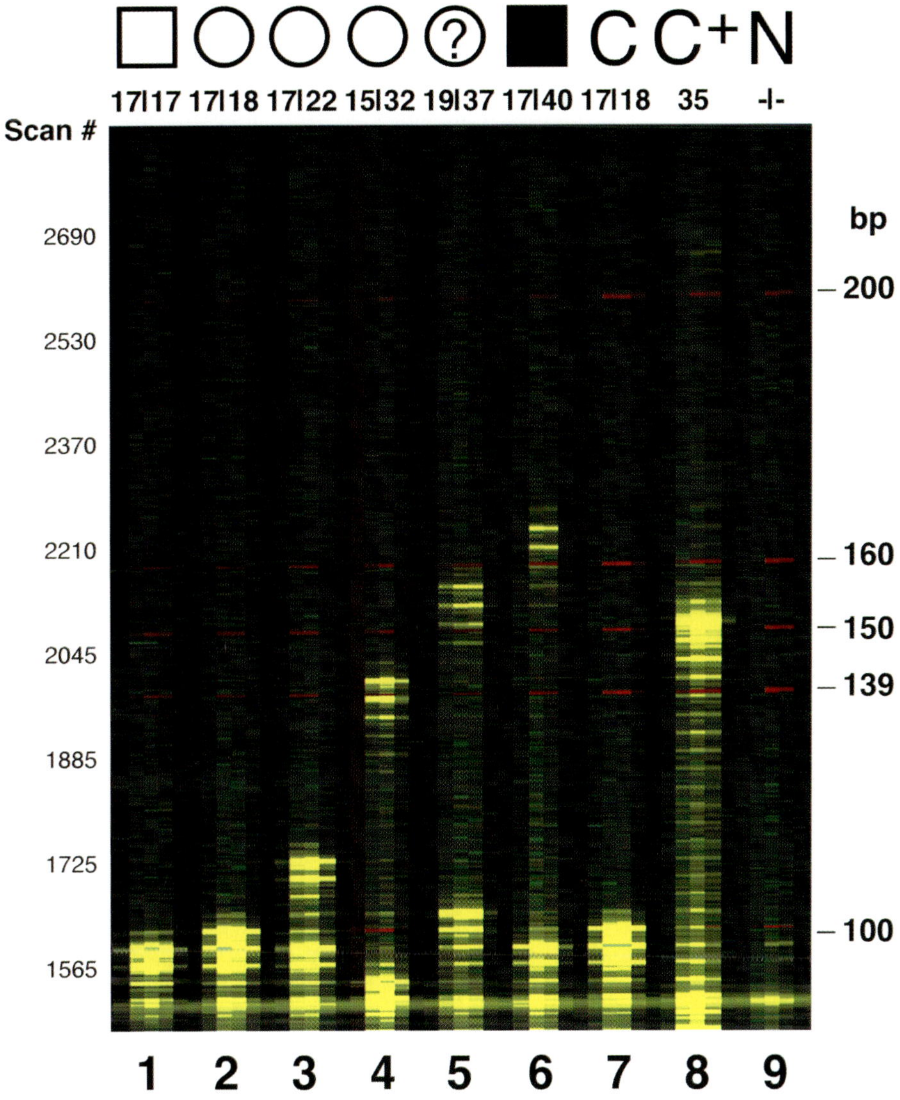

Color Plate 1, Fig. 1. (*see* discussion in Chapter 10, p 106.) Representative gel image for HD analysis using fluorescence PCR. Scan numbers are listed on the left. TAMRA size standard is listed as base pairs (bp) on the right. Phenotype and genotype of patients studied are listed at the top. Lanes 1–3, CAG repeats characteristic of normal alleles (17/17; 17/18; 17/22), Lane 4, a normal (15) and a mutable (32) allele. Lane 5, a normal (19) and a HD (37) allele with reduced penetrance. Lane 6: a normal (17) and a HD (40) allele; Lane 7, a normal control (C) with CAG repeats characteristic of normal (17/18) alleles. Lane 8, Cloned standard control (C+) with 35 CAG repeats (from M. R. Hayden, University of British Columbia). Lane 9, Negative control (N) reaction tube with all reagents except DNA.

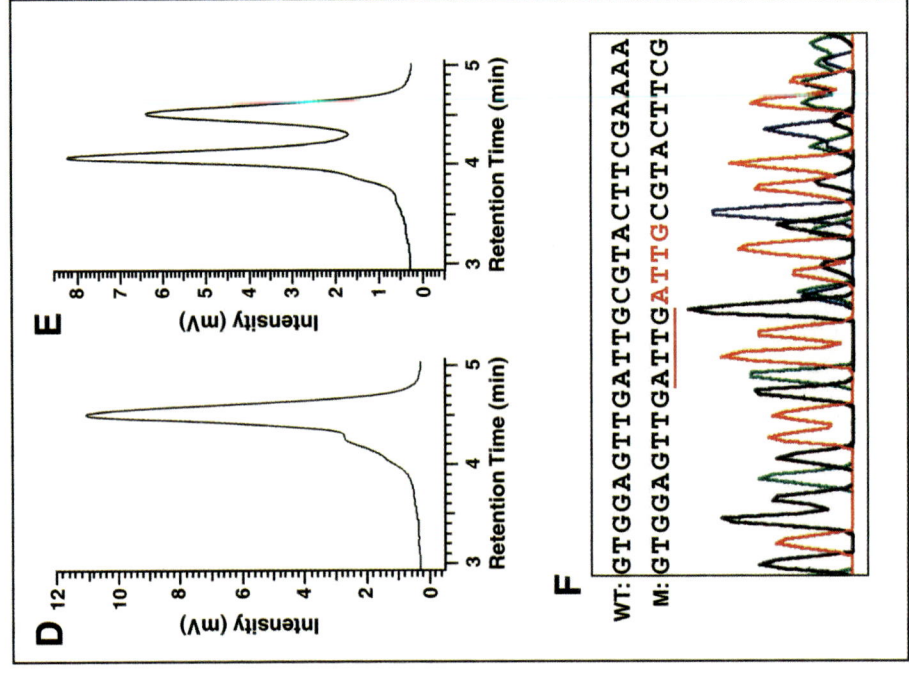
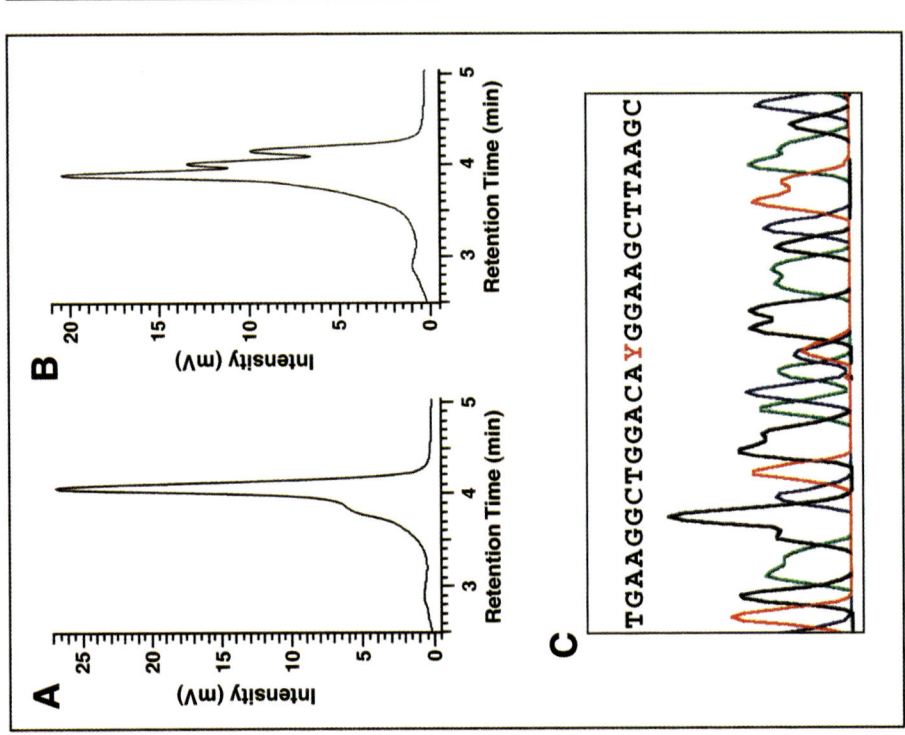

Color Plate 2, Fig. 2. (*see* full caption and discussion in Chapter 12, p. 127.) Illustration of representative DHPLC and direct sequencing data for an *MECP2* missense and insertion mutation.

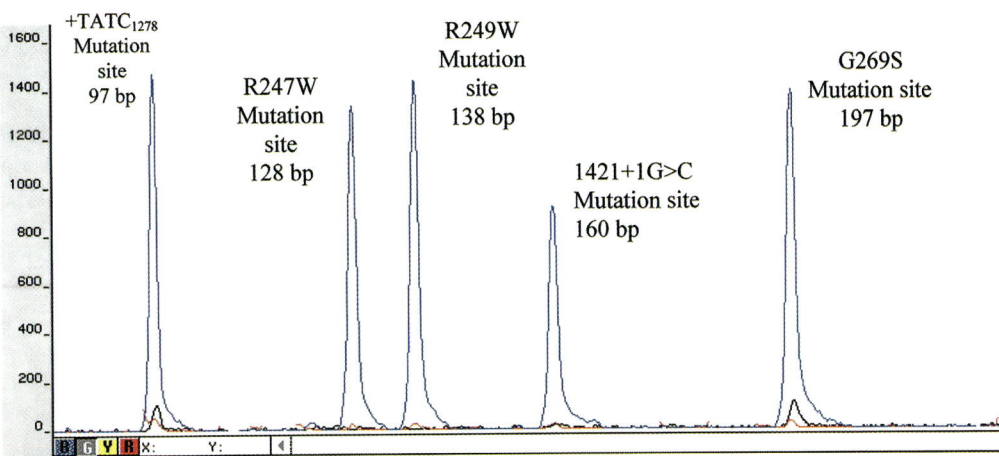

Color Plate 3, Fig. 2. (*see* discusson in Chapter 13, p 138). Results of the Tay-Sachs ASA assay for a normal sample (no mutations). Note that there are 5 blue peaks seen, which correspond to the normal allele at each of the mutation sites in the *HEXA* gene. The sizes of the peaks are as given, and are also listed in **Table 3**.

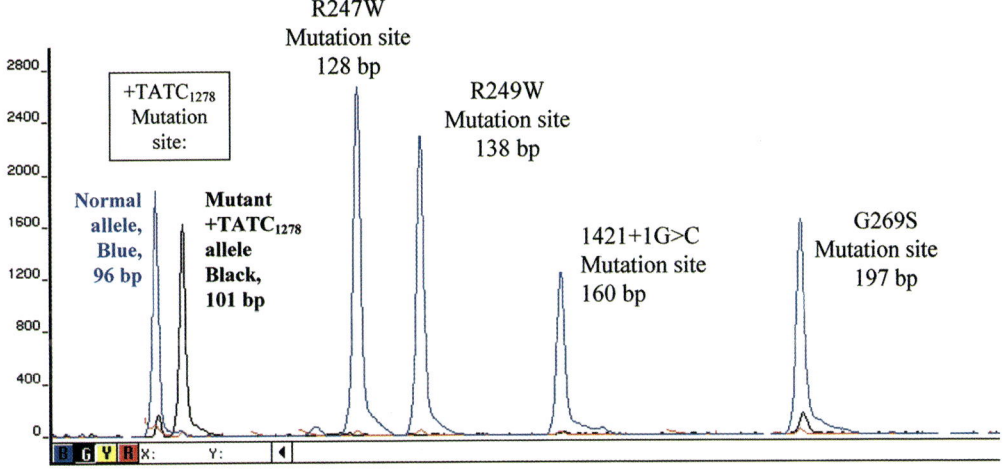

Color Plate 4, Fig. 3. (*see* discussion in Chapter 13, p 139.) Example of Tay-Sachs ASA assay results for a carrier of the +TATC$_{1278}$ mutation. The normal allele at the +TATC1278 site is indicated by a blue peak of 96 bp, while the mutant +TATC$_{1278}$ allele is indicated by the black peak of 101 bp. Note that the size difference between normal and mutant +TATC$_{1278}$ alleles is due the presence of the 4 bp insertion mutation.

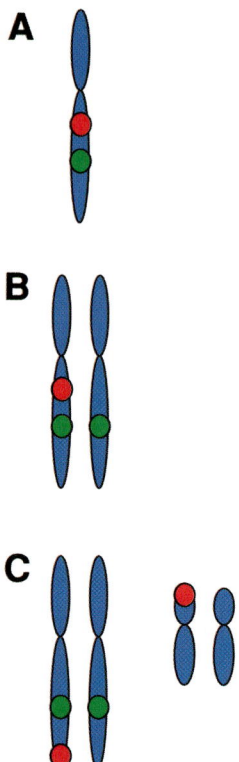

Color Plate 5, Fig. 1. (*see* full caption and discussion in Chapter 21, p 220.) Schematic representations to depict the typical hybridization patterns of test and control probes in the detection of microdeletions and cryptic translocations.

● Control fibroblasts
● EDMD fibroblasts

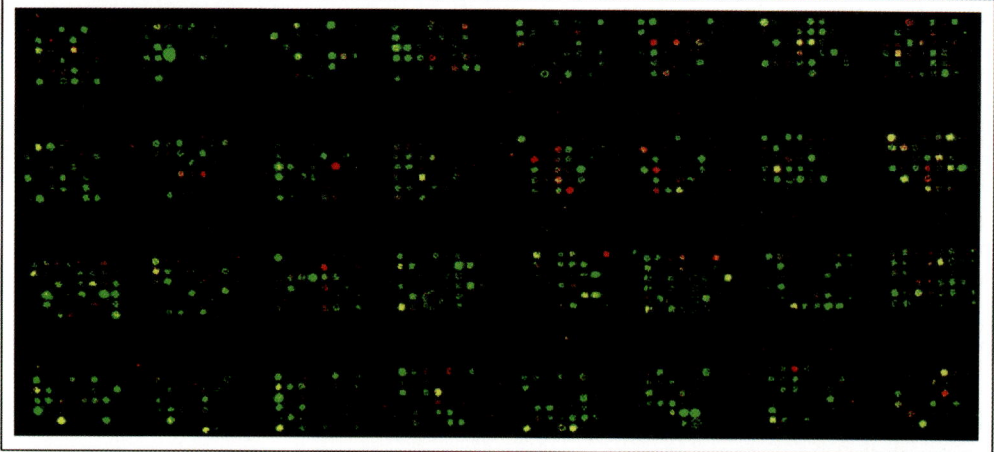

Color Plate 6, Fig. 2. (*see* discussion in Chapter 23, p 259.) Typical microarray image. One microgram of total RNAs from control fibroblasts and EDMD fibroblasts were labeled with biotin and FITC, respectively. Microarray hybridization and TSA detection was performed. The captured image was obtained by an overlay of both Cy3 and Cy5 images of the microarray. An in-house microarray on which 1536 different probes were spotted was used in this study. Our microarray showed low background and good resolution.

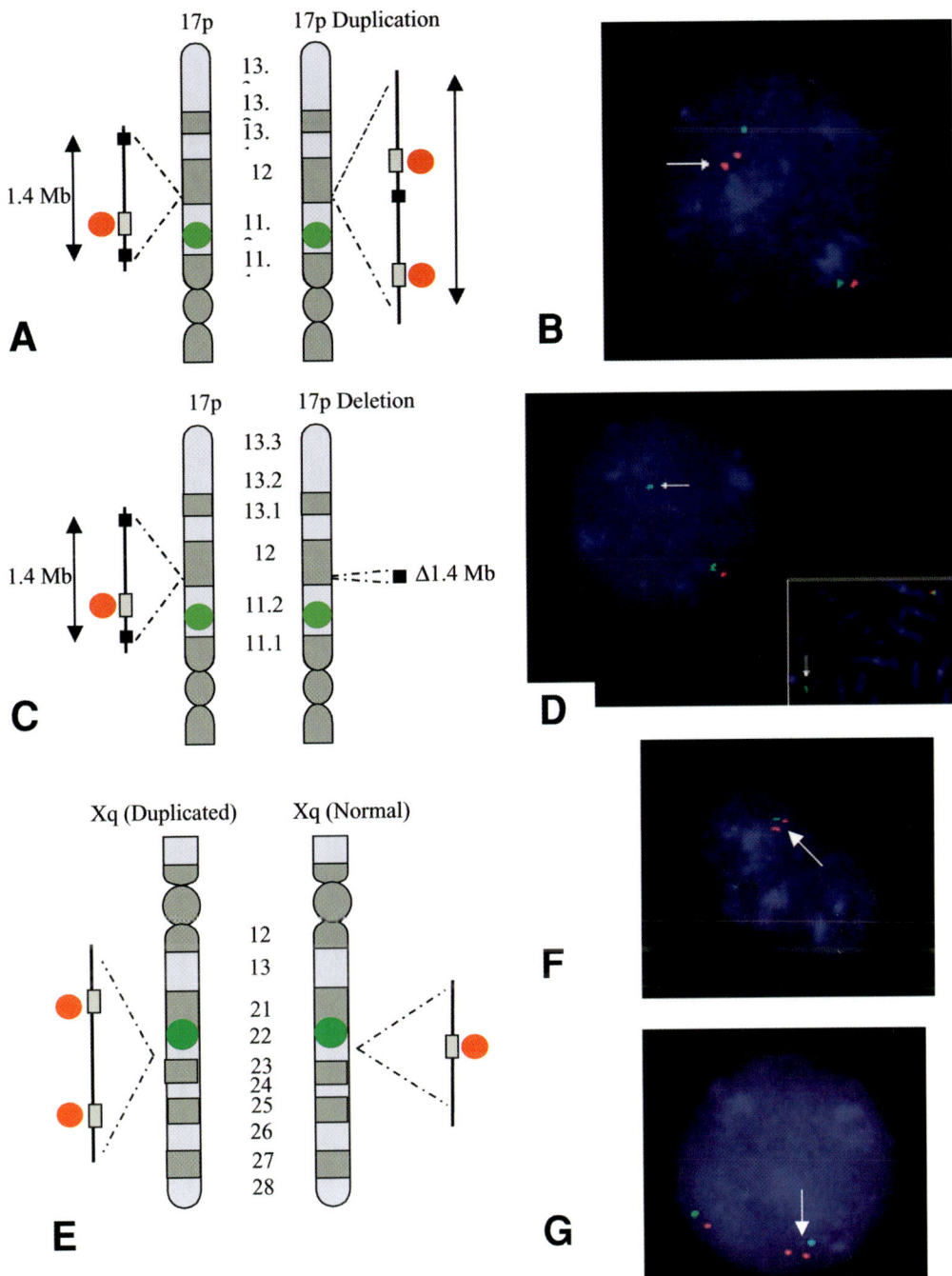

Color Plate 7, Fig. 2. (*see* full caption and discussion in Chapter 21, p. 224–225.) Schematic representation of CMT1A duplication, HNPP deletion; and PMD duplication with accompanying FISH images from representative patient test samples.

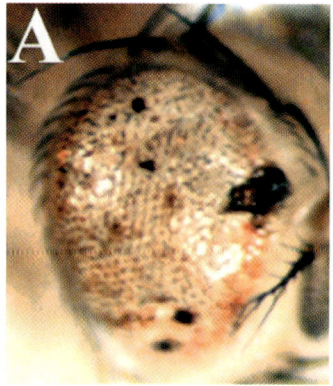 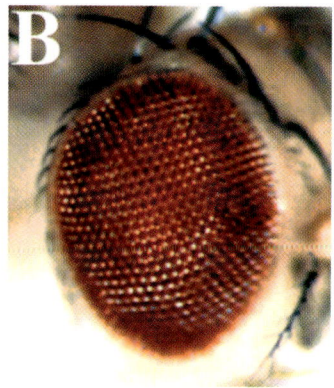

Color Plate 8, Fig. 1. (*see* discussion in Chapter 22, p 245.) Neurodegeneration induced by expression of expanded Machado-Joseph Disease (MJD) protein in the *Drosophila* retina, and its suppression by the molecular chaperone Hsp70. **(A)** Fly expressing the mutant disease form of the MJD protein with an expanded polyglutamine repeat (MJDtr-Q78), results in severe degeneration of the eye as indicated by the absence of red pigment in the eye. **(B)** Co-expression of the molecular chaperone Hsp70 with MJDtr-Q78 leads to dramatic suppression of degeneration, restoring eye structure back toward normal (*see* also **ref. 33**).

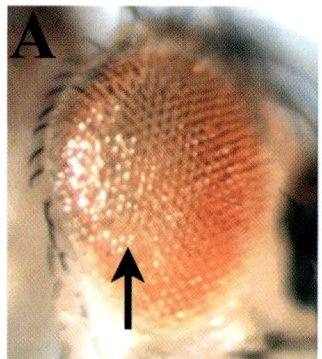 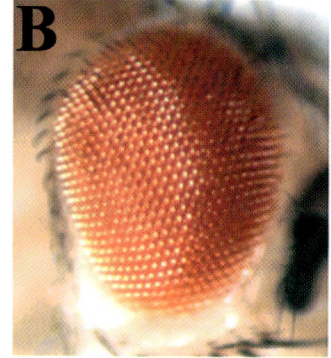 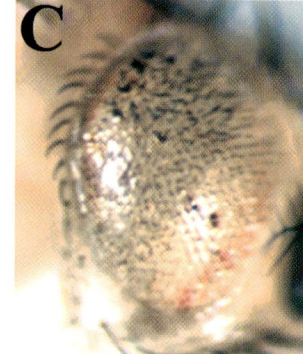

Color Plate 9, Fig. 2. (*see* discussion in Chapter 22, p 245.) Neurodegenerative phenotype caused by expression of pathogenic Machado-Joseph Disease (MJD) protein in the *Drosophila* retina and the modulatory effects of the molecular chaperone Hsp70. **(A)** Fly expressing the mutant disease form of the MJD protein with a moderately expanded polyglutamine repeat (MJDtr-Q61), exhibiting degeneration of the eye. Loss of pigmentation is noted by white patch in the eye (arrow). **(B)** Co-expression of the molecular chaperone Hsp70 with MJDtr-Q61 results in an eye that is phenotypically normal with the pigmentation restored. **(C)** Co-expression of a dominant negative mutant of the molecular chaperone Hsc70 with MJDtr-Q61 results in enhanced degeneration with severe loss of eye pigmentation. (*see* also **refs. 33** and **34**).

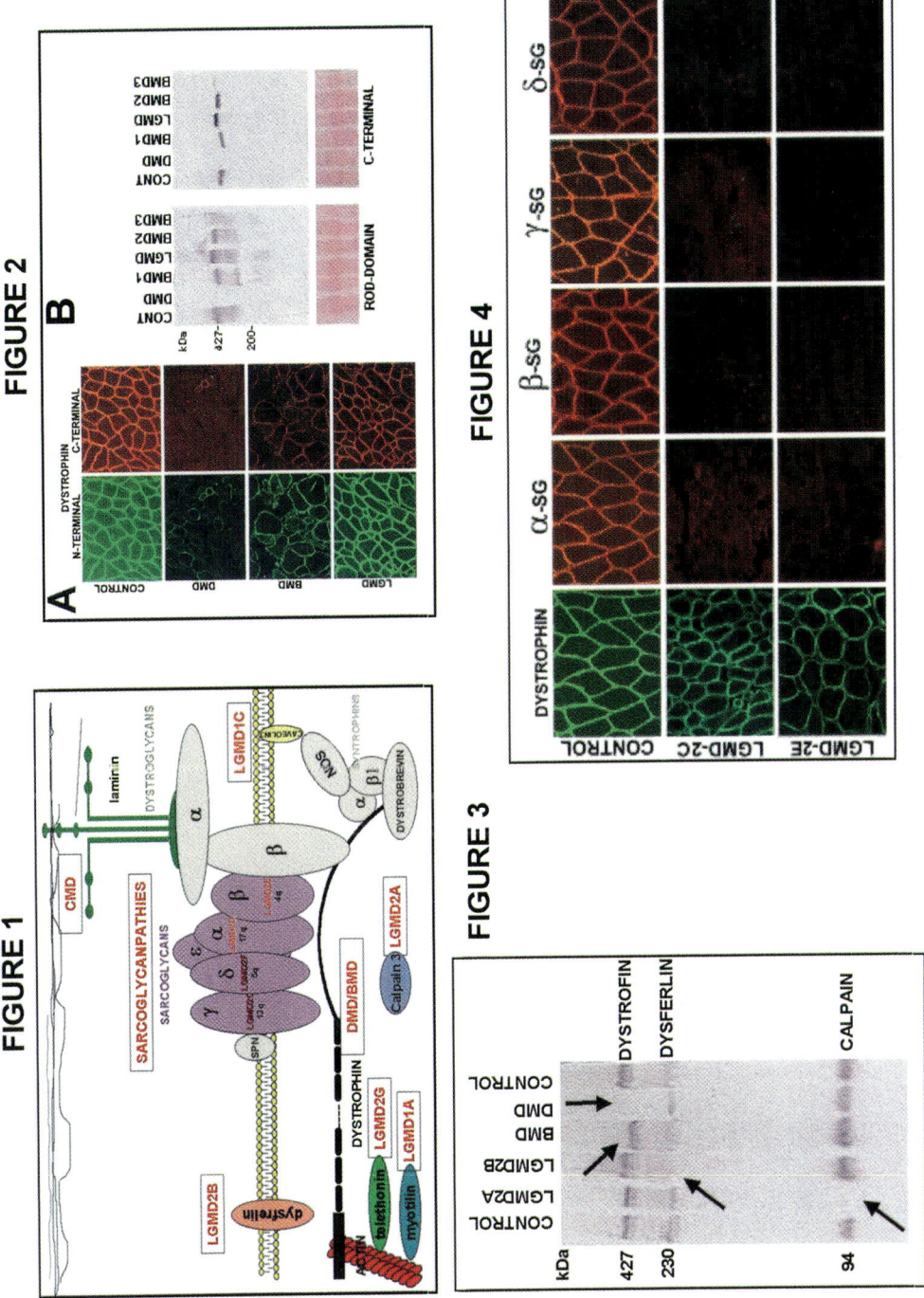

Color Plate 10, Figs. 1–4. (*see* discussion and captions in Chapter 32, pp 356–366.)

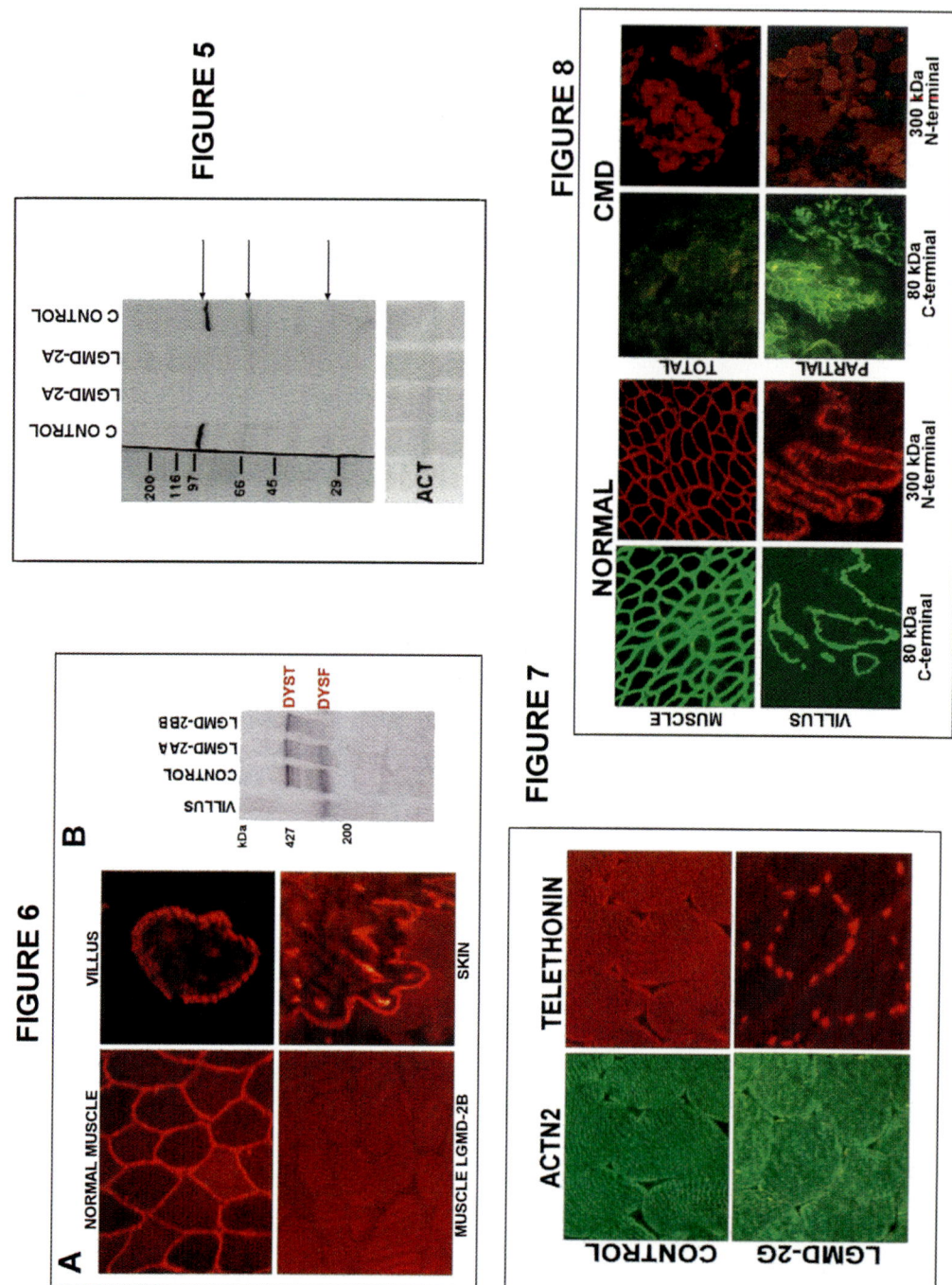

Color Plate 11, Figs. 5–8. (*see* discussion and captions in Chapter 32, pp 367–370.)

8

Antibody-Based Detection of CAG Repeat Expansion Containing Genes

Yvon Trottier

1. Introduction

Polyglutamine (polygln) expansion in specific proteins is one of the most intriguing pathogenic mechanisms causing adult-onset neurodegenerative disorders. In all the cases studied so far, the normal gene products tolerate a rather wide variation in size of a polygln tract (ranging typically between 10–35 glns) without any detectable adverse effect. However, beyond the threshold of about 35–42 glns those proteins acquire toxic properties, which correlate with a conformational change—revealed by the 1C2 antibody interaction—and a tendency of mutant proteins to aggregate in vitro and in vivo. To date nine polygln expansion disorders have been characterized, the list of which includes Huntington's disease (HD) the most frequent of them, six spinocerebellar ataxias (SCA1-3, SCA6, SCA7, and SCA17), spino-bulbar muscular atrophy (SBMA) and dentatorubral-pallidoluysian atrophy (DRPLA) (reviewed in **refs. *1*,*2***). Most of theses diseases present a strong inverse correlation between the length of the polygln tract and the age of onset of clinical symptoms. At the DNA level, the expanded CAG repeats, which code for the polygln stretch, are found to be unstable upon transmission from one generation to the other, with a clear tendency to expansion. These two features account for the anticipation phenomenon (increased severity and earlier onset of the disease in successive generations) that is observed at various degrees in polygln expansion diseases. Because anticipation is observed in many other neurological disorders (and suggested in some non-neurological diseases), for which the causative gene has not yet been cloned, it is suggested that polygln expansion could be involved in the disease process *(3,4)*.

We reported in 1995 that an anti-polygln monoclonal antibody (Mab), 1C2, selectively recognizes polygln expansion *(5)*. On Western blot (WB) analysis, the antibody was able to detect mutant proteins involved in HD, SCA1, and SCA3, but not the corresponding normal forms. Moreover we showed that the ability of 1C2 to detect the mutant HD protein, huntingtin, clearly increases with the polygln size: signal intensity increases by about 20–40 fold from mutant proteins with small pathological expansion (39–40 glns) to those with long expansion (60–85 glns) associated with juvenile onset *(5)*. Thus, the efficiency of detection of pathological alleles appears to parallel the severity of the disease.

Given its properties, we used the 1C2 antibody as a tool to identify new polygln expansion proteins in lymphoblastoid cell lines (LCL) derived from SCA2 and SCA7 patients *(5,6)*. Very recently, Nakamura et al. *(2)* have used the 1C2 antibody to show that polygln expansion in the TATA-binding protein (TBP) is responsible for SCA17. Other groups showed that 1C2 detects an unknown protein with a long polygln stretch in some patients with childhood onset schizophrenia *(7,8)*. In this report, we summarize the binding properties of 1C2 antibody, and outline a strategy using this antibody as a probe to detect new polygln expansion in neurodegenerative disorders, with proven or suggested anticipation.

1.1. Interaction Properties Between 1C2 and Polyglns

To analyze further how 1C2 selectively recognizes the expanded polygln stretches, a kinetic analysis of the interaction between the F(ab) fragment of 1C2 antibody and polygln tracts was performed in native conditions using a surface plasmon resonance biosensor (BIAcore™, Pharmacia) *(9)*. The F(ab) fragment showed a much stronger affinity for an expansion of 73 glns than for a normal polygln stretch (15 glns). Notably, the complex formed between the F(ab), having a single antigen binding site, and the polygln expansion is remarkably stable and dissociates 100 times more slowly than the one formed with a normal polygln stretch. These data indicate that 1C2 antibody recognizes a "new" conformation formed by the expanded polyglutamine.

1.2. Search for New Polygln Disorders Using 1C2 Antibody: Important Considerations

Since pathological tissues from patients are not always available, LCLs represent an unlimited source for proteins extract preparation and are very useful for initial screening. However, false-negative results may be obtained if the pathogenic protein is weakly or not expressed in LCL. Even though these cells are a convenient source of material, one should note, however, that the polyclonality of the LCL may generate some variability in the detection *(6–8)*, and that Epstein-Barr Virus (EBV) transformation of lymphocytes has been reported to be associated with loss of DNA methylation *(10)*, which in turn may alter gene expression.

In the search of a putative polygln expansion protein, it is strongly recommended to analyze several LCL samples from patients affected by the same pathology, and to study early-onset cases given the strong polygln length-dependent properties of the antibody. Furthermore, analysis of patients and normal relatives from the same family may help to confirm the specificity of polygln expansion detection as the signal intensity and protein size is expected to increase in the successive generation *(5,6)*.

When performing a WB analysis of LCL protein extracts, 1C2 typically reveals the general transcription factor TBP at about 49 kDa as well as an unknown ~230 kDa nuclear protein, yet with a much lower signal intensity than proteins with polygln expansion *(5)*. Therefore, it is helpful to have a positive control (e.g., protein extract from SCA3 LCL) in order to compare the signal intensity of any protein detected by 1C2. Be aware that TBP has also a polymorphic polygln stretch with the most common allelic form of 38 glns, and rare variants having 42 glns and this polymorphism could lead to variability in the signal of detection and the migration of the protein *(5)*.

Under some conditions of immunodetection, for instance, high concentration of antibody, very long exposure of the blot or low stringent washes, several additional proteins of different sizes are faintly detected in normal controls, and some of them even show a variability of detection, that could perhaps be explained by polymorphic polygln *(9)*. On WB, these "background" proteins were more commonly observed in the nuclear extract preparation. Indeed, polygln stretches are present in many eukaryotic proteins, particularly in several transcription factors *(11)*. If these proteins mask a specific signal or cause problems of interpretation, we recommend to perform a cellular fractionation and to analyze the cytoplasmic and nuclear proteins separately. Alternatively, a two dimension gel may allow a better evaluation of the signal specificity *(7)*.

2. Materials

2.1. Whole Protein Extract from Cultured Cells

1. $TGEK_{50}$ buffer: 50 mM Tris-HCl, pH 7,8, 10% glycerol, 1 mM EDTA, 50 mM KCl.
2. The serine protease inhibitor phenyl methyl sulfonyl fluoride (PMSF) 100X stock solution is prepared at 100 mM in isopropanol. PMSF is stored at –20°C and heated at 37°C before use.
3. The Protease Inhibitors Cocktail (PIC) 1000X consists of leupeptin 2.5 mg/mL, aprotinin 2.5 mg/mL, antipain 2.5 mg/mL, chymostatin 2.5 mg/mL, and pepstatin 2.5 mg/mL.
4. Before using the $TGEK_{50}$ buffer, add PMSF at 1/100 (v/v) and PIC at 1/1000 (v/v) and keep the buffer on ice.
5. Vibracell 72412 (Bioblock Scientific).
6. Bradford protein assay (BioRAD Laboratories, Cat. no. 500-0006).

2.2. Standard SDS-PAGE: Gel Casting

1. We recommend the use of commercial apparatus (such as the MINI-PROTEAN II from Bio-RAD Laboratories, Cat. no. 165-2940) to facilitate casting and handling of protein gels.
2. Sample loading buffer 4X: 8% (w/v) sodium dodecyl sulfate (SDS), 240 mM Tris-HCl, pH 6.8, 40% glycerol, 20% β-mercaptoethanol.
3. 30% acrylamide solution: 29 g of acrylamide (electrophoresis grade) and 0.8 g of N,N'-methylene-bisacrylamide are dissolved in 80 mL distilled water. Adjust the volume at 100 mL and store at 4°C.
4. Solution (5mL) for one 6 % acrylamide separating gel: mix 2.7 mL of H_2O, 1.0 mL of 30% acrylamide solution, 1.3 mL of 1.5 M Tris-HCl, pH 8.8, and 50 μL of SDS 10%.
5. Solution (5 mL) for one 12% acrylamide separating gel: mix 1.7 mL of H_2O, 2.0 mL of 30% acrylamide solution, 1.3 mL of 1.5 M Tris-HCl, pH 8.8, and 50 μL of SDS 10%.
6. For 5 mL of stacking solution, mix 3.4 mL of H_2O, 0.83 mL of 30% acrylamide solution, 0.63 mL of 1.0 M Tris-HCl, pH 6.8, and 50 μL of SDS 10%.
7. 10% (w/v) ammonium persulfate (APS).
8. TEMED (N,N,N',N'-tetrametyl-ethylenediamine) (Sigma).
9. Running buffer 10X: 60.6 g Tris base, 288 g glycine, 20 g SDS in 2 L H_2O.

2.3. Electrophoresis

1. To load the protein gels, use Hamilton syringe or 20-μL pipetteman fine cones.

2.4. Western Blotting

1. Nitrocellulose membrane (Schleicher & Schuell, Cat. no. 10 411180).
2. The transfer buffer consists of 1X running buffer, 15% methanol, and 0.1% SDS.

2.5. Immunodetection

1. Ponceau S red concentrate is available from Sigma (Cat. no. P7767).
2. 5% nonfat milk solution: 2.5 g nonfat dry milk in 50 mL PBS 1X.
3. 1C2 antibody is now available from Chemicon (Cat. no. MAB1574).
4. The secondary antibody is a horseradish peroxidase-conjugated affinity-purify goat anti-mouse IgG, Fc fragment specific (minimal cross-react with Hu, Sr) from Jackson-ImmunoResearch Laboratories (Cat. no. 111-035-046).
5. Substrate for peroxidase reaction is the SuperSignal West Pico luminol/enhancer solution from Pierce. (Cat. no. 34080ZZ).

3. Methods
3.1. Whole Protein Extract from Cultured Cells

1. For LCLs grown in suspension, collect the cells by centrifugation at 400g for 10 min at 4°C, resuspend the pellet in ice-cold PBS 1X, and centrifuge again. Remove the PBS and keep the pellet on ice or store at −80°C.
2. Resuspend the pelleted cells in 100 µL of chilled TGEK$_{50}$ buffer (containing protease inhibitors) per 10^6 cells, transfer the cells in a 1.5 mL eppendorf tube and leave on ice for 10 min.
3. Disrupt the cells by sonication using Vibracell with a probe of 13 mm: amplitude 27%, pulse 60% during 10 s at 4°C (other techniques for cells disruption are reported in **ref. 12**).
4. Remove the debris by centrifugation at 12,000g for 15 min at 4°C.
5. Transfer the supernatant (protein homogenate) in a fresh eppendorf tube and keep on ice.
6. Determine the protein concentration using a Bradford assay as recommended by the manufacturer.
7. Make aliquots of 50 micrograms of protein per tube, freeze samples in liquid nitrogen and store at −80°C (*see* **Note 1**).

3.2. Standard SDS-PAGE: Gel Casting

1. Clean the plates with water and ethanol (avoid the use of detergents), and assemble the gel plates as indicated on the manufacturer's instructions.
2. For initial analysis, we use to analyze the whole protein extracts on both a 6% and a 12% separating SDS-polyacrylamide gel electroporesis (PAGE) in order to obtain a good separation of the large and small size proteins, respectively (*see* **Note 2**).
3. Before pouring the acrylamide solutions, add 50 µL of 10% APS and 4 µL of TEMED to the 6% acrylamide solution, and add 50 µL of 10% APS plus 2 µL of TEMED to the 12% acrylamide solution. Mix well but avoid bubbles and pour between the plates (*see* **Note 3**). Immediately overlay the acrylamide solution with few drops of isobutanol. Allow the gel to polymerize for 30 min.
4. Discard the overlay solution, rinse with water and drain well with paper towel the rest of the water solution.
5. Before pouring the stacking gel, add 50 µL of 10% APS and 5 µL of TEMED. Pour the solution onto the separating gel and insert the clean comb (*see* **Note 4**). Allow the gel to polymerize for 15 min.

6. Remove carefully the combs, assemble the gels on the electrophoresis device as indicated by the manufacturer. Add running buffer 1X to top and bottom reservoir. The gels are ready for loading the protein samples.

3.3. Electrophoresis

1. Thaw the protein samples on ice. Add the sample loading buffer to get 1X final concentration. Mix well and denature samples in boiling water bath for 5 min.
2. To load precisely the samples on the bottom of the slots, use a Hamilton syringe (wash well between each sample) or a micropipetter with a dispensable long narrow tip.
3. Perform the electrophoresis at 200 V, until the dye front reaches the bottom of the gel.

3.4. Western Blotting

Transfer:
1. The proteins are transfer on 0.45 μm nitrocellulose membrane (*see* **Note 5**). With a pencil, mark the side of the membrane in order to orient it later on.
2. The transfer is performed using transfert buffer (*see* **Note 6**) as indicated on the manufacturer's instructions.
3. Western blotting is performed either at 200 mA for 1 h at room temperature with the presence of a ice block in the reservoir buffer or at 40 mA overnight at room temperature.

3.5. Immunodetection

1. After transfer, wash the membranes with 30 mL PBS 1X for 5 min.
2. Optional: using ponceau S, the proteins can be stained on the membrane to appreciate the uniformity of either the protein quantity, or quality or transfer. After staining, wash the membrane with distilled water to remove the excess of dye and the protein patterns will appear. To remove completely the ponceau red, wash several times with PBS 1X.
3. Block the nonspecific sites on the membranes by incubation in 50 mL of 5% non-fat dry milk solution for 1 h at room temperature or overnight at 4°C.
4. Remove the milk solution and rinse the membrane with PBS 1X.
5. Dilute the 1C2 antibody (*see* **Note 7**) at 1/2000–1/5000 in non-fat milk solution. Incubate by rocking for 1h at RT (*see* **Note 8**). 10 mL of antibody solution is enough to incubate one or two membranes of 9 cm × 6.5 cm in an appropriate plastic box (*see* **Note 9**).
6. Remove the antibody solution (*see* **Note 10**), and wash the membranes 3 times with 30 mL PBS 1X for 10 min at RT.
7. Incubate the membrane with the secondary antibody: 10 mL of diluted secondary antibody (*see* **Note 11**) in 0.5% nonfat dry milk is used for two membranes. Rock the membranes for minimum 1 h at RT.
8. Wash the membranes 5 times with 50 mL PBS 1X for 10 min (*see* **Note 12**).
9. Perform the chemiluminescent peroxidase reaction as indicated on the manufacturer's instructions. Place the membrane on glass plate, protein side up, and add 2 mL substrate solution, so that the reagents are held by surface tension on the surface membrane. Let stand for 5 min at RT. Drain off the excess of reagents by holding the glass plate vertically and touching the edge of the membrane against tissue paper. Cover the membrane with a saran wrap. Gently smooth out air pockets, with putting pressure on the membrane.
10. In a dark room, using red safelights, place an autoradiography film on the top of the membrane and expose 30 s. Remove and develop the film. During this process, place another film on the membrane and close the cassette. On the basis of the results obtained with the first film, estimate how long to continue the exposure of the second one. Comparing films with different exposure times help to evaluate the specificity of the signal (*see* **Note 13**).

4. Notes

1. It is recommended to thaw the protein samples only once, and discard the rest. Repeating thawing-freezing rounds may generate artifacts due to protein degradation.
2. We also use gradient gel 5–15% polyacrylamide that gives a good protein separation. Casting of such a gel may represent some difficulties. Precasted gradient and standard gels are also commercially available.
3. Caution: when pouring acrylamide solution, plan to have the separating gel at least 0.5 cm below the slots.
4. 10 teeth-comb (1 mm) allows the loading of maximum 25 µL or up to 200 µg of protein extract per slot, without aberrant migration of the proteins, while 15 teeth-comb (1 mm) allows loading of maximum 15 µL or up to 50 µg of protein extract per slot.
5. For the gel prepared with the MINI-PROTEAN II apparatus, nitrocellulose membranes of 9 cm × 6.5 cm will be large enough for the protein transfer. They should be carefully handled with gloves.
6. SDS is usually dispensable for the transfer, but we consistently observe a better immunodetection with 1C2 antibody when the transfer was performed with SDS.
7. The antibody concentration to be used on WB may vary from one batch to another and should be determined, but 1/2000 is a good dilution to start with.
8. Using incubation for 1 h at RT, we obtained better detection and less background than when the detection is performed overnight at 4°C. This is thought to be due to the instability of the 1C2 antibody once diluted.
9. The antibody solution should cover the membranes during rocking to keep them wet. Alternatively, the membranes and antibody solution can be sealed in a plastic bag.
10. In general, antibody solution can be reused immediately for incubation with other membranes or after being stored overnight at 4°C. Our experience indicates that repeatedly freezing and thawing the diluted 1C2 antibody decreases the polygln expansion detection and increases the background.
11. The secondary antibody should be diluted according to the supplier's instructions; too high a concentration of the secondary antibody would result in high background.
12. Given the very stable interaction of 1C2 with long polygln, intensive washes are strongly recommended to reduce the background as much as possible. Washing can even be extended to several hours without loss of signal.
13. If background is high the membrane may be rewashed and redetected, but this causes a slight loss of sensitivity.

Acknowledgments

I thank Y. Lutz and L. Tora for kindly providing the 1C2 antibody; G. Stevanin, A. Brice, D. Devys, G. Imbert, and J. L. Mandel for helpful discussion, G. Zeder-Lutz and M. H. V. Van Regenmortel for collaboration; and C. Weber for technical assistance. The work was supported by Grants from CNRS, INSERM, EEC (BMH4-CT96-0244), GREG, and CHRU of Strasbourg. YT was supported by Medical Research Council of Canada and Hereditary Disease Foundation.

References

1. Paulson, H. L. and Fischbeck, K. H. (1996) Trinucleotide repeats in neurogenetic disorders. *Annu. Rev. Neurosci.* **19**, 79–107.

2. Nakamura, K., Jeong, S. Y., Uchihara, T., Anno, M., Nagashima, K., Nagashima, T., et al. (2001) SCA17, a novel autosomal dominant cerebellar ataxia caused by an expanded polyglutamine in TATA-binding protein. *Hum. Mol. Genet.* **10** 1441–1448.
3. McInnis, M. G. and Margolis, R. L. (1998) Anticipation and psychiatric disorders, in *Genetic Instabilities and Hereditary Neurological Diseases* (Wells, R. D. and Warren, S. T., eds.), Academic Press, San Diego, CA, USA, pp. 401–408.
4. Paterson, A. D., Naimark, D. M. J., Vincent, J. B., Kennedy, J. L., and Petronis, A. (1998) Genetic anticipation in Neurological and other disorders, in *Genetic Instabilities and Hereditary Neurological Diseases* (Wells, R. D. and Warren, S. T. eds.), Academic Press, San Diego, CA, USA, pp. 413–424.
5. Trottier, Y., Lutz, Y., Stevanin, G., Imbert, G., Devys, D., Cancel, G., et al. (1995) Polyglutamine expansion as a pathological epitope in Huntington's disease and four dominant cerebellar ataxias. *Nature* **378,** 403–406.
6. Stevanin, G., Trottier, Y., Cancel, G., Dürr, A., David, G., Didierjean, O., et al. (1996) Screening for proteins with polyglutamine expansions in autosomal dominant cerebellar ataxias. *Hum. Mol. Genet.* **5,** 1887–1892.
7. Moriniere, S., Saada, C., Holbert, S., Sidransky, E., Galat, A., Ginns, E., et al. (1999) Detection of polyglutamine expansion in a new acidic protein: a candidate for chilhood onset schizophrenia. *Mol. Psychiatry* **4,** 58–63.
8. Joober, R., Benkelfat, C., Jannatipour, M., Turecki, G., Lal, S., Mandel, J. L., et al. (1999) Polyglutamine-containing proteins in Schizophrenia. *Mol. Psychiatry* **4,** 53–57.
9. Trottier, Y., Zeder-Lutz, G., and Mandel, J.-L. (1998) Selective recognition of proteins with pathological polyglutamine tracts by a monoclonal antibody in *Genetic Instabilities and Hereditary Neurological Diseases* (Wells, R. D. and Warren, S. T. eds.) Academic Press, San Diego, CA, USA, pp. 447–453.
10. Petronis, A., Gottesman, I. I., Crow, T. J., DeLisi, L. E., Klar, A. J., Macciardi, F., et al. (2000) Psychiatric epigenetics: a new focus for the new century. *Mol. Psychiatry* 5; 342–346.
11. Gerber, H. P., Seipel, K,. Georgiev, O., Hofferer, M., Hug, M., Rusconi, S., and Schaffner, W. (1994) Transcriptional activation modulated by homopolymeric glutamine and proline stretches. *Science* 263; 808-811.
12. Harlow, E. and Lane, D. (1999) *Using Antibodies: A Laboratory Manual.* Cold Spring Harbor Laboratory Press, Cold Spring Harbor, NY, USA.

9

Detection of Trinucleotide Repeat Containing Genes by Matrix-Assisted Laser Desorption/Ionization (MALDI) Mass Spectrometry

Chung-Hsuan Chen, Nicholas T. Potter, and Nelly T. Taranenko

1. Introduction

A conventional method used for DNA size measurement is gel electrophoresis. In this process, an electric field is applied to a gel medium and DNA molecules are separated by size. However, this process is slow and takes hours to separate different sizes of DNA fragments. Other physical and chemical properties of gel, such as viscosity, temperature, and temperature gradients, tend to influence the migration of DNA fragments. Alternatively, mass spectrometry has been used for several decades to measure the molecular weight of gaseous samples by mass-to-charge ratio (M/Z). In order to use a mass spectrometer for molecular weight determination, molecules of interest need to be produced in a gas phase. Owing to the practically zero vapor pressure of biomolecules at room temperature, a method to convert biomolecules from solid or liquid form into gaseous form is required. Desorption of biomolecules from solid phase using a laser beam has been pursued for decades. However, large biomolecules tend to fragment when laser desorbed, due to their strong absorption of laser photons. Results of laser desorption experiments indicated that the upper mass limit was ~5000 Daltons. Matrix-assisted laser desorption/ionization (MALDI) was developed to circumvent the problem of laser absorption-induced dissociation (*1*). With MALDI, laser energy is absorbed by matrix molecules instead of the biomolecule so that no fragmentation of the biomolecule occurs. Because mass spectrometry is based on the detection of gas phase ions, a means to produce ions is also needed. In the MALDI process, positive or negative biomolecular ions are produced by attaching or detaching a proton to/from a biomolecule. Once the ions have been produced, various sizes of ions may be separated by electric or magnetic fields. The mass resolved ions are then accelerated to impinge on a charged particle detector such as an electron multiplier or a microchannel plate (MCP). When one ion impacts an electron multiplier, as many as 10^6 electrons can be produced after several stages of multiplication, thus making it possible to detect even a single charged particle.

Owing to its relatively simple principle of operation, which is somewhat analogous to a gel electrophoresis device, a time-of-flight mass spectrometer (TOF-MS) is most often used for biomolecule detection. A TOF-MS is a device in which ions of different masses are given the same energy and are allowed to traverse a field-free distance. Because of their different velocities, the ions of differing masses arrive at the end of the field-free distance at different times. In order to achieve high resolution in mass spectra, the ions need to be produced in a very short time interval and in a very small volume. Thus, ionization by a short-duration pulsed laser is an ideal choice for a TOF-MS. One major advantage of a TOF-MS over other types of mass spectrometers is its ability to measure all of the masses in a sample in a very short period of time. Another is its capability of measuring ions with very high molecular weights. In principle, there are no limitations on the size of molecules that can be measured by a TOF-MS.

In 1987, Hillenkamp and his coworkers *(1)* discovered that large protein molecular ions can be produced without much fragmentation by laser desorption, if these biomolecules are mixed with smaller organic compounds, which serve as a matrix for strong absorption of laser photons. This process is now called matrix-assisted laser desorption (MALD). The typical preparation technique for MALD is to dissolve biomolecular samples in solution, then prepare another solution of small organic molecules, such as 3-hydroxypicolinic acid. These two solutions are subsequently mixed and a small amount is placed on a metal plate to dry. After the crystallization of the sample, the sample plate is placed in the mass spectrometer for analysis. The molar ratio of matrix to analyte is typically more than 1000 to 1. During the MALD process, matrix molecules strongly absorb the laser energy and become vaporized, carrying the large interspersed biomolecules along with them during the fast vaporization process. Large gas-phase biomolecules can be produced without fragmentation, which is probably due to minimal direct absorption of laser energy; thus, "soft" desorption can be achieved. The meaning of "soft" here indicates no breaking apart of molecules. However, it was found that protein parent ions are also produced during the MALD process in addition to the expected neutral molecules. Thus, these desorbed ions can be directly detected by a mass spectrometer. The process involving ionization and matrix-assisted laser desorption at the same time is abbreviated MALDI (namely, matrix-assisted laser desorption and ionization). The mechanism of ion production has been speculated by many researchers to involve proton transfer. The proton attachment to a biomolecule leads to the production of a positive ion, namely $(M + H)^+$. The proton transfer from biomolecule to matrix molecule produces negative ions, namely $(M - H)^-$.

Since the discovery of MALDI, many research groups have succeeded in using it to measure various proteins and large organic compounds. MALDI has also been applied to DNA segments. Initially, success was limited to small DNA detection and gaining some understanding of the mechanism of the MALDI process. Then, Wu et al. *(2)* discovered 3-hydroxypicolinic acid is a good matrix for mixed-base oligonucleotide and succeeded in detecting oligonucleotides of 67 bases. With the development of an instrument to give high ion energy and the use of new matrices, we reported the first measurement of detecting longer oligonucleotides (500 bp) with MALDI *(3)*. That confirmed that MALDI can be used to measure long DNA for various applications.

MALDI is emerging as a new technology for the rapid, reliable, and inexpensive detection of genetic polymorphisms and mutations *(4)*. The separation time needed for MALDI can be shorter than a few hundred microseconds, compared to hours for conventional gel analysis and minutes for capillary gel electrophoresis. Since MALDI detection of biomolecules is based on molecular mass, there is no need to exogenously label biomolecules, which can significantly reduce both time and cost. In addition, there is no concern for identification ambiguity resulting from band compression, reiterated nucleotide sequences or high GC content. Collectively, these advantages make MALDI mass spectrometry an attractive methodology for automated, high-throughput DNA genotyping.

We have utilized MALDI for DNA sequencing *(5–7)*, genotyping for cystic fibrosis *CFTR* mutations *(8,9)*, single nucleotide polymorphism (SNP) detection *(9)*, short tandem repeat (STR) genotyping for forensic applications *(10)*, gender determination *(11)*, and multiplex hybridization detection *(12)*. Recently, we have utilized MALDI for the analysis of trinucleotide repeat (TNR) containing genes and demonstrated its validity for the accurate genotyping of expanded CAG alleles associated with both Huntington's disease (HD) and Dentatorubral pallidoluysian atrophy (DRPLA), two autosomal dominant neurodegenerative disorders *(13)*. In this chapter, we outline the methodology for the application of MALDI for the detection and quantitation of TNR containing genes and illustrate its utility for the accurate genotyping of this type of DNA mutation.

2. Materials

2.1. Genomic DNA Isolation

Genomic DNA was isolated from peripheral blood leukocytes utilizing Puregene® DNA isolation kits as described by the manufacturer. (Gentra Systems, Minneapolis, MN, USA).

2.2. PCR

1. Primers for amplification across the HD CAG repeat expansion are:
   HD-1: 5'-ATG AAG GCC TTC GAG TCC CTC AAG TCC TTC-3'
   HD-3: 5'-GGC GGT GGC GGC TGT TGC TGC TGC TGC TGC-3' *(14)*.
   Primers are aliquoted as 20 µM stocks in molecular biology grade water.
   Primers for amplification across the DRPLA (CTG-B37) repeat expansion are:
   DRPLA-F: 5'-CAC CAG TCT CAA CAC ATC-3'
   DRPLA-R: 5'-CCT CCA GTG GGT GGG GAA ATG CTC-3' *(15)*.
   Primers are aliquoted as 20 µM stocks.
2. 10X PCR buffer:500 mM KCl, 15 mM MgCl$_2$, and 100 mM Tris-HCl, pH 9.0 (Fisher Scientific, Fair Lawn, NJ, USA).
3. Taq DNA polymerase (5 U/µL Fisher Scientific).
4. 10 mM dNTP stocks (GeneAmp® dNTPs, Applied Biosystems, Foster City, CA, USA).
5. Molecular biology grade water (Sigma Chemical Co., St Louis, MO, USA).
6. DMSO, molecular biology grade (Sigma).

2.3. Post-PCR DNA Purification

Prior to MALDI analysis, PCR products are purified by column chromatography using the QIAquick PCR purification kit (Qiagen Inc, Valencia, CA, USA) following the manufacturer's instructions.

2.4. MALDI

A schematic of a typical TOF-MS meter is shown in **Fig. 1**. A UV laser—generally a nitrogen laser—is used for laser desorption and ionization of biomolecular samples,

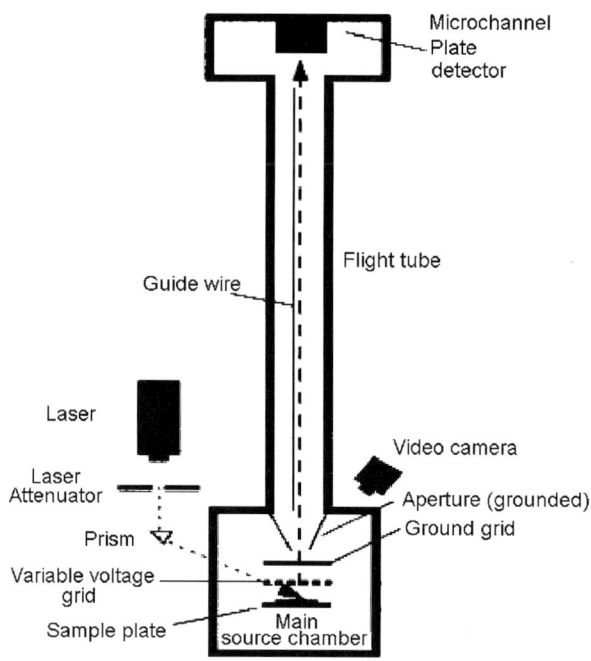

Fig. 1. Schematic of a typical MALDI time-of-flight mass spectrometer

which are typically prepared on a stainless steel substrate. The sample plate is often biased with a high voltage of 20,000–30,000 volts to provide ion energy for the desorbed biomolecular ions. For most commercial MALDI TOF-MS, an extraction region is often biased with a pulsed voltage to achieve ion focusing to improve mass resolution. This is called delayed ion extraction *(16,17)*. After ion extraction, the biomolecular ions travel in the drift tube (electric field free) region of the TOF-MS. Biomolecular ions with different masses are separated in the drift tube by their mass-to-charge ratio. In other words, ions with different masses take different lengths of time to pass through the drift tube. After passing through the drift tube, biomolecular ions are subsequently detected by an ion detector, which can be an electron multiplier, a microchannel plate, or a channeltron. Most commercial MALDI instruments are equipped with a microchannel plate detector (*see* **Note 1** and **2**).

For most DNA detection experiments, a mixture of 0.3 M 3-hydroxypicolinic acid (3-HPA), 0.5 M picolinic acid and 0.3 M ammonium fluoride with molar ratio of 9:1:1 was used as the matrix. All chemicals were purchased from Aldrich Inc. and used without further purification (*see* **Note 3** and **4**).

There are several commercial manufacturers of MALDI TOF-MS, most of whom utilize a sample plate which is computer controlled to position the sample with respect to the desorption laser beam. The sample plate can typically hold ~100 samples for analysis. Each mass spectrum is obtained by averaging the signal resulting from several laser shots (4~256), depending on the signal intensity. Most equipment is optimized for protein detection instead of oligonucleotide due to the difficulty of detecting large oligonulceotides and poor mass resolution of oligomers larger than 100 nucleotides.

3. Methods

3.1. Genomic DNA Isolation

Genomic DNA was isolated from peripheral blood leukocytes utilizing Puregene® DNA isolation kits and described by the manufacturer. (Gentra Systems).

Typical sample volumes were 3 mL and generally yielded a purified DNA product with a concentration of 0.5–1.0 µg/µL.

3.2. PCR

1. The PCR reaction cocktail was made as a 10X stock and 24 µL aliquots were dispensed into 10 × 0.5 mL thin-walled micro-amp PCR tubes (Applied Biosystems). The 10X reaction mix consists of the following:

   HD
25 µL	10X PCR buffer
5 µL	HD-1 (from 20 µM stock)
5 µL	HD-3 (from 20 µM stock)
5 µL	Each dNTP (dATP, dCTP, dGTP, and dTTP from 10 mM stock)
168 µL	Molecular biology grade water
25 µL	Dimetryl Sulfoxide (DMSO)
2 µL	Taq polymerase (5 U/µL)

   DRPLA
25 µL	10X PCR buffer
5 µL	DRPLA-F (from 20 µM stock)
5 µL	DRPLA-R (from 20 µM stock)
5 µL	Each dNTP (dATP, dCTP, dGTP, and dTTP from 10 mM stock)
180.5 µL	Molecular biology grade water
12.5 µL	DMSO
2 µL	Taq polymerase (5 U/µL)

2. 1 µL (50–100ng) genomic DNA was added to each 24 µL aliquot and PCR was performed in a Perkin-Elmer 9600 thermocycler using the following conditions:

   HD   94°C, 4 min; 35 cycles of 94°C, 65°C, 72°C for 1 min each; 72°C extension for 10 min; 4°C soak.

   DRPLA   94°C, 4 min; 30 cycles of 94°C, 62°C, 72°C for 1 min each; 72°C extension for 10 min; 4°C soak.

3.3. Post-PCR DNA Purification

Prior to MALDI analysis, PCR products were purified by column chromatography using the QIAquick PCR purification kit (Qiagen Inc) following the manufacturers instructions. After the final elution step, DNAs were precipitated with alcohol, centrifuged, and resuspended in 5 µL of molecular biology grade water for subsequent MALDI analysis.

3.4. MALDI

1. The sample was prepared for mass spectrometry analysis by mixing an equal volume of aqueous analyte solution with matrix solution. Typically, 1 µL of the mixed solution was spotted on a stainless-steel sample plate and dried by a forced nitrogen gas jet at ambient temperature. The sample was then crystallized. It was found that better MALDI signals

Fig. 2. A typical needle-shaped crystalline sample with 3-HPA matrix.

are obtained when 3-hydroxypicolinic acid is used as matrix and the matrix forms needle-like crystals (see **Fig. 2**) The sample plate can accommodate up to one hundred samples a time. The typical spot size of a sample is 1–3 mm. The laser spot size ranges from 50 to 100 µm.

2. The dried sample was immediately loaded into the TOF-MS. A mechanical pump pumped the pressure of the chamber of the mass spectrometer down to a few microns. Then, a turbo pump pumped down the pressure to 10^{-6} Torr or lower. The high voltages were then provided for ion acceleration. The polarity of biased voltage of the target was determined by the type of ions to be detected. When positive ions are to be detected, the voltage should be positive. On the other hand, negative voltage should be provided for negative ion detection. The sample plate was moved by the servo system, which was controlled by the computer. A small lamp was used to illuminate the sample plate so that the spot of the selected sample and the position of the laser beam hitting the sample could be constantly monitored by a video camera.

3. A linear TOF-MS (Voyager, PerSeptive Biosystems, Framingham, MA, USA) equipped with a nitrogen laser was used to obtain mass spectra. The laser beam was directed onto the sample surface. In general, better signals were obtained when the laser beam was aimed at the edge of the sample where better crystals were formed. The typical laser fluence was between 45–65 mJ/cm^2. The acceleration voltage was set between 25000–30000 volts (V). The pressure of the chamber of the mass spectrometer was 1.5×10^{-7} Torr. New calibration is needed whenever the acceleration voltage was changed. We typically used insulin and myoglobin mixed with the sample for internal calibration.

4. The DNA ions were collected and their signal amplified by a microchannel plate. The amplitude of signals depends on the bias voltage on the microchannel plate. The signals were digitized by a digital oscilloscope (Tektronix 520). Delayed ion extraction *(16,17)* was also installed to improve mass resolution by time focusing for the same type of ions, which were produced at different locations during the desorption/ionization process. The amplitude of the pulse voltage was 1500V–2000V. The pulse duration was typically set at ~10 µsec.

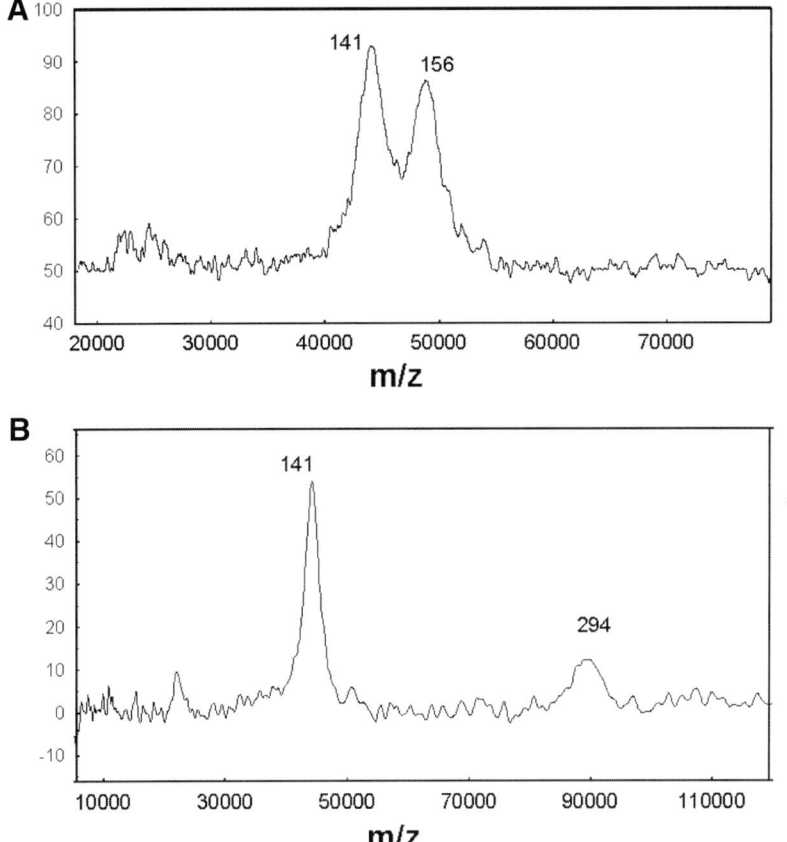

Fig. 3. MALDI analysis of the DRPLA CAG repeat. (**A**) Mass spectrum from a normal individual carrying 15 and 20 CAG repeats and (**B**) Mass spectrum from a DRPLA patient carrying 15 and 66 CAG repeats.

5. A mass spectrum for each laser pulse was recorded by the computer. The final mass spectrum was typically obtained by averaging 8–256 laser shots. The spectrum was saved in the computer and could be printed whenever desired. Computer software was developed to determine the mass resolution for each peak and the peak position could also be automatically located. Correlation for the electronic background was also included in the software to normalize mass spectra to improve mass resolution.
6. We have used MALDI to measure CAG expansions in both the DRPLA and HD genes. An example of MALDI mass spectrum for a normal individual and a DRPLA patient sample are shown in **Fig. 3**. The sample from the normal individual shows 15 and 20 CAG repeats while the patient sample shows 15 and 66 repeats. Both samples are heterozygous. A mass spectrum from a HD patient is shown in **Fig. 4**. It indicates the sample is heterozygous with CAG repeat of 17 and 40 (*see* **Note 5**). It is worth mentioning that all peaks in mass spectra correspond to the mass of single-stranded DNA. PCR products are dissociated into single-stranded DNA either during the process of mixing DNA with matrix compounds or the laser desorption process. The mass resolution is not high enough to separate the two peaks expected of each single-stranded DNA of slightly different mass.

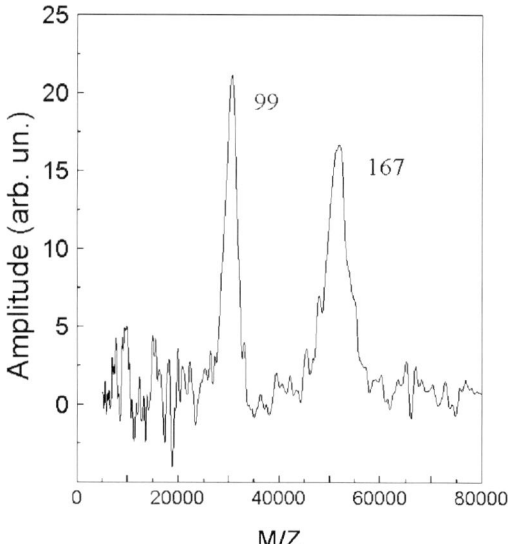

Fig. 4. MALDI analysis of the HD-CAG repeat. A mass spectrum of a HD patient carrying 17 and 40 CAG repeats.

4. Notes

1. Reflectron TOF-MS is often used to improve mass resolution since the velocity distribution that spoils mass resolution in linear time-of-flight mass spectrometers can be compensated. However, for large DNA fragment measurements, the use of reflectron time-of-flight does not significantly improve mass resolution. It does, however, reduce the signal level drastically due to the limited lifetimes of DNA ions *(18)* Thus, a linear time-of-flight mass spectrometer is still preferred.
2. In general, electrospray mass spectrometry can exhibit better mass resolution than TOF-MS. However, an electrospray mass spectrometer is not suitable for measuring several biomolecules simultaneously, since each molecule tends to give several peaks due to the multiply charged ions. On the other hand, when only one short tandem repeat allele is to be analyzed, an electrospray mass spectrometer can give better mass resolution for determining the sizes of DNAs. However, sample impurity (i.e., salt contamination) is very critical for using electrospray mass spectrometry for DNA measurements. If electrospray mass spectrometry is to be used, more extensive DNA purification will be necessary.
3. Instead of a mixture of 3-hydroxypicolinic acid, picolinic acid, and ammonium fluoride, it is also possible to use a mixture of 0.2 *M* 2,3,4-trihydroxyacetophenone, 0.2 *M* 2,4,5-trihydroxyacetophenone, and 0.3 *M* ammonium citrate dibasic of equal volume *(19)* as the matrix. In general, the mass resolution is better with this alternate matrix but the ion signal amplitude is less for large DNA fragments, compared to the 3-HPA solution.
4. Three characteristics of the matrix are important in successful MALDI-MS experiments: 1) The matrix must incorporate the analyte in a way that does not break bonds or otherwise compromise the integrity of the intra-DNA binding; 2) the matrix must have strong absorption at the chosen laser wavelength; and 3) the matrix must provide a ready source of charge, which can be readily transferred to the DNA molecules. This is critically important, as the interaction between analyte molecules and crystals might be the key that determines the usefulness of a matrix material for MALDI.

5. Our results of MALDI for both DRPLA and HD samples show that mass spectrometry can be used for the accurate determination of the number of CAG repeats. Since the time needed for mass spectrometric analysis can be much shorter than the gel electrophoretic method, MALDI has the potential to become a valuable tool for the quick diagnosis of these dominantly inherited neurodegenerative disorders as well as a useful methodology for the detection of repeat-containing genes in general. We expect the same technology can be applied to other repeat expansion diseases. Microsatellites, which often contain tandem repeats, are useful markers for many genetic loci. Thus, we expect that MALDI-TOF-MS will be useful for the analysis of microsatellite markers as well. However, expansion of MALDI mass spectrometry for DNA analysis critically depends on mass resolution and sensitivity.

Acknowledgment

Research was sponsored by the National Institute of Justice, grant number 97-LB-VX-A047, and in part by the Office of Biological and Environmental Research, US Department of Energy, under contract DE-AC05-00OR22725 with UT-Battelle, LLC. N. T. Potter acknowledges the funding provided by the State of Tennessee, Department of Health and Environment, Mental Health and Mental Retardation. We also acknowledge PerSeptive BioSystems (a subsidiary of Applied Biosystems Inc.) for providing a mass spectrometer for part of this work.

The submitted manuscript has been authored by a contractor of the US Government under contract number AC05-00OR22725. Accordingly, the US Government retains a nonexclusiove, royalty-free license to publish or reproduce the published form of this contribution, or allow others to do so, for US Government purpose.

References

1. Karas, M. and Hillenkamp, F. (1988) Laser desorption ionization of proteins with molecular masses exceeding 10000 Daltons. *Anal. Chem.* **60**, 2299–2301.
2. Wu, K. J., Steding, A., and Becker, C. H. (1993) Matrix-assisted Laser Desorption Time-of-Flight Mass Spectrometry of oligonucleotide using 3-hydroxypicolinic acid as an ultraviolet sensitive matrix. *Rapid Comm. Mass Spectrom.* **7**, 142–146.
3. Tang, K., Taranenko, N. I., Allman, S. L., Chang, L. Y., and Chen, C. H. (1994) Detection of 500-nucleotide DNA by Laser Desorption Mass Spectrometry. *Rapid Comm. Mass Spectrom.* **8**, 727–730.
4. Jackson, P. E., Scholl, P. F., and Groopman, J. D. (2000) Mass spectrometry for genotyping: an emerging tool for molecular medicine. *Mol. Med. Today* **6(7)**, 271–276.
5. Taranenko, N. I., Chung, C. N., Zhu, Y. F., Allman, S. L., Golovlev, V. V., et al. (1997) Matrix-assisted laser desorption/ionization for sequencing single-stranded and double-stranded DNA. *Rapid Comm. Mass Spectrom.* **11**, 386–392.
6. Taranenko, N. I., Allman, S. L., Golovlev, V. V., Taranenko, N. V., Isola, N. R., and Chen, C. H. (1998) Sequencing DNA using mass spectrometry for ladder detection. *Nucleic Acids Res.* **26**, 2488–2490.
7. Zhu, Y. F., Taranenko, N. I., Allman, S. L., Taranenko, N. V., Martin, S. L., Haff, L. A., and Chen, C. H. (1997) Oligonucleotide sequencing by fragmentation in matrix-assisted laser desorption/ionization time-of-flight mass spectrometry. *Rapid Comm. Mass. Spectrom.* **11**, 897–903.
8. Chang, L. Y., Tang, K., Schell, M., Ringelberg, C.; Matteson, K. J., Allman, S. L., and Chen, C. H. (1995) Detection of ΔF508 Mutation of the cystic fibrosis gene by matrix-

assisted laser desorption/ionization mass spectrometry. *Rapid Comm. Mass Spectrom.* **9,** 772–774.

9. Taranenko, N. I., Matteson, K. J., Chung, C. N., Zhu, Y.F., Chang,L.Y., Allman, S. L., et al. (1996) Laser desorption mass spectrometry for point mutation detection *Gene. Anal. Biom. Eng.* **13,** 87–94.
10. Taranenko, N. I., Golovlev, V. V., Allman, S. L.; Taranenko, N. V.,Chen, C. H., Hong, J., and Chang, L. Y. (1998) Matrix-assisted laser desorption/ionization for short tandem repeat loci. *Rapid Comm. Mass Spectrom.* **12,** 413–418.
11. Taranenko, N. I.; Potter, N. T.; Allman, S. L.; Golovlev, V. V.; and Chen, C. H. (1999) Gender identification by matrix-assisted laser desorption/ionization time-of-flight mass spectrometry. *Anal. Chem.* **71,** 3974–3976.
12. Isola, N. R., Allman, S. L., Golovlev, V. V., and Chen, C. H. (2001) MALDI-TOF Mass spectrometric method for detection of hybridized DNA oligomers. *Anal. Chem.* **73,** 2126–2131.
13. Taranenko, N. I., Potter, N. T., Allman, S. L., Golovlev, V. V., and Chen, C. H. (1999) Detection of trinucleotide expansion in neurodegenerative disease by matrix-assisted laser desorption/ionization time-of-flight mass spectrometry; *Gene. Anal: Biom. Eng.* **15,** 25–31.
14. Warner, J. P., Barron, L. H., and Brock, J. H. (1993) A new polymerase chain reaction (PCR) assay for the trinucleotide repeat that is unstable and expanded on Huntington's disease chromosomes. *Mol. Cell Probes* **7,** 235–239.
15. Nagafuchi, S., Yanagisawa, H., Sato, K., Shirayama,T., Ohsahki, E., Bundo, M.,et al. (1994) Dentatorubral and pallidoluysian atrophy expansion of an unstable CAG trinucleotide on chromosome 12p. *Nat. Genet.* **6,** 14–18.
16. Vestal, M. L., Juhasz, P., and Martin, S. A. (1995) Delayed extraction matrix-assisted laser desorption time-of-flight mass spectrometry. *Rapid Comm. Mass Spectrom,* **9,** 1044–1050.
17. Whittal, R. M., and Li, L. (1995) High resolution MALDI in a linear time-of-flight mass spectrometer. *Anal. Chem.* **67,** 1950–1954.
18. Jones, R. B., Allman, S. L., Tang, K., Garrett, W. R., and Chen, C. H. (1993) Neutralization of negatively charged oligonucleotides. *Chem. Phys. Lett.* **212,** 451–457.
19. Zhu, Y. F., Chung, C. N., Taranenko, N. I., Allman, S. L., Martin, S. A., Haff, L. and Chen, C. H. (1996) The study of 2,3,4-trihydroxyacetophenone and 2,4,6-trihydroxyacetophenone as matrix for DNA detection in matrix-assisted laser desorption/ionization time-of-flight mass spectrometry. *Rapid Comm. Mass Spectrom.* **10,** 383–388.

10

Fluorescence PCR and GeneScan® Analysis for the Detection of CAG Repeat Expansions Associated with Huntington's Disease

Cindy L. Vnencak-Jones

1. Introduction

Huntington's Disease (HD) is an autosomal dominant neurodegenerative disorder with an incidence of 1/10,000. The disease is characterized by involuntary choreic movements, psychiatric disorders, dementia, and death within 15–20 years. When the HD gene was cloned in 1993, it was discovered that the causative mutation was an expansion of a CAG trinucleotide repeat (1) and to date HD is one of 14 trinucleotide repeat diseases (2). The expanded CAG sequence in exon 1 of the HD gene, *IT15*, likewise encodes an expanded polyglutamine tract resulting in an aberrant huntingtin protein, which leads to neuronal specific death predominantly within the neostratum (3). Huntington is widely expressed in both the brain and nonneural tissues (4) and interestingly, N-terminal fragments of huntingtin, containing the elongated polyglutamine residues form aggregates, and can be visualized as cytoplasmic and nuclear inclusions (5).

Cloning of the HD gene eliminated the need for linkage analysis (6) and enabled rapid and accurate analysis using PCR to precisely determine the CAG repeat number in exon 1 of the *IT15* gene. The HD phenotype results from an expanded allele that contains 40 or greater repeats and the size of the repeat is inversely correlated with the age of onset (7). Repeats of 36 to 39 are considered HD alleles with reduced penetrance (8,9). Further, although normal alleles are characterized with CAG repeats less than or equal to 35, alleles with repeats between 27 and 35 are "mutable" in that they can exhibit meiotic instability (10).

Referrals for HD testing are diverse and include diagnostic testing for the adult or symptomatic child, presymptomatic testing for the consenting adult, prenatal testing of the fetus at risk, and preimplantation genetic testing for the couple at risk. For presymptomatic, prenatal, and pre-implantation testing, because there are significantly different implications for the patient with a CAG repeat number of 35 vs 36 or a repeat number of 39 vs 40, repeat size determination must be accurate. The method contained within this chapter describes the use of fluorescence polymerase chain reaction (PCR) coupled with denaturing gel electrophoresis and GeneScan® analysis to precisely and reproducibly determine CAG repeat lengths in the *IT15* gene for clinical testing for HD.

2. Materials
2.1. DNA Extraction: Peripheral blood

1. Puregene™ DNA Isolation Kit # D-40K (Gentra Systems, Inc.) The kit contains: Red Blood Cell Lysis Solution, Cell Lysis Solution, Protein Precipitation Solution, and DNA Hydration Solution. All reagents should be stored at room temperature and properly discarded prior to the expiration date.
2. Isopropanol: stored at room temperature in a fire cabinet.
3. Ethanol (70%).
4. 1% agarose gel, electrophoresis chamber and 1X TBE buffer.
5. Size Standard: Lambda DNA digested with *Hind*III (Invitrogen, # 15612013).

2.2. Fluorescence PCR

1.Ampli*Taq* DNA polymerase; 10X PCR Buffer II; 25 m*M* $MgCl_2$ (Applied Biosystems N808-0172) stored at $-20°C$.
2. dNTP stock: 1.25 m*M* each of dATP, dCTP, dGTP and dTTP (NEB N0446S) stored at $-20°C$.
3. DMSO (Sigma D5879) stored as aliquots at room temperature and protected from light.
4. 10X PCR buffer minus Mg (Invitrogen, Y02028).
5. Oligonucleotide primers: 5'-ATG AAG GCC TTC GAG TCC CTC AAG TCC TTC-3' *(11)*. This customized fluorescent primer is available from Applied Biosystems (*see* Website: www.appliedbiosystems.com) and is labeled with HEX for a color display of yellow. Upon receipt, lyophilized primers are reconstituted in 500 µL of sterile 1X TE. Fluorescent tagged primers are light-sensitive and should be handled in minimum lighting. 5'-GGC GGT GGC GGC TGT TGC TGC TGC TGC TGC-3' *(11)*. This customized nonfluorescent primer is purchased upon request from ResGen Invitrogen Corporation (*see* Website: www.resgen.com) and is received in 1X TE. The optical density of the stock tubes are measured and the concentrations recorded. Primers are stored at $-20°C$ and are diluted as needed prior to using.

2.3. 5% Denaturing Polyacrylamide Gel and Electrophoresis

1. ABI Prism 377 DNA Sequence System.
2. Alconox Lab Detergent (Fisher Scientific 04-322-4).
3. Long Ranger Singel Packs (BioWhittaker Molecular Applications 50691). Store shielded from direct light at room temperature. Discard before expiration date.
4. Blue-Dextran-EDTA loading buffer (Applied Biosystems 402055).
5. TAMRA 500 size standard (Applied Biosystems 401733).
6. Resin, Amberlite Mixed Bed (Applied Biosystems 400665).
7. Formamide (Sigma F-7508).

2.4. GeneScan® Analysis and Interpretation

1. GeneScan® Software and User Manual.

3. Methods
3.1. DNA Extraction-Peripheral Blood

1. For a generous amount of high-quality DNA, begin with 3–5 mL of whole blood. For a normal white count, the expected yield will be 300–500 µg of DNA.
2. Using universal precautions, transfer the whole blood from the lavender top EDTA collection tube to a sterile, labeled 50 mL polypropylene tube. Follow the instructions as

outlined in the manufacturers manual adjusting the amount of reagents based on the initial volume used. The optional RNase treatment step in this protocol can be omitted (*see* **Notes 1** and **2**).
3. Based on the size of the DNA pellet, add 300–500 µL of Hydration Solution for a concentration of about 1 µg/µL.
4. To estimate the concentration and assess the quality of the extracted DNA, subject 2 µL of the homogenous DNA solution to gel electrophoresis on a 1% agarose gel. For comparison, simultaneously analyze 2 µL of Lambda DNA in an adjacent well.

3.2. Fluorescence PCR

1. Each assay should include: the patient in duplicate (varying the amount of template DNA that is added); a normal control, with previously characterized CAG repeats of normal size; a positive control, with previously characterized expanded CAG repeats or a cloned standard with a known expanded CAG repeat length and; a negative control, with all reagents added to the reaction tube except template DNA (*see* **Note 3**).
2. Thaw reagents on ice and vortex each well before using. In subdued lighting, set up the following 15 µL reaction adding each in the following order. 1 µM each of the forward and reverse primer; 1.5 µL 10X PCR buffer II; 2.4 µL 1.25 mM dNTP stock; 1.5 µL 25 mM MgCl$_2$; sterile dH$_2$O (amount needed to bring volume to 15 µL); 1.5 µL DMSO and 1 U *Taq* DNA polymerase (dilute 5 U/µL stock 1:5 with 10X PCR buffer minus Mg) add 1 µg of patient DNA (*see* **Notes 4–7**) and immediately place in 94°C preheated DNA thermal cycler.
3. PCR is performed using the following conditions (*see* **Note 8**).

1 cycle	Initial Denaturation	94°C, 6 min
25 cycles	Denaturation	94°C, 30 s
	Annealing	65°C, 30 s
	Extension	72°C, 45 s
1 cycle	Final Extension	72°C, 10 min

4. Store reaction tubes protected from light and at –20°C.

3.3. 5% Denaturing Polyacrylamide Gel

1. Rinse both glass plates with warm water. In a squirt bottle, dissolve Alconox in warm water and apply diluted soap solution to the plates and wash gently with a gauze pad. Rinse the plates well with warm water followed by dH$_2$O. Securely position plates in horizontal position to air dry.
2. Mount the clean plates in the gel cassette as described by the manufacturer.
3. Prepare the gel solution in the gel pack as described by the manufacturer.
4. Squeeze the gel solution into a 50-mL tube avoiding the use of great force which may introduce air into the solution. Cap the tube and invert gently to mix.
5. Pour the gel by first attaching a sterile disposable 50 mL pipet to an electric pipetter. Aspirate 30–40 mL of gel solution into the pipette. With the plate assembly at an angle, begin slowly injecting the gel solution across the top of the glass. As the gel solution travels down between the plates, tap the glass firmly to expel any air bubbles that may form. When complete, secure the comb in place to form wells. Allow the gel to polymerize 2 h before removing the comb (*see* **Notes 9** and **10**).
6. Carefully remove the comb. Remove excess gel material from the top of the plate assembly and from each well. Mount the plate assembly onto the instrument, add 1X TBE buffer to the chambers, perform the plate check, prerun, prepare the sample sheet and prerun the gel as described by the manufacturer (*see* **Notes 11** and **12**).

7. Prepare gel-loading solutions. Using a scoopula, dispense resin beads into a 0.5 mL microfuge tube filling to 1/3 the volume of the tube. To the resin, add 400 µL formamide, cap the tube, and invert several times to mix. Prepare a 1:4 TAMRA/Blue Dextran dye solution. Vortex well to mix.
8. Prepare amplicons for analysis. Thaw the reaction tubes and vortex well to ensure homogeneity. Prepare 1:2 dilution of each of the amplicons using 1X TE. Label a set of 0.5 mL microfuge tubes corresponding to each reaction. Into each labeled tube dispense 2 µL of the TAMRA/Blue Dextran dye solution and 4 µL of the freshly prepared formamide, mix well and quick spin. Add 1 µL of the diluted amplicon to the corresponding tube and mix well.
9. Heat gel-loading tubes to 94°C for 3 min to denature and place tubes on ice to prevent reannealing.
10. Using a 0.17 mm flat capillary pipette tip, aspirate 2.5 µL of the denatured amplicon/TAMRA/Blue Dextran/formamide solution and carefully dispense into the corresponding well of the gel. Proceed with electrophoresis according to the manufacturer at 3000 V for 2.5 h.

3.4. GeneScan Analysis and Interpretation

1. Perform the analysis of the gel using the GeneScan software version provided by the manufacturer (*see* **Notes 13** and **14**).
2. Interpret the CAG repeat number for each sample based on the size of the amplicon as generated in the tabular data. Verify that the CAG repeat number of all controls is as expected before reporting patient results (*see* **Note 15**).

4. Notes

1. PCR analysis for the determination of the CAG repeat number in exon 1 of the *IT15* gene can be performed on high molecular-weight DNA extracted from any source including fresh tissue or cultured fetal cells using the protocol outlined by the manufacturer.
2. For clinical testing, the laboratory should verify that informed consent was obtained and in the case of predictive testing assure that proper psychiatric evaluation and genetic counseling was performed prior to submission of the specimen *(12)*. Further, if the laboratory is performing this assay for clinical testing, the laboratory must document proficiency by participating in a proficiency program for HD analysis.
3. The positive control should be a previously well characterized proficiency specimen or a cloned standard. Standardization between labs is especially important for this assay, where the difference between one repeat can have significantly different implications.
4. The quality of each new shipment of fluorescent and nonfluorescent primer is tested with existing reagents prior to using for clinical specimens.
5. To minimize freezing and thawing and to preserve the integrity of the fluorescent primer, fluorescent primers are stored as 5–10-µL aliquots for one-time use and diluted prior to using. Further, aliquot tubes are placed in 50-mL tubes, wrapped in aluminum foil, and placed in a freezer that is not self-defrosting. The amount of primer dispensed into each aliquot should be sufficient for a typical number of reactions performed at one time by the laboratory but not in large excess since unused primers are subsequently discarded.
6. The nonfluorescent labeled primer is likewise stored as concentrated stock aliquots with 50 µL in each tube. Aliquots are used in sequential order and repetitive freezing and thawing does occur. The same primer stock tube is used until about 5–10 µL of undiluted primer remains. At that time, this aliquot is discarded and the subsequent stock aliquot is put into use.
7. For larger runs, it is easiest to prepare a "cocktail" that includes premixing of the appropriate amounts of dH_2O; 10X buffer II; dNTP; $MgCl_2$, dimethyl sulfoxide (DMSO), and

Taq polymerase. To prepare, determine the number of reactions, add 2 (to assure adequate amount), and multiply this number times the reagent volume per reaction for each cocktail reagent. After adding each component, vortex well. This cocktail can be added in one pipetting step to each reaction tube to which each of the primers have previously been added. Mix well and add the template DNA using sterile, cotton-plugged pipet tips. Because DMSO is a heavy liquid, vortex the cocktail frequently during the pipetting of the cocktail into each reaction tube.

8. Similar to amplification of other trinucleotide repeat disease sequences (i.e., CGG for Fragile X Syndrome) the efficiency of amplification of the larger CAG repeat alleles is greatly reduced. Under the conditions described here, amplicons generated from expanded CAG repeat alleles of 40–50 appear to represent only about 20–30% of the total product as opposed to the theoretic yield of 50%. Further, if the efficiency of the PCR assay is somehow greatly diminished, amplicons from the expanded allele may not be detected and the allele "drops out." This will result in a false-negative and an apparent homozygous normal pattern will be observed. This phenomenon is compounded if this assay is performed on DNA extracted from paraffin-embedded tissue which is inherently partially degraded and often results in decreased amplification efficiency. Requests for HD testing on archived material is not uncommon and must be carefully interpreted to prevent reporting of a false negative result. The addition of 3.4 M Betaine (Sigma B-2629) to the reaction tube can be used to increase the amplification efficiency and the percentage of amplicons representing the expanded allele. In contrast, other variations to this procedure such as increasing the cycle number, increasing the length of time of the 72°C extension within each cycle, or increasing the final extension time from 10–30 min did not appear to significantly enhance amplification of the expanded allele.

9. Plates are received as a set and are numerically labeled. A log is maintained on the instrument recording the usage of each plate set. With frequent usage, there is a propensity to generate numerous small bubbles within the freshly poured gel suggesting a build up of residue. Should this occur, the use of this plate set is discontinued until the plates are cleaned. This special washing includes washing the plates with diluted Alconox, rinsing with warm water and then rinsing with 3 M HCl, warm water, 3 M NaOH, warm water and followed by dH$_2$O and air drying.

10. Best results are obtained if the gel is used immediately after polymerization but can be stored over night. To prevent drying, place gauze dampened with 1X TBE electrophoresis buffer across the top of the gel and cover with plastic wrap. Do not remove the comb if storing the gel overnight.

11. Use a dry folded Kimwipe pressed at the top of the well and pulled back to remove excess gel material from the top of the glass plates. Repeat using a Kimwipe dampened with dH$_2$O. Attach a gel loading capillary pipet tip (0.17 mm flat) to a 1000 pipet tip using parafilm to secure the joint and fill the device with dH$_2$O. With the gel apparatus at an angle and facing downward, position the dH$_2$O- filled pipet tip into each well to expel the dH$_2$O. Use a Kimwipe to remove resulting debris. Wells must be thoroughly cleaned before loading.

12. After removing excess gel debris and urea and prior to mounting the gel cassette onto the instrument, prepare a 30% Blue-Dextran-EDTA dye solution using dH$_2$O. To enhance visibility of the well during loading, using the same pipet tip device described above to flush out the wells, now fill the tip with the diluted blue loading dye. With the gel apparatus flat, force the dye solution into each well. Use a Kimwipe to blot excess dye that did not enter the well.

13. A representative gel image from the HD assay is shown in **Fig. 1**. Retain the gel file on disk or print a copy of the image and the corresponding sample sheet as a record for the laboratory.

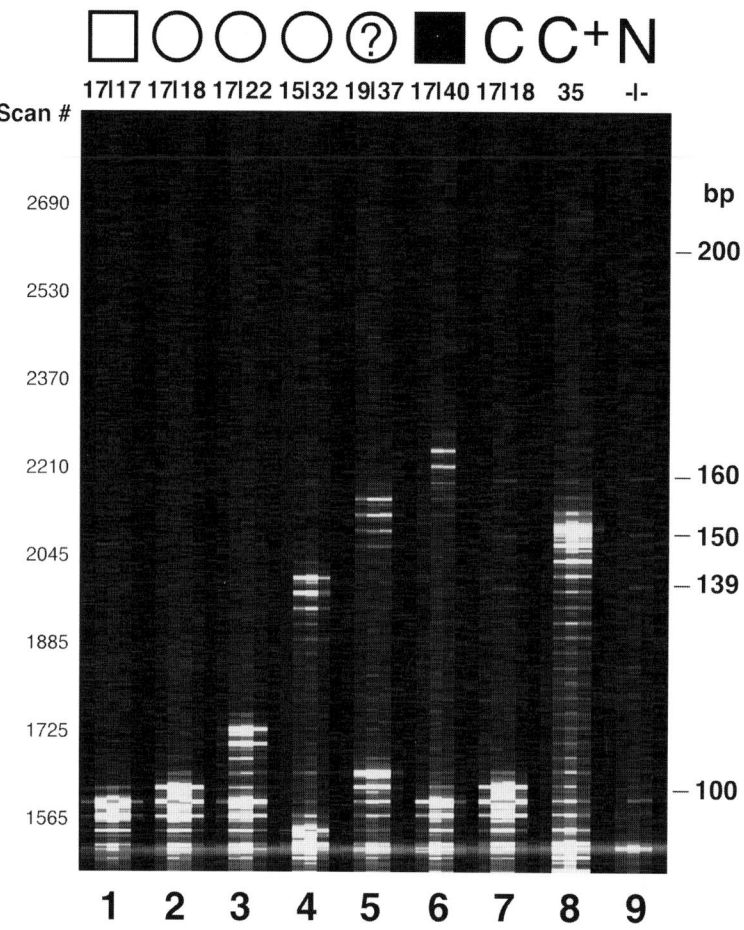

Fig. 1. Representative gel image for HD analysis using fluorescence PCR. Scan numbers are listed on the left. TAMRA size standard is listed as base pairs (bp) on the right. Phenotype and genotype of patients studied are listed at the top. Lanes 1–3, CAG repeats characteristic of normal alleles (17/17; 17/18; 17/22), Lane 4, a normal (15) and a mutable (32) allele. Lane 5, a normal (19) and a HD (37) allele with reduced penetrance. Lane 6: a normal (17) and a HD (40) allele; Lane 7, a normal control (C) with CAG repeats characteristic of normal (17/18) alleles. Lane 8, Cloned standard control (C+) with 35 CAG repeats (from M. R. Hayden, University of British Columbia). Lane 9, Negative control (N) reaction tube with all reagents except DNA. (See color plate 1 appearing in the insert following p. 82)

14. A representative electropherogram from the HD assay is shown in **Fig. 2.** Since large expanded repeats are less efficiently amplified, when a homozygous normal pattern is observed (lane 1) it is necessary to modify the peak detection window under analysis parameters to 20 to increase the sensitivity of the analysis, ensure accurate results and prevent a false negative result.
15. Regardless of whether a gel (ABI 377) or capillary (ABI 310, 3100 or 3700) format was used for fragment separation or which version of GeneScan analysis was used for the analysis, the CAG repeat number is directly correlated with the size of the amplicon generated in this assay. **Table 1** represents the most commonly observed CAG repeat numbers and the size of their corresponding amplicons.

Detection of CAG Repeat Expansions

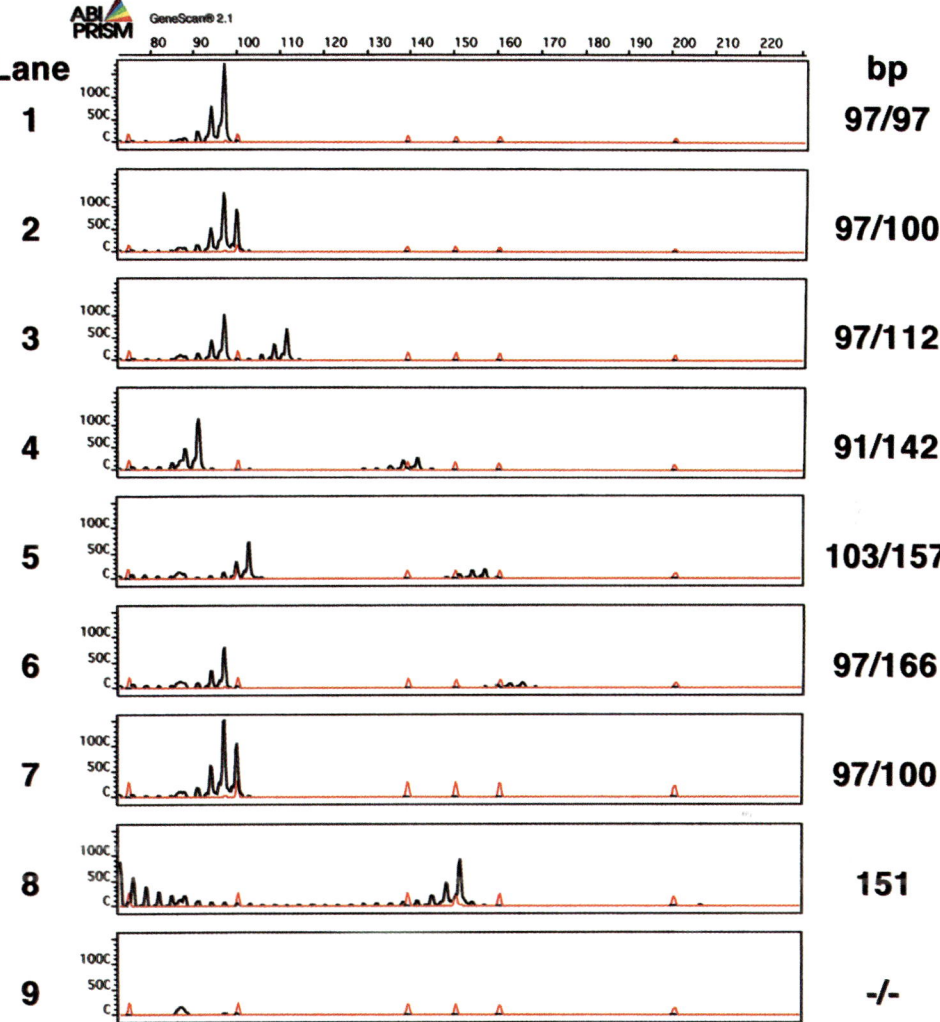

Fig. 2. Electropherogram generated from gel image shown in **Fig. 1.** Corresponding gel lane numbers listed on the left. Amplicon sizes in bp listed on the right.

Table 1
Interpretation of CAG Repeat Number Based on Size of Amplicon

Fragment size	88	91	94	97	100	103	106	109	112
CAG repeat #	14	15	16	17	18	19	20	21	22
Fragment size	115	118	121	124	127	130	133	136	139
CAG repeat #	23	24	25	26	27	28	29	30	31
Fragment size	142	145	148	151	154	157	160	163	166
CAG repeat #	32	33	34	35	36	37	38	39	40
Fragment size	169	172	175	178	181	184	187	190	193
CAG repeat #	41	42	43	44	45	46	47	48	49

References

1. Huntington's Disease Collaborative Research Group (1993) A novel gene containing a trinucleotide repeat that is expanded and unstable on Huntington's disease chromosomes. *Cell* **72,** 971–983.
2. Cummings, C. J. and Zoghbi, H. Y. (2000) Fourteen and counting: unraveling trinucleotide repeat diseases. *Hum. Mol. Gene.* **9,** 909–916.
3. Reiner, A., Albin R. L., Anderson K. D., et al. (1988) Differential loss of striatal projection neurons in Huntington disease. *Proc. Natl. Acad. Sci. USA* **85,** 5733–5737.
4. Davies, S. W., Turmaine M., Cozens, B. A., et al. (1997) Formation of neuronal intranuclear inclusions underlies the neurological dysfunction in mice transgenic for the HD mutation. *Cell* **90,** 537–548.
5. Strong, T. V., Tagle, D. A., Valdes, J. M., et al. (1993) Widespread expression of the human and rat Huntington's disease gene in brain and nonneural tissues. *Nat. Gene.* **5,** 259–265.
6. Meissen, G. J., Myers, R. H., Mastromauro, C. A., et al. (1988) Predictive testing for Huntington's disease with use of a linked DNA marker. *N. Engl. J. Med.* **318,** 535–542.
7. Duyao, M., Ambrose, C., Myers, R., et al. (1993) Trinucleotide repeat length instability and age of onset in Huntington's disease. *Nat. Gene.* **4,** 387–392.
8. McNeil, S. M., Novelletto, A., Srinidhi, J., et al. (1997) Reduced penetrance of the Huntington's disease mutation. *Hum. Mol. Gene.* **6,** 775–779.
9. Rubinsztein, D. C., Leggo, J., Coles, R., et al. (1996) Phenotypic characterization of individuals with 30–40 CAG repeats in the Huntington Disease (HD) gene reveals HD cases with 36 repeats and apparently normal elderly individuals with 36–39 repeats. *Am. J. Hum. Gene.* **59,** 16–22.
10. Goldberg, Y. P., McMurray, C. T., Zeisler J., et al. (1995) Increased instability of intermediate alleles in families with sporadic Huntington disease compared to similar sized intermediate alleles in the general population. *Hum. Mol. Gene.* **4,** 1911–1918.
11. Warner, J. P., Barron, L. H. Brock, D. J. H., et al. (1993) A new polymerase chain reaction (PCR) assay for the trinucleotide repeat that is unstable and expanded on Huntington's disease chromosomes. *Mol. Cell. Probes* **7,** 235–239.
12. (1994) Guidelines for the molecular genetics predictive test in Huntington's disease. International Huntington Association (IHA) and the World Federation of Neurology (WFN) Research Group on Huntington's Chorea. *Neurology* **44,** 1533–1536.

III

SEQUENCE-BASED MUTATION DETECTION

11

Molecular Detection of Galactosemia Mutations by PCR-ELISA

Kasinathan Muralidharan and Wei Zhang

1. Introduction

1.1. Galactosemia

Classic galactosemia is an autosomal recessive disorder caused by the deficiency of galactose 1-phosphate uridyltransferase, GALT (EC 2.7.7.12) *(1)*. It presents with vomiting and diarrhea in neonates within a few days of milk intake. Most patients develop jaundice and hepatic failure. If untreated, it is potentially lethal. Newborn screening for galactosemia is routine in all states in the USA and in many other countries. Elimination of dietary galactose is the main treatment. However, this does not prevent secondary complications such as growth retardation, mental retardation, dyspraxia, cataracts, ataxia, and ovarian failure later in life *(2)*.

The main dietary source of galactose in humans is from lactose in milk. Lactose is hydrolyzed to glucose and galactose in the intestine. Galactose-1-phosphate is produced from galactose by the action of galactokinase. Galactose 1-phosphate uridyltransferase (GALT) catalyzes the conversion of uridine diphosphoglucose and galactose-1-phosphate to glucose-1-phosphate and uridine diphosphogalactose. Absence of GALT activity leads to intracellular accumulation of galactose 1-phosphate.

1.2. Mutations in GALT Gene

Over 130 mutations have been identified in human GALT gene *(3)*. These range from missense, nonsense, and splice site mutations to small and large deletions.

The most common classic galactosemic (G) allele is the Q188R mutation. A glycine to arginine change at codon 188 is caused by an A to G missense mutation. It is present in all ethnic groups to varying extents. It is the most common mutation in white galactosemic patients in the United States.

K285N is the second most common among white patients and is most prevalent among patients derived from the area of Southern Germany, Croatia, and Austria. K285N is owing to G to T change that converts a lysine to asparagine at codon 285.

Y209C mutation is due to A to G change that substitutes cystine for tyrosine at codon 209. L195P is caused by a T to C change resulting in leucine to proline at codon 195. These two mutations account for about 2% of white galactosemic patients.

S135L mutation is the most common G allele in blacks in the US. A C to T change substitutes a leucine instead of serine at codon 135. It has so far been exclusively found in African Americans *(4)*. Q188R is the next most common G allele in the black population.

The Duarte variant (N314D) is found in all populations ranging from 6–20% of the defective alleles. This is caused by an A to G transition that leads to an asparagine to aspartate at codon 314. Duarte variant shows about a 25% reduction in enzymatic activity.

The six common mutations (Q188R, N314D, S135L, K285N, L195P, and Y209C) account for 87.5% of the mutant alleles in US galactosemic population *(3)*. We routinely test for these mutations in our galactosemia patients. The other mutations occur at a low frequency and are detectable by single-strand conformation polymorphism analysis and sequencing *(5)*.

1.3. PCR-ELISA

The method described here is based on allele-specific oligonucleotide (ASO) hybridization. It can simultaneously detect the six common mutations in GALT gene. This method is rapid and suitable for batch analysis of large number of samples.

Three regions of the GALT gene that harbor these six mutations are amplified by multiplex polymerase chain reaction (PCR). The PCR product is labeled with Digoxigenin (DIG) by the incorporation of DIG labeled nucleotides in the PCR reaction. The DIG moiety is later used to detect the presence of hybrids with sequence specific oligonucleotide capture probes.

Figure 1 illustrates the various steps involved in the enzyme-linked immunosobent assay (ELISA) detection of GALT mutations. The PCR product is denatured and hybridized to biotinated sequence-specific oligonucleotide probes. A separate hybridization reaction is used to detect the normal and mutant sequence at each of the loci. The specificity of the hybridization is optimized by the design of oligonucleotides and the temperature of hybridization. Further, nonbiotinylated blocking oligos are used to reduce background hybridization. An oligonucleotide containing the normal sequence is used as a blocking agent in the case of hybridization to detect the mutation and vise versa. Blocking oligos compete with specific oligos for hybridization and reduce nonspecific hybridization. Further, in a heterozygote sample, cross-hybridization of normal sequence to mutant detection probe is reduced by the specific hybridization with the normal blocking oligo. Because the specific oligo is biotinylated, only the specific hybrids are detectable in subsequent steps. The hybrids formed with blocking oligos are not detectable. The hybrids with biotinylated oligos are then immobilized on streptavidin-coated microtiter plates. Unbound hybrids and PCR products are washed away. The plates are then treated with anti-DIG antibodies conjugated with peroxidase. These antibodies react to the DIG on the PCR product bound and immobilized by the specific oligonucleotide. Unbound antibodies are washed away. The presence of hybrids is detected by activity of peroxidase enzyme on a colorimetric substrate.

2. Materials

2.1. Reagents for DNA isolation (6)

1. TKM1: 10 mM Tris-HCl, pH 7.6. 10 mM KCl, 10 mMMgCl$_2$, 2 mM EDTA, and 2.5% Nonidet P-40.

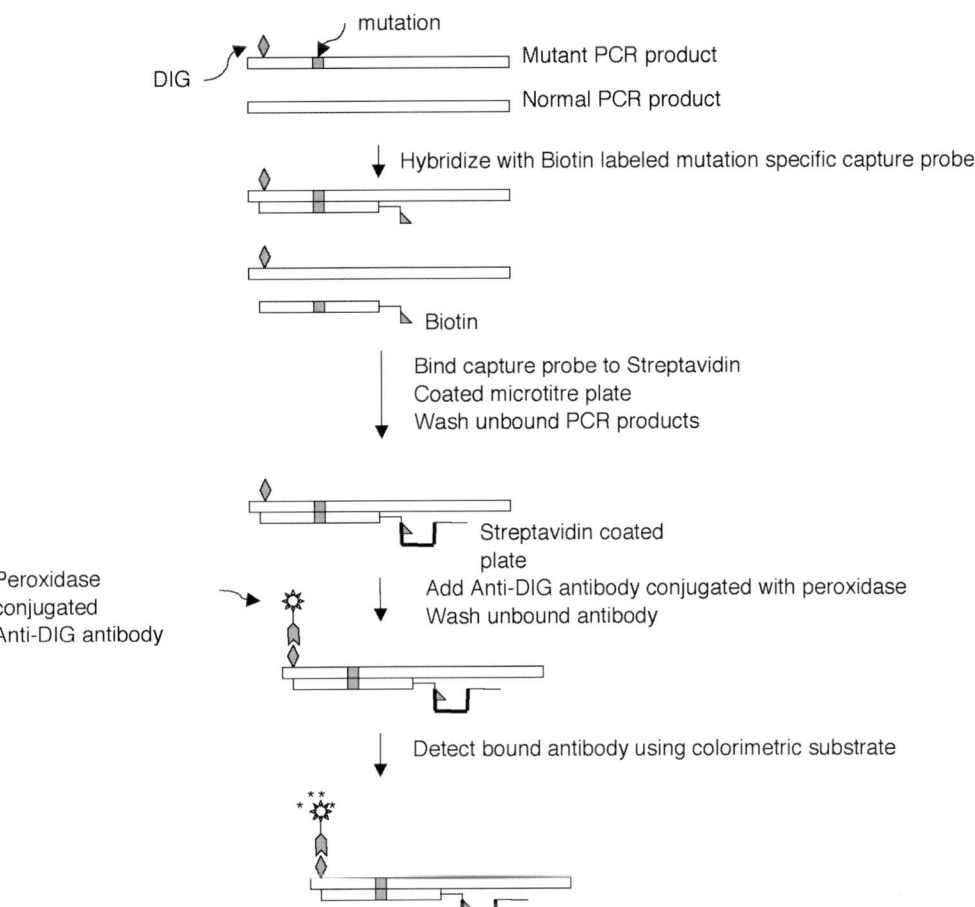

Fig. 1. PCR-ELISA detection of GALT gene mutations. Three regions of the GALT gene are amplified by multiplex PCR. The PCR product is labeled with Digoxigenin (DIG) by the incorporation of DIG-labeled nucleotides in the PCR reaction. The PCR product is denatured and hybridized to biotinylated sequence-specific oligonucleotide probe. A separate hybridization reaction is used to detect the normal and mutant sequence at each of the loci. The specificity of the hybridization is optimized by the design of oligonucleotides and the temperature of hybridization. Further, nonbiotinylated blocking oligonucleotides are used to reduce background hybridization. The hybrids with biotinylated oligonucleotides are then immobilized on streptavidin-coated microtiter plates. Unbound hybrids and PCR products are washed away. The plates are then treated with anti-DIG antibodies conjugated with peroxidase. Unbound antibodies are washed away. The presence of allele specific hybridization is detected by activity of peroxidase enzyme on a colorimetric substrate.

2. TKM2: 10 mM Tris-HCl, pH 7.6. 10 mM KCl, 10 mM MgCl$_2$, 2 mM EDTA, 0.4 M NaCl, 0.05% sodium dodecyl sulfate (SDS).
3. 6 M NaCl in water.
4. Absolute Ethanol.

Table 1
PCR Primers

Primer Sequence (5'→3')	Mutation	Size of PCR product
GALT 9: GGT CAG CAT CTG GAC CCC AGG **INJR:** GGG GTC GAC GCC TGC ACA TAC TGC ATG TGA	K285N, Y209C, N314D	668
GALT6: AGG AGG GAG TTG ACT TGG AGT **IN7R:** GGG GAC ACA GGG CTT GGC TCT CTC CCA	Q188R, L195P	428
S135LF: GAG TGA TAC TCC TTT ACC TCA GGA CCC AGT **S135LR:** GGA CCG ACA TGA GTG GCA GCG TTA CAT TC	S135L	252

2.2. PCR Amplification

1. Roche Biochemicals PCRELISA DIG labeling kit (Cat. no. 1636 120) may be used for the DIG labeling of PCR products (*see* **Note 1**).
2. PCR primers: The primer sequences are provided in **Table 1**. Stock solutions at 10 pmol/ µL in water are used.
3. Taq DNA polymerase (5 U/ µL) (Roche Molecular Biochemicals).

2.3. Reagents for DIG ELISA

1. PCR ELISA detection kit (Cat. no. 1636111) was purchased from Roche Biochemicals. This kit provides streptavidin-coated microtiter wells, the reagents necessary for hybridization, washing, anti-DIG antibodies, and substrate for detection. Kit components are diluted and used according to the manufacturers instructions. Storage conditions of diluted anti-DIG antibodies and substrate are critical. Control DIG-labeled PCR product and biotin-labeled capture probe are provided with the kit. These can be useful in troubleshooting (*see* **Note 2**).
2. Biotin-labeled detection oligonucleotides. The sequences of the oligos are provided in **Table 2**.
3. Blocking oligos: Blocking oligos are used in the hybridization reaction to enhance specificity. For example in the hybridization reaction for the detection of the 188R mutation, an oligonucleotide corresponding to this mutation is biotin labeled and nonbiotin labeled oligo corresponding to the wild-type 188Q is used to block the nonspecific binding of 188R to 188Q target sequence (*see* **Note 2c**).

3. Methods

3.1. DNA Isolation (6)

1. Collect whole blood in a Vacutainer tube (purple-stopper) containing EDTA as anticoagulant.
2. Transfer 5 mL of blood into a 15-mL centrifuge tube and add 5 mL of TKM1. Mix well.
3. Centrifuge at 1200–1500*g* for 10 min at room temperature in a tabletop centrifuge.
4. Slowly pour off the supernatant and save the nuclear pellet
5. Wash the pellet in 5 mL of TKM1 buffer and centrifuge as in step 3.
6. Gently resuspend the pellet in 0.8 mL of TKM2. Mix thoroughly by pipetting back and forth. Place on rocker overnight or at 55°C for 30 min.
7. Add 0.3 mL of 6 *M* NaCl to the tube and mix well.

Table 2
Oligonucleotides for Hybridization[a]

Probe name	Sequence 5'→3'	Allele
Q188RN3	CCC CCA CTG CCA GGT AAG GGT	Q188 (normal)
Q188RM3	CCC CCA CTG CCG GGT AAG GGT	R188 (mutant)
N314DN2	GGG CCA ACT GGA ACC ATT GGC A	N314 (normal)
N314DM2	GGG CCA ACT GGG ACC ATT GGC A	D314 (mutant)
S135LN2	CTT CCA CCC CTG GTC GGA TGT A	S135 (normal)
S135LM	CTT CCA CCC CTG GTT GGA TGT A	L135 (mutant)
L195PN	AGC AGT TTC CTG CCA GAT ATT G	L195 (normal)
L195PM	AGC AGT TTC CCG CCA GAT ATT G	P195 (mutant)
Y209CN	CAG CAG GCC TAT AAG AGT CAG	Y209 (normal)
Y209CM	CAG CAG GCC TGT AAG AGT CAG	C209 (mutant)
K285NN	TCT TGA CCA AGT ATG ACA ACC T	K285 (normal)
K285NM	TCT TGA CCA ATT ATG ACA ACC T	N285 (mutant)

[a] Oligonucleotides when used as sequence specific detection probes are end-labeled with Biotin. Oligonucleotides used for blocking nonspecific binding in hybridization reactions are not labeled with biotin.

8. Centrifuge at 1200–1500g for 10 min.
9. Transfer the supernatant using a transfer pipet to a new tube. Add 2.2 mL absolute ethanol. Mix. Spool DNA precipitate to a wand.
10. Rinse DNA precipitate in 70% ethanol. Air dry.
11. Suspend DNA in 10 mM Tris-HCl, 1 mM EDTA, and pH 8.0.

3.2. Multiplex PCR

1. For each assay, include water blank, positive and negative controls.
2. Each PCR reaction has a final volume of 50 µL. Each reaction contains 8.0 µL dNTP with DIG UTP, 5 µL 10XPCR buffer, 5 µL of MgCl$_2$ (10 mM stock), 1.5 µL each of the three forward and three reverse primers, 5 µL of DNA (concentration of 30 ng/µL) and 0.25 µL Taq polymerase. Water is added to bring the total volume to 50 µL.
3. The PCR thermal cycling parameters included an initial denaturation step of 94°C for 5 min, 35 cycles each of 45 s of denaturation at 94°C, 1 min annealing at 65°C, and 1 min extension at 72°C. The final extension at 72°C was for 3 min.

3.3. Allele Specific Hybridization and ELISA

1. Alkali denaturation of PCR products: Add 3 µL of denaturation buffer from the PCR ELISA Detection kit to 3 µL of PCR product. Incubate at room temperature for 10 min. Add 19 µL of ddH$_2$O to each tube.
2. Stock probe solutions are prepared according to the concentrations given in **Table 3** (*see* **Note 2c**).
3. Prepare hybridization probe. Each sample requires 2 µL of probe stock mixed with 28 µL of hybridization buffer.
4. Each hybridization reaction consists of 150 µL of hybridization solution mixed with 30 µL of the above probe solution and 20 µL of denatured PCR product. The hybridization is performed in the streptavidin-coated microtiter plates. The plates are covered with self-adhesive cover foils. Twelve wells are needed for each sample to assay for the 6 mutations and the corresponding normal sequence.

Table 3
Hybridization Probes

Allele detected	Biotin-labeled probe (0.75 pico mole)	Blocking Oligonucleotide (unlabeled)	Blocking Oligo (pico mole)
Q188 (normal)	BQ188RN3	Q188RM3	0.19
R188 (mutant)	BQ188RM3	Q188RN3	3.0
N314 (normal)	BN314DN2	N314DM2	7.5
D314 (mutant)	BN314DM2	N314DN2	16.5
S135 (normal)	BS135LN2	S135LM	0.75
L135 (mutant)	BS135LM	S135LN2	0.75
L195 (normal)	BL195PN	L195PM	7.5
P135 (mutant)	BL195PM	L195PN	7.5
Y209 (normal)	BY209CN	Y209CM	7.5
C209 (mutant)	BY209CM	Y209CN	7.5
K285 (normal)	BK285NN	K285NM	3.0
N285 (mutant)	BK285NM	K285NN	3.0

5. The plates are placed in a shaking water bath in contact with water. The water bath is set at 65°C. The shaker is set at 50 rpm. After about 15 min, the temperature of the water bath is set to 37°C. The temperature is allowed to drop to 37°C. This takes about 3 h if the lid of the water bath is left open. The lid can be left closed and let the hybridization continue overnight. This method of starting at a high temperature and allowing slow cooling allows for optimal hybridization of multiple probes with different annealing temperatures (*see* **Note 2d**).
6. Pour off the hybridization solution. Wash the plate by adding 250 µL of wash solution to each well and shaking out the solution. This is repeated three times.
7. Add 200 µL of anti-DIG-antibody working stock solution to each well. Incubate in a waterbath at 37°C with shaking (*see* **Note 2e**).
8. Wash three times as in **step 6**.
9. Add 200 µL substrate solution and incubate at 37°C. Color development will be evident in 10–30 min. The results can be scored by visually as positive or negative for color reaction. However, colorimetric reading of the plates provides a more secure interpretation. The plate is read on a microtiter plate reader at 405 nm (**Fig. 2**).
10. Interpretation: The readings from the water blank must be less than 0.1 OD units. The signal from specific hybridization must be more than three times that in the blank. If the background is high, refer to **Note 3**. The negative and positive controls should behave as expected. Genotype of a sample is scored as normal at a locus if no hybridization is detected with the mutant probe. If hybridization is detected only in the mutant probe and not in the normal probe, the individual is homozygous for that mutation. If hybridization is detected with both normal and mutant probes, the individual is heterozygous for that mutation. Individuals heterozygous for two mutations are compound heterozygotes (*see* **Notes 2–4**).

4. Notes

1. Instead of using the DIG-PCR labeling kit, one can purchase DIG labeled UTP (Roche Biochemicals, Digoxigenein-11-2'-deoxy-uridine triphosphate [alkali stable], Cat. no. 1093 088) and use it with a regular PCR kit. DIG-UTP is used at 0.1 mM in a dNTP solution containing 2 mM each of dATP, dGTP, and dCTP and 1.9 mM dTTP.

Galactosemia Mutations

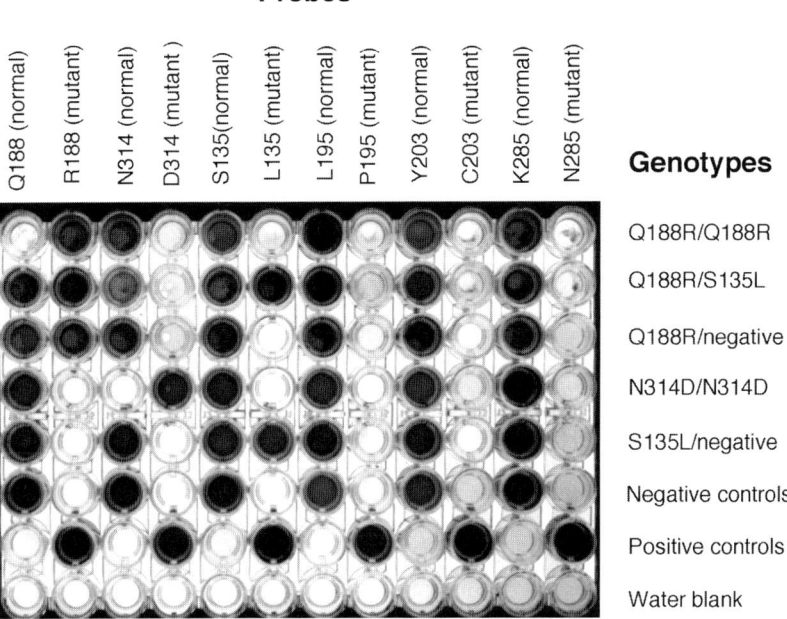

Fig. 2. PCR-ELISA detection of GALT gene mutations. Photograph of an ELISA plate illustrating the results obtained by OLA hybridization. Each horizontal row corresponds to a patient. Each column corresponds to the biotinylated probe used for hybridization. Genotype of a sample is scored as normal at a locus if no hybridization is detected with the mutant probe. If hybridization is detected only in the mutant probe and not in the normal probe, the individual is homozygous for that mutation. If hybridization is detected with both normal and mutant probes, the individual is heterozygous for that mutation. Individuals heterozygous for two mutations are compound heterozygotes.

2. Low or no signals in all reactions: This could be owing to problems in either amplification of target sequences or in hybridization and detection steps.
   a. To determine if the problem is with PCR, check PCR products by electrophoresis in agarose gel. Poor DNA quality or low DNA quantity is the common cause of poor PCR amplification. This can be corrected by improving the quality and quantity of DNA. Insufficient mixing of the sample and reagents is a common problem.
   b. If PCR amplification is normal, check reagents used for denaturation, hybridization, and detection. Make sure appropriate reagents used for denaturation and hybridization.
   c. The concentration of blocking oligonucleotide influences sensitivity and specificity of the hybridization. It can be adjusted up or down to adjust the signal to noise ratio.
   d. Temperature is a critical component of hybridization. Check temperature of the water bath using an external thermometer and adjust temperature settings if necessary.
   e. Diluted anti-DIG antibodies and substrate solutions have limited viability. Store these reagents according to kit instructions. Make fresh dilution of antibodies and substrate if the signals are low or absent in positive hybridization controls.
3. High reading in blanks can be caused by low stringency of the hybridization reaction. Check concentrations of the hybridization and detection reagents, and temperature of the water baths.

4. False-negatives and false-positives: Diagnostic methods based on PCR and ASO hybridization are susceptible to fail due to unexpected changes in the intended primer and oligonucleotide binding sites.

   Polymorphisms, mutations, or deletions in the PCR primer binding sites can result in poor PCR amplification or fail to amplify an allele. This can result in falsely scoring the other allele as being homozygous.

   Polymorphisms and other mutations in the binding site for the detection oligo will result in lack of hybridization. In such a case, that allele is not detected. This results in falsely scoring the other allele of the locus as being homozygous.

Acknowledgments

The authors wish to thank Dr. Louis J. Elsas for support and encouragement on this project.

References

1. Segal, S., Berry, G.T. (1995) Disorders of galactose metabolis, in The Metabolic Basis of Inherited Disease 1, Scriver, D., Beaudet, A., Sly, W., Valle, D. (eds.), McGraw-Hill, New York, USA, pp. 967–1000.
2. Segal, S. (1995) Galactosemia unsolved. *Eur. J. Pediatr.* **154,** S97–102.
3. Elsas, L. J., 2nd and Lai, K. (1998) The molecular biology of galactosemia. *Genet. Med.* **1,** 40–48.
4. Lai, K., Langley, S. D., Singh, R. H., Dembure, P. P., Hjelm, L. N., and Elsas, L. J., 2nd. (1996) A prevalent mutation for galactosemia among black Americans. *J. Pediatr.* **128,** 89–95.
5. Elsas, L. J., Langley, S., Steele, E., Evinger, J., Fridovich-Keil, J. L., Brown, A., et al. (1995) Galactosemia: a strategy to identify new biochemical phenotypes and molecular genotypes. *Am. J. Hum. Genet.* **56,** 630–639.
6. Lahiri, D. K. and Nurnberger, J. I., Jr. (1991) A rapid non-enzymatic method for the preparation of HMW DNA from blood for RFLP studies. *Nucleic Acids Res.* **19,** 5444.

12

Denaturing High-Performance Liquid Chromatography and Sequence Analyses for *MECP2* Mutations in Rett Syndrome

Inge M. Buyse and Benjamin B. Roa

1. Introduction

The detection of abnormal variations in a DNA sequence is a challenge for genetic research and clinical diagnostic applications. Among the different mutation detection methods currently available, DNA sequence analysis is largely considered the gold standard because it provides complete information about the nature and location of a particular sequence variant. The development of polymerase chain reaction (PCR)-based direct sequencing protocols using automated fluorescence detection systems has greatly facilitated DNA sequencing. In spite of numerous improvements in sequencing technologies, however, full-scale DNA sequence analysis remains relatively costly and labor-intensive in many laboratory settings. Unknown sequence variants in a region of interest may also be detected using a variety of mutation-scanning methods *(1)*. These scanning methods have inherent strengths and limitations, and typically require confirmatory sequence analysis to identify the DNA alteration as a mutation or a non-disease-associated variant. Technical strengths may include speed, ease of use, and lower assay costs, but the critical limitation for most scanning methods involves a suboptimal mutation-detection rate.

Denaturing high-performance liquid chromatography (DHPLC) is a very sensitive PCR-based mutation scanning method for detecting nucleotide variants and small insertions/deletions in an efficient and automated manner *(2–4)*. DHPLC relies on the principle of DNA duplex separation by ion-pair reverse-phase liquid chromatography under partially denaturing conditions. DNA regions of interest are initially amplified by PCR, and the resulting PCR products are then subjected to heat denaturation and re-annealing. This induces heteroduplex formation between complementary wild-type and variant strands of PCR products derived from heterozygous DNA templates. In contrast, perfectly matched homoduplex PCR products result from homozygous wild-type or homozygous variant DNA templates. Heteroduplex DNA molecules exhibit differential elution profiles compared to homoduplex DNA upon analysis on a DHPLC column under partially denaturing conditions (*see* **Note 1**). The design of DHPLC assay conditions is facilitated by computer software analysis of the DNA sequence of interest

(Stanford DHPLC, Website: http://insertion.stanford.edu; Wavemaker Utility Software by Transgenomic, Inc.). The software analyzes the melting profile of a given DNA sequence to aid in PCR design, and recommends assay parameters such as buffer gradients and analytical temperatures for DHPLC analysis. The predicted DHPLC assay conditions can be optimized empirically by analyzing positive controls for each fragment at different temperatures, within an interval above and below the predicted melting temperature (*see* **Note 3**). The use of multiple analytical temperatures is especially important if the sequence contains more than one melting domain (*see* **Note 4**). Certain features should be considered in designing PCR primers for DHPLC analysis (*see* **Notes 5–7**). Major practical advantages of DHPLC technology include high sensitivity and process automation. Identification of heterozygous sequence variants is the most straightforward application of DHPLC, which is amenable to detection of homozygous, hemizygous, and low-level nucleotide variants as well (*see* **Notes 8–10**). Multiple studies have documented DHPLC detection rates of 95–100% as applied to different disease genes such as phosphomannomutase 2 (carbohydrate-deficient glycoprotein syndrome type IA [CDGS type 1A]) *(5)*, protein C (protein C deficiency) *(6)*, Tuberous sclerosis genes *TSC1 (7)* and *TSC2 (8)*, *hMLH1* (hereditary nonpolyposis colon cancer, HNPCC) *(9)*, and Factor VIII (hemophilia A) *(10)*. Our laboratory's experience with DHPLC technology involves clinical diagnostic testing for mutations in the *MECP2* gene causing Rett syndrome *(11)*.

Rett syndrome (RTT) is an X-linked dominant neurodevelopmental disorder that is a leading genetic cause of mental retardation in females, with a prevalence of ~1/10,000–1/15,000 girls *(12–14)*. Patients appear to develop normally until 6–18 mo of age, followed by regression of motor and language development. Typical features include loss of speech and purposeful hand use, development of stereotypic hand movements, microcephaly, intermittent hyperventilation, ataxia, seizures, and autistic features. The condition subsequently stabilizes with patients usually surviving into adulthood *(13)*. Atypical variants of RTT range from a severe phenotype with absence of early normal development, congenital hypotonia, and infantile spasms, to the milder "forme fruste" phenotype with less severe regression, milder mental retardation *(15)*, and preserved speech in some cases *(16)*.

The Rett syndrome gene was mapped to chromosome Xq28 and identified as *MECP2*, which encodes methyl-CpG-binding protein 2 *(17)*. MeCP2 is a ubiquitously expressed 486 amino acid nuclear protein that functions as a transcriptional repressor and contains the functionally important methyl-binding domain (MBD) and the transcriptional repression domain (TRD) *(18–20)*. The *MECP2* gene contains 4 exons, of which exons 2–4 contribute to the coding region (20). *MECP2* mutations have been identified in up to 80% of classic sporadic RTT patients *(17,21,22)*. Most mutations occur *de novo* on the paternally inherited X chromosome *(23)*. Different types of *MECP2* mutations include frameshift, nonsense, missense, and splice-site mutations. Truncating frameshift mutations consist of insertions, deletions, or complex insertion/deletions mostly in *MECP2* exon 4, which contains quasi-palindromic sequences that predispose the region to DNA rearrangements.

Mutations in *MECP2* can cause a variable spectrum of clinical phenotypes *(22,24–26)*. The X-inactivation (XCI) pattern is a modulating factor of the phenotype *(22,24,27)*. Although RTT was long considered to be a male-lethal disorder, *MECP2* mutations have been identified in clinically affected males with variable phenotypes

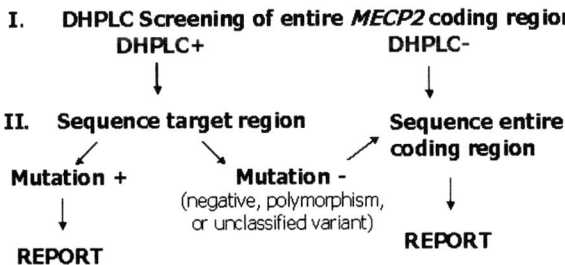

Fig. 1. Illustration of the two-tiered diagnostic strategy for *MECP2* testing in female Rett patients. Female patients are initially screened for variants in the *MECP2* coding region by DHPLC analysis. DHPLC positive samples are further analyzed by targeted sequencing analysis to identify the specific heterozygous mutation. Samples whose sequence variant proves to be a polymorphism or an uncharacterized variant, and for samples that are DHPLC test negative, the entire *MECP2* coding region is reamplified by PCR and sequenced in both directions.

(22,27–31). Moreover, *MECP2* mutations were found in a significant proportion of patients with uncharacterized mental retardation, including patients who were previously diagnosed or evaluated for Angelman syndrome *(32) (33)*. Collectively, these data support *MECP2* mutation analysis in male and female patients with unexplained neurodevelopmental disorders and mental retardation. Given the spectrum of clinical phenotypes attributed to *MECP2* mutations, the utility of a highly sensitive molecular diagnostic strategy is evident.

Our laboratory has developed a two-tiered DNA testing strategy for Rett syndrome that combines DHPLC analysis (using a Transgenomic WAVE Nucleic Acid Fragment Analysis system) and automated fluorescent DNA sequencing (*see* flowchart in **Fig. 1**; *see* **Note 1**). The strategy involves initial DHPLC screening for heterozygous sequence variants in the *MECP2* coding region of female patients. DNA samples that test positive by DHPLC are re-amplified for the target region, and the specific PCR products are sequenced in both forward and reverse orientations to identify the change as a mutation or polymorphism. Patient DNA samples whose sequence change is not a definitive mutation (i.e., a polymorphism or unclassified variant), as well as samples that test negative by DHPLC analysis, are subsequently re-amplified by PCR and sequenced for the entire *MECP2* coding region (**Fig. 1**). This two-tiered diagnostic approach minimizes the risk for both false-positives and false-negatives, and has proven to be highly sensitive, robust, and efficient *(11)*. The combination of DHPLC and sequence analysis under optimized conditions provides a powerful tool for heterozygote detection in the molecular diagnostic laboratory setting, where the challenge is to maximize assay sensitivity. In terms of efficiency, screening for *MECP2* variants by DHPLC prior to targeted sequence analysis reduced the need for full gene sequencing in as much as 40% of our initial RTT caseload. In the case of male patients, bi-directional sequencing is performed on the entire *MECP2* coding region to detect hemizygous mutations in the *MECP2* gene (*see* **Note 8**). Overall, DHPLC is a powerful tool for diagnostic testing, particularly when utilized in conjunction with DNA sequence analysis (*see* **Notes 11** and **12**). Our protocol and observations concerning the strengths and limitations of DHPLC and sequencing technology are outlined in this chapter.

2. Materials

2.1. DNA Extraction from Whole Blood

1. Puregene DNA isolation kit (Gentra Systems Inc.).
2. Isopropanol.
3. 70% Ethanol.

2.2. PCR Amplification and Agarose Gel Electrophoresis

1. *Taq* DNA polymerase.
2. 10X PCR buffer: 500 mM KCl, 200 mM Tris-HCl, pH 8.4, 15 mM MgCl$_2$.
3. 2.5 mM PCR grade dideoxynucleotide triphosphate (dNTP) mix.
4. Oligonucleotide primers (**Tables 1** and **2**), 6.25 µM.
5. Agarose.
6. 10X TBE buffer: 890 mM Tris-borate, 20 mM EDTA.
7. 6X gel loading buffer: 0.25% bromophenol blue, 0.25% xylene cyanol FF, 15% Ficoll.
8. Ethidium bromide, 10 mg/mL. This is a mutagen and may be carcinogenic. It should be handled wearing gloves and should be disposed of properly.

2.3. DHPLC

1. Ion pairing reagent: 2 M triethylammoniumacetate (TEAA) in water, pH 7.4 (Transgenomic, Inc.).
2. HPLC-grade acetonitrile (Transgenomic, Inc.). Acetonitrile is toxic and needs to be handled wearing gloves.
3. Buffer A: 0.1 M TEAA; Buffer B: 0.1 M TEAA/25% acetonitrile; Buffer C: 8% acetonitrile; Buffer D: 75% acetonitrile. Buffers A and B have a shelf life of 1 wk.
4. DHPLC System: WAVE Nucleic Acid Fragment Analysis system (Transgenomic, Inc.), with column cleaning module for rapid column cleaning and re-equilibration, and Wavemaker Utility Software (initially version 3.4, since upgraded to version 4.1; Transgenomic, Inc.).

2.4. Direct Sequencing Analysis

1. QIAquick PCR purification kit (Qiagen).
2. Ethanol.
3. BigDye Terminator Cycle Sequencing Ready Reaction kit (ABI).
4. Sequencing buffer: 80 mM Tris-HCl, pH 8.7, 2 mM MgCl$_2$.
5. Oligonucleotide primers (**Table 2**), 3.2 µM.
6. AGTC Gel Filtration Cartridge (Edge BioSystems).
7. Sequencing analysis: 3700 DNA analyzer (ABI) and Sequencing analysis version 3.6.1 software (ABI).
8. SEQUENCHER version 4.0 software (Gene Codes Corporation).

3. Methods

3.1. DNA Extraction from Whole Blood

1. Whole blood collected in sterile tubes containing sodium EDTA anticoagulant is used for DNA isolation. The blood sample should be less than 5 d old and stored at 4°C.
2. DNA isolation from whole blood using the Puregene kit (Gentra Systems Inc.) is performed according to the manufacturer's instructions. *See* **Note 2** for alternative extraction protocols.
3. The DNA sample is quantitated, and its concentration adjusted to 50 ng/µL. DNA samples are stored at 4°C.

Table 1
PCR Primer Sequences, Buffer Gradients, and Run Temperatures for DHPLC Analysis

Exon	Name	Primer sequence	Buffer B gradient	DHPLC temperature
2	EX2 FWD	TAA GCT GGG AAA TAG CCT AG	55–64%	59°C
	EX2 REV	TTA TAT GGC ACA GTT TGG CA		
3	EX3 FWD	AGG ACA TCA AGA TCT GAG TG	60–69%, 55–64%	61°C, 63°C
	EX3 REV	GGT CAT TTC AAG CAC ACC TG	53–62%, 52–61%	66°C, 67°C
4(A)	EX4A-FWD1	CGC TCT GCC CTA TCT CTG	61–70%, 57–66%	61°C, 64°C
	EX4A-REV	ACA GAT CGG ATA GAA GAC TC	55–64%	66°C
4(B)	EX4B-FWD	CCA CCC AGG TCA TGG TGA TC	58–67%, 56–65%	64°C, 65°C
	EX4B-REV1	TGA GTG GTG GTG ATG GTG GT		
4(C/D)	EX4C-FWD	GGA AAG GAC TGA AGA CCT GT	58–67%, 57–66%	65°C, 66°C
	EX4D-REV	GCT CTC CCT CCC CTC GGT GT		
4(E)	EX4E-FWD	GGA GAA GAT GCC CAG AGG AG	57–66%, 57–66%	58°C, 60°C
	EX4E-REV	CGG TAA GAA AAA CAT CCC CA	55–64%, 54–63%	63°C, 65°C

Table 2
Primer Sequences fro PCR-Direct Sequencing Analysis

Exon		Name	Primer sequence
2	PCR	EX2 FWD	TAA GCT GGG AAA TAG CCT AG
		EX2 REV	TTA TAT GGC ACA GTT TGG CA
	SEQ	EX2 FWD	TAA GCT GGG AAA TAG CCT AG
		EX2 REVSEQ	CTA AAA AAA AAA AAA GGA AGG TTA C
3	PCR/SEQ	EX3 FWD	AGG ACA TCA AGA TCT GAG TG
		EX3 REV	GGT CAT TTC AAG CAC ACC TG
4	PCR	EX4A-FWD2	CGA GTG AGT GGC TTT GGT GA
		EX4E-REV	CGG TAA GAA AAA CAT CCC CA
	SEQ	EX4A-FWD2	CGA GTG AGT GGC TTT GGT GA
		EX4A-REV	ACA GAT CGG ATA GAA GAC TC
		EX4B-FWD	CCA CCC AGG TCA TGG TGA TC
		EX4B-REV2	TGA GTG GTG GTG ATG GTG GT
		EX4C-FWD	GGA AAG GAC TGA AGA CCT GT
		EX4D-REV	GCT CTC CCT CCC CTC GGT GT
		EX4E-FWD	GGA GAA GAT GCC CAG AGG AG
		EX4E-REV	CGG TAA GAA AAA CAT CCC CA

3.2. PCR Amplification

1. PCR amplification of the *MECP2* coding region (exons 2, 3, and 4) is performed for DHPLC and direct sequencing analysis. Exons 2 and 3 are amplified using flanking intronic primers, and the same PCR products are used for DHPLC and sequencing analyses (**Tables 1** and **2**). The large exon 4 is amplified two different ways: 1) four overlapping PCR products spanning the exon 4 coding region are amplified for DHPLC analysis (primers are listed in **Table 1**); 2) one large exon 4 PCR product is amplified as a template for dye-terminator sequencing analysis using nested sequencing primers (**Table 2**).
2. PCR reactions are carried out in 50 µL reaction volumes containing 100 ng genomic DNA, 1X PCR buffer, 0.2 m*M* dNTP, 0.25 µ*M* of each primer, and 1.25 units of *Taq* polymerase.
3. PCR cycling conditions for all exons include an initial denaturation at 94°C for 2 min 30 s, followed by 10 "step-down" cycles of 30 s at 94°C, 30 s at 65°C (decreasing 1.5°C per cycle), and 1 min 45 s at 72°C, followed by 28 cycles of 30 s at 94°C, 30 s at 51°C and 1 min 30 s at 72°C, and a final extension step at 72°C for 15 min.
4. PCR products are stored at 4°C until further use.

3.3. Agarose Gel Electrophoresis

1. PCR products to be used as sequencing templates are analyzed by electrophoresis on a 0.8% agarose TBE gel containing 50 µg/100 mL ethidium bromide. A mixture of 5 µL PCR product and 1 µL 6X gel loading buffer is loaded for agarose gel analysis.
2. The 0.8% gel is electrophoresed in 1X TBE running buffer at 125V for 45 min. The PCR products are visualized by UV light illumination. PCR product quality and fragment sizes are compared to controls to detect deletion or insertion mutations (particularly for *MECP2* exon 4).

3.4. DHPLC

1. DHPLC analysis is performed on a WAVE Nucleic Acid fragment analysis system (Transgenomic, Inc.), with column cleaning module for rapid column cleaning and re-equilibration, according to the manufacturer's recommendations and assay conditions optimized for each fragment of interest. The WAVEMAKER Utility Software (Transgenomic, Inc.) was used to analyze the *MECP2* melting profile, and to predict assay parameters such as buffer gradient values and the fragment melting temperatures. Analytical buffer gradients and temperatures for individual PCR fragments were empirically optimized using available positive controls. Multiple analytical temperatures were implemented for the analysis of individual fragments (*see* **Table 1**).
2. Heteroduplex formation is induced by heat denaturation of PCR products (entire reaction volume) at 94°C for 5 min followed by gradual cooling from 94°C to 25°C over 45 min.
3. Denatured PCR products are loaded on the DHPLC instrument, and specific volumes are injected (8 µL for exon 3 and 4A fragments, 10 µL for all other fragments). PCR products are eluted from the column at a flow rate of 0.9 mL buffer per minute. The buffer A/B gradient runs for 4.5 min and consists of a linear (2% per minute) decrease and increase of Buffer A and buffer B, respectively. Implemented run temperatures and buffer B gradient values for each of the PCR fragments are listed in **Table 1**. Loading and re-equilibration buffer values are 5% higher and 5% lower than the gradient start values for buffer A and buffer B, respectively (**Table 1**).
4. Data analysis is based on visual inspection of the chromatogram and comparison with normal controls included in each run. Heterozygous profiles are detected as distinct elution peaks when compared to homozygous wild-type peaks (*see* **Fig. 2**).

3.5. Direct Sequencing Analysis

1. Sequencing of PCR products is performed using BigDye Terminator reactions (ABI) in both forward and reverse orientations. The sequencing primers used for exons 2, 3, and 4 (4A, 4B, 4C/D and 4E, FWD and REV) are listed in **Table 2**.
2. PCR product templates are purified using the QIAquick PCR purification kit (Qiagen) according to the manufacturer's instructions.
3. Two microliters of each purified PCR product is used as sequencing template in a final reaction volume of 20 µL containing 3.2 pmol primer and 0.5X BigDye Ready Reaction buffer (1:1 of BigDye Terminator Ready Reaction:sequencing buffer).
4. PCR cycling conditions for the dye-terminator sequencing reactions include 25 cycles of 10 s at 96°C, 5 s at 50°C and 4 min at 60°C.
5. Sequencing products are purified using the AGTC Gel Filtration Cartridge (Edge Biosystems) according to manufacturer's instructions.
6. Samples are loaded without further manipulation and analyzed on the 96-capillary ABI 3700 DNA Analyzer (ABI) according to manufacturer's instructions.
7. Sequencing chromatograms are assembled and analyzed using Sequencher 4.0 analysis software (Gene Codes Corporation). Comparison of the aligned patient data (forward and reverse strands) with the wild-type sequence facilitates the identification of sequence variants.

4. Notes

1. DHPLC detection of sequence variants is based on column chromatography resolution of DNA homoduplex and heteroduplex molecules under partially denaturing conditions. Our laboratory developed a two-tiered diagnostic strategy for *MECP2* that uses DHPLC screening for heterozygous variants, coupled with bi-directional DNA sequence analysis to confirm positive and negative results (**Fig. 1**). Representative data for two types of mutations are shown in **Fig. 2**. Normal and mutant DHPLC chromatograms illustrate a heterozygous point mutation in *MECP2* exon 2 (nucleotide change of 316 C-T, predicting an amino acid substitution of R106W), and a heterozygous insertion mutation in *MECP2* exon 4 (418ins4). Sequence confirmation of these respective mutations is also shown in **Fig. 2**. Mutant DHPLC patterns can vary depending on a number of factors. A distinct heteroduplex peak is observed in some cases, while alterations in the profile of the normally occurring peak may be seen in other cases.
2. The DNA samples used for DHPLC analysis should be pure and free of protein and membranous remnants that could irreversibly damage the DHPLC column and decrease peak resolution. Alternatives to the described Puregene DNA extraction protocols include Qiagen blood extraction kits, and standard DNA extraction protocols involving phenol/chloroform extraction and ethanol DNA precipitation. Crude DNA preparations such as alkaline lysates should be avoided. PCR products for use in DHPLC analysis should be free of mineral oil, and should not be subjected to spin column purification.
3. The sensitivity of DHPLC is dependent on assay conditions designed around the melting profile of a specific DNA sequence. The design of PCR amplicons that ideally contain a single melting domain is facilitated by software analysis, which also predicts DHPLC assay conditions such as buffer gradients and analytical temperatures. However, the predicted DHPLC assay conditions should be empirically validated using available positive controls for each fragment.
4. The use of multiple analytical temperatures for each fragment significantly enhances DHPLC sensitivity. PCR fragment sizes are typically restricted for DHPLC analysis (<600 bp), since longer fragments are likely to contain multiple melting domains. In the case of *MECP2*, the large exon 4 was subdivided into 4 overlapping PCR fragments for

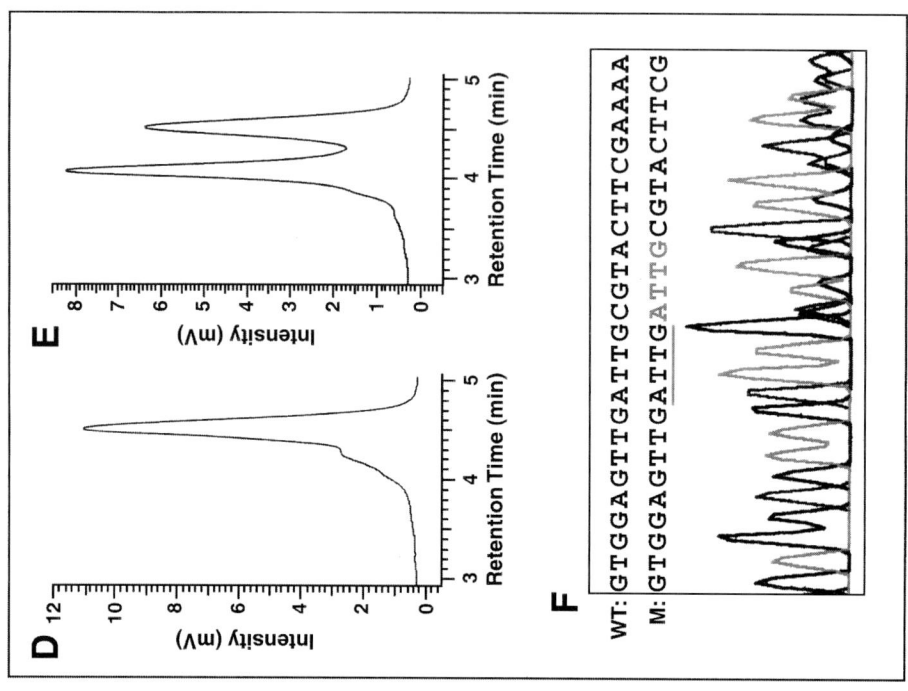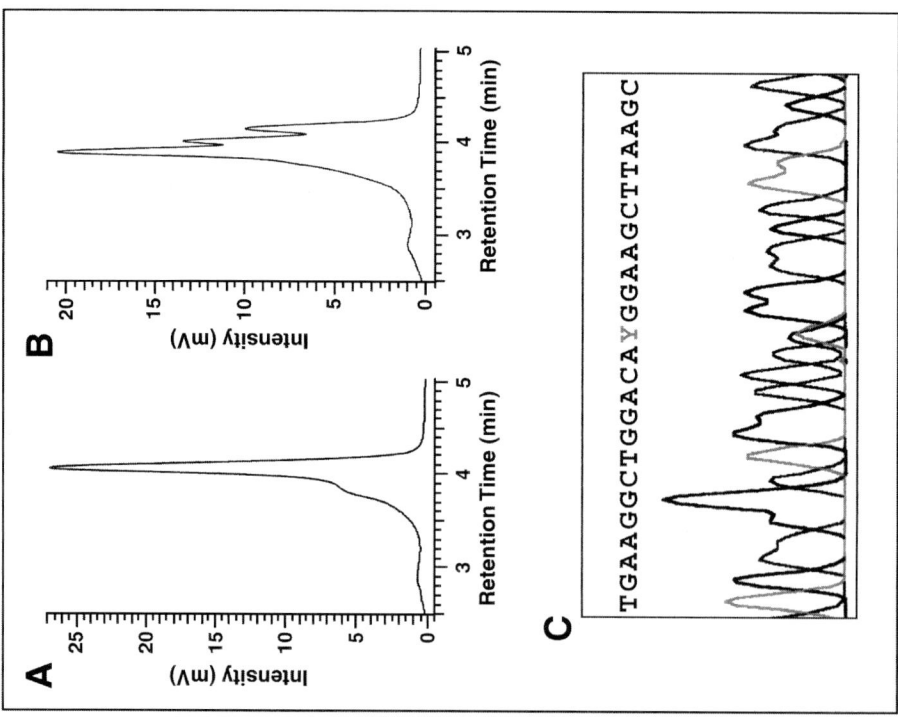

DHPLC analysis (**Table 1**). However, even short fragments can contain more than one melting domain depending on the DNA sequence. Moreover, algorithms for melting profile predictions are not 100% accurate, as exemplified by sequence variants in *MECP2* (*11,34*) and the *MET* proto-oncogene (*35*) that were missed at the software-predicted temperatures.

5. The use of nonmodified PCR primers is preferable for DHPLC analysis. Our initial protocol for *MECP2* analysis was streamlined using primers with universal M13 tails at the 5' end to generate PCR products for both DHPLC and dye-primer sequencing analyses (*11*). Although the DHPLC detection rate was not affected, we noticed a relative broadening of elution peaks for some PCR fragments at particular temperatures. Our current protocol uses nonmodified PCR products, which yield sharper DHPLC elution peaks under similar analytical conditions.

6. PCR primer design should avoid commonly occurring polymorphisms within the amplicon that would yield false-positives on DHPLC analysis. In the case of the *MECP2* gene, four benign polymorphic sites were identified 74, 90, 109, and 115 bp upstream of the intron 3/exon 4 splice junction. Different forward primers were designed for DHPLC and sequence analyses. The primer EX4A-FWD1 (**Table 1**) located 55 bp upstream of exon 4 is used to generate the PCR product for DHPLC (excluding the polymorphisms). A second primer, EX4A-FWD2 (**Table 2**) located 150 bp upstream of exon 4 can be used to generate the PCR template for sequencing (including the polymorphic sites).

7. DHPLC analysis can be supplemented by other strategies to maximize mutation detection. In addition to point mutations, large intragenic deletions in *MECP2* have been reported that can abolish internal primer binding sites in exon 4. To reduce the risk of false-negatives due to PCR non-amplification, we incorporated a long exon 4 PCR product (~1.45 kb). This long PCR fragment is analyzed for size differences by agarose gel electrophoresis, and then used as a template for BigDye-terminator sequencing reactions using a series of nested exon 4 sequencing primers (**Table 2**). The use of distinct PCR amplicons for DHPLC and sequencing analyses adds to the robustness of our diagnostic assay. Indeed, several patients with *MECP2* intragenic deletions were identified using this comprehensive approach.

8. DHPLC is ideal for screening heterozygous variants, such as *MECP2* mutations in Rett syndrome females. With the identification of *MECP2* mutations in affected males, however, male patients are increasingly submitted for testing. Detection of homozygous or hemizygous sequence variants by DHPLC would require mixing PCR products from test samples with known controls. Our current policy is to analyze male DNA samples using bi-directional sequencing, which readily detects hemizygous mutations in *MECP2*, and does not involve mixing of clinical patient samples.

9. DHPLC can be useful for detecting low-level mosaicism. We identified an apparent mosaic nonsense mutation in *MECP2* whose allele representation appears to be detectable by DHPLC but not by standard sequencing analysis (manuscript in preparation). Similar findings were reported for *TSC1* and *TSC2* gene analysis in tuberous sclerosis patients (*36*).

Fig. 2. *(opposite page)* Illustration of representative DHPLC and direct sequencing data for a *MECP2* missense and insertion mutation. (**A**) and (**B**) show DHPLC chromatograms (eluted at 66°C) corresponding to the wild-type exon 3 (**A**) and the 316 C-T mutation (R106W) (**B**). Direct sequencing analysis of the exon 3 PCR fragment confirmed the heterozygous 316 C-T mutation encoding the amino acid substitution of Arginine to Tryptophan, R106W (**C**). (**D**) and (**E**) show DHPLC chromatograms (eluted at 61°C) corresponding to the wild-type exon 4A (**D**) and the 418ins4 mutation (**E**). Direct sequencing analysis confirmed the heterozygous frameshift insertion mutation 418ins4 (**F**). (See color plate 2 appearing in the insert following p. 82)

10. DHPLC analysis can detect heterozygous sequence variants that may not be apparent on sequencing analysis. Sequencing artifacts have been known to occur, particularly in strings of repetitive nucleotides. We identified several unrelated patients with abnormal DHPLC patterns indicative of a heterozygous sequence variant, but whose sequencing chromatograms show only the variant base (and not the wild-type allele). Further investigations confirmed these patients to be heterozygous for the variant allele, consistent with the initial DHPLC data (unpublished observations).
11. Specific sequence variants may exhibit characteristic DHPLC elution profiles. Although DHPLC pattern recognition might be useful for mutation identification, great caution should be exercised in using pattern recognition alone without DNA sequence confirmation. Certain factors such as condition and age of the column, and subtle fluctuations in temperature, could influence the specific DHPLC patterns. In addition, different base substitutions within certain sequence contexts may exhibit similar elution profiles. Confirmatory sequencing is therefore recommended for mutation identification, particularly in a diagnostic setting.
12. Additional features may facilitate DHPLC analysis for different applications. These include hardware features such as UV detection systems, automated fraction collection, high-throughput modes, and improved oven temperature controls. Developments in software analysis may also facilitate automated data analysis of normal and variant DHPLC elution profiles. While no single mutation detection method currently provides 100% detection under all conditions, DHPLC is a powerful mutation scanning method that is applicable to many research and diagnostic settings.

Acknowledgments

We thank Drs. David Stockton, Karla Bowles, and Dani Bercovich for reviewing the manuscript.

References

1. Cotton, R. G. (1997) Slowly but surely towards better scanning for mutations. *Trends Genet.* **13,** 43–46.
2. O'Donovan, M. C., Oefner, P. J., Roberts, S. C., Austin, J., Hoogendoorn, B., Guy, C., et al. (1998) Blind analysis of denaturing high-performance liquid chromatography as a tool for mutation detection. *Genomics* **52,** 44–49.
3. Oefner, P. J. and Underhill, P. A. (1998) DNA mutation detection using denaturing high-performance liquid chromatography (DHPLC), in *Current Protocols in Human Genetics* (Dracopoli, N.C., Haines, J., Korf, B.R., Morton, C., Seidman, C.E., Seidman, J.G., et al. eds.), Wiley & Sons, New York, NY, USA, pp. 7.10.11–17.10.12.
4. Liu, W., Smith, D. I., Rechtzigel, K. J., Thibodeau, S. N., and James, C. D. (1998) Denaturing high performance liquid chromatography (DHPLC) used in the detection of germline and somatic mutations. *Nucleic Acids Res.* **26,** 1396–1400.
5. Erlandson, A., Stibler, H., Kristiansson, B., Wahlstrom, J., and Martinsson, T. (2000) Denaturing high-performance liquid chromatography is a suitable method for PMM2 mutation screening in carbohydrate-deficient glycoprotein syndrome type IA patients. *Genet. Test* **4,** 293–297.
6. Taliani, M. R., Roberts, S. C., Dukek, B. A., Pruthi, R. K., Nichols, W. L., and Heit, J. A. (2001) Sensitivity and specificity of denaturing high-pressure liquid chromatography for unknown protein C gene mutations. *Genet. Test.* **5,** 39–44.
7. Roberts, P. S., Jozwiak, S., Kwiatkowski, D. J., and Dabora, S. L. (2001) Denaturing high-performance liquid chromatography (DHPLC) is a highly sensitive, semi-

automated method for identifying mutations in the TSC1 gene. *J. Biochem. Biophys. Methods* **47,** 33–37.
8. Choy, Y. S., Dabora, S. L., Hall, F., Ramesh, V., Niida, Y., Franz, D., et al. (1999) Superiority of Denaturing High Performance Liquid Chromatography over single-stranded conformation and conformation-sensitive gel electrophoresis for mutation detection in TSC2. *Ann. Hum. Genet.* **63,** 383–391.
9. Holinski-Feder, E., Muller-Koch, Y., Friedl, W., Moeslein, G., Keller, G., Plaschke, J., et al. (2001) DHPLC mutation analysis of the hereditary nonpolyposis colon cancer (HNPCC) genes hMLH1 and hMSH2. *J. Biochem. Biophys. Methods* **47,** 21–32.
10. Oldenburg, J., Ivaskevicius, V., Rost, S., Fregin, A., White, K., Holinski-Feder, E., et al. (2001) Evaluation of DHPLC in the analysis of hemophilia A. *J. Biochem. Biophys. Methods* **47,** 39–51.
11. Buyse, I. M., Fang, P., Hoon, K. T., Amir, R. E., Zoghbi, H. Y., and Roa, B. B. (2000) Diagnostic testing for Rett syndrome by DHPLC and direct sequencing analysis of the *MECP2* gene: identification of several novel mutations and polymorphisms. *Am. J. Hum. Genet.* **67,** 1428–1436.
12. Rett, A. (1966) [On a unusual brain atrophy syndrome in hyperammonemia in childhood]. *Wien. Med. Wochenschr.* **116,** 723–726.
13. Hagberg, B., Aicardi, J., Dias, K., and Ramos, O. (1983) A progressive syndrome of autism, dementia, ataxia, and loss of purposeful hand use in girls: Rett's syndrome: report of 35 cases. *Ann. Neurol.* **14,** 471–479.
14. Hagberg, B. (1985) Rett's syndrome: prevalence and impact on progressive severe mental retardation in girls. Acta Paediatr. Scand. 74, 405–408.
15. Hagberg, B. A. (1989) Rett syndrome: clinical peculiarities, diagnostic approach, and possible cause. *Pediatr. Neurol.* **5,** 75–83.
16. Zappella, M., Gillberg, C., and Ehlers, S. (1998) The preserved speech variant: a subgroup of the Rett complex: a clinical report of 30 cases. *J. Autism. Dev. Disord.* **28,** 519–526.
17. Amir, R. E., Van den Veyver, I. B., Wan, M., Tran, C.Q., Francke, U., and Zoghbi, H. Y. (1999) Rett syndrome is caused by mutations in X-linked *MECP2*, encoding methyl-CpG-binding protein 2. Nat Genet. 23, 185–188.
18. Jones, P. L., Veenstra, G. J., Wade, P. A., Vermaak, D., Kass, S. U., Landsberger, N., et al. (1998) Methylated DNA and *MECP2* recruit histone deacetylase to repress transcription. *Nat. Genet.* **19,** 187–191.
19. Nan, X., Ng, H. H., Johnson, C. A., Laherty, C. D., Turner, B. M., Eisenman, R. N., and Bird, A. (1998) Transcriptional repression by the methyl-CpG-binding protein *MECP2* involves a histone deacetylase complex. *Nature* **393,** 386–389.
20. Reichwald, K., Thiesen, J., Wiehe, T., Weitzel, J., Poustka, W. A., Rosenthal, A., et al. (2000) Comparative sequence analysis of the *MECP2*-locus in human and mouse reveals new transcribed regions. *Mamm. Genome* **11,** 182–190.
21. Dragich, J., Houwink-Manville, I., and Schanen, C. (2000) Rett syndrome: a surprising result of mutation in *MECP2*. *Hum. Mol. Genet.* **9,** 2365–2375.
22. Hoffbuhr, K., Devaney, J. M., LaFleur, B., Sirianni, N., Scacheri, C., Giron, J., et al. (2001) *MECP2* mutations in children with and without the phenotype of Rett syndrome. *Neurology* **56,** 1486–1495.
23. Trappe, R., Laccone, F., Cobilanschi, J., Meins, M., Huppke, P., Hanefeld, F., and Engel, W. (2001) *MECP2* mutations in sporadic cases of rett syndrome are almost exclusively of paternal origin. *Am. J. Hum. Genet.* **68,** 1093–1101.
24. Amir, R. E., Van den Veyver, I. B., Schultz, R., Malicki, D. M., Tran, C. Q., Dahle, E. J., et al. (2000) Influence of mutation type and X chromosome inactivation on Rett syndrome phenotypes. *Ann. Neurol.* **47,** 670–679.

25. Cheadle, J. P., Gill, H., Fleming, N., Maynard, J., Kerr, A., Leonard, H., et al. (2000) Long-read sequence analysis of the *MECP2* gene in Rett syndrome patients: correlation of disease severity with mutation type and location. *Hum. Mol. Genet.* **9,** 1119–1129.
26. Huppke, P., Laccone, F., Kramer, N., Engel, W., and Hanefeld, F. (2000) Rett syndrome: analysis of *MECP2* and clinical characterization of 31 patients. *Hum. Mol. Genet.* **9,** 1369–1375.
27. Wan, M., Lee, S.S., Zhang, X., Houwink-Manville, I., Song, H. R., Amir, R. E., et al. (1999) Rett syndrome and beyond: recurrent spontaneous and familial *MECP2* mutations at CpG hotspots. *Am. J. Hum. Genet.* **65,** 1520–1529.
28. Villard, L., Kpebe, A., Cardoso, A. K., Chelly, P. J., Tardieu, P. M., and Fontes, M. (2000) Two affected boys in a rett syndrome family: clinical and molecular findings. *Neurology* **55,** 1188–1193.
29. Orrico, A., Lam, C., Galli, L., Dotti, M. T., Hayek, G., Tong, S. F., et al. (2000) *MECP2* mutation in male patients with non-specific X-linked mental retardation. *FEBS Lett.* **481,** 285–288.
30. Meloni, I., Bruttini, M., Longo, I., Mari, F., Rizzolio, F., D'Adamo, P., et al. (2000) A mutation in the rett syndrome gene, *MECP2*, causes X-linked mental retardation and progressive spasticity in males. *Am. J. Hum. Genet.* **67,** 982–985.
31. Clayton-Smith, J., Watson, P., Ramsden, S., and Black, G. C. (2000) Somatic mutation in *MECP2* as a non-fatal neurodevelopmental disorder in males. *Lancet* **356,** 830–832.
32. Watson, P., Black, G., Ramsden, S., Barrow, M., Super, M., Kerr, B., and Clayton-Smith, J. (2001) Angelman syndrome phenotype associated with mutations in *MECP2*, a gene encoding a methyl CpG binding protein. *J. Med. Genet.* **38,** 224–228.
33. Couvert, P., Bienvenu, T., Aquaviva, C., Poirier, K., Moraine, C., Gendrot, C., et al. (2001) *MECP2* is highly mutated in X-linked mental retardation. *Hum. Mol. Genet.* **10,** 941–946.
34. Nicolao, P., Carella, M., Giometto, B., Tavolato, B., Cattin, R., Giovannucci-Uzielli, M. L., et al. (2001) DHPLC analysis of the *MECP2* gene in Italian Rett patients. *Hum. Mutat.* **18,** 132–140.
35. Nickerson, M. L., Warren, M. B., Zbar, B., and Schmidt, L. S. (2001) Random mutagenesis-PCR to introduce alterations into defined DNA sequences for validation of SNP and mutation detection methods. *Hum. Mutat.* **17,** 210–219.
36. Jones, A. C., Sampson, J. R., and Cheadle, J. P. (2001) Low level mosaicism detectable by DHPLC but not by direct sequencing. *Hum. Mutat.* **17,** 233–234.

13

Multiplexed Fluorescence Analysis for Mutations Causing Tay-Sachs Disease

Tracy L. Stockley and Peter N. Ray

1. Introduction

Tay-Sachs disease is a severe, neurodegenerative disease fatal in childhood that is caused by deficiency of the enzyme β-hexosaminidase A (Hex A) *(1)*. Tay-Sachs is most common in the Ashkenazi Jewish population, with an incidence of 1/3600 affected individuals and a carrier rate of approx 1 in 30 *(1)*. Owing to the severity of the disease and the high incidence, carrier screening for Tay-Sachs disease has been available to Ashkenazi Jewish individuals since the 1970s, which has greatly reduced the number of affected children in this population. The standard method of carrier testing is by biochemical analysis for reduced Hex A activity in serum *(2)*.

Although biochemical testing for Tay-Sachs is an established population screening method, there are several difficulties with the biochemical screening. One difficulty is the presence of an 'inconclusive' range of Hex A activity, in which it cannot be precisely determined if a person is a carrier or noncarrier of Tay-Sachs *(3)*. Another difficulty is that women who are pregnant or using birth control pills can show an artificially reduced Hex A level when serum is tested, necessitating the use of biochemical testing of leukocytes for these women, which may also produce an inconclusive result *(2,3)*. Inaccurate Hex A levels may also result in some individuals from the presence of pseudodeficiency alleles, which metabolize the natural Hex A substrate G_{M2} ganglioside appropriately but do not metabolize the artificial substrate 4-methylumbelliferone (4MUG) used in the biochemical enzyme assay analysis *(4,5)*. Thus, these individuals may be labeled as Tay-Sachs carriers although they do not carry a pathogenic mutation.

Due to these limitations with biochemical testing for Tay-Sachs disease, there is interest in providing molecular analysis of the *HEXA* gene for Tay-Sachs as a complement to biochemical testing. In the Jewish population, there are three common mutations in the *HEXA* gene causing Tay-Sachs disease. An insertion mutation in exon 11, +TATC$_{1278}$, accounts for ~82% of mutations *(6)*, a splice error in intron 12, 1421+1G>C, for an additional 11% of mutations *(7,8)* and a missense mutation 805G>A (G269S) in exon 7 for 3% of mutations *(9,10)*. Together, these three mutations account for a total of ~95% of the mutations in HEXA causing Tay-Sachs disease in the Jewish population.

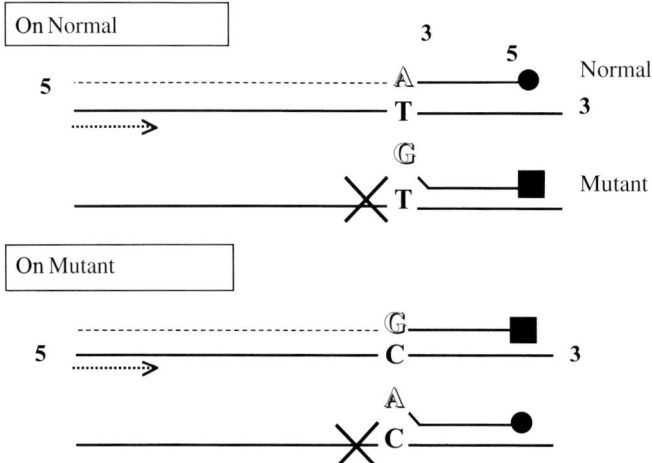

Fig. 1. Theory of the ASA assay to detect mutations causing Tay-Sachs disease. For each mutation site, a normal and mutant primer were designed that differ at the 3' base, and correspond at this base to either the normal or mutant allele sequence. The primers are labeled with different fluorescent dyes (blue dye for the normal primer, black dye for the mutant primer, as indicated by either a circle or square respectively in the figure above), so that the resultant PCR products can be discriminated by dye colors. The assay was designed so that if only the normal sequence is present, amplification from only the normal primer is seen, and if only the mutant sequence is present, amplification from only the mutant primer is seen. If both normal and mutant sequences are present, amplification from both primers occurs.

Two common nonpathogenic mutations, 739C>T (R247W) and 745C>T (R249W), in exon 7 of *HEXA* also exist, although at higher frequency in the non-Jewish population *(4,5)*. However, due to the importance of detecting pseudodeficiency alleles and the increasing rate of intermarriage between Jewish and non-Jewish individuals, molecular testing should include testing for these two pseudodeficiency alleles.

Most current molecular diagnostic tests for Tay-Sachs *HEXA* mutations are labor-intensive and costly and thus are not suitable for testing large numbers of individuals as required in a population-screening situation. The current molecular tests commonly rely on restriction-enzyme differences between normal and mutant alleles, with the assay involving polymerase chain reaction (PCR) amplification of relevant regions of the *HEXA* gene and digestion with appropriate restriction enzymes for each mutation to be detected *(11)*.

We have developed an alternate rapid method to simultaneously detect the three pathogenic and two pseudodeficiency *HEXA* mutations in one tube. This assay is based on allele specific amplification (ASA), which relies on the fact that primers that have a mismatch at the 3' terminal nucleotide will not function in PCR *(12,13)*. In this test, fluorescent dye labeled primers specific to either normal or mutant DNA sequences are used to determine an individual's genotype for Tay-Sachs *HEXA* mutations or pseudodeficiency alleles, as shown in **Fig. 1**. For each mutation site, a normal and mutant primer are designed that differ at the 3' base, and correspond at this base to either the normal or mutant allele sequence. The primers are labeled with different flourescent dyes, so that the resultant PCR products can be discriminated by dye colors. The assay was designed so that if only the normal sequence is present, amplification from only the normal primer is seen, and if only the mutant sequence is present, amplification from only the mutant primer is seen. If both normal and mutant sequences are present, amplification from both primers occurs. In some primers, additional base pair mismatches besides the mismatch at the 3' end were required to achieve specific amplification of normal and mutant alleles.

This chapter will present methods for the application of the ASA method to detect the three common pathogenic mutations and the two pseudodeficiency alleles in the *HEXA* gene. This method has been adapted to simultaneously detect multiple mutations in other disorders for which molecular analysis is required, with appropriate design of primers for specific allele amplification.

2. Materials
2.1. Genomic DNA Isolation

1. DNA should be isolated from blood or leukocyte pellets using standard procedures (*see* **Note 1**). Materials required will depend on the extraction method used.

2.2. PCR

1. Primers required are listed in **Table 1**, labeled with the correct phosphoramidite dye 6-FAM or HEX, or unlabeled as specified (*see* **Notes 2** and **3**). Note that for the +TATC$_{1278}$ and 1421+1G>C mutations, there are three primers each: one primer for the mutant allele (M or MA), one primer for the normal allele (N or NA), and a common reverse primer (B or no letter). However, for the 805G>C (G269S), 739C>T (R247W), and 745C>T (R249W) mutations, there are specific mutant and normal primers for each, but the reverse primer (1559U59B) is common for all three mutations due to the close proximity of these mutations.
2. 10X 1.5 m*M* MgCl$_2$ PCR buffer: 500 m*M* KCl, 15 m*M* MgCl$_2$, 100 m*M* Tris-HCl, pH 8.8 (Perkin Elmer, Foster City, CA, USA). Store at –20°C in a non-frost free freezer.
3. Ampli*Taq*® DNA polymerase, 5 U/ µL, (Applied Biosystems, Foster City, CA, USA). Store at –20°C in a non-frost free freezer.
4. dNTP stocks: 100 m*M* stocks of each of dATP, dCTP, dGTP, and dTTP (Amersham Pharmacia, Piscataway, NJ, USA). Store at –20°C in a non-frost free freezer, and avoid freeze-thaw cycles.

2.3. Polyacrylamide Gel Electrophoresis on Fluorescence Detection Apparatus

1. Fluorescence detection sequencing apparatus, ABI 377 model (*see* **Note 4**).
2. 6% polyacrylamide gel solution: 7 *M* urea (ultrapure grade urea, Amresco, Solon, OH, USA), 6% acrylamide/bisacrylamide 19:1 (prepared from a 40% acrylamide/bisacrylamide solution; ultrapure grade acryl/bis 19:1; Amresco), 1XTBE buffer. Acrylamide and bisacrylamide are neurotoxins. Avoid inhalation and skin contact, and use in a fume hood.
3. 10% ammonium persulfate (Amresco), prepared fresh from powder.
4. TEMED (N,N,N,N'-tetra-metylethylenediamine; Bio-Rad, Hercules, CA, USA). Store at 4°C. TEMED is extremely flammable, and should be used in a fume hood. Avoid inhalation and contact with skin, eyes and clothing.
5. 10X Tris-boric acid-EDTA (TBE) buffer: 890 m*M* Tris base, 890 m*M* boric acid, 20 m*M* EDTA, pH 8.0.

Table 1
List of Primer Sequences for Tay-Sachs ASA Assay[a]

Mutation site	Primer name	Primer specific to mutant or normal allele	Primer sequence	Final concentration in 25 μL reaction
+TATC$_{1278}$ (Exon 11)	2659 U29M 2659 U30N 2729 L26	Mutant allele Normal allele	5' HEX CTG CCC CCT CGT ACC TGA ACC GTG TAT CT 3' 5' 6-FAM CTG CCC CCT CGT ACC TGA ACC GTG TAT CCT 3' 5' GCT CTC TGC TTT CAC CTT CAA ATG CC 3'	4 pmoles 3 pmoles 4 pmoles
1421+1G>C (Intron 12)	2915 L24M 2915 L24N 2779 U25	Mutant allele Normal allele	5' HEX CAC CTC CCC CCC GAA AAC CCT TAG 3' 5' 6-FAM CAC CTC CCC CCC GAA AAC CCT TAC 3' 5' AGT TAC CCC ACC ATC ACC AGA CTG T 3'	6 pmoles 6 pmoles 6 pmoles
805G>A (G269S) (Exon 7)	1726 L30M 1726 L30NA	Mutant allele Normal allele	5' HEX GTC CCT CTC GTC CCA GAC ATC ATT CTT ACT 3' 5' 6-FAM GTC CCT CTC GTC CCA GAC ATC ATT CTC ACC 3'	10 pmoles 7 pmoles
739C>T (R247W) (Exon 7)	1660 L25MA 1660 L27N	Mutant allele Normal allele	5' HEX CAA GCA CAC GGA TAC CCC CGA GCC A 3' 5' 6-FAM TGC AAG CAC ATG GAT ACC CCG GAG TCG 3'	5 pmoles 7 pmoles
745C>T (R249W) (Exon 7)	1666 L31M 1666 L31NA	Mutant allele Normal allele	5' HEX TGT CAA ACT CTG TAA GCA CAC GGA TAC CTC A 3' 5' 6-FAM TGT CAA ACT CTG CAA GCA CAC AGA TAC TCC G 3'	18 pmoles 10 pmoles
Common primer for 805G>A, 739C>T and 745C>T	1559 U29B		5' AAG TGT GAA CCT GAA GAG TGT CTT GTG CC 3'	12 pmoles

[a] Primers should be labeled with HEX phosphoramidite dye (HEX), 6-FAM, phosphoramidite dye (6-FAM), or unlabeled.

Tay-Sachs Disease Mutations

6. Denaturing loading buffer: 1 mL deionized formamide (Fisher Scientific, Fair Lawn, NJ, USA) (*see* **Note 5**), 170 µL of 50 mg/mL blue dextran (Sigma Chemical, St. Louis, MO, USA), 20 µL 0.5 *M* EDTA pH 8.0 (Bio-Rad). Store at 4°C.
7. Size standard: GS 2500 ROX size standard (Applied Biosystems, Perkin Elmer Cetus). Store at 4°C.

3. Methods

3.1. Genomic DNA Isolation

1. Genomic DNA should be isolated from blood or leukocyte pellets using standard procedures (*see* **Note 1**). The final DNA concentration should be 50–200 ng/µL.

3.2. PCR

1. The reagents for each 25 µL Tay Sachs ASA assay reaction are as follows:

17.94 µL	ddH$_2$O	
2.5 µL	10X 1.5 MgCl$_2$ buffer	
0.05 µL	each dNTP (dATP, dCTP, dTTP, dGTP), 100 m*M* concentrations (200 µ*M* dNTPs final concentration)	
0.08 µL	primer 2659U29M, 50 pmoles/µL	(4 pmoles total/ 25 µL reaction)
0.06 µL	primer 2659U30N, 50 pmoles/µL	(3 pmoles total/ 25 µL reaction)
0.08 µL	primer 2729L26, 50 pmoles/µL	(4 pmoles total/ 25 µL reaction)
0.12 µL	primer 2915L24M, 50 pmoles/µL	(6 pmoles total/ 25 µL reaction)
0.12 µL	primer 2915L24N, 50 pmoles/µL	(6 pmoles total/ 25 µL reaction)
0.12 µL	primer 2779U25, 50 pmoles/µL	(6 pmoles total/ 25 µL reaction)
0.20 µL	primer 1726L30M, 50 pmoles/µL	(10 pmoles total/ 25 µL reaction)
0.14 µL	primer 1726L30NA, 50 pmoles/µL	(7 pmoles total/ 25 µL reaction)
0.10 µL	primer 1660L25MA, 50 pmoles/µL	(5 pmoles total/ 25 µL reaction)
0.14 µL	primer 1660L27N , 50 pmoles/µL	(7 pmoles total/ 25 µL reaction)
0.36 µL	primer 1666L31M, 50 pmoles/µL	(18 pmoles total/ 25 µL reaction)
0.20 µL	primer 1666L31NA, 50 pmoles/µL	(10 pmoles total/ 25 µL reaction)
0.24 µL	primer 1559U29B, 50 pmoles/µL	(12 pmoles total/ 25 µL reaction)
0.4 µL	Taq polymerase (5 U/µL)	
2.0 µL	100 ng genomic DNA plus ddH$_2$O to bring volume to 2.0 µL	

    25.0 µL total

2. We typically prepare a large reaction kit that will test 100 samples to reduce the set up time needed for each reaction (*see* **Note 6**). The kit is prepared as shown in **Table 2**, and includes ddH$_2$O, buffer, dNTPs, and all primers. The kits are prepared and stored in aliquots of approx 40 single reactions at –20°C in a non-frost free freezer-protected from light. When a test is required, an appropriate number of aliquots are thawed, and in each PCR tube the following is prepared:

22.6 µL	Tay Sachs ASA kit
0.4 µL	Taq polymerase
2.0 µL	100 ng genomic DNA plus ddH$_2$O to bring volume to 2.0 µL

    25.0 µL

3. For each PCR reaction, appropriate control samples should be prepared. Controls required include a normal sample, positive controls for each mutation tested (*see* **Note 7**) as well as a negative control containing all reagents but no DNA.

Table 2
Preparation of PCR Kit for Tay Sachs ASA Reactions[a]

Reagents for 100 Tay Sachs ASA reactions		Volume (µL)
DdH$_2$O		1794
10X PCR buffer, containing 1.5 mM MgCl$_2$		250
dNTPs: use 100 mM stocks of dNTPs		
dATP		5
dCTP		5
dTTP		5
dGTP		5
Primers	Stock concentration	Volume (µL)
2659 U29M	50 pmole/µL	8
2659 U30N	50 pmole/µL	6
2729 L26	50 pmole/µL	8
2915 L24M	50 pmole/µL	12
2915 L24N	50 pmole/µL	12
2779 U25	50 pmole/µL	12
1726 L30M	50 pmole/µL	20
1726 L30NA	50 pmole/µL	14
1660 L25MA	50 pmole/µL	10
1660 L27N	50 pmole/µL	14
1666 L31M	50 pmole/µL	36
1666 L31NA	50 pmole/µL	20
1559 U29B	50 pmole/µL	24
	Total =	2260 µL
	Number of reactions =	100

[a]Reagents should be added in the order given, and mixed between additions. Aliquots of the final kit should be made and stored at –20°C in a non-frost free freezer protected from light.

4. PCR is performed using the following conditions in a Perkin-Elmer Model 9600 thermocycler: Initial denaturation, 94°C, 10 min; cycle (25 cycles), 94°C, 30 s denaturing; 60°C, 30 s annealing; 72°C, 30 s extension; final extension, 72°C, 10 min.
5. Before running on a gel, samples can be stored for 1–2 d at 4°C if necessary, in a covered rack to protect the samples from light.

3.3. Polyacrylamide Gel Electrophoresis on Fluorescence Detection Apparatus

1. Prepare 1X TBE buffer by diluting the 10X TBE stock with water.
2. Prepare the 6% polyacrylamide gel solution by mixing 9 g of urea, 3.75 mL of 40% acrylamide/bis-acrylamide 19:1 solution, 2.5 mL 10X TBE and water to a total volume of 25 mL. Stir until the urea is dissolved, and filter and degas the solution for 10 min.

3. Prepare the plates for the gel solution. For the ABI 377 model, the gel thickness should be 0.2 mm, with a plate length of 12 cm. Combs used should be 0.2 mm thickness, with 34 flat-bottomed wells.
4. Mix 50 mL 6% polyacrylamide gel solution with 250 µL 10% APS and 35 µL TEMED. Swirl to mix the reagents, and dispense the gel solution into the prepared plate apparatus for the ABI GeneScan appartus. Be careful to avoid introducing bubbles into the gel as it is dispensed between the glass plates. Allow the gel to polymerize for at least 2 h.
5. Once the gel is polymerized, remove the comb and assemble the gel and plates onto the ABI GeneScan. Add 1X TBE buffer to upper and lower chambers.
6. Set up the ABI GeneScan run with the following specifications:
    a. Scan approx 900–2000.
    b. Run 1 h, 1.2 kV, 60 mA, 200W.
    c. Matrix: matrix appropriate for dye set containing 6-FAM, HEX, and ROX dyes.
    d. Filter set: B.
    e. Pre-run: 15 min, or until gel temperature reaches 51°C.
7. Prepare the PCR samples by diluting 1:1 with 1X TBE.
8. Combine the following in a 0.5 mL eppendorf tube for each sample:

    1.0 µL diluted PCR product
    0.5 µL TAMRA 2500 standard (ABI)
    3.5 µL denaturing loading buffer
    ―――――――――――――――――
    5.0 µL total

9. Prior to loading samples, denature the mix above at 95°C for 5 min and snap-freeze on ice.
10. Load 2 µL of the mix onto the prepared and pre-run gel on the ABI GeneScan apparatus.

3.4. Analysis of Results

1. After the gel run is complete, the data for each lane should be printed with peak size and peak heights.
2. The interpretation of results depends on the color and expected fragment lengths of the peaks seen. **Table 3** provides a summary of results expected for normal and carriers of each of the 5 Tay-Sachs mutations. For example, an individual who is a carrier of the +TATC$_{1278}$ mutation will show a yellow/black (HEX dye) peak of 100 bp corresponding to the +TATC$_{1278}$ mutation and a blue (6-FAM dye) peak of 96 bp corresponding to the normal allele at the mutation site, as well as 4 additional blue peaks of 128 bp, 138 bp, 160 bp, and 197 bp corresponding to normal alleles at the other 4 mutation sites.
3. Examples of results expected for normal samples and carriers of the +TATC$_{1278}$ mutation are shown in **Figs. 2** and **3**.
4. Controls should be checked to ensure that the expected results are obtained. Negative controls should show complete absence of product, and all positive controls should show the correct peak sizes and colors as expected.
5. Peak heights on carriers should have yellow/black (HEX) and blue (6-FAM) peaks for the particular mutation site that are approx equal, with less than two-fold difference between the HEX and 6-FAM peak heights (*see* **Notes 3** and **8**).

3.5. Quality Control

1. Peak heights should be at least 500 U and less than 4000 U for all peaks seen.
2. If positive or normal controls do not amplify, the analysis should be repeated. If amplification product is seen in the blank, discard the water, buffers, and primers and repeat the analysis.

Table 3
Summary of Results Expected for Normal or Carrier Samples for the Three Pathogenic and Two Pseudodeficiency Mutations in the *HEXA* Gene[a]

Mutation	Normal (++)	Carrier (+/M)
+TATC$_{1278}$ (Exon 11)	Blue peak, size = 96 bp	Blue peak, size = 96 bp; Yellow peak, size = 100 bp
739C>T (R247W) (Exon 7)	Blue peak, size = 128 bp	Blue peak, size = 128 bp; Yellow peak, size = 126 bp
745C>T (R249W) (Exon 7)	Blue peak, size = 138 bp	Blue peak, size = 138 bp; Yellow peak, size = 138 bp
1421+1G>C (Intron 12)	Blue peak, size = 160 bp	Blue peak, size = 160 bp; Yellow peak, size = 160 bp
805G>A (G269S) (Exon 7)	Blue peak, size = 197 bp	Blue peak, size = 197 bp; Yellow peak, size = 197 bp

[a]Results for normal and mutant alleles are distinguished by peak color differences for all mutations, and by peak color and peak size differences in the case of the +TATC$_{1278}$ mutation and the 745C>T (R249W) pseudodeficiency allele. The size differences are due to a 4 bp insertion in the case of the +TATC$_{1278}$ mutation, and differences in lengths of the normal and mutant primers in the case of the 745C>T (R257W) allele.

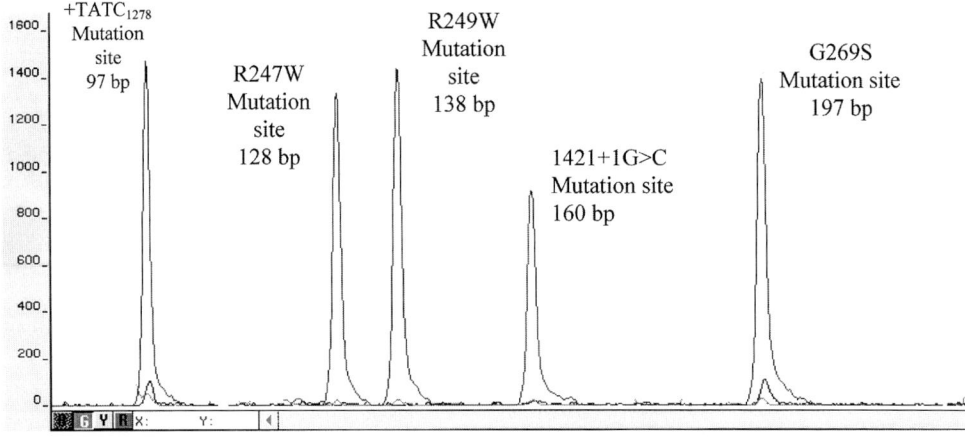

Fig. 2. Results of the Tay Sachs ASA assay for a normal sample (no mutations). Note that there are 5 blue peaks seen, which correspond to the normal allele at each of the mutation sites in the *HEXA* gene. The sizes of the peaks are as given, and are also listed in **Table 3**. (See color plate 3 appearing in the insert following p. 82)

4. Notes

1. We have used the Tay-Sachs ASA test successfully on genomic DNA prepared from blood using phenol-chloroform techniques and Qiagen Spin-Amp columns (Qiagen, Mississauga, Ontario, Canada). As well, we have also used this assay on genomic DNA extracted from leukocyte pellets using a single-strand DNA extraction method with Chelex 100 resin (Bio-Rad) *(14)*.

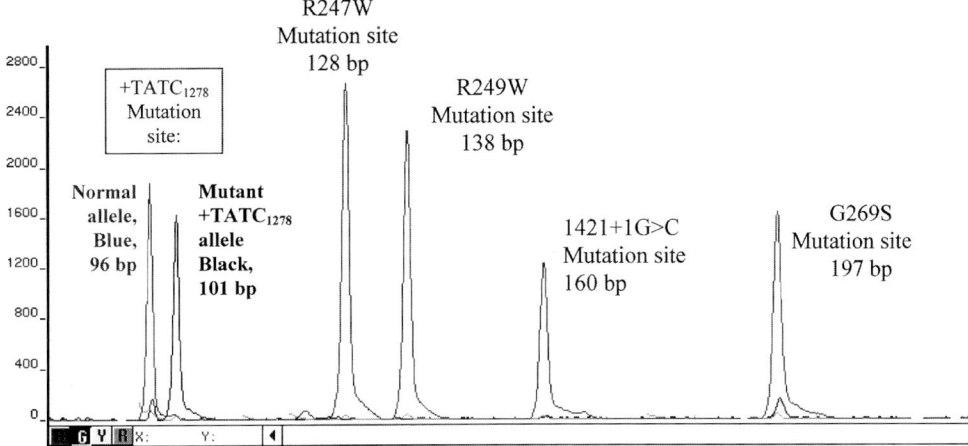

Fig. 3. Example of Tay Sachs ASA assay results for a carrier of the +TATC$_{1278}$ mutation. The normal allele at the +TATC$_{1278}$ site is indicated by a blue peak of 96 bp, while the mutant +TATC$_{1278}$ allele is indicated by the black peak of 101 bp. Note that the size difference between normal and mutant +TATC$_{1278}$ alleles is due the presence of the 4 bp insertion mutation. (See color plate 4 appearing in the insert following p. 82)

2. All 6-FAM and HEX fluorescent-dye labeled primers should be stored in a –20°C freezer and protected from light. Prior to use, new primers should be accurately quantitated and a trial on samples with known mutations performed on all new kits prior to use in routine testing.
3. The primer concentrations as listed in **Table 1** should produce a reliable discrimination of normal and mutant alleles with low background false priming. However, we have noticed that when new primer lots are received, an accurate quantitation of the primer concentration and trial of new primers on known control samples is required. If discrimination of any of the normal or mutant alleles is not ideal when using new primers, adjustments can be made to the primer concentrations and tested on samples with known mutations to improve discrimination as necessary. Primer concentration adjustments should start with small changes to the primer concentrations so as to not unbalance the other primers in the multiplex reaction.
4. The Tay-Sachs ASA assay can also be run on an ABI 373 model with the following specifications: 24-cm plates, scan approx 375–900, run 2 h, 2500 V, 60 mA, 40 W, filter set B, and Amidite matrix.
5. To deionize formamide, mix 10 mL of formamide and 1 g of Amberlite MB150 ion-exchange resin (Applied Biosystems). Stir for 1 hour at room temperature. Check that the pH is greater than 7.0. If the pH is not greater than 7.0, decant the formamide into a beaker containing another 5 g of ion-exchange resin and repeat 30-min stirring. When the pH is greater than 7.0, filter twice through No. 1 Whatman paper. Dispense into aliquots and store for up to 3 m at –15°C to –25°C protected from light.
6. When preparing the kit for 100 Tay-Sachs ASA reactions, all reagents should be added in the order listed in **Table 2** and mixed well by vortexing after each addition to the kit. Small aliquots of the kits should be made, and stored in a –20°C freezer protected from light. Avoid freeze-thaw cycles of the kit, which can decrease the stability of the primers and nucleotides.

7. As controls for Tay-Sachs mutations, we use DNA extracted from cell lines obtained from Coriell Cell Repositories (Camden, NJ, USA, *see* Website: locus.umdnj.edu/ccr/). Cell line GM03770 is a compound heterozygote with the 1421+1 G>C mutation and the 805 G>C (G269S) mutation. Cell line GM02968B is a compound heterozygote with the +TATC$_{1278}$ mutation and the 1421+1 G>C mutation. Controls for the pseudodeficiency alleles 739C>T (R247W) and 745C>T (R249W) were previously obtained from patient samples.

8. In some cases, we have seen some bleed-through of one dye color into another that produces small shadow peaks under the larger correct peaks. This is commonly seen with bleed-through of the yellow/black (HEX) dye peak under a blue (6-FAM) peak. However, the bleed-through is typically less than 20% of the height of the larger peak, and so is not mistaken for a real mutation, which will show approx equal peak heights of HEX and 6-FAM. To minimize bleed-through, overall peak heights should be less than 4000, and samples should be diluted and re-run on another gel if peaks are too high. All samples with potential mutations should be carefully checked to ensure that normal and mutant peak heights are of approx equal heights, and that controls analyzed on the same gel produce appropriate results. If bleed-though persists, primer concentrations may need to be adjusted using samples with known mutations as discussed in **Note 3**.

Acknowledgments

T. L. Stockley was supported by a fellowship from the Ontario Ministry of Health during development of the Tay-Sachs ASA assay. We thank B. Triggs-Raine for providing genomic DNA for the 745C>T (R249W) pseudodeficiency allele.

References

1. Gravel, R. A., Clarke, J. T. R., and Kaback, M. M. (1995) The G$_{M2}$ gangliosidoses, in *The Metabolic and Molecular Bases of Inherited Disease* (Scriver, C.R., Beaudet, A.L., Sly, W.S., and Valle, D., eds.), McGraw-Hill, New York, USA, pp. 2839–2879.
2. O'Brien, J. S., Okada, S., Chen A., and Fellerup, D. L. (1970) Tay-Sachs disease. Detection of heterozygotes and homozygotes by serum hexosaminidase assay. *N. Engl. J. Med.* **283,** 15–20.
3. Natowicz, M. R. and Prence, E. M. (1996) Heterozygote screening for Tay-Sachs disease: past successes and future challenges. *Curr. Opin. Ped.* **8,** 625–629.
4. Triggs-Raine, B. L., Mules, E. H. Kaback, M. M., Lim-Steele, J. S. T., Dowling, C. E., Akerman, B. R., et al. (1992) A pseudodeficiency allele common in non-Jewish Tay-Sachs carriers: Implications for carrier screening. *Am. J. Hum. Genet.* **51,** 793–801.
5. Cao, Z., Natowicz, M. R., Kaback, M. M., Lim-Steele, J. S. T., Prence, E. M., Borwn, D., et al. (1993) A second mutation associated with apparent β-hexosaminidase A pseudodeficiency: identification and frequency estimation. *Am. J. Hum. Genet.* **53,** 1198–1205.
6. Myerowitz, R. and Costigan, F. C. (1988) The major defect in Ashkenazi Jews with Tay-Sachs disease is an insertion in the gene for the α-chain of β-hexosaminidase. *J. Biol. Chem.* **263,** 18,587–18,589.
7. Myerowitz, R. (1988) Splice junction mutation in some Ashkenazi Jews with Tay-Sachs disease: evidence against a single defect within this ethnic group. *Proc. Natl. Acad. Sci. USA* **85,** 3955–3959.
8. Arpaia, E., Dumbrille-Ross, A., Maler T., Neote, K., Tropak, M., Troxel, D., et al. (1988) Identification of an altered splice site in Ashkenazic Tay-Sachs disease. *Nature* **333,** 85–86.

9. Navon, R. and Proia, R. L. (1989) The mutations in Ashkenazi Jews with adult G_{M2} gangliosidosis, the adult form of Tay-Sachs disease. *Science* **243,** 1471–1474.
10. Paw, B. H., Kaback, M. M., and Neufeld, E. F. (1989) Molecular basis of adult-onset G_{M2} gangliosidoses in patients of Ashkenazi Jewish origin: Substitution of serine for glycine at position 269 of the α-subunit of β-hexosaminidase. *Proc. Natl. Acad. Sci. USA* **86,** 2413–2417.
11. Triggs-Raine, B. L., Feigenbaum, A. S. J., Natowicz, M., Skomorowski, M. A., Schuster, S. M., Clarke, J. T. R., et al. Screening for carriers of Tay-Sachs disease among Ashkenazi Jews: a comparison of DNA-based and enzyme-based tests. *N. Engl. J. Med.* **323,** 6–12.
12. Wu, D. Y., Ugozzoli, L., Pal, B. K., and Wallace, R. B. (1989) Allele-specific enzymatic amplification of beta-globin genomic DNA for diagnosis of sickle cell anemia. *Proc. Natl. Acad. Sci. USA* **86,** 2757–2760.
13. Howard, T. D., Bleecker, E. R., and Stine, O. C. (1999) Fluorescent allele-specific PCR (FAS-PCR) improves the reliability of single nucleotide polymorphism screening. *BioTechniques* **26,** 380–381.
14. Walsh, P. S., Metzger, D. A., and Higuchi, R. (1991) Chelex 100 as medium for simple extraction of DNA for PCR-based typing from forensic material. *Biotechniques* **10,** 506–513.

14

Single-Strand Conformational Polymorphism Analysis (SSCP) and Sequencing for Ion Channel Gene Mutations

Kylie A. Scoggan and Dennis E. Bulman

1. Introduction

Single-strand conformational polymorphism (SSCP) analysis is a technique used to screen for the presence of sequence variations in short DNA fragments. This technique relies on the ability of single-stranded DNA molecules to fold into unique secondary structures, the conformations for which are based on their primary nucleotide sequence. Changes in the nucleotide sequence, owing to a polymorphism or a mutation, are expected to alter the secondary structure of the molecule resulting in a shift in mobility through a nondenaturing polyacrylamide gel. It is this aberrant migration pattern that indicates the presence of a DNA sequence alteration. SSCP analysis is capable of detecting single nucleotide differences and, since its first use by Orita et al. *(1,2)* this technique has been extremely successful and widely used to detect disease-causing mutations *(3–8)*. Although SSCP analysis has experienced wide spread use, its sensitivity is variable and has been reported to range from 35–100 % *(9)*. Overall, however, the sensitivity of SSCP analysis reported in the literature usually ranges from 75–98 % with the most critical parameter being the size of the DNA fragment being evaluated. There are numerous reports stating that the sensitivity of SSCP analysis decreases as the size of the polymerase chain reaction (PCR) products become larger than 200 base pairs *(10,11)*. Other parameters such as electrophoresis temperature, buffer concentration, gel concentration, cross-linker concentration, and the addition of compounds to the gel matrix, have also been altered to optimize SSCP sensitivity.

Although SSCP analysis will not identify a mutation 100% of the time, its main advantages are its simplicity and its capacity for examining multiple samples simultaneously. SSCP analysis is only capable of detecting sequence variations and not localizing these changes to a particular nucleotide. Therefore, DNA sequencing must be used to determine the specific alteration. Although the sensitivity of DNA sequencing for identifying a sequence change is 100% and can be used to screen genes for mutations, it is usually more cost efficient to screen initially by SSCP analysis. This is especially true if one is screening a large number of samples, the gene of interest has many exons, and the majority of mutations are novel. For example, ion channel genes such as the α_{1A}-subunit of the P/Q-type calcium channel gene, CACNA1A, and the skeletal muscle α_1-subunit of the sodium channel gene, SCN4A, consist of 47 and 24 exons, respectively *(12,13)*. Mutations in the CACNA1A gene are usually novel and have been implicated in episodic ataxia type-2 (EA-2) *(3,12,14–16)*, familial hemiplegic migraine (FHM) *(12,17–19)*, severe progressive cerebellar ataxia *(20)*, and spinocerebellar ataxia type-6 (SCA6) *(21)*. Similarly, missense mutations in SCN4A have been found to cause hyperkalemic periodic paralysis, paramyotonia congenita, myotonia fluctuans, myotonia permanens, acetazolamide-responsive myotonia, or hypokalemic periodic paralysis (reviewed in **ref.** *(4)*). Because mutations in CACNA1A, as well as in SCN4A have been demonstrated to cause a number of distinct clinical disorders, large numbers of patients tend to be screened for mutations in these genes. It is important to note that highly variable sequences such as those which contain polymorphic repeats are not amenable to SSCP analysis. Thus it would be difficult to detect mutations in exon 47 of the CACNA1A gene, which contains a polymorphic CAG repeat, whose expansion is responsible for SCA6. In order to screen for an EA-2 or FHM causing mutation in this exon, one would have to PCR amplify exon 47 and then directly sequence the resulting products.

Once a sequence variation is detected, the onus is on the investigator to prove that it is a mutation and not a polymorphism or rare sequence variant. Sequence alterations that result in a nonsense codon, change in transcriptional reading frame, or alteration in a splice site are easily proved to be mutations. On the other hand, changes that result in amino acid substitutions may not necessarily be a mutation. Mutations that result in genetic disease are distinguished from polymorphisms by their functional consequences either in vivo or in vitro. Supporting evidence for a mutation includes the cosegregation of the mutation with affected family members, its absence in unaffected family members, as well as its absence in the general population. Without a functional test, loss of amino acid sequence conservation at a particular residue may be used to suggest that the resulting amino acid change is a mutation. In general, if an amino acid at a given position is conserved through evolution, the particular amino acid is thought to be of functional importance at that position.

The basic SSCP analysis method consists of PCR amplifying and radiolabeling wild-type and mutant target DNA, denaturing the PCR products and electrophoresing them side-by-side through a nondenaturing polyacrylamide gel. The single-stranded DNA molecules form three-dimensional conformations (conformers) based on their primary sequence. If there is a difference between the wild-type and mutant sequence, then one or both of the mutant strands may migrate with a mobility that is different from that of the wild-type strands. The DNA strand with an aberrant mobility can either be directly isolated from the gel, amplified, and sequenced, or alternatively, the exon that displayed the aberrant conformer can be amplified and sequenced directly to determine the precise alteration.

There are more advanced protocols for performing SSCP analyses using current technologies such as denaturing high performance liquid chromatography (DHPLC) *(22,23)*, and capillary and microchip electrophoresis *(24)* for automated high throughput mutation screening. Although most SSCP techniques incorporate radioactivity into the DNA fragments, there are also nonradioactive SSCP analysis methods using minigels and silver staining of DNA *(5,25)* or ethidium bromide staining *(26)*. Here we describe the basic SSCP protocol that is technically easy and can be used to efficiently screen large numbers of samples.

2. Materials

2.1. SSCP Analysis

2.1.1. Primer Design

1. Primer designing software available (*see* Website: http://www-genome.wi.mit.edu/cgi-bin/primer/primer3www.cgi).

2.1.2. Optimization of PCR Amplification Conditions

1. Genomic DNA (100 ng/µL).
2. *Taq* DNA polymerase (Boehringer Mannheim), 5U/µL. Store at −20°C (not in a frost-free freezer).
3. 10 X PCR buffer with varying Mg^{2+} concentrations: 100 mM Tris-HCl, pH 8.3, 500 mM KCl, 10, 15, or 20 mM $MgCl_2$, 100 µg/mL BSA, 2.0 mM of each dNTP (Amersham Pharmacia).
4. Intronic oligonucleotide primers; 17 µM.
5. Agarose (molecular biology grade).
6. 10X TBE buffer: 890 mM Tris-boric acid, 25 mM EDTA.
7. 6X loading buffer: 0.25% bromophenol blue, 30% glycerol in TE (10 mM Tris-HCl, 1 mM EDTA, pH 8.0).
8. Ethidium bromide (10 mg/mL). This fluorescent chemical is light-sensitive. It is a mutagen and may be carcinogenic and/or teratogenic, therefore, it should be handled wearing gloves and ethidium bromide-containing solutions should be disposed of appropriately.

2.1.3. Radiolabeling PCR Products for SSCP Analysis

1. Genomic DNA (100 ng/µL).
2. *Taq* DNA polymerase (Boehringer Mannheim), 5 U/µL. Store at −20°C (not in a frost-free freezer).
3. 10 X PCR buffer (low dCTP): 100 mM Tris-HCl, pH 8.3, 500 mM KCl, 10, 15, or 20 mM $MgCl_2$, 100 µg/mL BSA, 2.0 mM of each dGTP, dATP, dTTP, and 4 µM dCTP (Amersham Pharmacia).
4. Intronic oligonucleotide primers; 17 µM.
5. [α-^{32}P]-dCTP (10 mCi/mL or 3000 Ci/mmol).

2.1.4. Polyacrylamide Gel Electrophoresis (SSCP Analysis)

1. Loading dye: 98% formamide, 10 mM EDTA, 0.1% bromophenol blue, 0.1% xylene cyanol FF.
2. Running buffer: 0.5X TBE at 4°C.
3. SSCP gel: a nondenaturing gel (55 cm × 30 cm) consisting of 7.5% acrylamide (49:1 acrylamide:bisacrylamide) and 0.5 X TBE at 4°C.
4. 40% stock acrylamide with 2% crosslinker (bisacrylamide). Acrylamide is light-sensitive and a neurotoxin. It should be handled with appropriate protective gear. The powder is preferably weighed out in a fumehood, wearing a face mask, gloves, and eye protection.
5. Nalgene green collar filter (CN, 0.2 µm diameter pore size).
6. 10X TBE buffer: 890 mM Tris-boric acid, 25 mM EDTA.
7. 10% APS (1.0 g ammonium persulphate in 10 mL deionized distilled (dd) H_2O, stored at 4°C).
8. TEMED.
9. Electrophoresis and gel apparatus: glass plates (55 cm × 30 cm), 0.40 mm spacers, shark's tooth comb.
10. X-ray film (AGFA Curix Ortho HT-G Ecopac).

2.2. DNA Sequencing

2.2.1. DNA Template Preparation

1. Exonuclease I (USB), 10 U/μL. Store at −20°C (not in a frost-free freezer).
2. Shrimp alkaline phosphatase (Roche Diagnostics GmbH), 1 U/μL. Store at −20°C (not in a frost-free freezer).

2.2.2. Polyacrylamide Gel Electrophoresis (DNA Sequencing)

1. Thermo Sequenase Radiolabeled Terminator Cycle Sequencing Kit (USB Corp).
2. Radiolabeled dideoxy-terminators: [α-^{33}P]-ddNTPs (450μCi/mL).
3. The primers that are used for PCR amplification can be used for DNA sequencing (*see* **Subheading 2.1.1.**).
4. DNA sequencing gel: 6% acrylamide gel (SequaGel-6, National Diagnostics, Atlanta, GA, USA).
5. 10% APS.
6. DNA sequencing electrophoresis and gel apparatus: glass plates (55 cm × 30 cm), 0.40 mm spacers, shark's tooth comb.
7. X-ray film (AGFA Curix Ortho HT-G Ecopac).

3. Methods

3.1. SSCP Analysis

3.1.1. Primer Design

1. Primers are designed within the intronic sequence flanking each exon (see **Notes 1–3**).
2. 50% GC content, at least 20 nucleotides in length.

3.1.2. Optimization of PCR Amplification Conditions

1. Dilute genomic DNA samples to 100 ng/μL in 10 m*M* Tris-HCl, pH 8.3. Analyze both affected and unaffected individuals simultaneously.
2. Use control genomic DNA to optimize PCR conditions for each primer set by varying the magnesium concentration present in the buffer and the annealing temperature used during thermocycling (*see* **Note 4**).
3. To ensure that the genomic DNA concentrations between samples are similar, amplify all samples with an optimized primer set in the absence of [α-^{32}P]-dCTP and visualize the intensity of the PCR products on an agarose gel (*see* **Note 5**).

4a. For a single sample, set up a 50.0 μL reaction by adding the following reagents in this order: 38.0 μL dd H$_2$O, 5 μL 10X PCR buffer, 2.0 μL forward primer, 2.0 μL reverse primer, 1.0 μL *Taq* DNA polymerase, 2.0 μL genomic DNA sample.

4b. For multiple samples prepare a master mix of the aforementioned reaction by multiplying the volumes of the individual components, excluding the genomic DNA, by the number of samples plus one (n + 1). Mix well and aliquot 48.0 μL of the master mix into separate tubes containing 2.0 μL of each genomic DNA sample.

5. PCR is performed using the following parameters and an MJ Research Tetrad Model PTC-225 Thermocycler:

Initial Denaturation	94°C, 2 min
30 cycles	94°C, 45 s denaturation
	X°C, 45 s optimized annealing temperature
	72°C, 45 s extension
Final primer extension	72°C, 10 min
End	4°C, hold

6. 2 µL of 6 X loading buffer is added to 10 µL of each PCR product and loaded onto a 1.2 % agarose gel. Electrophoresis is performed at constant voltage of 100 V in 1X TBE buffer. Visualize band intensities and adjust the stock (100 ng/µL) genomic DNA concentrations accordingly.
7. To prepare a 1.2 % agarose gel: weigh 1.2 g agarose in an erlenmyer flask. Add 10 mL of 10X TBE and 90 mL of dd H_2O to the flask and heat in a microwave for 2 min on high or until all of the agarose is melted. Bring the final volume to 100 mL by adding dd H_2O if necessary. Add 5 µL of ethidium bromide, stir, and pour into gel casting apparatus.

3.1.3. Radiolabeling PCR Products for SSCP Analysis

1a. For a single sample, set up a 50 µL reaction by adding the following reagents in this order: 37.9 µL dd H_2O, 5 µL 10X PCR buffer (low dCTP), 2.0 µL forward primer, 2.0 µL reverse primer, 0.1 µL [α-^{32}P]-dCTP, 1.0 µL *Taq* DNA polymerase, 2.0 µL genomic DNA sample (*see* **Note 6**).
1b. For multiple samples prepare a master mix of the above reaction by multiplying the volumes of the individual components, excluding the genomic DNA, by the number of samples plus one (n + 1). Mix well and aliquot 48.0 µL of the master mix into separate tubes (or wells) containing 2.0 µL of each genomic DNA sample.
2. PCR is performed using the same parameters as in **Subheading 3.1.2, step 5**.

3.1.4. Polyacrylamide Gel Electrophoresis (SSCP Analysis)

1. To make 40% stock acrylamide with 2% crosslinker (bisacrylamide 2 g/100 mL) weigh out 490 g acrylamide and 10 g bisacrylamide into a 2 L beaker. Add dd H_2O to 1250 mL and stir for 60 min or until dissolved. Filter solution using Nalgene green collar filter and store in a dark bottle at 4°C.
2. To make an SSCP gel, pipette the following solutions into a 100 mL erlenmyer flask: 20.6 mL 40% stock acrylamide with 2 % crosslinker, 82.7 mL dd H_2O, 5.5 mL 10X TBE, 1.2 mL 10 % APS, and 35.0 µL TEMED. Swirl the solution and pour onto glass plates.
3. After the gel has polymerized at room temperature, transfer it to a cold room (4°C).
4. Dilute each radiolabeled 50 µL PCR product with 10 µL loading dye.
5. Heat denature samples at 95°C for 5 min, immediately place samples on ice, and subsequently load a 4 µL aliquot of each denatured PCR product onto a nondenaturing gel (55 cm × 30 cm) consisting of 7.5% acrylamide (49:1 acrylamide:bisacrylamide) and 0.5X TBE.
6. Pre-cool running buffer and gel to 4°C (*see* **Note 7**).
7. Perform gel electrophoresis at 4°C (preferably in a cold room) and at 50 W.
8. Transfer gel to filter paper and expose to X-ray film at –80°C overnight.
9. Detect abnormally migrating conformers by visual inspection of the autoradiograph (*see* **Note 8**).

3.2. DNA Sequencing

3.2.1. DNA Template Preparation

1a. To sequence only the mutant allele, excise the aberrantly migrating conformer from the SSCP gel using a razor blade by lining up the autoradiograph with the actual gel. Place the gel slice into a 1.5 mL eppendorf tube containing 200 µL of 10 m*M* Tris-HCl, pH 8.3. Carefully remove any filter paper from the gel slice, vortex, and freeze/thaw twice. Use 5 µL of this mixture as the DNA template for PCR amplification (*see* **Subheading 3.1.2, step 4a**). Proceed to **step 2** or (*see* **Note 9**).
1b. To sequence both alleles simultaneously, exons containing aberrantly migrating conformers are amplified from genomic DNA without radionucleotides (*see* **Subheading 3.1.2, step 4a**).

2. To 5–10 µL of PCR product add 1 µL of exonuclease I and 1 µL of shrimp alkaline phosphatase and adjust the final volume to at least 10 µL with dd H$_2$O if necessary (*see* **Note 10**).
3. Incubate samples at 37°C for 20 min followed by an incubation at 70°C for 10 min.
4. Proceed to DNA sequencing reaction.

3.2.2. Polyacrylamide Gel Electrophoresis (DNA Sequencing)

1. DNA sequence analysis of the treated PCR products is then performed manually using the Thermo Sequenase Radiolabeled Terminator Cycle Sequencing Kit (USB Corp) according to the manufacturer's recommended conditions (*see* **Note 11**).
2. Denatured sequencing reactions are immediately loaded and electrophoresed (constant power at 90 watts) through a 6% acrylamide gel in 1X TBE buffer.
3. To make the DNA sequencing gel: add 1.2 mL 10% APS to 100 mL SequaGel-6 in a 100-mL erlenmyer flask. Swirl the solution and pour onto DNA sequencing gel apparatus.
4. Gels are transferred onto filter paper and dried for 2 h at 80°C.
5. Sequencing gels are exposed to X-ray film for 18 h at –80°C, and autoradiographs are examined for sequence changes (*see* **Notes 12** and **13**).

4. Notes
4.1. SSCP Analysis
4.1.1. Primer Design

1. Primers should be designed at a distance from the intron/exon boundary in order to visualize the entire intron/exon splice sites during DNA sequence analysis. As a general rule, primers are designed at least 20 nucleotides away from the intron/exon boundaries.
2. In general, neither primer should be highly similar to DNA repeats. Many primer-designing programs will test whether or not the primer falls within a repeated sequence of DNA (*see* Website: http://www-genome.wi.mit.edu/cgi-bin/primer/primer3www.cgi).
3. In cases where the intronic sequence of the candidate gene is not available, primers must be designed based on the cDNA sequence of that gene.

4.1.2. Optimization of PCR Amplification Conditions

4. A single DNA fragment observed on an agarose gel must be obtained before proceeding to SSCP analysis. In some cases a single PCR product may never be achieved; therefore, a different set of primers will be required.
5. Genomic DNA concentrations should be adjusted to obtain equivalent PCR product loading among samples.

4.1.3. Radiolabeling PCR Products for SSCP Analysis

6. For SSCP analysis, incorporation of ^{32}P radioactive nucleotides within the PCR product was found to increase the resolution and reduce the number of artifacts compared to the use of end-labeled oligonucleotide primers or ^{35}S-dATP incorporation (reviewed in **ref.** *[27]*).

4.1.4. Polyacrylamide Gel Electrophoresis (SSCP Analysis)

7. Both the gel and the buffer should be cooled to 4°C before running samples, in order to prevent the gel from heating and thereby providing enough energy to alter the three dimensional conformation and the migration pattern of the conformer.
8. Aberrantly migrating conformers may be very obvious or very subtle. The bands with altered mobility may be bright and sharp or faint and blurry. They may be well-separated from the normally migrating bands or they may be close in proximity and appear as a blur. All alterations, clear or subtle, should be investigated for a mutation. Sample autoradio-

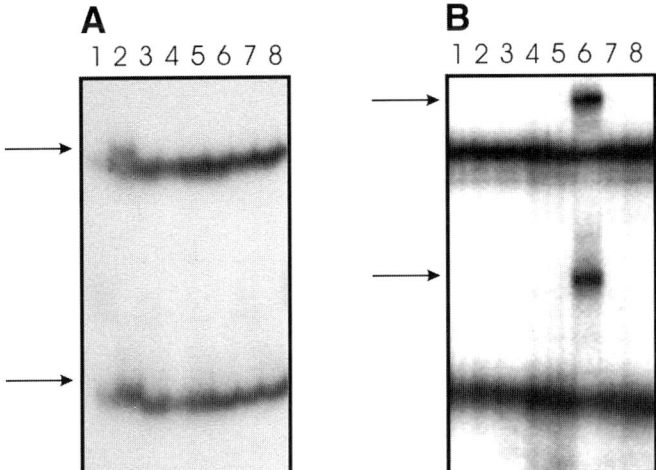

Fig. 1. SSCP analysis of the CACNA1A gene illustrating the detection of mutations by the presence of a subtle change in conformer gel mobility (arrows), panel (**A**) lane 2, and an obvious shift in gel mobility (arrows), panel (**B**) lane 6, compared to normally migrating conformers.

graphs of SSCP analysis of the CACNA1A gene are shown in **Fig. 1**. More subtle changes may be detected by using different electrophoresis conditions. The use of normal controls, nondenatured controls, and positive controls when available are extremely helpful for interpretation of band patterns.

4.2. DNA Sequencing

4.2.1. DNA Template Preparation

9. Isolating the aberrantly migrating conformer from the SSCP gel to use as DNA template for PCR amplification of the exon and subsequent direct DNA sequencing allows only the mutant allele to be sequenced. This approach is advantageous since deciphering the mutant DNA sequence is much easier in the absence of the normal allele. If the aberrantly migrating conformer cannot be isolated from the SSCP gel then the exon can be PCR amplified using genomic DNA. This method is slightly faster than the previous approach and is suitable for identifying simple sequence changes since both the mutant and normal alleles are sequenced simultaneously.

10. Exonuclease I and SAP treatment without PCR product purification can only be done using PCR products optimized as a single band on an agarose gel. Exonuclease I and SAP treatment degrade single-stranded DNA and dephosphorylate the ends of DNA molecules, respectively. These enzymes are used to eliminate the presence of unwanted PCR primers from the PCR products so that the amplification primers cannot interfere with the DNA sequencing reaction.

4.2.2. Polyacrylamide Gel Electrophoresis (DNA Sequencing)

11. Using end-labeled terminators instead of end-labeled primers increases the signal to noise ratio (decreases the presence of background bands) during DNA sequencing analysis.

12. DNA sequencing of PCR products derived directly from the amplification of genomic DNA will demonstrate heterozygosity at the position of the mutation; displaying both a wild-type and a mutant allele in the case of an autosomal dominant disorder.

13. DNA sequencing of an aberrantly migrating conformer that has been excised from the SSCP gel will not demonstrate heterozygosity at any base, including the site of the mutation.

Acknowledgments

This work was supported by a grant from the Canadian Institutes of Health Research. K. A. Scoggan is a recipient of a Canadian Institutes of Health Research Fellowship Award. D. E. Bulman is a Scholar of the Canadian Institutes of Health Research. The authors thank Dr. R. Kothary for reviewing this manuscript.

References

1. Orita, M., Iwahana, H., Kanazawa, H., and Sekiya, T. (1989) Detection of polymorphism of human DNA by gel electrophoresis as single strand conformation polymorphism. *Proc. Nat. Acad. Sci. USA* **86,** 2766–2770.
2. Orita, M., Suzuki, Y., Sekiya, T., and Hayashi, K. (1989) A rapid and sensitive detection of point mutations and genetic polymorphism using polymerase chain reaction. *Genomics* **5,** 874–879.
3. Scoggan, K. A., Chandra, T. C., Nelson, R., Hahn, A. F., and Bulman, D. E. (2001) Identification of two novel mutations in the CACNA1A gene responsible for episodic ataxia type-2. *J. Med. Genet.* **38,** 249–253.
4. Bulman, D. E., Scoggan, K. A., van Oene, M. D., Nicolle, M. W., Hahn, A. F., Tollar, L. L., and Ebers, G. C. (1999) A novel sodium channel mutation in a family with hypokalemic periodic paralysis. *Neurology* **53,** 1932–1936.
5. Ainsworth, P. J., Surth, L. C., and Couler-Mackie, M. B. (1991) Diagnostic single strand conformational polymorphism, (SSCP): A simplified non-radioisotopic method as applied to a Tay-Sachs B1 variant. *Nucleic Acids Res.* **19,** 405–406.
6. Mashiyama, S., Murakami, Y., Yoshimoto, T., and Sekiya, T. (1991) Detection of p53 gene mutations in human brain tumors by single-strand conformation polymorphism analysis of polymerase chain reaction products. *Oncogene* **6,** 1313–1318.
7. Hurk, J. A. J. M., Pol, T. J. R., Molloy, C. M., Brunsmann, F., Ruether, K., Zrenner, E., et al. (1992) Detection and characterization of point mutations in the chloroderemia candidate gene by PCR-SSCP analysis and direct DNA sequencing. *Am. J. Hum. Genet.* **50,** 1195–1202.
8. Dean, M., White, M. D., Amos, J., Gerrard, B., Stewart, C., Khaw, K. T., and Leppert, M. (1990) Multiple mutations in highly conserved residues are found in mildly affected cystic fibrosis patients. *Cell* **61,** 863–870.
9. Glavac, D. and Dean, M. (1993) Optimization of the single-strand conformation polymorphism (SSCP) technique for detection of point mutations. *Hum. Mutat.* **2,** 404–414.
10. Michaud, J., Brody, L. C., Steel, G., Fontaine, G., Martin, L. S., Valle, D., Mitchell, G. (1992) Strand-separating conformational polymorphism analysis: efficacy of detection of point mutations in the human ornithine delta-aminotransferase gene. *Genomics* **13,** 389–394.
11. Sheffield, V. C., Beck, J. S., Kwitek, A. E., Sandstrom, D. W., and Stone, E. M. (1993) The sensitivity of single-strand conformational polymorphism analysis for the detection of single base substitutions. *Genomics* **16,** 325–332.
12. Ophoff, R. A., Terwindt, G. M., Vergouwe, M. N., van Eijk, R., Oefner, P. J., Hoffman, S. M., et al. (1996) Familial hemiplegic migraine and episodic ataxia type-2 are caused by mutations in the Ca2+ channel gene CACNL1A4. *Cell* **87,** 543–552.

13. McClatchey, A. I., Lin, C. S., Wang, J., Hoffman, E. P., Rojas, C., and Gusella, J. F. (1992) The genomic structure of the human skeletal muscle sodium channel gene. *Hum. Mol. Genet.* **1,** 521–527.
14. Yue, Q., Jen, J. C., Thwe, M. M., Nelson, S. F., and Baloh, R W. (1998) De novo mutation in CACNA1A caused acetazolamide-responsive episodic ataxia. *Am. J. Med. Genet.* **77,** 298–301.
15. Denier, C., Ducros, A., Vahedi, K., Joutel, A., Thierry, P., Ritz, A., et al. (1999) High prevalence of CACNA1A truncations and broader clinical spectrum in episodic ataxia type 2. *Neurology* **52,** 1816–1821.
16. Jen, J., Yue, Q., Nelson, S. F., Yu, H., Litt, M., Nutt, J., and Baloh, R. W. (1999) A novel nonsense mutation in CACNA1A causes episodic ataxia and hemiplegia. Neurology. **53,** 34–37.
17. Ducros, A., Denier, C., Joutel, A., Vahedi, K., Michel, A., Darcel, F., et al. (1999) Recurrence of the T666M calcium channel CACNA1A gene mutation in familial hemiplegic migraine with progressive cerebellar ataxia. *Am. J. Hum. Genet.* **64,** 89–98.
18. Carrera, P., Piatti, M., Stenirri, S., Grimaldi, L. M., Marchioni, E., Curcio, M., Righetti, P. G., Ferrari, M., and Gelfi, C. (1999) Genetic heterogeneity in Italian families with familial hemiplegic migraine. Neurology. **53,** 26–33.
19. Battistini, S., Stenirri, S., Piatti, M., Gelfi, C., et al. (1999) A new CACNA1A gene mutation in acetazolamide-responsive familial hemiplegic migraine and ataxia. *Neurology* **53,** 38–43.
20. Yue, Q., Jen, J. C., Nelson, S. F., and Baloh, R. W. (1997) Progressive ataxia due to a missense mutation in a calcium-channel gene. *Am. J. Hum. Genet.* **61,** 1078–1087.
21. Zhuchenko, O., Bailey, J., Bonnen, P., Ashizawa, T., Stockton, D. W., Amos, C., et al. (1997) Autosomal dominant cerebellar ataxia (SCA6) associated with small polyglutamine expansions in the alpha 1A-voltage-dependent calcium channel. *Nat. Genet.* **15,** 62–69.
22. Gross, E., Arnold, N., Goette, J., Schwarz-Boeger, U., and Kiechle, M. (1999) A comparison of BRCA1 mutation analysis by direct sequencing, SSCP and DHPLC. *Hum. Genet.* **105,** 72–78.
23. Choy, Y. S., Dabora, S. L., Hall, F., Ramesh, V., Niida, Y., Franz, D., et al. (1999) Superiority of denaturing high performance liquid chromatography over single-stranded conformation and conformation-sensitive gel electrophoresis for mutation detection in TSC2. *Ann. Hum. Genet.* **63,** 383–391.
24. Tian, H., Jaquins-Gerstl, A., Munro, N., Trucco, M., Brody, L.C., and Landers, J.P. (2000) Single-strand conformation polymorphism analysis by capillary and microchip electrophoresis: a fast, simple method for detection of common mutations in BRCA1 and BRCA2. *Genomics* **63,** 25–34.
25. Sougakoff, W., Lemaitre, N., Cambau, E., Szpytma, M., Revel, V., and Jarlier, V. (1997) Nonradioactive single-strand conformation polymorphism analysis for detection of fluoroquinolone resistance in mycobacteria. *Eur. J. Clin. Microbiol. Infect. Dis.* **16,** 395–398.
26. Yap, E. P. and McGee, J. O. (1992) Nonisotopic SSCP detection in PCR products by ethidium bromide staining. *Trends Genet.* **8,** 49.
27. Hayashi, K. and Yandell, D. W. (1993) How sensitive is PCR-SSCP? *Hum. Mutat.* **2,** 338–346.

15

Pulse Field Gel Electrophoresis for the Detection of Facioscapulohumeral Muscular Dystrophy Gene Rearrangements

Luciano Felicetti and Giuliana Galluzzi

1. Introduction

Facioscapulohumeral muscular dystrophy (FSHD) is an autosomal dominant myopathy with a frequency of 1 in 20,000 and a penetrance of 95% by the age of 20. It affects specific muscle groups (facial, upper girdle, upper arm, pelvic girdle, and foot extensor) and displays a variety of phenotypic expression, ranging from almost asymptomatic forms to more severe wheelchair-bound cases. Linkage and physical mapping strategies have identified a polymorphic EcoRI locus on human chromosome 4q35 *(1)* that is composed of a variable number of 3.3 kb KpnI tandem repeat units (D4Z4) and appears to be tightly linked to FSHD. In normal subjects the p13E-11 EcoRI fragment containing the KpnI repeats varies in size between 50 and 300 kb, while in sporadic and familial cases of FSHD the disease cosegregates with a fragment widely below this range, i.e., between 10 and 34 kb *(2–7)*. It has been demonstrated that the 4q35 rearrangement involves the deletion of an integral number of 3.3 kb repeat units *(8)*. In the human genome the 3.3 kb repeats are present on several chromosomes other than 4q, specifically the short arms of acrocentric chromosomes, 1q12 and 10qter, as shown by *in situ* hybridization experiments *(9,10)*. The spreading of KpnI repeat sequences on human chromosomes generates artifacts in the interpretation of DNA analysis in normal subjects and FSHD patients, because multiple EcoRI fragments are observed after hybridization with p13E-11 probe and time-consumimg linkage analysis with 4q35 and 10qter markers is required to assign the chromosomal origin of the alleles *(5,11)*. We found that the 10qter locus shows a high degree of homology with the 4q35 locus as shown by restriction mapping and *in situ* hybridization experiments *(12)*. However restriction site differences between the two loci can be exploited for cleaving the interfering 10qter alleles: double digestion of genomic DNA with EcoRI and BlnI resulted in a marked improvement of molecular diagnosis and genetic counseling for the disease *(13)* (**Fig. 1**). When the 10qter sequences are compared with the homologous 4q35 sequences published by other laboratories *(14,15)* the degree of homology ranges between 98% and 100%; this facilitates interchromosomal exchanges resulting in the

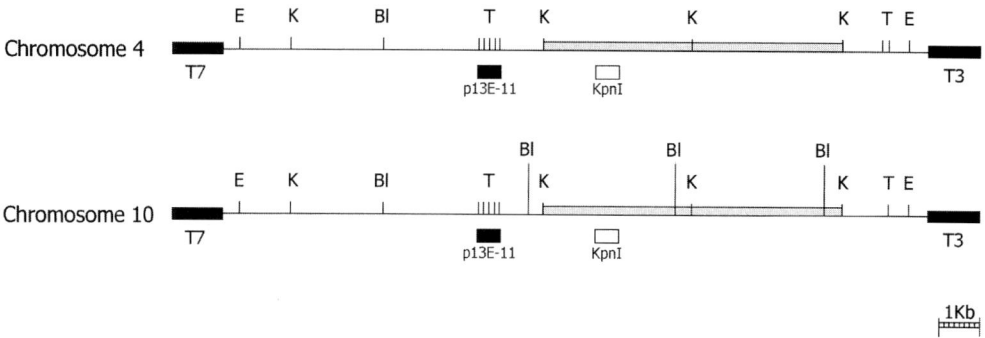

Fig. 1. Restriction sites and probes of the EcoRI polymorphic fragments derived from 4qter and 10qter loci. The restriction map has been originally constructed by using phage clones derived from FSH patients carrying short fragments of different chromosomal origin and has been confirmed by extended studies on genomic DNA. Additional polymorphisms at the EcoRI and Tru9I sites are very rare.

E, EcoRI; Bl, BlnI; T, Tru9I. The bars refer to the single copy probe p13E-11 (closed) and cloned KpnI sequences (open).

Chromosomes		p13E-11 alleles			Allele configuration
4	10	Eco RI	BlnI^R	BlnI^S	
		4	2	2	Disomy (canonical pattern)
		4	3	1	Trisomy
		4	4	0	Tetrasomy
		4	1	3	Monosomy (spurious)
		4	0	4	Nullisomy (spurious)
		4	2	2	False disomy
		3	1	2	Deletion of p13E-11 and KpnI repeats
		5	3	2	Somatic mosaicism (4q)

Fig. 2. Variations of the number and size of p13E-11 BlnI-resistant alleles as an effect of inter chromosomal 4q-10q exchanges and other DNA rearrangements. The subtelomeric

reshuffling of 4q BlnI-resistant and 10q BlnI-sensitive repeats from one chromosome to the other, with the appearance of hybrid chromosomes *(16,17)* (**Fig. 2**). These rearrangements may have implications for the accuracy of molecular diagnosis and genetic counseling: a short 4q BlnI-sensitive fragment in a FSHD patient can be misinterpreted as an allele of 10q origin, unrelated to the disease and viceversa a short 10q BlnI-resistant fragment in an unaffected subject can be erroneously taken as an allele of 4q origin that carries the mutation (nonpenetrant gene carrier). Although in 98% of FSHD patients *(18,19)* molecular diagnosis is based on the presence of p13E-11 BlnI-resistant fragments in the range 10–30 kb that are rarely involved in translocations, conventional agarose gel electrophoresis does not clearly separate higher MW alleles in the range 30–300 kb that represent the majority of p13E-11 alleles of 4qter and 10qter origin and could play a role in the etiology of the disease. Taking into account these considerations, we decided to use pulsed field gel electrophoresis (PFGE) as a routine test for molecular diagnosis of 4q35 rearrangements in family members of FSHD families, both sporadic and familial (**Fig. 3**). Other advantages of PFGE analysis are: 1) a more accurate sizing of the smaller BlnI-resistant fragments implicated in the disease, especially in the range 28 to 40 kb; 2) complete allele assignment to 4qter and 10qter in family trees, without using other polymorphic markers (**Fig. 4**); 3) identification of BlnI-resistant fragments wherever they are dislocated, using KpnI cloned sequences as probe (**Fig. 5**). The use of Tru9I and Tru9I-BlnI are recommended to clear up the hybridisation pattern of 4q and 10q alleles by cleaving fragments derived from other chromosomes *(17)*; 4) detection of somatic mosaicism in normal and affected individuals; 5) study of deletions of p13E-11 associated with removal of various lengths of KpnI repeat array. Finally, being concerned about the poor quality of genomic DNA extracted by the high salt procedure from frozen blood samples, we further modified the PFGE procedure to minimize degradation of high molecular-weight DNA: density-gradient purified leukocytes from venous blood (preferably fresh) were included into agarose plugs and all the steps of DNA extraction and restriction with EcoRI, True 9I, and BlnI were performed directly within the plugs.

Fig. 2. *(continued)* regions of 4q and 10q chromosomes have been magnified: the closed symbol refers to 4q BlnI-resistant repeats, the open symbol to 10q BlnI-sensitive epeats. BlnI^R: fragments resistant to BlnI digestion (4q-type). BlnI^S : fragments sensitive to BlnI digestion (10q-type). Trisomy and tetrasomy originate from transfer of the whole set of BlnI-resistant repeats to one or both 10q chromosomes, respectively. Monosomy and nullisomy are due to the insertion of BlnI-sensitive repeats between p13E-11 and BlnI-resistant repeats of one or both 4q chromosomes, respectively. In these cases, size and distribution of BlnI-resistant repeats can be monitored by hybridization with cloned KpnI repeats as a probe. False disomy is due to a reciprocal translocation of BlnI-sensitive repeat array onto a 4q chromosome and of BlnI-resistant repeat array onto a 10q chromosome. This particular configuration is not distinguishable from the canonical pattern, unless studying segregation of PFGE alleles through the family. Somatic mosaicism: the 4q chromosome marked with an asterisk corresponds to a supernumerary 4q BlnI-resistant fragment present in a sub-population of lymphocytes. A subject with a short extra fragment carries a *de novo* FSHD mutation: he can be affected or unaffected, depending on the proportion of somatic cells involved in the rearrangement.

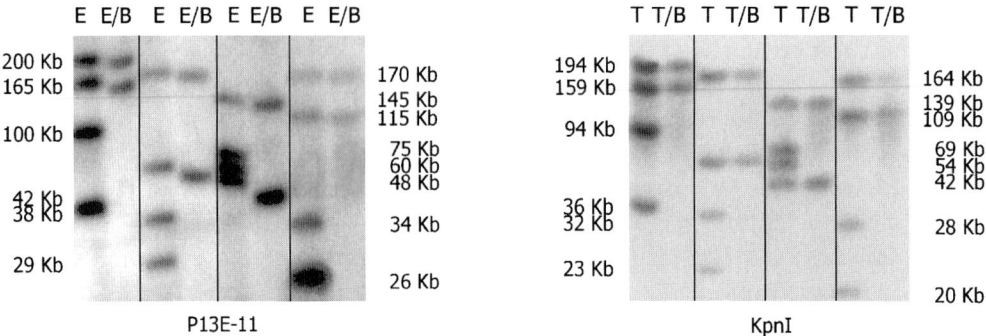

Fig. 3. Canonical pattern of 4q and 10q alleles in normal subjects. The size of alleles ranges between 26 and 200 kb and their homogeneity is confirmed by hybridization with KpnI sequences after Tru9I and Tru9I/BlnI digestions. It is worth noting that two individuals carry fragments lower than 30 kb that disappear after BlnI digestion: this demonstrates that they contain BlnI-sensitive sequences of 10q origin.

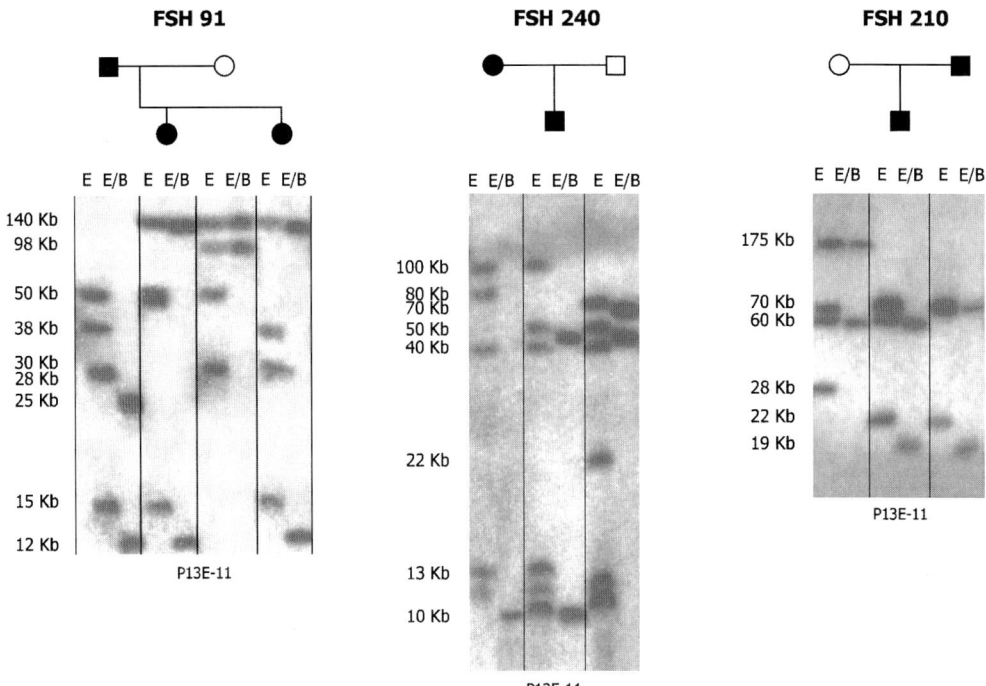

Fig. 4. Segregation of 4qter and 10qter alleles in FSHD families. All the patients carry an EcoRI BlnI-resistant fragment ranging between 15 and 22 kb: after BlnI digestion the fragment is reduced of 3 kb. The number of KpnI repeats spared by the deletion is approx three in Fam. 91, two in Fam. 240, and five in Fam. 210. Segregation of 4q–10q alleles through the family members can be easily followed. The affected mother in Fam. 240 is monosomic.

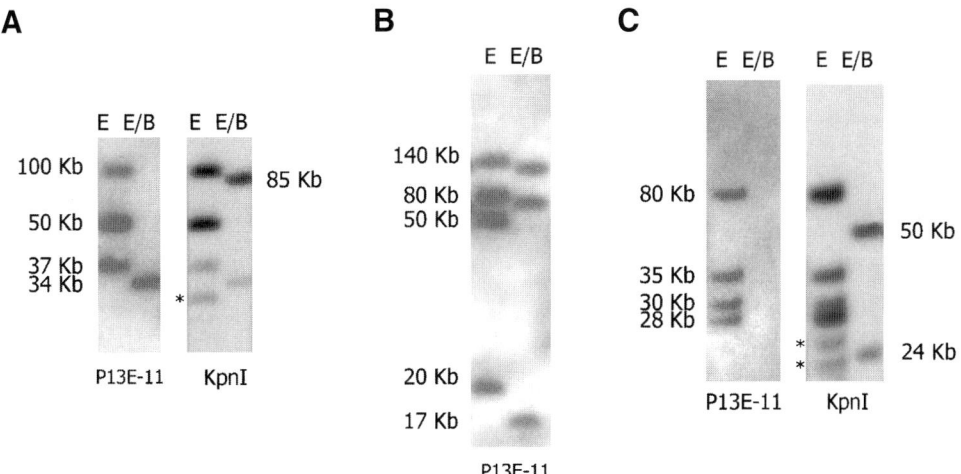

Fig 5. 4q–10q subtelomeric exchanges in FSHD patients. (**A**) Spurious monosomy in a patient with a border line fragment (37 kb). After hybridization with KpnI sequences an additional 85 kb BlnI-resistant fragment appears: it derives from the 100 kb fragment. (**B**) Trisomy in a patient with an FSHD fragment of 20 kb. (**C**) Nullisomy in a patient showing four p13E-11 bands after EcoRI digestion and none after BlnI digestion. After hybridization with KpnI sequences, two BlnI-resistant fragments appear: the 50 kb fragment originates from the 80 kb fragment; the shortest (24 kb) must derive from one of the smaller EcoRI alleles. Nullisomy is due to partial translocations of 10q-type repeats on both 4q chromosomes: in this patient one of the translocation events involves the FSHD fragment directly. The asterisk marks BlnI-sensitive fragments probably deriving from other chromosomes and revealed by hybridization with KpnI sequences only.

2. Materials

2.1. Lymphocytes Isolation from Whole Blood

1. 10 mL of fresh peripheral blood in EDTA-tubes for each subject to be analyzed.
2. 1X PBS (Dulbecco's Phosphate-Buffered Saline [PBS]) (Oxoid, Basingstoke, UK). Sterilize by autoclaving.
3. Lympholyte-H (Cedarlane, Hornby, Ontario, Canada). Light-sensitive, store at 4°C.

2.2. Inclusion of Cells in Agarose Blocks

1. Low Melt Agarose, preparative grade (Bio-Rad, Hercules, CA, USA).
2. 1X PBS.
3. Insert moulds containing rectangular holes of $10 \times 5 \times 2$ mm (Amerham Pharmacia Biotech, Uppsala, Sweden). Moulds are stored in 0.1 M HCl.

2.3. DNA Isolation in Agarose Blocks

1. Sarcosyl (Sigma, St. Louis, MO, USA).
2. Proteinase K (Roche, Mannheim, Germany).
3. 1000X Phenyl-methyl-sulphonyl-fluoride (PMSF) fresh stock solution: 40 mg/1 mL Ethanol 99% (Sigma). This solution is highly toxic by inhalation and in contact with skin. It should be handled wearing lab coat, gloves, and eyes/face protection.

4. TE 10/1 pH 8.0: 10 mM Tris-HCl, 1 mM EDTA, pH 8.0, sterilized by autoclaving. 0.5 M EDTA pH 8.0.

2.4. Restriction Endonuclease Digestion of DNA in Agarose Blocks

1. Restriction enzymes:
    a. EcoRI (Roche)/ 10X buffer H: 50 mM Tris-HCl, 100 mM NaCl, 10 mM $MgCl_2$, 1 mM dithiothreitol (DTT), pH 7.5.
    b. BlnI (Amersham Pharmacia Biotech, Buckinghamshire, UK)/10X buffer K: 200 mM Tris-HCl, pH 8.5, 100 mM $MgCl_2$, 10 mM DTT, 1 M KCl.
    c. Tru9I (Roche)/10X buffer M: 10 mM Tris-HCl, pH 7.5, 50 mM NaCl, 10 mM $MgCl_2$, 1 mM DTE.
2. 0.5 M EDTA pH 8.0.
3. Sterile bidistilled water.

2.5. Pulse Field Gel Electrophoresis

1. High strength analytical grade agarose (Bio-Rad).
2. 5X TBE: 54 g Tris base, 27.5 g boric acid, 3.72 g EDTA per liter.
3. Bidistilled water.
4. Ethidium bromide (Sigma) 10 mg/mL. This flourescent dye is stored at 4°C and is light-sensitive. It is mutagen and may be carcinogenic. It should be handled wearing gloves and solutions containing EtBr should be disposed of properly.
5. Low Melt Agarose, preparative grade (Bio-Rad).
6. 50 kb ladder: λ DNA can be ligated to produce a mixture of monomers, dimers, trimers, etc. On PFGE these concatemers produce a ladder of bands, each successive rung corresponding to a 50 kb increment.
7. DNA Molecular-weight Marker XV (Roche).

2.6. Southern Blot

1. Denaturing solution: 0.4 M NaOH, 0.6 M NaCl.
2. Nylon transfer membrane Hybond N^+ (Amersham Pharmacia Biotech).
3. 0.2 M Tris-HCl, pH 7.5.
4. Whatman 3 MM chromatography paper.
5. Paper towels.
6. Bidistilled water.

2.7. Prehybridization and Hybridization

1. 20X SSC: 3 M NaCl, 0.3 M sodium citrate.
2. Deionized Formamide (Sigma).
3. Ficoll 400 (Amersham Pharmacia Biotech AB, Uppsala, Sweden); PVP Polyvinylpirrolidone (BDH, Poole, UK); BSA-Bovine serum albumin (Sigma). Use a 100X stock solution (Denhardt's) containing 2% Ficoll, 2% PVP, and 2% bovine serum albumin (BSA).
4. Sonicated salmon sperm DNA, 2 mg/mL (Sigma).
5. Dextran sulfate sodium salt (Pharmacia Biotech AB).
6. Sodium dodecyl sulfate (SDS) (Lauryl Sulfate, Sigma).
7. Multiprime DNA labelling system (Amersham Pharmacia Biotech).
8. [α-^{32}P] dNTP (Amersham Pharmacia Biotech).

2.8. Special Equipment: PFGE Apparatus

The apparatus can be constructed or purchased commercially. The protocol described in methods was developed for use with Gene Navigator System (Amersham Pharmacia Biotech AB) consisting of:

1. 1 (or 2) Gene Navigator electrophoresis unit: includes pump for buffer circulation, cooling coil, safety lid with cable, insert moulds, gel staining tray, silicone tubing, hose connectors, and hose clamps.
2. 1 Gene Navigator control unit. A controller for switching the electric field and programming up to 9 run procedures.
3. Power supply for electrophoresis applications requiring a maximum of 600 V and 400 mA.
4. HEX electrode kit: it includes a Hexagonal electrode, gel supporting tray and gel casting frame, and double combs 1- and 2-mm thick and preparative combs.
5. Thermostatic cooler circulating constant temperature liquid to external electrophoresis equipment.

3. Methods

3.1. Lymphocytes Isolation from Whole Blood

1. Collect 10 mL of fresh blood into a tube containing EDTA as anticoagulant (*see* **Note 1**).
2. Dilute the blood by addition of an equal volume of 1X PBS.
3. Carefully layer 20 mL of the diluted blood over 10 mL of Lympholyte-H in a 50 mL Falcon tube, at room temperature. Alternatively Lympholyte-H can be underlayered. Avoid mixing of blood and separation fluid.
4. Centrifuge at 800g for 30 min at room temperature in a G8 swing-out rotor without brake.
5. After centrifugation the mononuclear cells form a distinct band at the sample/medium interface. Remove the cells from the interface using a sterile Pasteur pipet, without collecting any of the clear Ficoll layer.
6. Transfer the cells to a new Falcon tube and dilute with 1X PBS up to 30 mL.
7. Centrifuge at 500g for 5 min.
8. Rinse the pellet twice with 10 mL of 1X PBS.

3.2. Inclusion of Cells in Agarose Blocks

1. Resuspend the pellet gently in 10 mL of 1X PBS.
2. Count cells in one large square (i.e., 16 small squares) of a haemocytometer.
3. Spin down cells and resuspend the pellet in PBS to a homogeneous concentration of 2×10^6 per 50 µL. Avoid the formation of air bubbles.
4. Wash moulds thoroughly in distilled water, dry, and cover one side with a tape. Add equal volume of 1% Low Melt Agarose in PBS (kept between 40–42°C) to the cell suspension. Mix gently with a Pasteur pipet to ensure that the cells are evenly dispersed throughout the agarose. Dispense the mixture immediately into moulds (precooled in ice). Place the moulds in ice and allow to set for 20–30 min. Each plug has a standard volume of 100 µL and contains 2×10^6 cells (i.e., if you resuspend 12×10^6 cells in 300 µL of PBS, and add an equal volume of 1% Low Melting Agarose, you obtain a mixture of 600 µL to be dispensed into 6 plugs: each of the plugs will contain 2×10^6 cells).

3.3. DNA Isolation in Agarose Blocks

1. Remove the tape. Using a bent Pasteur pipet, gently transfer the plugs to a Falcon tube containing the extraction solution (0.5 M EDTA, 1% Sarcosyl, 1 mg/mL Proteinase K; at least 1 mL per plug).

2. Incubate at 50°C for 20–36 h, with gentle shaking.
3. Put the tube on ice for 5–10 min to allow blocks to set.
4. Discard the supernatant and put the blocks on a Petri dish: wash 15 min with TE pH 8.0. Repeat washing 4 times (see **Note 2**).
5. Transfer the plugs in a new Falcon tube: add 1X PMSF solution (stock solution diluted up to 40 μg/mL with TE pH 8.0–1 mL/plug) and incubate at 50°C for 30 minutes with gentle shaking (see **Note 3**).
6. Put the tube on ice for 5–10 min to allow blocks to set.
7. Discard PMSF solution and add TE, pH 8.0 (at least 1 mL per plug).
8. Repeat washing 4 times.
9. Store plugs in 0.5 M EDTA, pH 8.0, at 4°C until the subsequent steps: agarose plugs can be stored in this solution for several years.

3.4. Restriction Endonuclease Digestion of DNA in Agarose Blocks

1. Pick up the plugs to be digested with a spatula and wash twice in sterile bidistilled water for 20 min using Petri dishes.
2. Use 1/4 of plug (corresponding to 0.5×10^6 cells) for each DNA digestion. Cut slices on a sterile surface using a sterile scalpel: unused parts can be stored again in 0.5 M EDTA, pH 8.0.
3. Transfer each slice in a 2 mL microfuge tube and add water, appropriate restriction buffer 10X and 80–100 U of restriction enzyme added in three equal portions at 1-h intervals (final volume 400 μL). Incubate overnight at the optimal temperature for each restriction enzyme (37°C for EcoRI and BlnI; 50°C for Tru9I) (see **Note 4**).
4. When the DNA is to be digested with more than one restriction enzyme (i.e., EcoRI/BlnI or Tru9I/BlnI double digestions), remove as much as possible of the first restriction enzyme buffer from each tube and replace it with 1 mL of TE. Gently remove the TE and replace with the appropriate second 10X restriction buffer, restriction enzyme, and water: repeat the digestion as described in step 3.
5. After digestion, place on ice for 15–30 min. Remove all buffer and wash with sterile water for 30 min at 4°C.
6. The slices may be either used directly or stored in 0.5 M EDTA, pH 8.0 at 4°C for later use.

3.5. Pulse Field Gel Electrophoresis

1. For a 15×15 cm gel, dissolve 1.4 g of High Strength Analytical Grade Agarose in 115 mL of 0.5X TBE (1.2%). Cool to 37°C.
2. Pour the agarose into the casting tray: allow the gel to harden for 1 h at room temperature (better at 4°C).
3. Prepare 2.5 L of 0.5X TBE: pour into the Electrophoresis unit connected to a thermostatic cooler preset at 12°C and allow to cool.
4. Sample plugs should be loaded before the gel is submerged. Embed the DNA plugs in individual wells of the gel with a spatula. Carefully push the plugs with a bent Pasteur pipet and suck down with a syringe. The slices should be placed in contact with the front wall. Make sure that the sample slices are not wider than the well (see **Note 5**).
5. Seal the wells with a few drops of 0.5% low melting point agarose in 0.5X TBE.
6. Submerge the gel-supporting tray, with the gel attached, in the electrophoresis tank and place the HEX electrode.
7. Load the marker(s) into the gel. Heat Marker XV at 65°C for 5 min before loading.

8. Set the Program Controller and the Power Supply and run the electrophoresis. After 5–10 min start the buffer circulating (*see* **Note 6**).
9. After electrophoresis, stain the gel in a 1 µg/mL solution of EtBr for 30 min at room temperature. Destain the gel in water for 30 min and photograph the gel under UV light.

3.6. Southern Blot

The following procedures require little or no modification for use with PFGE.

1. After photography, incubate the gel in 300 mL of denaturing solution with gentle shaking for 20 min.
2. Prepare the capillary transfer apparatus, wrapping a piece of Whatman 3 MM paper around a piece of Plexiglas to form a support longer and wider than the gel. Place the wrapped support inside a large baking dish. Fill the dish with denaturing solution until the level of the liquid reaches almost to the top of the support: smooth out all air bubbles with a glass rod.
3. Using a paper cutter, cut a piece of nylon membrane about 1 mm larger than the gel in both dimensions. Float the membrane on the surface of a dish of deionized water until it wets completely and then transfer the membrane in the denaturing solution for at least 5 min.
4. Remove the gel from the denaturing solution and place it on the support so it is centered on the wet 3 MM paper. Make sure that there are no air bubbles between the 3 MM paper and the gel.
5. Place the wet nylon membrane on top of the gel so that the corners are aligned. Make sure that there are not air bubbles between the membrane and the gel.
6. Wet three sheets of 3 MM paper (cut to exactly the same size as the gel) in denaturing solution and place them on top of the wet nylon membrane. Smooth out any air bubbles with a glass rod. Cover the wet 3 MM paper with three sheets of dry 3 MM paper.
7. Cut a stack of paper towels (8–10 cm) just smaller than the 3 MM papers. Place the towels on the 3 MM papers. Put a glass plate on top of the stack and weigh down with a 500 g weight. This procedure is aimed at setting up a flow of liquid from the reservoir through the gel and the nylon membrane: the fragments of denatured DNA are eluted from the gel and are immobilized on the nylon membrane.
8. Allow the transfer of DNA to proceed for 16–24 h.
9. Remove the paper towels and the 3 MM papers: mark the position of the gel slots on the membrane with a pencil.
10. Soak the membrane in 0.1 M Tris-HCl, pH 7.5, for 20 min.
11. Place the membrane on a 3 MM paper and allow to dry for 30 min at room temperature.
12. Store the membrane between two sheets of 3 MM paper for later use.

3.7. Prehybridization and Hybridization

1. Wet the membrane with 3X SSC for 5–15 min.
2. Prepare the prehybridization mix (50% deionized formamide; 3X SSC; 5X Denhardt's; 200 µg/mL Salmon Sperm). Heat at 65°C for 15 min before use.
3. Place the wet membrane into a heat-sealable bag and seal three sides. Add 5–10 mL of preheated prehybridization mix, squeeze air from the bag and seal the open side.
4. Incubate 6–16 h at 42°C.
5. Prepare the appropriate labeled probe using Amersham's Multiprime DNA labeling system according to the instructions of the manufacturer: 20–40 ng of linear ds DNA are suitable for each hybridization (*see* **Note 7**).
6. Replace the prehybridization mix in the bag with 4–5 mL of a new mix containing 50% deionized formamide; 3X SSC; 5X Denhardt's; 200 µg/mL Salmon Sperm; 10% Dextran

Sulfate, and 20–40 ng of denatured labeled probe. Squeeze air from the bag and seal the open side.
7. Incubate at 42°C for 16–20 h.
8. Remove the membrane from the bag and wash in 300 mL of 2X SSC for 15–20 min at room temperature.
9. Repeat washing with 0.5X SSC/0.1% SDS at 58°C for 15–30 min (*see* **Note 8**).
10. Cover the wet membrane with a Saran Wrap. Expose the membrane to X-ray film (at −70°C) to obtain an autoradiographic image. Exposure time varies from 8–48 h.

4. Notes

1. Venous blood can be stored at room temperature and used within 2–3 d.
2. To avoid nuclease activity, keep TE 10/1, pH 8.0, as an autoclaved stock solution ready to use.
3. Addition of 40 µg/mL PMSF is used to reduce protease activity.
4. Optimal DNA digestion depends on the specific activity of the enzyme and on quality of DNA included in agarose plugs. Partial digestion leads to abnormal hybridization patterns, complicating PFGE analysis significantly.
5. Mobility of fragments depends on the quantity of DNA applied on each slot: overloading with high DNA concentrations decreases fragment mobility.
6. Selection of pulse time depends on the size range of interest. In general, shorter pulse times are used to separate smaller DNA fragments and longer pulse times are used to separate larger fragments. To resolve DNA molecules in the range 7–200 kb (size range of interest in FSHD) the following running parameters are recommended:

Phase	Time (h)	Pulse (s)	Voltage
1	1:30	0.3	300
2	3:30	0.5	300
3	4:00	1	300
4	2:30	5	300

According to the suggestions of manufacturers of PFGE apparatus, running parameters can be modified in order to obtain the maximum resolution of DNA fragments in the size range of interest.

7. Both the single copy p13E-11 probe and KpnI repeat sequence were obtained from a 4.9 kb KpnI fragment subcloned in PUC21 *(17)*. p13E-11 was amplified by PCR as a 680 nt fragment using primers 1F (5'-GCTATAGATTCTAGGTGCTTCA-3') and 2R (5'-TTG CCTGTGAGTTCGAATGCAC-3'). The 3.3 kb KpnI repeat unit was digested with BamHI and the restricted fragments (1,387; 737; and 1,179 nt) subcloned in PUC21. The 1,179 nt fragment was used as probe.
8. When the FSHD specific fragment is smaller than 15 kb, it comigrates in the gel with similar fragments derived from other chromosomes and its identification can be difficult. In this case it is convenient to increase the stringency of hybridization conditions.

Acknowledgments

We are most grateful to Luca Colantoni and Monica Rossi for their help in editing the Materials and Methods section. We are also indebted to Dr. Enzo Ricci and Prof. Pietro Tonali for clinical and diagnostic evaluation of FSHD patients. This work has been financially supported by Telethon Italy grant No. 1296.

References

1. Sarfarazi, M., Wijmenga, C., Upadhyaya, M., Weiffenbach, B., Hyser, C., Mathews, K. D., et al. (1992) Regional mapping of facioscapulohumeral muscular dystrophy gene on 4q35. Combined analysis of an international consortium. *Am. J. Hum. Genet.* **51,** 396–403.
2. Wijmenga, C., Hewitt, J. E., Sandkuijl, L. A., Clark, L. N., Wright, T. J., Dauwerse, H. G., et al. (1992) Chromosome 4q DNA rearrangements associated with facioscapulohumeral muscular dystrophy. *Nature Genet.* **2,** 26–30.
3. Passos-Bueno, M. R., Wijmenga, C., Takata, R. E., Marie, S. K. N., Vainzof, M., Pavanello, R. C., et al.(1993) No evidence of genetic heterogeneity in Brazilian facioscapulohumeral muscular dystrophy families (FSHD) with 4q markers. *Hum. Mol. Genet.* **2,** 57–562.
4. Upadhyaya, M., Lunt, P., Sarfarazi, M., Broadhead, W., Farnham, J., Harper, P. (1992) The mapping of chromosome 4q markers in relation to Facioscapulo humeral muscular dystrophy (FSHD) *Am. J. Hum Genet.* **51,** 404–410.
5. Weiffenbach, B., Dubois, J., Storvick, D., Tavil, R., Jacobsen, S. J., Gilbert, J., et al. (1993) Mapping the facioscapulohumeral muscular dystrophy gene is complicated by 4q35 recombination events. *Nat. Genet.* **4,** 165–169.
6. Wijmenga, C., van Deutekom, J. C. T., Hewitt, J. E., Padberg, G. W., van Ommen, G-J. B., Hofker, M. H., and Frants, R. R. (1994) Pulse field gel electrophoresis of the D4F104S1 locus reveals the size and the paternal origin of the FSHD-associated deletions. *Genomics* **19,** 21–26.
7. Cacurri, S., Deidda, G., Piazzo, N., Novelletto, A., La Cesa, I., Servidei, S., et al. (1994) Chromosome 4q35 haplotypes and DNA rearrangements segregating in affected subjects of 19 Italian families with facioscapulohumeral musculatur dystrophy (FSHD). *Hum. Genet.* **94,** 367–374
8. Van Deutekom, J. C. T., Wijmenga, C., Van Tienhoven, E. A. E, Gruter, A. M., Hewitt, J. E., Padberg, G. W., et al. (1993) FSHD associated DNA rearrangements are due to deletions of integral copies of a 3.2 kb tandemly repeated units. *Hum. Mol. Genet.* **2,** 2037–2042.
9. Deidda, G., Cacurri, S., Grisanti, P., Vigneti, E., Piazzo, N., and Felicetti, L. (1995) Physical mapping evidence for a duplicated region on chromosome 10qter showing high homology with the FSHD locus on chromosome 4qter. *Eur. J. Hum. Genet.* **3,** 15–167.
10. Winokur, S. T., Bengtsson, U., Vargas, J. C., Wasmuth, J. J., and Altherr, M. R. (1996) The evolutionary distribution and structural organization of the homeobox-containing repeat D4Z4 indicates a functional role for the ancestral copy in the FSHD region. *Hum. Mol. Genet.* **5,** 1567–1575.
11. Deidda, G., Cacurri, S., La Cesa, I., Scoppetta, C., Felicetti, L. (1994) 4q35 molecular probes for the diagnosis and genetic counseling of facioscapulohumeral muscular dystrophy. *Ann. Neurol.* **36,** 117–118.
12. Deidda, G., Cacurri, S., Grisanti, P., Vigneti, E., Piazzo, N., and Felicetti, L. (1995) Physical mapping evidence for a duplicated region on chromosome 10qter showing high homology with the FSHD locus on chromosome 4qter. *Eur. J. Hum. Genetics.* **3,** 155–167.
13. Deidda, G., Cacurri, S., Piazzo, N., and Felicetti, L. (1996) Direct detection of 4q35 rearrangements implicated in facioscapulohumeral muscular dystrophy (FSHD). *J. Med. Genet.* **33,** 361–365.
14. Hewitt, J. E., Lyle, R., Clark, L. N., Valleley, E. M., Wright, T. J., Wijmenga, C., et al. (1994) Analysis of the tandem repeat locus D4Z4 associated with facioscapulohumeral muscular dystrophy. *Hum. Mol. Genet.* **3,** 1287–1295.

15. Lee, J. E. H., Goto, K., Sahashi, K., Nonaka, I., Matsuda, C., Arahata, K. (1995) Cloning and mapping of a very short (10 kb) EcoRI fragment associated with facioscapulohumeral muscular dystrophy (FSHD). *Muscle Nerve* **Suppl 2,** S27–S31.
16. Van Deutekom, J. C. T., Bakker, E., Lemmers, R. J., van der Wielen, M. J. R, Bik, E., Hofker, M. H., et al. (1996) Evidence for subtelomeric exchange of 3.3 kb tandemly repeated units between chromosomes 4q35 and 10q26: implications for genetic counselling and etiology of FSHD1. *Hum. Mol. Genet.* **5,** 1997–2003.
17. Cacurri, S., Piazzo, N., Deidda, G., Vigneti, E., Galluzzi, G., Colantoni, L., et al. (1998) Sequence homology between 4qter and 10qter loci facilitates the instability of subtelomeric KpnI repeats units implicated in facioscapulohumeral muscular dystrophy. *Am. J. Hum. Genet.* **63,** 181–190.
18. Galluzzi, G., Deidda, G., Cacurri, S., Colantoni. L., Piazzo, N., Vigneti, E., et al. (1999) Molecular analysis of 4q35 rearrangements in facioscapulohumeral muscular dystrophy (FSHD): application to family studies for a correct genetic advice and a reliable prenatal diagnosis of the disease. *Neuromusc. Disord.* **9,** 190–198.
19. Ricci, E., Galluzzi, G., Deidda, G., Cacurri, S., Colantoni, B., Merico, B., et al. (1999) Progress in the molecular diagnosis of facioscapulohumeral muscular dystrophy (FSHD) and correlation between the number of KpnI repeats at the 4q35 locus and clinical phenotype. *Ann. Neurol.* **45,** 751–757.

16

Denaturing Gradient Gel Electrophoresis (DGGE) for Mutation Detection in Duchenne Muscular Dystrophy (DMD)

Luciana C. B. Dolinsky

1. Introduction

Duchenne muscular dystrophy (DMD) is an X-linked recessive neuromuscular disorder caused by mutations in the dystrophin gene at Xp21. Approximately two-thirds of the mutations are intragenic deletions of one or more of the 79 exons that constitute the 2.4 Mb dystrophin gene, 5% are duplications, and the remaining 30% are mutations that are very difficult to identify by current diagnostic screening strategies *(1–3)*. The great majority of deletions can be detected by polymerase chain reaction (PCR) multiplex approach *(4,5)* or Southern blot analysis probed with dystrophin cDNA *(1)*.

The identification of the molecular lesion in the 30% of patients that do not have a detectable deletion or duplication represents a great obstacle in genetic analysis of these patients and genetic counseling of their relatives. Many different techniques such as single-strand conformation polymorphism (SSCP), hydroxylamine and osmium tetroxide (HOT) chemical cleavage, heteroduplex analysis (HA), protein truncation test (PTT), are available to identify these single nucleotide variations *(6–9)*. However, there is not a very high mutation detection rate when we use each one of these techniques for Duchenne muscular dystrophy analysis. As such, in nondeletion cases diagnostic laboratories use dystrophin muscle biopsy staining and/or haplotype analysis to assess risk status. However, even both these strategies may be inconclusive because there is a high intragenic recombination rate (~10%) and high mutation rate (1/3 of all cases) using haplotype analysis, and technical difficulties associated with muscle biopsy staining. Recently, a highly sensitive SSCP assay showed the highest mutation detection rate (90%), suggesting that DNA analysis might be better than muscle biopsy staining for diagnostic purposes *(10)*. But even using highly sensitive SSCP, some families are excluded from carrier detection and prenatal diagnosis because detection rate is not 100%.

In this work, we have used denaturing gradient gel electrophoresis (DGGE) *(11,12)*, the most sensitive electrophoretic method (detection rate close to 100%) *(13)*, to scan for DMD mutations. Mutation identification in the majority of DMD cases is important because it can then provide empiric diagnostic and risk information for the patient and other family members.

DGGE is a methodology that is conceptually similar to heteroduplex analysis *(8,9)*. The method is based on the electrophoretic mobility of a double-strand DNA molecule through linearly increasing concentrations of a denaturing agent (formamide and urea). As the DNA fragment proceeds through the gradient gel, it will reach a position where the concentration of the denaturing agent equals the melting temperature (Tm) of its lowest melting domain causing denaturation and the consequent marked retardation of its electrophoretic mobility. As the Tm of a melting domain is dependent on its nucleotide sequence DNA fragments differing by a single nucleotide change in their lowest melting domain, when electrophoresed through denaturing gradient gels, will suffer branching and consequent retardation of their mobility at different positions along the gel, allowing their separation *(14)*. The melting behavior of a DNA fragment can be simulated by computer analysis *(15)*.

However, because of complete strand dissociation, DGGE cannot resolve DNA fragments differing by nucleotide changes in the highest melting domain. This problem is overcome by introducing a GC clamp tail as short as 40bp to serve as a high Tm domain and prevent complete dissociation of the DNA fragment. The introduction of a GC clamp tail increases the percentage of mutation detection by DGGE to close to 100% *(13)*. The PCR-DGGE combination is extremely powerful when applied to heterozygous nucleotide variants because of continuous denaturation and reannealing of single-strand molecules during PCR, allowing for the formation of homoduplex and heteroduplex molecules. The presence of a single nucleotide change within heteroduplexes decreases their melting temperature allowing separation from the homoduplexes and easier visual detection of the mutants.

DGGE methodology does not require the use of radioisotopes, is less labor- and time-consuming, is generally cheaper than other point mutation detection techniques, and has a very high mutation detection rate (close to 100%) *(16,17)*.

For the analysis of dystrophin mutations we utilize DGGE with the following modifications: 1) All primers are designed to include intronic sequences flanking each exon *(16)*. This increases even more the mutation detection rate *(16)*. 2) PCR assays for male specimens require a 1:1 mixture of patient and male control DNA to allow for the detection of heteroduplexes, as DMD is an X-linked disease. 3) Segments are amplified in multiplex combinations of up to 6 exons to reduce workload. 4) The choice of the gel denaturant range is based upon the Tm (melting temperature) of the fragment to be analyzed. 5) Electrophoretic runs are performed at a constant temperature (58°C) that exceed the Tm of an A-T rich DNA fragment in the absence of denaturing agents. 6) Amplicons are detected using polyacrylamide, urea, and formamide gel electrophoresis with ethidium bromide staining. 7) Employing these modifications DGGE permits analysis and identification of point mutations, *de novo* mutations and obligate carriers of the dystrophin gene mutations *(17)*.

2. Materials

2.1. DNA Isolation from Lymphocytes (18)

1. 1X Blood lysis buffer (1 L): 155 mM NH$_4$Cl, 10 mM KHCO$_3$, 1 mM EDTA, pH 7.4. This solution should be stored at 4°C up to 6 mo. Blood should be handled using universal precautions. Take care to discard all blood contaminated materials in appropriate biological waste containers.

DGGE Analysis for DMD

2. 1X Nucleus lysis buffer (1L): 10 mM Tris-HCl, 400 mM NaCl, 2 mM EDTA, pH 8.2. Store at 4°C up to 1 y.
3. 10% SDS Ultra PURE (Gibco BRL). Store at room temperature up to 1 y.
4. Pronase 20 mg/mL (Sigma). Store at −20°C up to 1 y.
5. 6 M NaCl. Store at room temperature up to 6 mo.
6. Absolute ethanol.
7. TE-4 (0.5L): 10 mM Tris-HCl, 0.1 mM EDTA, pH 7.5. Store at room temperature up to 1 y.

2.2. PCR

1. Ampli*Taq* DNA polymerase (Perkin Elmer), 5 U/µL. Store at −20°C (not in a frost-free freezer).
2. 10X Super *Taq* PCR buffer (Sphero-Q): 100 mM Tris-HCl. pH 9.0, 500 mM KCl, 15 mM MgCl$_2$, 0.1% gelatin and Triton X-100. Store at −20°C.
3. dNTP mix (Perkin Elmer): 25 mM of each dNTP. Store at −20°C.
4. 25 mM MgCl$_2$ solution (Perkin Elmer). Store at −20°C.
5. Oligonucleotide primers (**Table 1**, *see* **Note 1**); 12.5 pmol/µL.

2.3. Polyacrylamide Gel Electrophoresis

1. 12% PAA, 80% urea/formamide (1L): 12% acrylamide, 32% formamide (deionized), 5.6 M urea (ultra-pure) in 0.5X TAE buffer. Polyacrylamide, urea, and formamide are all toxic agents; always wear disposable gloves when handling these solutions. Store at 4°C up to 4 mo.
2. 9% PAA, 0% urea/formamide (1L): 9% acrylamide in 0.5X TAE buffer. Store at 4°C up to 4 mo.
3. 50X Tris-acetate (TAE) buffer: 400 mM Tris-acetate, 10 mM EDTA, pH 8.0. Store at room temperature up to 6 mo.
4. 10% ammonium persulfate (APS). Store at 4°C up to 1 mo.
5. TEMED (Bio-Rad). Electrophoresis grade reagent. TEMED should be handled wearing disposable gloves in a laminar flow. Store at room temperature up to 1 y.
6. 6X DGGE gel loading buffer (100 mL): 0.25% bromophenol blue (Sigma), 0.25% xylene cyanol (Sigma), 15% ficoll. Store at 4°C up to 6 mo.
7. Ethidium bromide (10 mg/mL). This fluorescent dye is light sensitive. It is a mutagen and may be carcinogenic/teratogenic. Always wear disposable gloves when handling ethidium bromide-containing solutions. Store at 4°C up to 6 mo.

2.4. DNA Sequencing Reactions

1. ABI PRISM™ BigDye™ Terminator Cycle Sequencing Ready Reaction Kit with Ampli*Taq*® DNA Polymerase (Perkin Elmer). The DNA double strand sequencing reactions were made using the above reaction kit for ABI PRISM model 377. Store at −20°C. Safety warnings on the reagents bottles and in the manufacturers' Material Safety Data Sheets (MSDS). Dispose of waste in accordance with all local, state, and federal health and environmental regulations and laws.

3. Methods
3.1. DNA Isolation from Lymphocytes

1. Starting material is 5 mL blood, decoagulated with EDTA (*see* **Note 2**).
2. Dilute 5 mL blood with 15 mL blood lysis buffer.
3. Lysis of the red blood cells takes place for at least 10 min.
4. Centrifuge for 10 min at 500g.

Table 1
Oliginucleotide Primers[a]

DGGE1F-AGCCTACTGGAGCAATAAAG	DGGE1R[b]-AGCTTGTCACAAACTAAACG
DGGE2F[b]-ATGGAAAGTTACTTTGGTTG	DGGE2R-CAGGTACATAGTCCATTTTG
DGGE3FA[b,c]-TAAATTGAGTGTATTTTTTT	DGGE3RA-TTTTGCCCTGTCAGGCCTTC
DGGE3FB[c]-CTTCAGTGACCTACAGGATG	DGGE3RB[b]-TTAACTTTCTTAAAAATAAG
DGGE4F[b]-CTCTCTGCTGGTCAGTGAAC	DGGE4R-GTGTGTCACAGCATCCAGAC
DGGE5F[b]- GCATTTGGTCTCTTACCTTC	DGGE5R- TTTGTTTCACACGTCAAGGG
DGGE6FA[b]-GGTTCTTGCTCAAGGAATGC	DGGE6RA-GCTCAGGAGAATCTTTTCAC
DGGE6FB-AAATATCATGGCTGGATTGC	DGGE6RB[b]-ATCAGAGTCTAAATCACCAC
DGGE7F-GCGCGTTTGTCTTTGTATGTGTG	DGGE7R[b]-TACTAAAAGCAGTGGTAGTC
DGGE8F-GCGCGCCAAAAATTGATGTGTAGTG	DGGE8R[b]-ATCTTGAATAGTAGCTGTCC
DGGE9F- TAATACGACTCACTATAGGGCCCTCCTCTCTATCCACTCC	DGGE9R[b]- GGAAGCAGTTCTCTGGTTTG
DGGE10F-GCGCGGACATTAATTGTGTAACACC	DGGE10R[b]-GGATGACTTGCCATTATAAC
DGGE11F-CACCGATTTACCTAGAGTTC	DGGE11R[b]-TCCAAAACTTGTTAGTCTTC
DGGE12F[b]-AAGTTGCTTTCAAAGAGGTC	DGGE12R-GTACTATACACAGAGTTTGC
DGGE13FA[b]-CATTTCAACACACATGTAAG	DGGE13RA-AGCAGCAGTTGCGTGATCTC
DGGE13FB-CTCTCACTCACATGGTGGTG	DGGE13RB[b]-CTTTTAAAGGACATATTTAG
DGGE14F-GCGCGGTTTGCTGATGCTTGATTG	DGGE14R[b]-CCCCGTGTCTTTTACAGCTA
DGGE15F-TATGTTTATTTATTCCTTGG	DGGE15R[b]-TACTGGGTTTTTATAAGACC
DGGE16F[b]-TGCAACCCAGGCTTATTCTG	DGGE16R-GCGCGTCTCTGAGATAGTCTGTAGC
DGGE17F-CCAAGCAGTCTTTACTGAAG	DGGE17R[b]-AGTTGCCTTCCTTCCGAAAG
DGGE18FA[b]-AGAAGAAAGAGATAATCAAG	DGGE18RA-AGTTGCCTTCCTTCCGAAAG
DGGE18FB-CTGGATTACTCGCTCAGAAG	DGGE18RB[b]-GCAGCACAAAATGAGTACAG
DGGE19F-GCGCGGATTCACAGTCCTTGTATTG	DGGE19R[b]-GTGTTTATCAAATCCCTAAG
DGGE20FA[b]-TCAGTCTGTGGGTTCAGGGG	DGGE20RA-CAGTAGTTGTCATCTGCTCC
DGGE20FB-GGCTGGAGTATCAGAACAAC	DGGE20RB[b]-GGAATGCCAAGAAATACCTA
DGGE21F-ATGTATGCAAAGTAAACGTG	DGGE21R[b]-TGGAAAATGTCAAGTTAGCC
DGGE22F[b]-CAATTAAGTGATTCTCATTC	DGGE22R-GCGTGCTTTATTGTTTTGAC
DGGE23FA[b]-TTCATCAATTAGGGTAAATG	DGGE23RA-AGGAGAGCTTCTTCCAGCGT
DGGE23FB-CGCCCTCTGAAATTAGCCGG	DGGE23RB[b]-GTAAATAAAAATGAGGGTAG
DGGE24F-AATCAGCACACCAGTAATGC	DGGE24R[b]-TTATTCATATTAAAGGCATC
DGGE25F[b]-TTTTCAGCTGGCTTAAATTG	DGGE25R-TTAAGTACGTTGAGGCAAGC
DGGE26FA[b]- CATCACTGTCAATAATCGTG	DGGE26RA-TAATACGACTCACTATAGGGTCAAG ATACTCTTCTTCAGC
DGGE26FB- TAATACGACTCACTATAGGGCTATCAGAGATGCACGAATG	DGGE26RB[b]- TTTACCTTCATCTCTTCAAC
DGGE27F[b]-TTTCATGCTATTAAGAGAGC	DGGE27R-AACTATGACCATGTATTGAC
DGGE28F[b]-TTCACATTTACTTTTCTACC	DGGE28R-CTGCATATAAATCATCATCA
DGGE29FA[b]- CATTTGCTGATAATCCAATG	DGGE29RA- TAATACGACTCACTATAGGGATTAGCTCATCCATGACTCC
DGGE29FB- TAATACGACTCACTATAGGGATTCGCATATTGGCACAGAC	DGGE29RB[b]- ATGCAAATTAGATTAAAGAG
DGGE30F[b]-TACAGAAAAGCTATCAAGAG	DGGE30R-CAAACAAAAAGAATGGAAGC
DGGE31F[b]-GTTAGTTGTTCTTTGTAGAG	DGGE31R-GCCCAACGAAAACACGTTCC
DGGE32F[b]-GGCAAAATTAAATCAGTGCC	DGGE32R-TAATGAGGAAAGTCAAGGGG
DGGE33F[b]-TAACTCTACTGATTATCATG	DGGE33R-GTGGAAAGAAGTGTTTGTGG
DGGE34F[b]-ATTTGAATTAAAGAGTAAAC	DGGE34R-ATGTGTTTTCACGTATGTTC
DGGE35F[b]-CAAGACATTACTTGAAGGTC	DGGE35R-CGCGCACAGAGAAGGGTGTAAAAGC
DGGE36F-GTCTAACCAATAATGCCATG	DGGE36R[b]-GTAACTGGTGTACAATTTGG
DGGE37F[b]-CTAACTTCAAGTCCTATCTC	DGGE37R-CGCAAGAGACCATTTAGCAC
DGGE38F[b]-CTTTTGGTTTATGTTTCTTA	DGGE38R-TTTATTTCCACTCCTAGTTC
DGGE39F-CTCTGTTAACAATGTACAGC	DGGE39R[b]-AAACCACAGGCAAGGTATAT
DGGE40F-GCGCGATTTTAATAATGTCTGCACC	DGGE40R[b]-AAAATCTGGTATTGACATTC

Table 1 *(continued)*
Oliginucleotide Primers[a]

DGGE41FA[b]-ATGTGGTTAGCTAACTGCCC	DGGE41RA-GCTGGATCTGAGTTGGCTCC
DGGE41FB-GAAGAAAGAGGAGCTGAATG	DGGE41RB[b]-AGTTGCAAACACATACGTGG
DGGE42F[b]-TAATGGAGGAGGTTTCACTG	DGGE42R-GCGCCACCTTGTAAAATACGAATG
DGGE43FA[b]-CTATAGACAGCTAATTCATT	DGGE43RA-GCTGCTGTCTTCTTGCTATG
DGGE43FB-AGTCTACAACAAAGCTCAGG	DGGE43RB[b]-TCATTTCTGCAAGTATCAAG
DGGE44F[b]-CGCTATATCTCTATAATCTG	DGGE44R-TAAAGAGTCCAGATGTGCTG
DGGE45F[b]-GACATGGGGCTTCATTTTTG	DGGE45R-GTTTTCATTCCTATTAGATC
DGGE46F[b]-GTTTGAGAACTATGTTGGAA	DGGE46R-ACTTCTTTATGCAAGCAGGC
DGGE47F-AGGTAGTTGGAATTGTGCTG	DGGE47R[b]-GCACCCAGGAAACAAAATAC
DGGE48F[b]-GGCTTATGCCTTGAGAATTA	DGGE48R-CAGTGATATTGCCATTTTTC
DGGE49F[b]-ACTGTGAAGTTAATCTGCAC	DGGE49R-CAACAGGCGAAGCATAACCC
DGGE50F[b]-CGAATAAGTAATGTGTATGC	DGGE50R-TTGAACAAATAGCTAGAGCC
DGGE51FA[b]-TCTTCTTTTTTCCTTTTTGC	DGGE51RA-GATCAAGCAGAGAAAGCCAG
DGGE51FB-GATTTCAACCGGGCTTGGAC	DGGE51RB[b]-AAAGAAAAACTTCTGCCAAC
DGGE52F-GAAGAACCCTGATACTAAGG	DGGE52R[b]-CATCTTGCTTTGTGTGTCCC
DGGE53FA[b]-TCCTCCAGACTAGCATTTAC	DGGE53RA-AGACCTGCTCAGCTTCTTCC
DGGE53FB-TTCAGAACCGGAGGCAACAG	DGGE53RB[b]-CAAATGTAACCAGTATTTTA
DGGE54FA[b]-AAAAACTGACATTCATTCTC	DGGE54RA-TAATCCCGGAGAAGTTTCAG
DGGE54FB-CCAGTGGCAGACAAATGTAG	DGGE54RB[b]-CCCATTATTACAGCCAACAG
DGGE55F[b]-TTTATGAGTTCACTAGGTGC	DGGE55R-ACAAATGCTGAGAATTGTTC
DGGE56F-TCTGCACATATTCTTCTTCC	DGGE56R[b]-ATTTGTGGCCTTTTTGCTCC
DGGE57F-GATATTCTGACATGGATCGC	DGGE57R[b]-GTCACTGGATTACTATGTGC
DGGE58F-CCACAAGCCTTTCTTAGCAC	DGGE58R[b]-TCACCACTGATCCTTCTATC
DGGE59F[b]-AAAGAATGTGGCCTAAAACC	DGGE59R-GCGGCCGCCAAAGGGCCCTGAAGCAA
DGGE60F-ATTCTCATCTTCCAATTTGC	DGGE60R[b]-TTACTGTAACAAAGGACAAC
DGGE61FA[b]-GAATGAGAGAACATAATTTC	DGGE61RA-TGAGATGCTGGACCAAAGTC
DGGE61FB-GTCGAGGACCGAGTCAGGCA	DGGE61RB[b]-AGATGCAATAAAGTTAAGTG
DGGE62F-ATTTGACCTCCTTGCCTTTC	DGGE62R[b]-GTTAGTCACAATAAATGCTC
DGGE63F[b]-GTTTTCTTGACTACTCATGG	DGGE63R-CCTGTCATTTAACTTGGAGG
DGGE64F-GGCAAATCACTGGGCGTCGG	DGGE64R[b]-AGCAAAGACATAGTATCAAG
DGGE65F[b]-GGAAGGTTTTACTCTTTGAG	DGGE65R-CTAAGCCTCCTGTGACAGAG
DGGE66F-CCTCTAGGAAAGGGTCTAGT	DGGE66R[b]-CTAGGGTAATTAGCCAACAT
DGGE67F-TTGGATGTCAGGTTCTGCTG	DGGE67R[b]-GAAAGAATAAATATGTTACC
DGGE68FA[b]-CTTTCCTTTCATCCTTTGGC	DGGE68RA-TTGGACACTCTTTGCAGATG
DGGE68FB-AGAAACTGCCAAGCATCAGG	DGGE68RB[b]-AGGAGATAAAAGATCAAGTC
DGGE69F[b]-CGAAGAAATACATACGTGTT	DGGE69R-TGAACTAACTCTCACGTCAG
DGGE70F[b]-AGTTTTGAAATCATCCTGTC	DGGE70R-CATCAAACAAGAGTGTGTTCTG
DGGE71F-AGAAAGCGTGTGTCTCCTTC	DGGE71R[b]-AGCGAATGTGTTGGTGGTAG
DGGE72F-GATGGTATCTGTGACTAATC	DGGE72R[b]-TCAATCAATATTTGCCTGGC
DGGE73F[b]-TCACATAAGTTTTAATGAGC	DGGE73R-ACCTCTAAATCCCTCAAAGC
DGGE74F[b]-TGAGTCCCTAACCCCCAAAG	DGGE74R-GCACTCTGCATACCAATGAC
DGGE75FA[b]-TTTACTTTTTTGATGCC	DGGE75RA-CCTTTGTGTTGACGCAGTAG
DGGE75FB-CGCCGCCGGCCAATAGGAATCTGCAAGC	DGGE75RB[b]-GTTTGTAAAAATCCCATCTC
DGGE76F[b]-TGAAAAGTAATTCTGTTTTC	DGGE76R-TCTACCTTTCTTCAGACAAC
DGGE77F[b]-ATGTAATTTTCCATTATTTG	DGGE77R-GATCCCAGCAAATCTGAGTC
DGGE78F[b]-CCCTTTCTGATATCTCTGCC	DGGE78R-AAGCAGGATGAGACAGACAG
DGGE79F[b]-TGATGCTATCTATCTGCACC	DGGE79R-TTCTGCTCCTTCTTCATCTG

[a]Primer sequences are 5' to 3'. Name of the primer represents used technique (DGGE) and exon number in human DMD gene. F: forward primer; R: reverse primer.

[b]All primers marked with [b] have a 5'GC clamp tail as short as 40bp designed this way: 5' cgcccgccgcg ccccgcgcccggcccgccgccgcggccgc 3' (*see* **Note 1**).

[c]Some exons had to be split in two partly overlapping segments (*see* **Note 1**)

5. Discard the supernatant (*see* **Note 3**).
6. Resuspend the pellet in 5 mL blood lysis buffer.
7. Centrifuge for 10 min at 500g.
8. Discard the supernatant.
9. Check if the lysis was complete before continuing, the pellet should be white and clearly visible.
10. Resuspend the pellet in 1mL nucleus lysis buffer.
11. Add 35 µL of pronase (*see* **Note 4**).
12. Add 50 µL of 20%SDS.
13. Incubate overnight at 37°C. The solution should get clear and viscous.
14. Add 350 µL saturated 6 M NaCl.
15. Shake vigorously for 15 s.
16. Centrifuge for 15 min at 1000g.
17. Transfer supernatant into a clean tube. When supernatant is cloudy repeat this step.
18. Add two volumes of absolute ethanol to precipitate the DNA.
19. Dissolve DNA in 300–1000 µL TE-4.
20. Incubate 65°C for 30 min to get rid of any DNAse contamination.
21. Dissolve overnight.
22. Store the DNA sample at 4°C.

3.2. PCR

1. Set up the following reaction (30 µL) by adding the reagents in this order: 500 ng of a genomic male patient DNA, 500 ng of a male control DNA (*see* **Note 5**), H$_2$O, dNTP's mix (25 mM), 12.5 pmol/µL of each oligonucleotide primer (**Table 2**, *see* **Note 6**), 10X Super *Taq* buffer, 15 mM MgCl$_2$ (*see* **Note 7**) and 1.5 U Ampli*Taq* DNA polymerase (5 U/µL).
2. PCR is performed in a thermal cycler (DNA Engine Tetrad, MJ Research) using the following conditions (*see* **Note 8**):

Initial denaturation	94°C, 5 min
Stage I (5 cycles)	94°C, 30 s denaturation
	52°C, 30 s annealing
	72°C, 1 min extension
Stage II (5 cycles)	94°C, 30 s denaturation
	48°C, 30 s annealing
	72°C, 1 min extension
Stage III (25 cycles)	94°C, 30 s denaturation
	45°C, 30 s annealing
	72°C, 1 min extension
Stage IV (Final)	72°C, 5 min extension
	94°C, 10 min denaturation
	50°C, 1 min annealing
	4°C, soak

3. Store samples at 4°C.

3.3. Polyacrylamide Gel Electrophoresis (see *Note 9*)

1. Clean the glass plates with a specific glass detergent, rinse with 70% ethanol and dry them carefully.
2. Clean the spacers and comb with water and dry them carefully.

Table 2
DGGE Multiplex Exons

DGGE Multiplex[a]	DMD Exons[b]	Gel Denaturation (% U/F)[c]
A	38 - 43 A[d] - 47 - 50	15–30/15–50[e]
B	40 - 30 - 03 B	20–50
C	26 B - 20 A - 79	20–50
D	46 - 54 B - 13 B - 69	25–65/25–50[e]
E	15 - 29 B - 52	20–50
F	44[f] - 53 A - 25 - 65	25–65
G	43 A - 23 A - 56 - 71 - 41 B	25–65/25–60[e]
H	29 A - 26 A - 09	25–65/20–55[e]
I	34 - 20 B - 21 - 43 B - 60	25–65/25–50[e]
J	58 - 23 B - 51 A	20–40
K	16 - 61 B - 67	25–65
L	66 - 27 - 14	25–65
M	13 A - 70 - 41 A - 61 B	25–65
N	32 - 74 - 37 - 19	25–65/25–50[e]
O	08[f] - 77[f] - 72 - 68 A	25–65
P	11 - 31 - 64	15–50
Q	01[f] - 07 - 62 - 47	15–50
R	12 - 06 B - 51 B - 04	20–60/20–55[e]
S	02 - 73 - 06 A - 24[g] - 22 - 57	20–60
T	28 - 18 B - 49 - 54 A	25–50
U	05 - 33 - 17 - 45	20–50/15–50[e]
V	39 - 36 - 55	15–50

[a] Name of DGGE multiplex.
[b] Exon number in human DMD gene.
[c] Low and high % urea/formamide denaturation density gel solutions.
[d] Some exons had to be split in two overlapping fragments (see **Note 1**).
[e] Better % urea/formamide denaturation for the multiplex.
[f] Use 2 times more of this primer.
[g] Use 1.5 times more of this primer.

3. Assemble the DGGE gel sandwich and put it on the casting stand with the short glass plate forward. Check if the plates and spacers are aligned and even at the bottom.
4. Fill the electrophoresis tank with 7 L of 0.5X TAE buffer. Do not reuse the running buffer.
5. Preheat the buffer to the desired temperature (58°C). It may take up to 1.5 h.
6. Prepare the first gel (3 mL) with a 0% urea/formamide, 9% polyacrylamide solution. Add 80 µL 10% APS and 4 µL TEMED per 3 mL. Pour a layer of this solution on bottom of the gradient gel with a plastic pasteur pipet. Take care with air bubbles. Let it polymerize for 30 min.
7. Prepare a 11 mL solution with the low % urea/formamide and a 11 mL solution with the high % urea/formamide density gel solutions (**Tables 2** and **3**). In this system a 25 mL will just fill the plates. So you can do 11 mL of each one of the above solutions.
8. Add 60 µL 10% APS and mix by inverting the tubes several times.
9. Add 3 µL TEMED and mix again.
10. Put the solutions in the gradient former. The high density solution in the column with the outlet tube. Fill the gel sandwich until 5 mm below the comb. Let polymerize for one and a half hour.

Table 3
DGGE Density Gel Solutions

Gel (%)^a	0% Urea/Formamide^b (mL)	80% Urea/Formamide^b (mL)
0	11.0	0.0
5	10.3	0.7
10	9.6	1.4
15	8.9	2.1
20	8.3	2.8
25	7.6	3.4
30	6.9	4.1
35	6.2	4.8
40	5.5	5.5
45	4.8	6.2
50	4.1	6.9
55	3.4	7.6
60	2.8	8.3
65	2.1	8.9
70	1.4	9.6
75	0.7	10.3
80	0.0	11.0

^{a}% denaturation of DGGE gel.
^{b}Low and high % Urea/Formamide density gel solutions.

11. After polymerization wash the gel with water using a plastic pasteur pipet. Remove excess of water with a filter paper.
12. Prepare the stacking gel (5 mL) with a 0% U/F, 9% polyacrylamide solution. Add 70 µL 10% APS and 3.5 µL TEMED per 5 mL. Pour a layer of this solution on top of the gradient gel with a plastic pasteur pipet. Take care with air bubbles. Let it polymerize for 15–30 min.
13. After polymerization remove the gel sandwich from the casting stand and put it inside the electrophoresis tank. Check if the running buffer is on the desired temperature.
14. Remove the comb carefully and rinse the slots with running buffer.
15. An aliquot (10 µL) of each amplification mixture is combined with 6X DGGE gel loading buffer and pipetted into a gel lane.
16. Gel electrophoresis is performed at 100V constant voltage (20–40 mA for 1 gel), 58°C temperature for 16 h.
17. After the run, stain the gel for 15–20 min in 250 mL water containing 30 µL Ethidium Bromide solution on a gently shaking shaker.
18. Remove excess Ethidium Bromide rinsing the gel 3X with water.
19. Examine and photograph the gel under UV light (*see* **Notes 10–12**).

3.4. DNA Sequencing Reactions

1. Samples that have band shifts in the gel are re-amplified with the specific primer pair corresponding to that exon (*see* **Note 13**).
2. After PCR, products are sequenced in the ABI PRISM model 377, using ABI PRISM BigDye terminator Cycle Sequencing Ready Reaction Kit with Ampli Taq DNA Polymerase (Perkin Elmer) (*see* **Note 14**).

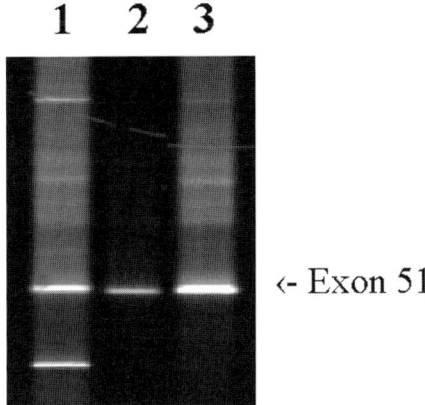

Fig. 1. *De novo* mutation detection. DGGE screen of DMD amplified exon 51. We can see a heteroduplex band in the DMD patient gene amplification corresponding to a 7547 C > T, Gln 2447 X mutation. Lane 1, DMD patient. Lane 2, DMD patient's mother. Lane 3, exon 51 normal sequence control.

4. Notes

1. DGGE segments have a GC clamp tail as short as 40 bp because it efficiently serves as a high TM domain preventing the complete denaturation of the fragment. The GC clamp tail increases the percentage of mutation detection to close to 100%. To improve detection of sequence changes some exons had to be split in two overlapping fragments because of the exon size, GC content or more than one Tm domain.
2. In the case of a blood clot, remove it with a Pasteur pipet and isolate DNA from the noncoagulated blood.
3. If after centrifugation a white pellet is not visible probably you have incomplete lysis. Resuspend and wait more 15 min to complete the lysis.
4. If you do not want to continue DNA isolation next day you can add Pronase and 20% SDS and leave samples in the room temperature for 72 h maximum, or you can add 20% SDS and keep this way for 15 d maximum. After this you can continue your regular DNA isolation.
5. It is not necessary to mix female DNA with a control DNA because females have two X chromosomes. In this case the heteroduplexes are naturally formed.
6. Equal amounts of each primer are used in the multiplexes. Some exons are not included in multiplexes and we do one primer pair PCR reaction for it. It is possible to combine these remaining segments, before loading on gel, to reduce workload.
7. For exons 03 A and 03 B and 75 B use 3.0 mM MgCl$_2$, for exons 18 A and 61 A use 8.5 mM MgCl$_2$ and for exon 63 use 6 mM MgCl$_2$ to improve DGGE analysis.
8. Multiplexes E and J shows better results when we change the annealing temperature of the first PCR stage to 53°C.
9. In our specific case we use DCODE™ System, Universal Mutation Detection System (Bio-Rad) to run the DGGE gels.
10. If a high background is observed we can introduce a destaining step of about 15 min in water.
11. Two photographs showing typical displays gels for different situations are shown in **Figs. 1** and **2**. **Figure 1** shows exon 51 mutation (7547 C>T, Gln 2447 X) only in the patient. The mother does not have the same mutation. It is a new mutation case. **Figure 2** shows exon

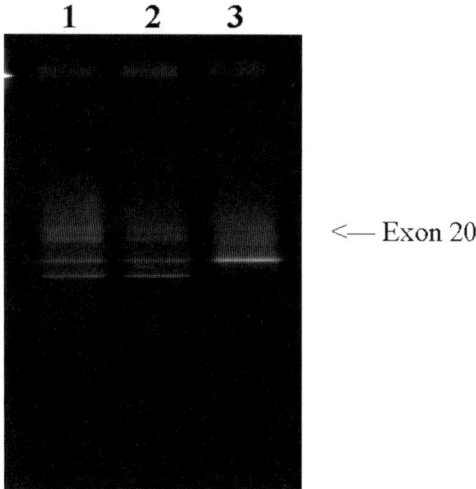

Fig. 2. Inherited mutation detection. DGGE screen of DMD amplified exon 20. We can see a heteroduplex band in both the mother and the DMD patient corresponding to a 2615 C > T, Gln 803 X mutation. Lane 1, DMD patient. Lane 2, DMD patient's mother. Lane 3, exon 20 normal sequence control.

20 mutation (2615 C>T, Gln 803 X) both in the patient and his mother. It is an inherited mutation case.
12. We could identify the DMD mutation in all patients we analyzed. Two other groups have also successfully identified mutations using the same system with some modifications *(10)*.
13. In this case samples are amplified using the non-GC-clamp primer and sequenced directly.
14. Other sequencing methods can be used.

Acknowledgments

This work was supported by CAPES, CNPq, and FUJB. The author greatly acknowledge the diagnostic and research group from Leiden and Groningen Universities (the Netherlands), especially Dr. Johan T. den Dunnen, Dr. Egbert Bakker and Dr. Robert M. W. Hofstra, for the important help in the beginning of this work.

References

1. Koenig, M., Hoffman, E. P., Bertelson, C. J., Monaco, A. P., and Kunkel, L. M. (1987) Complete cloning of the Duchenne muscular dystrophy (DMD) cDNA and preliminary genomic organization of the DMD gene in normal and affected individuals. *Cell* **50,** 509–517.
2. Den Dunnen, J. T., Grootscholten, P. M., Bakker, E., Blonden, L. A. J., Ginjaar, H. B., Wapenaar, M.C., et al. (1989) Topography of the Duchenne muscular dystrophy (DMD) gene: FIGE and cDNA analysis of 194 cases reveals 115 deletions and 13 duplications. *Am. J. Hum. Genet.* **45,** 835–847.
3. Hu, X., Ray, P. N., Murphy, E. G., Thompson, M.W., and Worton, R. G. (1990) Duplicational mutation at the Duchenne muscular dystrophy locus: its frequency, distribution, origin and phenotype genotype correlation. *Am. J. Hum. Genet.* **46,** 682–695.

4. Chamberlain, J. S., Gibbs, R. A., Ranier, J. E., Nguyen, P. N., and Caskey, C. T. (1988) Deletion screening of the Duchenne muscular dystrophy locus multiplex DNA amplification. *Nucleic Acids Res.* **16,** 11,141–11,156.
5. Beggs, A. H., Koenig, M., Boyce, F. M., and Kunkel, L.M. (1990) Detection of 98% of DMD/BMD gene deletions by polymerase chain reaction. *Hum. Genet.* **86,** 45–48.
6. Orita, M., Iwahana, H., Kanazawa, H., Hayashi, K., and Sekiya, T. (1989) Detection of polymorphisms of human DNA by gel electrophoresis as single-strand conformation polymorphism. *Proc. Natl. Acad. Sci. USA* **86,** 2766–2770.
7. Cotton, R. G. H. (1989) Detection of single base changes in nucleic acids. *Biochem. J.* **263,** 1–10.
8. Soto, D. M. and Sukumar, S. (1992) Improved detection of mutations in the p 53 gene in human tumors as single-strand conformation polymorphisms and double-strand heteroduplex DNA PCR. *Methods Appl.* **2,** 96–98.
9. White, M. B., Carvalho, M., Derse, D., O'Brien, S. J., and Dean, M. (1992) Detecting single base substitutions as heteroduplex polymorphisms. *Genomics* **12,** 301–306.
10. Mendell, J. R., Buzin, C. H., Feng, J., Yan, J., Serrano, C., Sangani, D. S., et al. (2001). Diagnosis of Duchenne dystrophy by enhanced detection of small mutations. *Neurology* **57,** 645–650.
11. Fischer, S. G. and Lerman, L. S. (1979). Lenght-independent separation of DNA restriction fragments in two-dimensional gel electrophoresis. *Cell* **16,** 191–200.
12. Fischer, S.G. and Lerman, L.S. (1983). DNA fragments differing by a single base-pair substitution are separated in denaturing gradient gels: correspondence with melting theory. *Proc. Natl. Acad. Sci. USA* **80,** 1579–1583.
13. Myers, R. M., Fischer, S. G., Lerman, L. S. and Maniatis, T. (1985). Nearly all single base substitutions in DNA fragments joined to a GC-clamp can be detected by denaturing gradient gel electrophoresis. *Nucleic Acids Res.* **13,** 3131–3145.
14. Myers, R. M., Maniatis, T., and Lerman, L.S. (1987). Detection and localization of single base changes by denaturing gradient gel electrophoresis, in Methods in Enzymology, vol. 155. (Wu, R., ed.), Academic Press, New York; pp. 501–527.
15. Lerman, L. S. and Silverstain, K. (1987). Computational simulation of DNA melting and its application to denaturing gradient gel electrophoresis, in Methods in Enzymology, vol. 155. (Wu, R., ed.), Academic Press, New York; pp. 482–501.
16. Hofstra, R. W. M. and den Dunnen, J. T. Personal communication. (*see* Website http://www.dmd.nl/dgge.html)
17. Dolinsky, L. C. B., Moura-Neto, R. S., and Falcão-Conceição, D. N. (2000) DGGE scan as a tool to look for new mutants and carriers of the DMD gene. *Am. J. Hum. Genet.* **67 (Suppl),** 4.
18. Miller, S. A., Dykes, D. D., and Polesky, H. F. (1988) A simple salting out procedure for extracting DNA from human nucleated cells. *Nucleic Acids Res.* **16,** 1215.

17

Genetic Diagnosis of Charcot-Marie-Tooth Disease

Frank Baas

1. Introduction

Over a century ago, the two French neurologists Charcot and Marie and the English neurologist Tooth described a peripheral neuropathy. They had defined a clinical entity, which is now known as Charcot-Marie-Tooth disease (CMT) or hereditary motor and sensory neuropathy (HMSN). Neuropathies are diseases of the peripheral nervous system (PNS). Many of them are disorders with a genetic basis (hereditary motor and sensory neuropathy, HMSN) but geneticists still use the term Charcot-Marie-Tooth disease (CMT). CMT has a prevalence of 1:2500 and is divided into several groups on the basis of clinical, electrophysiological and histological criteria. CMT 1 is a disorder with hypertrophic nerves, reduced nerve conduction velocities (NCVs) and signs of de- and remyelination. Type 2 is characterized by normal NCVs and a decreased number of large myelinated fibers on nerve biopsy, suggesting that the pathology of the axon is the major cause of disease. Type 3 is an autosomal recessive disease and usually a more severe form of CMT 1. CMT 4 is also an autosomal recessive disorder with axonal and demyelinating pathology. CMT-X is a dominant X-linked form of Charcot-Marie-Tooth disease, with severely affected males and, in many cases, females showing signs of a peripheral neuropathy.

To date, 10 genes (PMP22, peripheral myelin protein 22; MPZ, myelin protein zero; GJB1, connexin 32; EGR2, early growth response protein; NDRG1, N-myc down regulated gene 1, MTMR2, myotubularin related protein 2 kinase; GAN1, gigaxonin; NEFL, neurofilament light chain; KIF1B, kinesin 1Bβ; and PRX, periaxin) have been found mutated in CMT disease.

CMT 1A is the most frequent form of autosomal dominant demyelinating CMT and is caused by the presence of an additional copy of a 1.5 Mb region of chromosome 17, encompassing the PMP22 gene. This duplication results from unequal crossing-over of homologous chromosomes. The duplicated region is flanked by two highly homologous sequences, the CMT1A-REP *(1)*. The initial identification of the duplication suggested that a gene at the breakpoint or in the duplication causes CMT1A. However, after the identification of a point mutation in the PMP22 gene in a duplication negative patient with CMT1A, we showed that alterations in the sequence or gene dosage of the PMP22 gene are the cause of this disease *(2)*. The *Trembler* mouse, a model for CMT, has exactly the same mutation as we identified in a Dutch family. Subsequent studies identified more mutations in the PMP22 gene as well as other animal models for this disease. The reciprocal recombinational event of the unequal sister chromatid exchange, a deletion of 1.5 Mb encompassing the PMP22 gene, was identified in patients with hereditary neuropathy with liability to pressure palsies (HNPP) and we identified a stop codon in the PMP22 gene in a HNPP patient lacking this deletion, again implicating PMP22 as the gene responsible for HNPP *(3)*. Therefore, either overexpression or underexpression of the PMP22 gene can cause a peripheral neuropathy.

The myelin protein zero (MPZ) was identified as the gene for CMT 1 B *(4)*. This form is less frequent than CMT 1A and seems to have a more variable presentation. The phenotype can vary from severely slowed nerve conduction velocity with demyelination on nerve biopsies, to patients with (near) normal nerve conduction velocities suggesting mainly axonal involvement. Several cases of CMT 3 turned out to have either mutations of the PMP22 or MPZ gene or were *de novo* duplications suggesting recessive inheritance.

A severe form of CMT is called Dejerine-Sottas (DS) disease. Genetic studies have shown that DS can be due to mutations in several CMT genes. Initially DS was described as a recessive disorder, but we have shown that in fact the majority of the Dutch and Belgian DS cases are due to *de novo* duplication of the PMP22 gene *(5)*. This high frequency of *de novo* mutations is due to the high frequency for recombination surrounding the CMT1 region owing to the presence of the two highly homologous CMT1-REPS.

Diagnostic procedures for a dominant peripheral motor and sensory neuropathy are based on the detection of a change in allele dosage for the PMP22 gene. The common duplication encompassing PMP22 is 1.5 Mb long. Only when a change in PMP22 copy number or a rearrangement of chromosome 17 is not detected, mutation analyses for PMP22, MPZ, and GJB1 indicated in demyelinating forms of CMT. For axonal cases of CMT, analysis of GJB1 is warranted. The other CMT genes identified have thus far only been found mutated in a small proportion of the patients and hence do not warrant large-scale screening.

In principle a genetic test for the presence or absence of a duplication should give an unequivocal result. Methods for the detection of a gene dose alteration are preferably based on the presence or absence of specific DNA rearrangements. However, since not all PMP22 duplications result in the same DNA rearrangement, this approach is not feasible for all cases of CMT1A. We find that only 70% of the duplications can be detected by PCR-based approaches *(6)*. Therefore, we use a Southern blot based test to determine the allele dosage for PMP22, the gene for CMT1A, and a closely located marker, VAW409R3. The signals obtained are corrected for DNA loading by comparison to a reference marker E3.9, which is located on chromosome 22.

2. Materials

2.1. DNA Isolation

1. ACD buffer: 22.8 mM citric acid, 44.9 mM sodium citrate, 81.6 mM glucose (according to Becton Dickinson).
2. Buffers C1, G2, QBT, QC, QF (Qiagen Tip 100/G kit is supplied with buffers).
3. Ethanol 96% and ethanol 70%.
4. H$_2$O (Millipore or similar).
5. Isopropanol.

Genetic Diagnosis of CMT

6. Phosphate-buffered saline (PBS).
7. Proteinase K (20 µg/µL).
8. Qiagen Genomic Tip 100/G.
9. TE buffer: 10 mM Tris-HCl; 1 mM EDTA, pH 7.5.

2.2. DNA Digestion

1. EcoR1: store at –20°C.
2. EcoRI Restriction buffer: 50 mM Tris-HCl (pH 8.0), 10 mM MgCl$_2$, 100 mM NaCl.
3. Phenol mix, pH 8.0: (25:24:1), add: 25 vol Phenol, 24 vol Chloroform, 1 vol Isoamyl-alcohol. Store at 4°C in the dark.
4. 3 M NaAc, pH 5.2.
5. Ethanol 100 + 70%.
6. 5X Gel loading buffer: 50 mM Tris-HCl, 50 mM EDTA pH 8, 0.5% SDS, 1.46 M sucrose, 0.25% Orange G.

2.3. Agarose Gel Electrophoresis and Southern Blotting

1. Agarose: Multi purpose agarose.
2. DNA marker: λ-HindIII ladder (0.04 µg/µL).
3. Ethidium bromide: 5 µg/µL.
4. Neutralization buffer: 0.3 M Tris-HCl, 3 M NaCl, pH 7.0.
5. Denaturation buffer: 0.5 M NaOH, 1.5 M NaCl.
6. TAE buffer 50X: 2 M Tris-HCl, 5.7% glacial acetic acid, 50 mM EDTA, pH 8.0.
7. 20X SSC: 3 M NaCl, 0.3 M Sodium citrate, pH 7.0

2.4. Hybridization

1. Pre-hybridization buffer: 1 mM EDTA, 0.5 M Na$_2$HPO$_4$, pH 7.5, 7% SDS, 1% BSA fract.V.
2. RP mix (Random Primed Labeling Mix): 0.09 M HEPES, pH 6.9, 0.01 M MgCl$_2$, 0.05 mM dCTP, 0.05 mM dTTP, 0.05 mM dGTP, 0.01 M DTT, Random Primers (pd(N)$_6$): 0.0125 U/µL.
3. α-^{32}P-dATP: Specific activity: 3000 Ci/ mmol, 10 mCi/mL (9250 kBq/25 µL).
4. Column buffer (TES): 10 mM Tris-HCl, pH 7.5, 0.5 mM EDTA, 0.1% SDS.
5. Sephadex G50.
6. Wash buffers: 2X SSC/0.1% SDS; 1X SSC/0.1% SDS; 0.3X SSC/0.1% SDS.

2.5. Probe Information (PMP22/ VAW409R3a/E3.9) (see Note 1)

PMP22:
        Locus: 17p11.2
        Vector: pBluescript (3.0 kb)
        Selectable marker: Ampicillin (AMP)
        Cloningsite: EcoRI
        Insert size: 570 bp
        Genomic fragment detected on Southern blot : 11, 8, and 5 kb

VAW409R3a:
        Locus: 17p11.2 - 17p12 (D17S122)
        Vector: pUC18 (2.7 kb)
        Selectable marker: Ampicillin
        Cloning site: EcoRI - BamHI
        Insert size: 1.4 kb
        Genomic fragment detected on Southern blot: 2.1 kb

E3.9:
> Locus: 22q1.12-22q12 (CRYB3-gen)
> Vector: PSP65 (3.0 kb)
> Selectable marker: Ampicillin
> Cloning site: EcoRI
> Insert size: 3.9 kb
> Genomic fragment detected on Southern blot: 3,9 kb

In the probe mix use: 9 ng PMP22: 6 ng VAW409R3a, 9 ng E3.9

3. Methods

3.1. Genomic DNA Isolation from Blood

1. Add 1 mL ACD solution to 6 mL blood (EDTA) and mix gently (by hand). Store at −20°C (max. 6 mo).
2. Thaw blood/ACD mixture at room temperature.
3. Add to 5 mL blood/ACD mixture: 1X volume buffer C1 (cold) plus 3X volume dH_2O (cold).
4. Mix gently (by hand), then incubate on ice for 10 min.
5. Centrifuge at 1300g at 10°C for 15 min and discard supernatant gently.
6. Dissolve pellet in 1 mL buffer C1 (cold) plus 3 mL dH_2O (cold).
7. Mix and incubate on ice for 5 min.
8. Centrifuge at 1300g at 10°C for 15 min and discard supernatant gently.
9. Dissolve pellet in 5 mL buffer G2 and mix for 30 s.
10. Add 50 µL Proteinase K, mix and incubate at 50°C for 30–60 min.
11. Isolate DNA with Qiagen Genomic Tip 100/G. (*see* **Note 2**).

3.2. Restriction Digestion of Genomic DNA with EcoRI

The following steps should be performed in duplicate, since dosimetry will be done on two different gels and membranes.

1. For each gel: 3 normal samples (controls for 2 copies of PMP22), 1 HMSN positive control (duplicated samples, 3 copies of PMP22), 1 HNPP positive control (deleted sample, 1 copy of PMP22), should be prepared. This leaves sufficient space on the gel for the patients samples.
2. Incubate DNA samples at 50°C for 10 min to make sure that it is properly dissolved.
3. Combine 6 µg of genomic DNA, 20 µL 10X *Eco*RI Restriction buffer and H_2O to a final volume of 200 µL.
4. Add 1 µL EcoRI enzyme (50 U/µL).
5. Mix and incubate at 37°C for 16–18 h overnight.
6. Add 200 µL Phenol mix and mix.
7. Centrifuge at maximum speed for 15 min in microcentrifuge.
8. Take the (upper) DNA phase and add: 20 µL (3M) NaAc plus 500 µL ethanol 100%.
9. Mix and incubate at −20°C for 16–18 h (or at −80°C for 90 min).
10. Centrifuge at maximum 10,000g for 15 min in a microcentrifuge.
11. Discard supernatant carefully.
12. Wash pellet with 200 µL 70% Ethanol.
13. Centrifuge at maximum speed for 15 min in microcentrifuge, 10,000g.
14. Discard ethanol and dry pellet for 2 min in speedvac (Savant).
15. Dissolve pellet in 16 µL Tris/EDTA buffer.
16. Incubate at 55°C for 2 h.
17. Add 4 µL gel loading buffer and mix.

3.3. Agarose Gel Electrophoresis

1. Prepare 1X TAE running buffer from 50X stock.
2. Add 1.2 g agarose per 150 mL 1X TAE buffer use 150 mL per gel (make 2 gels).
3. Dissolve the agarose by heating in microwave oven, cool and pour into gel tray.
4. Incubate DNA samples with gel loading mix (and DNA markers) at 60°C for 10 min.
5. Centrifuge briefly (10–15 s) at maximum speed in microcentrifuge.
6. Place gel in 1X TAE running buffer and pipet samples carefully into gel lanes (do not forget a DNA marker).
7. Gel electrophoresis is performed at 1.3 V/cm gel for 16–18 h.
8. The gel is then stained with ethidium bromide :
    a. Place gel in 200 mL 1X TAE buffer
    b. Add 30 µL ethidium bromide
    c. Shake gently for 20 min
9. Visualize the DNA in the gel on a UV box (take a picture)
10. Fragment DNA products with UV light: 100 s (1.2 J/cm^2)

3.4. Southern Blotting

1. Transfer the DNA in the gel onto a positively charged nylon membrane (Hybound +) by Southern blotting.
2. First the gel is pretreated by soaking in a series of solutions to denature and neutralize the DNA and gel matrix:

    a. Place gel in a clean dish containing: 200 mL 0.5 M NaOH/1.5 M NaCl (denature).
    b. Shake slowly on a platform shaker for 20 min at room temperature.
    c. Replace with fresh denaturation buffer and shake for an additional 20 min.
    d. Pour off buffer and add: 200 mL 0.3 M Tris/3 M NaCl, pH 7.0 (neutralize).
    e. Shake as before for 20 min, replace with fresh neutralization buffer and shake another 20 min.
    f. The gel is then blotted overnight onto a membrane in neutralization buffer.

3. After overnight blotting remove the membrane from the gelstack.
4. Rinse the membrane briefly with 2X SSC, place it on a sheet of dry Whatman 3 MM paper and allow it to dry.
5. To immobilize the DNA, it is cross linked onto the membrane by UV in a UV cross-linker (use setting suggested by the manufacturer).

3.5. Hybridization with a Radio-Labeled DNA Probe

The membranes are hybridized with a mixture of the PMP22/VAW409R3a/E3.9 probe.

1. The probe is labeled with a-^{32}P-dATP (random-primed labeling).

    a. Mix PMP22/VAW409/E3.9 cocktail (21 ng, see Subheading 2.5) with 1 µL α-^{32}PdCTP, 1 µL Klenow polymerase and 1 µL 10 x RP-mix in a final volume of 10 µL and incubate for 60 min at 37°C. Stop the reaction by adding 100 µL column buffer (TES).
    b. Remove the free nucleotides from the probe over a Sephadex spin column.

2. Wet the membranes in 2X SSC.
3. Place them in a hybridization tube and add 30–40 mL hybridization buffer.
4. Place the tube in the hybridization oven and incubate > 2 h with rotation at 65°C.
5. Denature the probe by heating for 2 min at 100°C.
6. Place it on ice for 5 min.

7. Pour off the buffer from the hybridization tube and replace with 6 mL fresh buffer per membrane (buffer should be pre warmed at 65°C).
8. Add denatured probe and incubate o/n at 65°C.
9. Pour off the hybridization buffer/probe mix (into the radioactive waste!).
10. Place membranes in a clean dish containing 200 mL 2X SSC/0.1% SDS.
11. Shake for 20 min at 65°C, replace with fresh buffer and shake for another 20 min.
12. Replace the buffer with 200 mL 1X SSC/0.1% SDS and shake for 20 min, replace the buffer and shake for 20 min.
13. Replace the buffer with 200 mL 0.3X SSC/0.1% SDS and shake for 20 min, refresh the buffer and shake for further 20 min.
14. Pour off the final wash solution and air dry the membranes.
15. Wrap the membranes in plastic wrap and expose to a Phosphor Imager Screen.
16. After the exposure the results are analyzed with an Image Analysis program.

3.6. Dosimetry

1. The images from the exposed Phosphor Imager plates are analyzed with commercial image analysis software (e.g., ImageQuant from Pharmacia or Aida from Raytest, Germany). The ratio of the lowest PMP22 band, the E3.9 band and the VAW409R3a band are determined. We analyze only the 5 kb (lowest) PMP22 band since for dosimetry fragments of similar size should be analyzed. In case of (minimal) DNA degradation, small fragments are much less affected than large fragments. Indeed we see in some cases that the large PMP22 bands are slightly degraded. The DNA isolated by the Qiagen columns is of high molecular weight (> 50 kb) and perfectly suitable for this type of analysis. Degradation of the DNA is usually due to poor quality of the blood samples.
2. The ratio of the signals for PMP22 and VAW409 compared to the loading control E3.9 is determined for three normal controls (**Fig. 1**, lanes 2,3,4) and a mean for this group is calculated. This ratio is set at 100% and the patient samples are compared to this ratio *(7)*. Theoretically, a duplicated PMP22 sample should have 150% of the ratio of signals for PMP22/E3.9 and VAW409/E3.9 in the controls. The deleted samples (HNPP) have only 50% of the ratios of the controls. In practice, however, 150% is not always found. This is probably due to a contribution of the background signal noise to the hybridization signals. Generally we see ratios in the 130–150% range *(7)*. We have made a dataset of controls and duplicated samples to determine the range in which the duplicated samples should fall. Since each test is performed in duplicate and 2 different loci are measured in each test, the amount of false-positives is negligible. In rare cases there is a discrepancy between the two duplicate samples, most likely due to poor quality of the DNA (degradation). In that case the test is repeated. Some times the patients DNA must be isolated again.
3. **Figure 1** gives a representative example of a duplication test. The densitometric analysis is given in **Table 1**. The ratios of the signals of the bands for PMP22 and VAW3.9 over the E3.9 signal are given. Lanes 1, 8, and 12 are CMT duplications and lanes 5 and 6 are HNPP deletions. Lane 10 shows partially degraded DNA and rare polymorphisms of the PMP22 gene are shown in lanes 16 and 19. In both cases a restriction fragment length polymorphism results in alternatively sized PMP22 bands. In lane 16 the 4 kb PMP22 band is polymorphic, resulting in altered gene dose for this band when analyzed with this test. In this case the VAW409 signal should be used to determine whether a duplication is present. In lane 19 the 11 kb PMP22 band is polymorphic. This also does not affect the analysis of the VAW409 signal.
4. In very rare cases (< 1:500 duplications), we see a duplication of only PMP22 and not VAW409. PFGE analysis shows that this is due to alternatively sized duplications *(8)*. In that case only the PMP22 gene bands have an increased intensity.

Genetic Diagnosis of CMT

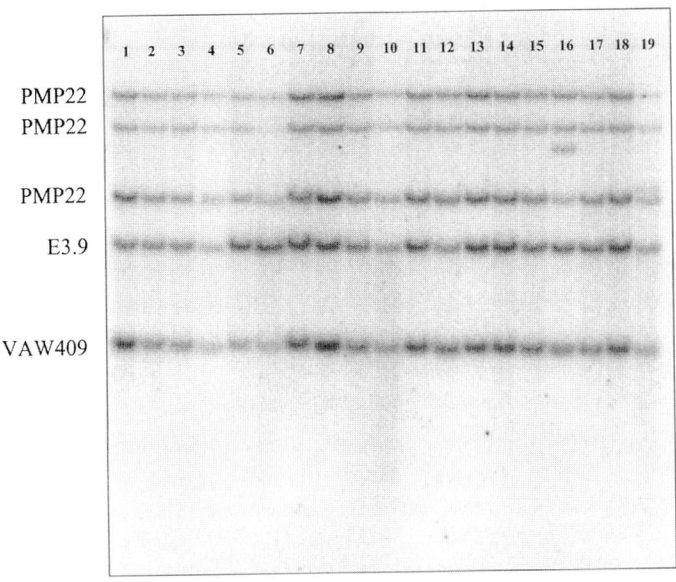

Fig. 1. Southern blot analysis for PMP22 gene copy number compared to the loading control E3.9.

Table 1
Densitometric Analysis of PMP22 Dosage Analysis

Lane	Sample	PMP22/E3.9[a]	VAW409/E3.9[a]	Result
1	Duplication	138	163	Duplication control
2	Normal	107	104	Control
3	Normal	98	101	Control
4	Normal	95	94	Control
5	Deletion	59	56	Deletion control
6	Patient	45	50	Deletion
7	Patient	95	89	Normal
8	Patient	150	140	Duplication
9	Patient	104	116	Normal
10	Patient	90	145	Degradation
11	Patient	106	109	Normal
12	Patient	142	148	Duplication
13	Patient	106	101	Normal
14	Patient	96	104	Normal
15	Patient	107	115	Normal
16	Patient	68	100	Polymorphism/normal
17	Patient	97	99	Normal
18	Patient	109	104	Normal
19	Patient	[b]	111	Polymorphism/normal

Densitometric analysis for PMP22 gene copy number.

[a]Ratio of PMP22/E3.9 is given as a percentage of the average signal ratios of the 3 normal control samples in lane 2, 3, and 4.

[b]This band could not be analyzed due to the presence of the polymorphism.

4. Notes

1. Probes are available upon request (*see* E-mail address: f.bass@amc.uva.nl).
2. DNA should be treated with care to avoid shearing. To analyze the DNA integrity 1 µg of undigested DNA can be run on a 0.6% agarose gel in 1X TAE at 1V/cm for 24 h The undigested DNA should be larger than 50 kb.

Acknowledgments

I thank Carin Sijmons for help in preparing this manuscript and Lourens Bordewijk for his assistance in developing the procedures described.

References

1. Reiter, L. T., Murakami, T., Koeuth, T., Pentao, L., Muzny, D. M., Gibbs, R. A., Lupski, J. R. (1996) A recombination hotspot responsible for two inherited peripheral neuropathies is located near a mariner transposon-like element. *Nat. Genet.* **3,** 288–297.
2. Valentijn, L. J., Baas, F., Wolterman, R. A., Hoogendijk, J. E., van den Bosch, N. H. A., Zorn, I., et al. (1992) Identical point mutations of the peripheral myelin protein gene PMP22 in Trembler-J mouse and Charcot-Marie-Tooth disease. *Nat. Genet.* **2,** 288–291.
3. Nicholson, G. A., Valentijn, L. J., Cherryson, A. K., Bragg, T. L., DeKroon, R. M., Ross, D. A., et al. (1994) A frame shift mutation in the PMP22 gene in hereditary neuropathy with liability to pressure palsies. *Nat. Genet.* **6,** 263–266.
4. Kulkens, T., Bolhuis, P. A., Wolterman, R., Kemp, S., te Nijenhuis, S., Valentijn, L. J., et al. (1993) Deletion of the codon for serine 34 from the major peripheral myelin protein P0 in Charcot-Marie-Tooth disease type 1B. *Nature Genet.* **5,** 35–39.
5. Hoogendijk, J. E., Hensels, G. W., Gabreëls-Festen, A. A. W. M., Gabreëls, F. J. M., Jansen, E. A. M., de Jonghe, P., et al. (1992) De novo mutation in hereditary motor and sensory neuropathy type 1. *Lancet* **339,** 1081–1082.
6. Stronach, E. A., Clarck, C., Bell, C., Lofgren, A., McKay, N. G., Timmerman, V., et al. (1999) A novel PCR based-based diagnostic tool for Charcot-Marie-Tooth Type 1A and Hereditary neuropathy with liability to pressure palsies. *J. Peripheral Nerv. Syst.* **4,** 117–122.
7. Hensels, G. W, Janssen, E. A. M., Hoogendijk, J. E., Valentijn, L. J., Baas, F., and Bolhuis, P. A. (1993) Quantitative measurement of duplicated DNA as a diagnostic test for Charcot-Marie-Tooth disease type 1A. *Clin. Chem.* **39,** 1845–1849.
8. Valentijn, L. J., Baas, F., Zorn, I., Hensels, G. W., de Visser, M., and Bolhuis, P. A. Alternatively sized duplication in Charcot-Marie-Tooth disease type 1A. *Hum. Mol. Genet.* **2,** 2143–2146.

18

Analysis of Human Mitochondrial DNA Mutations

Antonio L. Andreu, Ramon Martí, and Michio Hirano

1. Introduction

Mitochondria are the powerhouses of eukaryotic cells. These organelles generate energy in the form of adenosine triphosphate (ATP) from carbohydrates, fats, and proteins, via oxidative phosphorylation. By virtue of possessing their own genetic material—mitochondrial DNA (mtDNA)—mitochondria are unique mammalian organelles. Normal human mtDNA is a 16,569 base-pair (bp), double-stranded, circular molecule *(1)*. The molecules contain tightly compacted genes for 22 transfer (tRNAs), 13 polypeptides, and two ribosomal RNAs (rRNAs) (**Fig. 1**). All 13 polypeptides are subunits of the oxidative phosphorylation system: seven belong to Complex I (NADH-CoQ oxidoreductase), one to Complex III (CoQ-cytochrome *c* oxidoreductase), three to Complex IV (cytochrome *c* oxidase or COX), and two to Complex V (ATP synthase). These subunits are synthesized within the mitochondrion, where they are assembled together with a larger number of subunits encoded by the nuclear DNA (nDNA), that are synthesized in the cytoplasm and are transported into the mitochondrion *(2)*. Approximately 1,000 mitochondrial polypeptides are encoded in nDNA Complex II (succinate dehydrogenase-CoQ oxidoreductase), of which succinate dehydrogenase (SDH) is a component, is encoded entirely by nuclear genes; SDH thus serves as a marker for mitochondrial number and activity, independent of the mtDNA.

Since mitochondria are inherited only from the mother *(3)*, defects in mtDNA genes result in pedigrees exhibiting a pattern of solely maternal inheritance. Moreover, because there are hundreds or even thousands of mitochondria in each cell, with an average of 5 mtDNAs per organelle *(4,5)*, mutation in mtDNA may result in two populations of mtDNAs—mutated and wild-type—a condition known as heteroplasmy. The phenotypic expression of a mtDNA mutation is regulated by the threshold effect, that is, the mutant phenotype is expressed in the heteroplasmic cells only when the relative proportion of mutant mtDNAs reaches a certain value *(6)*. A respiratory chain deficiency may become manifest is some tissues, but not in others, if a number of mutant mtDNA exceeds a certain critical threshold. The threshold varies among tissues, depending on the oxidative energy requirements of that tissue: brain, heart, and skeletal muscle have extremely high energy requirements, and therefore it is no surprise that mitochondrial disorders frequently affect brain and muscle (i.e., mitochondrial encephalomyopathies and mitochondrial cardiomyopathies). Because both mitochon-

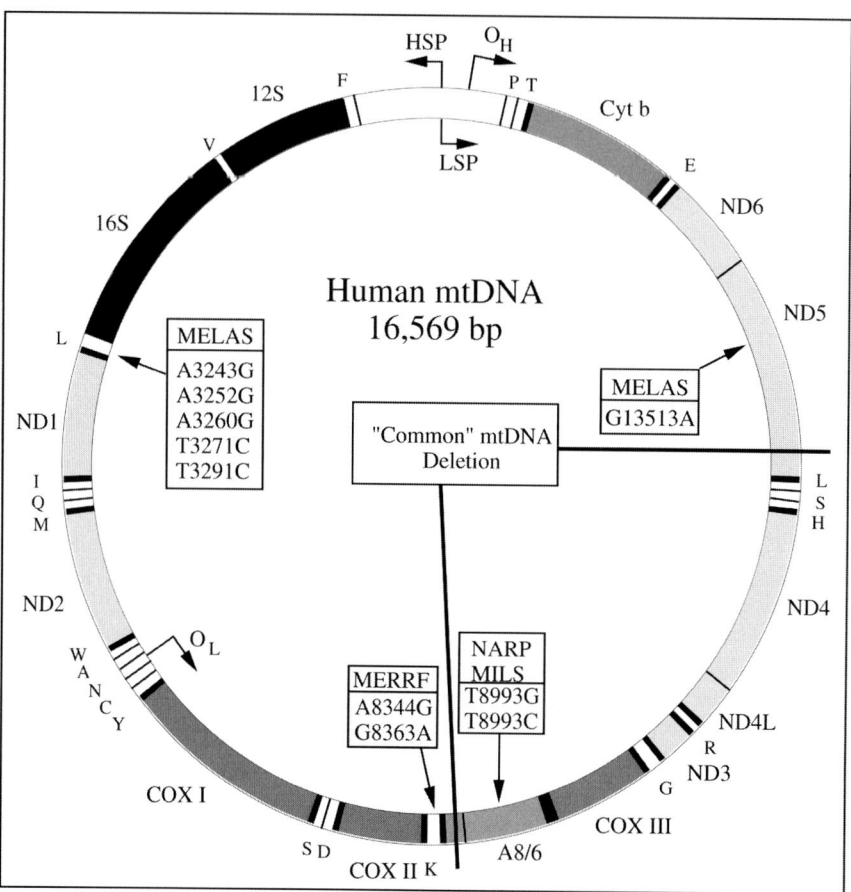

Fig. 1. Map of the human mitochondrial genome. The structural genes for the mtDNA-encoded 12S and 16S ribosomal RNAs, the subunits of NADH-coenzyme Q oxidoreductase (ND), cytochrome c oxidase (COX), cytochrome b (Cyt b), and ATP synthase (A), and 22 tRNAs (1-letter amino acid nomenclature), are shown. The origins of light-strand (O_L) and heavy-strand (O_H) DNA replication, and of the promoters for initiation of transcription from the light-strand (LSP) and heavy-strand (HSP), are shown by arrows. The "common" deletion, a mtDNA species often found in sporadic KSS/PEO is shown, as are common point mutations associated with maternally inherited encephalomyopathies (boxed).

drial division and mtDNA replication are random processes unrelated to cell division, a dividing cell will donate variable numbers of mitochondria and mtDNAs to its progeny. This process, known as mitotic segregation, can be important clinically if a patient harbors heteroplasmic populations or normal and mutated mtDNAs in his/her tissues. The phenotypic expression of a mutation may then vary among tissues or may change within a tissue over time.

From a practical point of view, the fact that mtDNA is polyplasmic presents serious problems when looking for point mutation with variable heteroplasmy in different tissues; therefore, it is best to analyze affected tissue. To date, the majority of mtDNA mutations have been identified in genes encoding tRNA, most notably, mutations associated with the clinical syndromes: mitochondrial encephalomyopathy, lactic acidosis, and stroke-like episodes (MELAS) and myoclonus-epilepsy and ragged-red fibers (MERRF) *(7–18)*. Our strategy for identifying mutations is based on polymerase chain reaction-restriction fragment length polymorphism (PCR-RFLP) analysis using, as a template, total DNA from patients. Because most of the mtDNA mutations present clinically as encephalomyopathies, skeletal muscle tissue typically harbors a high mutation load, thereby accounting for the greater likelihood of successfully identifying a pathogenic mutation in this tissue.

Another practical issue when performing mutational analysis of mtDNA is how to quantitate properly the percentage of mutated genomes. A PCR-based approach has a limitation when an accurate result is required. Typically, mutated and wild-type molecules co-exist and form heteroduplexes during PCR annealing steps. When the RFLP analysis is based on digestion of the mutant molecules, heteroduplex structures lead to underestimation of the amount of mutated mtDNA molecules. To avoid this artifact, a simple strategy is to add a radiolabeled nucleotide (i.e., ^{32}P-dATP) prior to the last PCR cycle, so that the percentage of mutated genomes will accurate represent the level in the tissue.

2. Materials

2.1. DNA Extraction

1. RSB buffer: 10 mM Tris-HCl, pH 7.4, 10 mM NaCl, 25 mM EDTA, 1% SDS.
2. Proteinase K solution: 5 µL of 100 µg/µL proteinase K stock solution in 0.5mL RSB buffer.
3. SEVAG: 4 mL of isoamylalcohol and 96 mL of chloroform (1:24 ratio).
4. TE buffer: 10mM Tris-HCl, pH 8.5, 1 mM EDTA.

2.2. Southern Blot Analyses

1. Depurination solution.   0.25 M HCl.
2. Denaturation solution.   0.5 M NaOH, 1 M NaCl.
3. Neutralization solution.   0.5 M Tris-HCl, 3 M NaCl, pH 7.4.
4. 10× SSC.   1.5 M NaCl, 0.15 M trisodium citrate.
5. Hybridization solution.   0.25 M sodium phosphate, pH 7.2, 7% SDS.
6. Washing solution I.   20 mM sodium phosphate, pH 7.2, 5% SDS.
7. Washing solution II.   20 mM sodium phosphate, pH 7.2, 1% SDS.

3. Methods

3.1. Isolation of Muscle DNA

The protocol presented here allows a high DNA recovery from an initial amount of 50 mg of muscle tissue (*see* **Note 1**). To isolate DNA from other sources (e.g., blood, skin, or hair follicles) commercially available kits can be used.

1. Crush 50 mg of frozen muscle into a fine powder (keep muscle in liquid nitrogen to prevent thawing). Keep mortar and pestle cold with liquid nitrogen.
2. Place the muscle powder in a 1.5 mL Eppendorf tube and add 0.5 mL of RSB buffer containing 1 mg/mL proteinase K.
3. Incubate overnight at 50°C or until solution is homogeneous.
4. Centrifuge tube and transfer supernatant to another tube.

Table 1
Analysis of Common mtDNA Point Mutations

Primers (5'-3')	PCR Fragment size	Restriction enzyme	Fragments
A3243G mutation			
(F) cctacttcacaaagcgcctt	402 bp	ApaI	307 and 95 bp
(R) cgatggtgagagctaaggtc			
A8344G mutation			
(F) ctaccccctctagagcccac	108 bp	BglI	73 and 35 bp
(R)[a]			
T8993G and T8993C mutations			
(F) ctataaacctagccatggcc	365 bp	HpaII/ AvaI[b]	194 and 171 bp
(R) agaggcttactagaagtgtg			

[a]The mismatch 41-mer primer is used: gtagtatttagttggggcatttcactgtaaagccgtgttgg.
[b]HpaII digests both the T-to-G and T-to-C mutations. AvaI digests only the T-to-G mutation.

5. Add 50 μL of 5 M NaCl.
6. Extract with 500 μL of phenol. Centrifuge at 14,000g for 10 min. Save the aqueous supernatant.
7. Extract with 250 μL of phenol and 250 μL of SEVAG. Centrifuge 10 min. Save supernatant.
8. Extract with 500 μL of SEVAG. Centrifuge 10 min. Save and transfer supernatant to a sterile tube.
9. Add 2 volumes of cold (−20°C) isopropanol and mix by inverting tube. DNA should precipitate and appear as white strands. Centrifuge 15 min.
10. Discard supernatant and rinse the pellet twice with ethanol 70%.
11. Resuspend the pellet in 50 μL of sterile water or TE buffer

3.2. PCR-RFLP Analysis of Common MtDNA Mutations

This PCR protocol can be used to study any putative mutation in mtDNA (*see* **Notes 2** and **3**). **Table 1** shows the oligonucleotide primers, PCR conditions, and RFLP strategy to study the most common mutations in the mtDNA. As a general rule, PCR conditions for mtDNA are less stringent than those for genomic DNA because the large number of mtDNA molecules provides many template targets for the primers. Accurate measurement of the proportion of mutated genomes is achieved by adding ^{32}P-dATP prior to the last PCR cycle (from this point, we are working with a radionuclide, therefore, safe laboratory practices for radioactive products must be applied). The PCR product is then digested with an appropriate restriction enzyme and the products of the digestion are run in a 12% nondenaturing polyacrylamide gel. The gel is then dried and subjected to autoradiography or exposed in a phosphor imager for quantification.

To minimize effort, we try to design primers 20-mer long with a G + C content of 50%. This will allow us to use annealing temperatures between 55–60°C.

1. Mix in a PCR tube.
   a. 20 ng of total DNA (in 1 μL).
   b. Water 25.5 μL.
   c. 10X PCR buffer 5 μL.
   d. 5 mM dNTP mixture 8 μL.

e. forward primer (2 pmol/μL) 5 μL.
f. reverse primer (2 pmol/μL) 5 μL.
g. Taq polymerase (5 U/μL) 0.5 μL.

2. We typically perform 30 cycles of PCR amplification; each cycle consists of three steps: 94°C for 15 s, 55°C for 15 s, 72°C for 1 min (Perkin-Elmer GeneAmp Thermal System 9700). After the initial PCR, we add 0.1 μL of ^{32}P-dATP (specific activity 3000 Ci/mmol) and perform a "last" cycle using the same conditions described earlier. If this "hot" PCR cycle is completed within 5 h of the first PCR, then additional Taq polymerase need not be added. However, if the hot PCR is performed the following day, then it is necessary to add 0.1 μL of Taq polymerase prior to the last cycle.

The sample is ready for digestion with an appropriate restriction enzyme. The product of the digestion is then run in a polyacrylamide gel (12% is an appropriate concentration for most of the RFLPs presented here) and the restriction pattern is analyzed.

3.3. Detection of Common Point Mutations of MtDNA by PCR-RFLP

Here, we present protocols for screening of four of the most common mtDNA mutations: A3243G, the A8344G, T8993G, and T8993C (number refer to the nucleotide (nt) in the original Cambridge sequence *(1,7,17,19,20)*. To date, the A3243G mutation has been the most prevalent mutation of mtDNA and is usually associated with MELAS syndrome; however, it can present with other clinical phenotypes such as diabetes and deafness (DAD) and maternally inherited progressive external ophthalmoplegia. The A8344G mutation is the most common cause of MERRF syndrome, but other mutations in the tRNA^{Lys} gene may cause a MERRF-like phenotype. Both the T8993G and T8993C mutations are associated with two phenotypes: neuropathy ataxia with retinitis pigmentosa (NARP) and maternally inherited Leigh syndrome (MILS) *(21,22)*. **Table 1** shows PCR-RFLP based strategies to screen for these mutations.

3.4. Detection of Point Mutations of MtDNA by Direct DNA Sequencing

Frequently, patients suspected of harboring a mtDNA molecular defect based on clinical, histological, and biochemical data, do not have one of the common mtDNA mutations. In this situation, one should screen for the presence of another point mutation. If the patient's clinical phenotype conforms to a well-recognized mitochondrial encephalomyopathy syndrome, e.g., MELAS, one can screen for less common mtDNA point mutations associated with that syndrome. If known mtDNA mutations are excluded, then direct sequencing of mtDNA provides the most simple and direct approach to identify a novel mtDNA mutation.

The next issue to resolve is the selection of mitochondrial genes to study. We base our decision on clinical, biochemical, and morphological criteria. For example, in patients with maternally inherited disorders with ragged-red fibers in a skeletal muscle biopsy and multiple defects of respiratory-chain enzymes with normal or elevated succinate dehydrogenase, we begin by sequencing the mtDNA-encoded tRNA genes. Alternatively, in a patient with a maternally inherited disorder with low activity of a single respiratory chain enzyme, we screen the mtDNA-encoded subunits of that enzyme complex. As most of the mutations described to date are in tRNA genes, we present a simple, fast, and reliable strategy to sequence all tRNAs genes, minimizing both effort and cost. **Table 2** shows primers and PCR conditions

Table 2
Primers and PCR Conditions to Amplify and Sequence the 22 mtDNA-Encoded tRNA Genes

Primer	5' site	Sequence (5'–3')	Tm (°C)	G +C% content
Fragment 1[a]				
M531F	531	taaccccatacccgaacca	62°	55
M718R	718	tcactggaacggggatgctt	62°	55
M1492F	1492	accctcctcaagtatattc	58°	45
M1803R	1803	tttcatctttcccttgcggt	58°	45
Fragment 2[b]				
M3087F	3087	ccaggtcggtttctatctac	60°	50
M3472R	3472	agagttttatggcgtcagcg	60°	50
M4181F	4181	acttcctaccactcacccta	60°	50
M4370F	4370	ttctccgtgccacctatcac	62°	55
M4388R	4388	tgataggtggcacggagaat	60°	50
M4556R	4556	ggtaaaaaatcagtgcgagc	58°	45
Fragment 3[c]				
M5454F	5454	ctcatcgcccttaccacgct	64°	60
M5651F	5651	cccttactagaccaatggga	60°	50
M5739R	5739	gcgggagaagtagattgaag	60°	50
M6022R	6022	tcggctcgaataaggaggct	62°	55
Fragment 4[d]				
M7361F	7361	agaaccctccataaacctgg	60°	50
M7667R	7667	gggcgtgatcatgaaaggtg	62°	55
M8123F	8123	aaccaaaccactttcaccgc	60°	50
M8490R	8490	atgggctttggtgagggagg	64°	60
Fragment 5[e]				
M9917F	9917	cgccgcctgatactggcatt	64°	60
M10149R	10149	tgtagccgttgagttgtggt	60°	50
M10347F	10347	atcctagccctaagtctggc	62°	55
M10568R	10568	taggcatagtagggaggata	58°	45
Fragment 6[f]				
M12057F	12057	aaaacaccctcatgttcata	58°	45
M12408R	12408	aacgagggtggtaaggatgg	62°	55
Fragment 7[g]				
M14587F	14587	accccataaataggagaag	58°	45
M14840R	14840	ttcatcatgcggagatgttg	58°	45
M15756F	15756	gaatcggaggacaaccagta	60°	50
M16119R	16119	tcatggtggctggcagtaat	60°	50

[a]Amplified with primers M531F and M1803R. Fragment size: 1086 bp (covers the following tRNAs: Phe and Val).

[b]Amplified with primers M3087F and M4553R. Fragment size: 1467 bp (covers the following tRNAs: Leu(UUR), Ile, Gln, and Met).

[c]Amplified with primers M5454F and M6022R. Fragment size: 568 bp (covers the following tRNAs: Trp, Ala, Asn, Cys, and Tyr).

[d]Amplified with primers M7361F and M8490R. Fragment size: 1129 bp (covers the following tRNAs: Ser(CUN), Asp, and Lys).

[e]Amplified with primers M9917F and M10568R. Fragment size: 670 bp (covers the following tRNAs: Gly and Arg).

[f]Amplified with primers M12057F and M12408R. Fragment size: 349 (covers the following tRNAs: His, Ser, and Leu[CUN]).

[g]Amplified with primers M14587F and M16119R. Fragment size: 1533 (covers the following tRNAs: Glu, Thr, and Pro).

to amplify and sequence all 22 tRNA genes with just seven PCR-amplified fragments. The GC% content of the primers have been adjusted such that all of the PCR reactions use the same annealing temperature (55°C).

3.5. Interpretation of mtDNA Sequencing Results

Due to its genetic characteristics (lack of introns and histones, and incomplete repair system), the mutation rate of mtDNA has been estimated to be 10-fold higher than that of nDNA *(23)* Consequently, one will find a high number of nucleotide changes when compared to the reference sequence. Most, but not all, these base-pair alterations will likely be previously identified neutral polymorphism. In this setting, establishing the putative pathogenic role of a nucleotide change can be a difficult task. The first step is to determine whether a particular nucleotide change has been previously reported as neutral or pathogenic. A comprehensive list of polymorphism can be found in MITOMAP: the human mitochondrial genome database (*see* Website: http://infinity.gen.emory.edu/mitomap.html). In general, to assign a pathogenic role to a nucleotide change the following canonical rules should be present:

1. The mutation should be heteroplasmic in the affected tissue.
2. The mutation can not be found in a series of at least 100 control DNAs from the same ethnic origin.
3. In the case of mutations in tRNA genes, the mutated nucleotide has to be well-conserved and, in addition, alters the secondary structure of the tRNA.
4. In the case of mutations in protein-coding genes, the resulting change in the translation has to affect an amino acid, that is well-conserved over evolution, indicating an important functional role.

3.6. mtDNA Depletion and Rearrangements

Patients with single large-scale deletions of mtDNA typically present with one of three clinical syndromes: Kearns-Sayre syndrome, sporadic progressive external ophthalmoplegia with ragged-red fibers, and Pearson marrow/pancreas syndrome *(19,24,25)*. The deletion range in length from 2–8.5 kb and are largely confined to an 11 Kb region in the major arc between the two origins of mtDNA replication (O_H and O_L; *see* **Fig. 1**). Regardless of the clinical presentation, 30–40% of all deletions are identical, that is called "common deletion," which spans 4,977 bp from the ATPase 8 gene to the ND5 gene. The clinical phenotype of the patients is determined by the distribution and relative abundance of the deletion in different tissues. Multiple deletions may present with a wide range of clinical phenotypes, but usually is associated with progressive external ophthalmoplegia, exercise intolerance, and muscle weakness *(26–29)*.

The standard technique to study mtDNA rearrangements is Southern blot analysis (**Fig. 2A**). Although long-PCR based strategies have been used to study deletions in mtDNA, the high number of false positives makes that technique difficult to interpret. Southern blot analysis is performed using a PCR-generated probe for mtDNA (**Table 3**). This probe is used to check for the presence of deleted molecules, which can be identified by conventional hybridization. Because the vast majority of mtDNA deletions are in the major arc between the origins of DNA replication, we use a probe in the minor arc between O_H and O_L (**Table 3**).

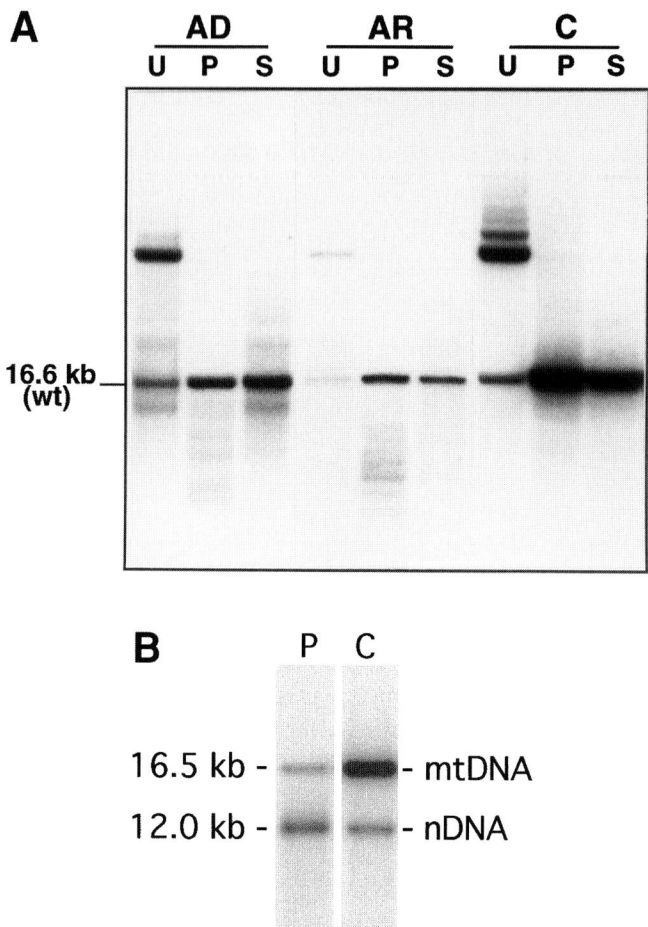

Fig. 2. **(A)** Southern blot analysis of total muscle DNA from an autosomal dominant progressive external ophthalmoplegia (AD) patient, autosomal recessive progressive external ophthalmoplegia (AR) patient, and a normal control (C). The DNA was digested with *Pvu*II (P), and *Sna*BI (S). In addition, uncut DNA (U) is included in the Southern blot analysis. The wild-type mtDNA molecule (wt) is 16.6 kb in length while multiple deletion of mtDNA are visible as faint bands below the wt-band. (Fig. 2A reprinted from Fig. 2 in **ref.** *[39]* with permission.) **(B)** Southern blot analysis of total muscle DNA from a patient with mtDNA depletion (P) and from a normal control (C). The nDNA band serves as a internal reference to account for the amount of total DNA loaded. (Fig. 2B reprinted with permission from Fig. 2 in **ref.** *[32]*).

Unlike point mutations and deletions, which are qualitative defects of mtDNA, depletion of mtDNA is a quantitative abnormality; there is a paucity of mtDNA, however, the residual mtDNA is qualitatively normal. Primary mtDNA depletion syndromes are presumed to be autosomal recessive disorders and typically manifest as myopathy, hepatopathy, or both in young children less than 2 y old *(30–32)*. Secondary depletion of mtDNA can be seen in disorders such as inclusion body myositis *(33)*. Finally, iatrogenic mtDNA depletion has been described in patients treated with nucleoside analogs (e.g., zidovudine, dideoxycytidine, and dideoxyinosine) that inhibit mtDNA replica-

Table 3
Probes for Southern Blot Analysis

Probe	Sequence (5'–3')
Mitochondrial DNA probe: The amplified fragment spans part of 16S-ND1 (nucleotides 1690–4707)[a]	(F): ccactccaccttactaccagac (R): gtaatgctagggtgagtggtagg
Nuclear DNA probe: The amplified fragment spans part of the nuclear DNA-encoded 18S rRNA gene (nucleotides 3657-5527)[b]	(F): tacctggttgatcctgccagtagcat (R): taatgatccttccgcaggttcacctac

[a]PCR conditions: one cycle of 94°C for 2 min; thirty cycles of 94°C for 30 s, 60°C for 30 s and 68°C for 6 min; and one final cycle of 68°C for 7 min (Perkin-Elmer 9700 thermocycler)

[b]PCR conditions: one cycle of 94°C for 2 min; 30 cycles of 94°C for 30 s, 60°C for 30 s and 68°C for 2 min; and one final cycle of 68°C for 7 min (Perkin-Elmer 9700 thermocycler).

tion *(34–36)*. Again, Southern blot analysis is the most reliable technique using hybridizations with two probes (one for mtDNA and another for nDNA).

3.7. Analysis for mtDNA Depletion

We start with 10–20 µg of total DNA that is digested with *Pvu*II or *Bam*HI (each enzyme has a single restriction site in mtDNA, therefore, they linearize normal mtDNA molecules) (*see* **Note 4**). The product of the digestion is run using conventional electrophoresis methods in a 1% agarose gel and then transferred to a nylon membrane. Hybridization is performed using probes listed in **Table 3**. Depletion of mtDNA is identified by measuring the ratio between the nuclear and the mitochondrial signals (**Fig. 2B**).

1. Extract total DNA as described previously (*see* **Subheading 3.1.**) and determine the DNA quality and concentration by spectrophotometry.
2. Digest 5–10 µg of total DNA from all the samples, including normal controls, with *Pvu*II.
3. Make a 1% agarose gel 20 cm in length using 1X Tris-Acetate EDTA (TAE) buffer containing 0.5 µg/liter ethidium bromide (EtBr). Load the total volume of each digestion product with an appropriate loading buffer in individual wells. In additional wells, load an apropriate molecular weight marker and at least one undigested DNA sample.
4. Run at 2–3 volts/cm (distance between electrodes) overnight in 1X TAE with 0.5 µg/mL EtBr. If necessary, continue the electrophoresis by inspecting the separation of the bands of the molecular weight marker.
5. Remove the gel from the electrophoresis apparatus and place it on an ultraviolet (UV) transilluminator. Put a fluorescent ruler on the gel aligning the beginning of the ruler with the sample wells, in order to determine the distance of DNA migration. Take a photograph of the UV illuminated gel.
6. Treat the gel with the following solutions in a shaker:
7. Depurination solution (15 min).
8. Denaturation solution (30 min).
9. Neutralizing solution (30 min).
10. Transfer the DNA from the gel to a nylon membrane as follows:

11. Fill one tray with 10X SSC.
12. Put a flat support tray on top of the tray with 10X SSC and lay a sheet of Whatman 3 Qualitative paper on it, with both ends submerged in the 10X SSC buffer.
13. Eliminate all air bubbles between the support tray and the Whatman paper.
14. Mark the gel for orientation by cutting off the upper left corner. Place the gel upside-down onto the wet Whatman paper surface.
15. Cut a sheet of nylon membrane (e.g., ZetaProbe GT, Bio-Rad) exactly the same size as the gel and soak in water.
16. Place the membrane on the gel. Place a sheet of Whatman 3 qualitative paper, moistened with 2X SSC, on the nylon membrane, then put three additional sheets dry Whatman paper on top. Avoid air bubbles in all the surfaces of contact.
17. Put about 10 cm of dry paper towels on top of the Whatman sheets.
18. Lay another flat tray onto the paper towels and put a 500 g weight on the top. Allow the DNA to transfer overnight.
19. Remove the gel and the membrane. Check with an UV transilluminator for the absence of DNA in the gel. Wash the membrane with 5X SSC for 5 min, then fix the DNA onto the membrane by baking the membrane between two Whatman 3 sheets in a vacuum oven at 80°C for 1 h.
20. Make the radiolabeled probes. For detection of mtDNA depletion, both mitochondrial and nuclear probes should be used. The primers and PCR conditions are described in the **Table 3**. Label the probe with ^{32}P using Random Primed DNA Labeling kit (Roche).
21. The membrane should be prehybridized 30 min at 65°C with the hybridization buffer. Then hybridize the membrane with both labeled probes simultaneously (ratio mitochondrial: nuclear should be 10^6 cpm/mL: 3×10^6 cpm/mL) in hybridization buffer, overnight at 65°C.
22. Wash the membrane at 65°C twice with washing solution I (30 min each), and twice with washing solution II (30 min each). After the last wash, check the membrane with Geiger counter. If the radioactive signal is too strong, make an additional wash with washing solution II.
23. Seal the membrane in a plastic bag, and expose to an X-ray film. The exposing time could be from 2 h to several days, depending on strength of the signals from the DNA probes.
24. By densitometry of the bands on the X-ray film or by direct measurement of the radiolabeled bands in a phosphor imager, calculate the ratio of the signals from the mtDNA band (16.5 kb) and the nuclear band (12 kb). Detect mtDNA depletion by comparing the ratio of mtDNA/nDNA bands in patients and controls.

3.8. Analysis for Single and Multiple Deletions of mtDNA

Single and multiple deletions of mtDNA are detected using the Southern blot analysis described earlier (*see* **Subheading 3.7.**) with modifications (*see* **Note 5**). The probe for nDNA is not required. Deletions of mtDNA are detected as bands shorter than the full-length 16.5 kb normal band.

Duplications of mtDNA often co-exist with single deletions. Duplicated mtDNA molecules produce Southern blot analysis results that are identical to heteroplasmic deleted mtDNA. Each duplicated mtDNA correspond to a wild-type molecule plus the nondeleted segment of the mtDNA *(37,38)*. For example, in a patient harboring both the 4,977 bp "common" deletion and the corresponding duplication, the individual would also have a 28,161 bp duplicated specie of mtDNA (corresponding to the 16,569 bp wild-type plus a 11,592 bp segment of mtDNA). If the duplicated molecule is cut with an restriction enzyme that recognizes a site within the duplicated segment, then the RFLP analysis will produce two bands, one identical to the wild-type mtDNA molecule and another identical to the deleted mtDNA specie *(37,38)*. Therefore, if the duplicated mtDNA molecule, corresponding to the "common" deletion, is digested with *Pvu*II (nt 2652) or *Bam*HI (nt 14,258), Southern blot analysis will reveal bands, identical in size, to the bands produced by a heteroplasmic population of the "common" mtDNA deletion. To resolve this problem, one can use a restriction enzyme that recognizes a unique site within the undeleted segment of mtDNA. In the case of the common deletion/duplication, *Sna*BI restriction enzyme can be used because there is a unique cutting site (nt 10,736) in the duplicated molecule. Thus, *Sna*BI digestion of a "common" duplication mtDNA molecule will produce a single linear band of approximately 28kb. By contrast, the "common" deletion molecule will remain uncut.

4. Notes

1. We generally obtain 50–100 μg of total DNA from 50 mg of skeletal muscle tissue. We quantitate the amount of DNA by measuring spectrophotometric absorbance at 260nm. Mitochondrial DNA constitutes about 1% of the total DNA.
2. In general, we prefer to identify mtDNA point mutations using a "gain-of-site" analysis, by employing a restriction enzyme that cuts the mutant sequence and not the wild-type sequence. There are at least two advantages of a gain-of-site analysis. First, we avoid false positive results due to other DNA polymorphisms within the enzyme recognition site. Second, we avert false positives caused by inadequate restriction enzyme cutting. For example, in a "loss-of-site" analysis, if the wild-type DNA is not completely cut because the restriction enzyme is degraded, then the result will be identical to that of a heteroplasmic mutation.
3. For an up-to-date listing of pathogenic mtDNA point mutations, readers are referred to the latest issue of the journal *Neuromuscular Disorders*, which maintains a comprehensive section entitled "Mitochondrial encephalomyopathies: gene mutations" edited by Dr. Serenella Servidei.
4. Because the amount of mtDNA in a tissue can vary widely from individual to individual, we generally use at least DNA samples from at least four controls for each Southern blot analysis for mtDNA depletion screening.
5. Multiple deletions of mtDNA may appear as faint bands, therefore, we often expose the membrane to X-ray film for extended periods of time. Degradation of DNA will appear as a smear rather than discrete bands below the 16.6 kb mtDNA band.

Acknowledgments

This work was supported by NIH grants HL59657 and HD37529 and by grants from the Muscular Dystrophy Association and Spanish Fondo de Investigación Sanitaria (FIS 00/797). Ms. Saba Tadesse is acknowledged for her excellent technical assistance.

References

1. Anderson, S., Bankier, A. T., Barrel, B. G., DeBruijin, M., Coulson, A. R., Drouin, J., et al. (1981) Sequence and organization of the human mitochondrial genome. *Nature* **290,** 457–465.
2. Attardi, G. and Schatz, G. (1988) Biogenesis of mitochondria. *Annu. Rev. Cell Biol.* **4,** 289–333.
3. Giles, R. E., Blanc, H., Cann, R. M., and Wallace, D. C. (1980) Maternal inheritance of human mitochondrial DNA. *Proc. Natl. Acad. Sci. USA* **77,** 6715–6719.

4. Bogenhagen, D. and Clayton, D. A. (1974) The number of mitochondrial deoxyribonucleic acid genomes in mouse L and human HeLa cells. *J. Biol. Chem.* **249,** 7991–7995.
5. Satoh, M. and Kuroiwa, T. (1991) Organization of multiple nucleotides of DNA molecules in mitochondria of a human cell. *Exp. Cell Res.* **196,** 137–140.
6. Wallace, D. C. (1992) Diseases of the mitochondrial DNA. *Annu. Rev. Biochem.* **61,** 1175–1212.
7. Goto, Y., Nonaka, I., and Horai, S. (1990) A mutation in the tRNA(Leu)(UUR) gene associated with the MELAS subgroup of mitochondrial encephalomyopathies. *Nature* **348,** 651–653.
8. Goto, Y. I., Nonaka, I., and Horai, S. (1991) A new mtDNA mutation associated with mitochondrial myopathy, encephalopathy, lactic acidosis and stroke-like episodes (MELAS). *Biochem. Biophys. Acta* **1097,** 238–240.
9. Morten, K., Cooper, J., Brown, G., Lake, B., Pike, D., and Poulton, J. (1993) A new point mutation associated with mitochondrial encephalomyopathy. *Hum. Mol. Genet.* **2,** 2081–2087.
10. Goto, Y., Tsugane, K., Tanabe, Y., Nonaka, I., and Horai, S. (1994) A new point mutation at nucleotide pair 3291 of the mitochondrial tRNALeu(UUR) gene in a patient with mitochondrial myopathy, encephalopathy, lactic acidosis, and stroke-like episodes (MELAS). *Biochem. Biophys. Res. Comm.* **202,** 1624–1630.
11. Manfredi, G., Schon, E. A., Moraes, C. T., Bonilla, E., Berry, G. T., and DiMauro, S. (1995) A new mutation associated with MELAS is located in a mitochondrial DNA polypeptide-coding gene. *Neuromusc. Disord.* **5,** 391–398.
12. Manfredi, G., Schon, E. A., Bonilla, E., Moraes, C. T., Shanske, S., and DiMauro, S. (1996) Identification of a mutation in the mitochondrial tRNA^{Cys} gene associated with mitochondrial encephalopathy. *Hum. Mutat.* **7,** 158–163.
13. Nishino, I., Komatsu, M., Kodama, S., Horai, S., Nonaka, I., and Goto, Y.-I. (1996) The 3260 mutation in mitochondrial DNA can cause mitochondrial myopathy, encephalopathy, lactic acidosis, and strokelike episodes (MELAS). *Muscle Nerve* **19,** 1603–1604.
14. Taylor, R. W., Chinnery, P. F., Haldane, F., Morris, A. A., Bindoff, L. A., Wilson, J., and Turnbull, D. M. (1996) MELAS associated with a mutation in the valine transfer RNA gene of mitochondrial DNA. *Ann. Neurol.* **40,** 459–462.
15. Santorelli, F. M., Tanji, K., Kulikova, R., Shanske, S., Vilarinho, L., Hays, A. P., and DiMauro, S. (1997) Identification of a novel mutation in the mtDNA ND5 gene associated with MELAS. *Biochem. Biophys. Res. Comm.* **238,** 326–328.
16. Hanna, M. G., Nelson, I. P., Morgan-Hughes, J. A., and Wood, N. W. (1998) MELAS: a new disease associated mitochondrial DNA mutation and evidence for further genetic heterogeneity. *J. Neurol. Neurosurg. Psychiatry.* **65,** 512–517.
17. Shoffner, J. M., Lott, M. T., Lezza, A. M. S., Seibel, P., Ballinger, S. W., and Wallace, D. C. (1990) Myoclonic epilepsy and ragged-red fiber disease (MERRF) is associated with a mitochondrial DNA tRNALys mutation. *Cell* **61,** 931–937.
18. Silvestri, G., Moraes, C. T., Shanske, S., Oh, S. J., and DiMauro, S. (1992) A new mtDNA mutation in the tRNALys gene associated with myoclonic epilepsy and ragged-red fibers (MERRF). *Am. J. Hum. Genet.* **51,** 1213–1217.
19. Holt, I. J., Harding, A. E., Petty, R. K., and Morgan Hughes, J. A. (1990) A new mitochondrial disease associated with mitochondrial DNA heteroplasmy. *Am. J. Hum. Genet.* **46,** 428–433.
20. de Vries, D., van, E. B., Gabreels, F., Ruitenbeek, W., and van Oost, B. (1993) A second missense mutation in the mitochondrial ATPase 6 gene in Leigh's syndrome. *Ann. Neurol.* **34,** 410–412.
21. Tatuch, Y., Christodoulou, J., Feigenbaum, A., Clarke, J. T. R., Wherret, J., Smith, C., et al. (1992) Heteroplasmic mtDNA mutation (T->G) at 8993 can cause Leigh disease when the percentage of abnormal mtDNA is high. *Am. J. Hum. Genet.* **50,** 852–858.

22. Santorelli, F. M., Shanske, S., Sciacco, M., Ciafaloni, E., Bonilla, E., De Vivo, D. C., and DiMauro, S. (1992) A new etiology for Leigh's syndrome: mitochondrial DNA mutation in the ATPase 6 gene. *Ann. Neurol.* **32,** 467–468.
23. Brown, W. M., George, M., Jr., and Wilson, A. C. (1979) Rapid evolution of animal mitochondrial DNA. *Proc. Natl. Acad. Sci. USA* **76,** 1967–1971.
24. Zeviani, M., Moraes, C. T., DiMauro, S., Nakase, H., Bonilla, E., Schon, E. A., and Rowland, L. P. (1988) Deletions of mitochondrial DNA in Kearns-Sayre syndrome. *Neurology* **38,** 1339–1346.
25. Rotig, A., Cormier, V., Blanche, S., Bonnefont, J. P., Ledeist, F., Romero, N., et al. (1990) Pearson's marrow-pancreas syndrome. A multisystem mitochondrial disorder in infancy. *J. Clin. Invest.* **86,** 1601–1608.
26. Zeviani, M., Servidei, S., Gellera, C., Bertini, E., DiMauro, S., and DiDonato, S. (1989) An autosomal dominant disorder with multiple deletions of mitochondrial DNA starting at the D-loop region. *Nature* **339,** 309–311.
27. Servidei, S., Zeviani, M., Manfredi, G., Ricci, E., Silvestri, G., Bertini, E., et al. (1991) Dominantly inherited mitochondrial myopathy with multiple deletions of mitochondrial DNA: clinical, morphologic, and biochemical studies. *Neurology* **41,** 1053–1059.
28. Hirano, M., Silvestri, G., Blake, D. M., Lombes, A., Minetti, C., Bonilla, E., et al. (1994) Mitochondrial neurogastrointestinal encephalomyopathy (MNGIE): clinical, biochemical, and genetic features of an autosomal recessive mitochondrial disorder. *Neurology* **44,** 721–727.
29. Suomalainen, A., Majander, A., Wallin, M., Setälä, K., Kontula, K., Leinonen, H., et al. (1997) Autosomal dominant progressive external ophthalmoplegia with mitochondrial deletions of mtDNA: clinical, biochemical, and molecular genetic features of the 10q-linked disease. *Neurology* **48,** 1244–1253.
30. Moraes, C. T., Shanske, S., Tritschler, H. J., Aprille, J. R., Andreetta, F., Bonilla, E., et al. (1991) MtDNA depletion with variable tissue expression: A novel genetic abnormality in mitochondrial diseases. *Am. J. Hum. Genet.* **48,** 492–501.
31. Vu, T. H., Sciacco, M., Tanji, K., Nichter, C., Bonilla, E., Chatkupt, S., et al. (1998) Clinical manifestations of mitochondrial depletion. *Neurology* **50,** 1783–1790.
32. Hirano, M. and Vu, T. H. (2000) Defects of intergenomic communication: where do we stand? *Brain Pathol.* **10,** 451–461.
33. Santorelli, F., Sciacco, M., Tanji, K., Shanske, S., Vu, T., Golzi, V., et al. (1996) Multiple mitochondrial DNA deletions in sporadic inclusion body myositis. *Ann. Neurol.* **39,** 789–795.
34. Dalakas, M. C., Illa, I., Pezeshkpour, G. H., Laukaitis, J. P., Cohen, B., and Griffin, J. L. (1990) Mitochondrial myopathy caused by long-term zidovudine therapy. *N. Engl. J. Med.* **322,** 1098–1105.
35. Arnaudo, E., Dalakas, M., Shanske, S., Moraes, C. T., DiMauro, S., and Schon, E. A. (1991) Depletion of muscle mitochondrial DNA in AIDS patients with zidovudine-induced myopathy. *Lancet* **337,** 508–510.
36. Lewis, W. and Dalakas, M. C. (1995) Mitochondrial toxicity of antiviral drugs. *Nature Med.* **1,** 417–422.
37. Poulton, J., Deadman, M. E., and Gardiner, R. M. (1989) Duplications of mitochondrial DNA in mitochondrial myopathy. *Lancet* **1,** 236–240.
38. Manfredi, G., Vu, T. H., Bonilla, E., Schon, E. A., DiMauro, S., Arnaudo, E., et al. (1997) Association of myopathy with large-scale mitochondrial DNA duplications and deletions: which is pathogenic? *Ann. Neurol.* **42,** 180–188.
39. Carrozzo, R., Hirano, M., Fromenty, B., Casali, C., Santorelli, F. M., Bonilla, E., et al. (1998) Multiple mtDNA deletions features in autosomal dominant and recessive diseases suggest distinct pathogeneses. *Neurology* **50,** 99–106.

19

Detection of Mitochondrial DNA Mutations Associated with Leber Hereditary Optic Neuropathy

Kasinathan Muralidharan

1. Introduction
1.1. Mitochondrial Genetics

Human genetic disease linked to mitochondrial DNA (mtDNA), the other genome in our cells, was first recognized in 1988 *(1,2)*. Mitochondrial genetics has certain unique features (reviewed in **ref. 3**): 1) Maternal inheritance: Mitochondria are maternally inherited and therefore mtDNA mutations are transmitted through maternal lineage. 2) Heteroplasmy: The presence mutant mtDNA in a population of normal molecules is termed heteroplasmy. The proportion of mutant molecules may be different in different tissues and can change with cycles of cell division. 3) Replicative segregation: Mitochondria are partitioned along with the cytoplasm during cell division. The distribution of mutant and normal molecules in the daughter cells of heteroplasmic cell may be unequal. In the course of development and differentiation, different parts of the body may have different proportion of mutant molecules. 4) Threshold effect: The percentage of mutant mtDNA molecules and the energetic needs of the tissue influence the penetrance of mutations. The effect of mtDNA mutations may become evident when either the number of mutant molecules or the deleterious effect exceeds a threshold. 5) High mutation rate: The mtDNA has a much higher mutation rate than nuclear DNA. This results in the recurrence of some mutations in different haplotype backgrounds. The high mutation rate also contributes to polymorphisms.

1.2. Leber Hereditary Optic Neuropathy (LHON)

Leber Hereditary Optic Neuropathy (LHON) is a maternally inherited disease that presents as mid-life acute or sub-acute painless loss of central vision. Ophthalmic examination commonly reveals a central scotoma, peripapillary telangiectasis, microangiopathy, disk pseudoedema, and vascular toruosity. Patients and maternal relatives may also manifest ancillary symptoms such as cardiac conduction defects, neurological manifestations including altered reflexes, ataxia, sensory neuropathy, and multiple sclerosis (MS).

The onset can range from sudden and complete loss of vision to progressive decline over the course of two years. The eyes may be simultaneously or sequentially affected. The age of onset is variable and has a mean of 27–34 y. The progression is also vari-

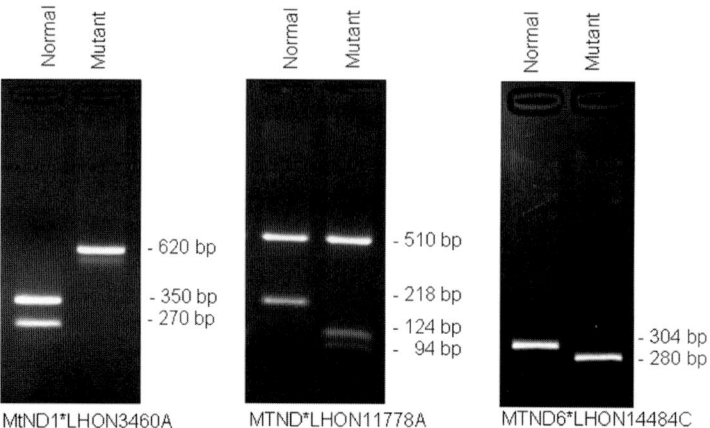

Fig. 1. Molecular detection of LHON mutations. Normal and mutant PCR-restriction digestion patterns for the three primary LHON mutations are shown. The mutation assayed is indicated at the bottom of the gel. The molecular sizes of the fragments are indicated in base pairs. Details of the PCR primers conditions and restriction enzymes are provided in **Table 2**.

able. The final visual acuity can range from 20/50 to no light perception. The probability of visual recovery also is variable.

The mitochondrial etiology of the LHON was demonstrated by Wallace et al. *(2)* who recognized the MTND4*11778A mutation in families with LHON. To date 27 mutations in mitochondrial DNA (mtDNA) have now been associated with LHON. Although all these mutations have LHON as a common feature, they differ with respect to other neurological features associated with them. The mitochondrial basis for LHON explains the maternal transmission of the disease. The variable penetrance and male bias in expression in LHON, however, remains enigmatic. Environment factors and other genetic factors have also been noted to play a role in LHON expression.

1.2.1. LHON Mutations

Over 27 mutations have been associated with LHON. Many of them are rare and their association with LHON remains to be confirmed.

Three of these mutations, 3460A, 11778A, and 14484C are generally agreed to be the primary mutations (*see* **Fig. 1** and **Table 1**). These mutations occur in multiple unrelated LHON, and have not been detected in a large number of control mtDNAs. The presence of these mutations greatly increases the probability of blindness. Together these account for over 90% of LHON cases reported (*see* **Notes 1** and **2**).

These and other mutations that primarily result in LHON alter mtDNA-encoded intrinsic membrane proteins that contribute to NADH: ubiquinone oxidoreductase or mitochondrial oxidative phosphorylation Complex I. 3460A and 14459A cause a severe defect in Complex I enzyme function. 11778A mutation show moderate impairment and 14484C only causes mild perturbation of Complex I.

The 14484C mutation occurs on a particular European mtDNA haplotype (haplotype J defined by MTND5*LHON13708A, MTND1*LHON4216C, and often MYCYB*LHON15257A). 11778A and 3460A are associated with a variety of haplotypes indicating that these mutations have arisen multiple times independently.

The secondary mutations have been found in increased frequencies in LHON patients, but generally in conjunction with one of the primary mutations. LHON mutations may also be classified into those that cause LHON alone and those that present with LHON and other neurological symptoms. For example MTND6*LHON14484C almost always only presents with LHON. MTND6*LDYTI14459A on the other hand, presents with LHON and dystonia and other neurological symptoms.

2. Materials

2.1. DNA Isolation (4)

1. TKM1: 10 mM Tris-HCl, pH 7.6. 10m M KCl, 10 mM MgCl$_2$, 2 mM EDTA, and 2.5% Nonidet P-40.
2. TKM2: 10 mM Tris-HCl, pH 7.6. 10m M KCl, 10 mM MgCl$_2$, 2 mM EDTA, 0.4 M NaCl, 0.05% sodium dodecyl sulfate (SDS).
3. 6 M NaCl.
4. Ethanol.

2.2. PCR Amplification

1. PCR primers: The primer sequences are provided in **Table 2**. Stock solution at 10 pmol/μL in water are used.
2. PCR Buffer 10X buffer: 100 mM Tris-HCl, pH 8.3, 15 mM MgCl$_2$, 500 mM KCl.
3. dNTPs: 10 mM each of dATP, dCTP, dGTP, and dTTP (Roche Biochemicals).
4. *Taq* DNA polymerase (5 U/μL) (Roche Molecular Biochemicals).

2.3. Restriction Digestion

1. Restriction enzymes BsaHI, MaeIII, and BstN1. Purchased from New England Biolabs. Reaction buffers provided by the manufacturer are used (*see* **Note 3**).

2.4. Agarose Electrophoresis

1. 2% and 3% Agarose gels in 1X TBE.
2. 1X TBE: 89 mM Tris-borate, 2 mM EDTA, pH 8.3.

3. Methods

3.1. DNA Isolation (4)

1. Collect whole blood in a Vacutainer tube (purple-stopper) containing EDTA as anticoagulant.
2. Transfer 5 mL of blood into a 15-mL centrifuge tube and add 5 mL of TKM1. Mix well.
3. Centrifuge at 1200–1500g for 10 min at room temperature in a tabletop centrifuge.
4. Slowly pour off the supernatant and save the nuclear pellet.
5. Wash the pellet in 5 mL of TKM1 buffer and centrifuge as before.
6. Gently resuspend the pellet in 0.8 mL of TKM2. Mix thoroughly by pipetting back and forth. Place on rocker overnight or at 55°C for 30 min.
7. Add 0.3 mL of 6 M NaCl to the tube and mix well.
8. Centrifuge at 1200–1500g for 10 min.
9. Transfer the supernatant using a transfer pipet to a new tube. Add 2.2 mL absolute ethanol. Mix. Spool DNA precipitate to a wand.
10. Rinse DNA precipitate in 70% ethanol. Air-dry.
11. Suspend DNA in 10 mM Tris-HCl, 1 mM EDTA, pH 8.0.

Table 1
Primary LHON Mutations

Mutation	Nucleotide change	Amino acid change	Control frequency (%)	Homo	Hetero	Visual recovery (%)	Penetrance in male relatives (%)	Penetrance in female relatives (%)	Occurrence
MTND4*LHON11778	G to A	Arg to His	0	+	+	4	33–60	82	70% of cases in Europe 90% of cases in Japan
MTND1*LHON3460	G to A	Ala to Thr	0	+	–	22	14–75	40–80	35% of cases in Europe
MTND6*LHON14484	T to C	Met to Val	0	+	+	37	27–80	68	68% of cases in Europe

Table 2
Molecular Analysis of Primary LHON Mutations

Mutation	Primers (5'- 3')	Annealing (°C)	Fragment size (bp)	Restriction enzyme (incubation °C)	Normal fragments (bp)	Abnormal fragments (bp)	References
MTND1*LHON3460A	Forward: TTCAAATTCCTCCCTGTACG Reverse: GTGACGCTCGTCATCGG (tail) TTCGAACACC	51	620	BsaHI (37°C)	350, 270	620	(8) (9)
MTND4*LHON11778	Forward: CCCACCTTGGCTATCATC Reverse: GCGATTGGAGCGGAATGG	51	728	MaeIII (55°C)	510, 218	510, 124, 94	(2)
MTND6*LHON14484	Forward: AAACAATGGTCAACCATGAAC[a] Reverse: GTCCGGGGATTTATTTAATTTTTT[a]	45	319	BstNI (60°C)	304, 15	280, 24, 15	(7)

[a]Nucleotides in italics altered to create MvaI site.

Table 3
Secondary LHON Mutations (3)

No.	Mutation	Status
1	MTATP6*LHON9101C	Primary
2	MTCO3*LHON9804A	Primary[a]
3	MTND2*LHON5244A	Primary[a]
4	MTND3*LHON10663C	Primary
5	MTND5*LHON13730A	Primary[a]
6	MTND6*LDYT14459A	Primary[a]
7	MTCO3*LHON9438A	Secondary[a]
8	MTCOI*LHON7444A	Secondary[a]
9	MTCOXIII*LHON9738	Secondary
10	MTCYB*LHON15257A	Secondary[a]
11	MTCYB*LHON15812A	Secondary[a]
12	MTND1*LHON3394C	Secondary[a]
13	MTND1*LHON4160C	Secondary[a]
14	MTND1*LHON4216C	Secondary[a]
15	MTND2*LHON4917G	Secondary[a]
16	MTND5*LHON13708A	Secondary[a]

[a]Mutations tested in extended LHON mutation analysis.

3.2. PCR Amplification

In each analysis a positive control, negative control, and water blank are included.

1. A 50 µL reaction is set with 8.0 µL of dNTP stock (final concentration 1.25 mM of each dNTP), 5 µL 10XPCR buffer, and 1.5 µL each of forward and reverse primer and 0.25 µL Taq polymerase. About 100 ng of isolated DNA was used per reaction.
2. The PCR thermocycling parameters included an initial denaturation step of 94°C for 4 min, 34 cycles of 1 min each of denaturation at 94°C, annealing at 51°C or 53°C according to the primers in the reaction (*see* **Table 2**), extension at 72°C, and final extension at 72°C for 3 min.

3.3. Restriction Digestion

1. 15 µL of the PCR product is digested in a 30 µL reaction. This volume contains 15 µL of PCR produce, 3 µL of the appropriate 10X restriction digestion buffer, and 10 U of restriction enzyme, and water to make up to 30 µL. The restriction digestion is carried out at the temperature optimum for the particular restriction enzyme (*see* **Table 2**).

3.4. Agarose Electrophoresis

1. The restriction digests are resolved on an agarose gel. MTND1*LHON3460 and MTND4*LHON11778 assays are resolved on 2% agarose TBE gels. MTND6*LHON 14484 digestion products is resolved on 3% agarose TBE gels.

4. Notes

1. The three mutations MTND1*LHON3460C, MTND4*LHON11778 and MTND6*LHON14484 account for about 90% of the cases of LHON in patients of European ancestry. Our laboratory also routinely tests for MTND6*LDYT14459A mutation which presents with dystonia. We also test for 12 other primary and secondary mutations when further testing is requested on patient who has tested negative on the first four mutations (**Table 3**). These additional mutations may contribute to 2–3% increased detection rate. The determination to carry out further testing is made by the physician based on the clinical presentation and pedigree information. Sequencing of the mtDNA on a research basis may be offered to patients where the clinical diagnosis strongly points to LHON.

2. Genetic counseling of maternal relatives of a patient with a LHON mutation is recommended. Counseling is complicated by the presence of heteroplasmy, sex of the individual, and other genetic and environmental factors.

3. Polymorphisms in the restriction enzyme recognition sites are a well-recognized problem in molecular diagnostic methods. A false-positive rate of 2–7% is estimated for primary LHON mutations *(5,6)*. This should be kept in mind as source of false positive or negatives in the analysis. Alternative restriction enzymes or sequencing of the region of the mtDNA is occasionally necessary.

    a. MTND4*LHON11778 is some times assayed by digestion with SfaNI restriction enzyme. This assay recognizes MTND4*LHON11778 by loss of digestion by SfaNI. A polymorphic loss of site is known for this mutation. Digestion with Mae III, which is a gain of site with the mutation, is useful if another method is needed.

    b. MTND6*LHON14484 can be assayed by Sau3AI using a mismatch primer *(7)*. A polymorphism at nucleotide 14485 creates a loss of this site.

    c. A polymorphism at nucleotide 3459 can result in a loss of BsaHI recognition site in the assay for MTND1*LHON3460C. MvaI is another enzyme that can be used to detect this mutation.

References

1. Holt, I. J., Harding, A. E., and Morgan-Hughes, J. A. (1988) Mitochondrial DNA polymorphism in mitochondrial myopathy. *Hum. Genet.* **79**, 53–57.
2. Wallace, D. C., Singh, G., Lott, M. T., Hodge, J. A., Schurr, T. G., Lezza, A. M., et al. (1988) Mitochondrial DNA mutation associated with Leber's hereditary optic neuropathy. *Science* **242**, 1427–1430.
3. Wallace, D. C., Brown, M.D., and Lott, M.T. (1996) *Mitochondrial Genetics. Emery and Rimoin's Principles and Practice of Medical Genetics*. Churchill Livingstone, London, UK.
4. Lahiri, D. K. and Nurnberger, J. I., Jr. (1991) A rapid non-enzymatic method for the preparation of HMW DNA from blood for RFLP studies. *Nucleic Acids Res.* **19**, 5444.
5. Johns, D. R. and Neufeld, M. J. (1993) Pitfalls in the molecular genetic diagnosis of Leber hereditary optic neuropathy (LHON). *Am. J. Hum. Genet.* **53**, 916–920.
6. Yen, M. Y., Wang, A. G., Chang, W. L., Hsu, W. M., Liu, J. H., and Wei, Y. H. (2000) False positive molecular diagnosis of Leber's hereditary optic neuropathy. *Zhonghua Yi Xue Za Zhi (Taipei)* **63**, 864–868.
7. Johns, D. R., Heher, K. L., Miller, N. R., and Smith, K. H. (1993) Leber's hereditary optic neuropathy. Clinical manifestations of the 14484 mutation. *Arch. Ophthalmol* **111**, 495–498.

8. Howell, N., Bindoff, L. A., McCullough, D. A., Kubacka, I., Poulton, J., Mackey, D., et al. (1991) Leber hereditary optic neuropathy: identification of the same mitochondrial ND1 mutation in six pedigrees. *Am. J. Hum. Genet.* **49,** 939–950.

9. Huoponen, K., Vilkki, J., Aula, P., Nikoskelainen, E. K., and Savontaus, M. L. (1991) A new mtDNA mutation associated with Leber hereditary optic neuroretinopathy. *Am. J. Hum. Genet.* **48,** 1147–1153.

IV

MOLECULAR DETECTION OF IMPRINTED GENES

20

PCR-Based Strategies for the Diagnosis of Prader-Willi/Angelman Syndromes

Milen Velinov and Edmund C. Jenkins

1. Introduction

Imprinting is the naturally occurring functional inequality of alleles of a given gene reflecting their parental origin. Only one of the alleles (either maternal or paternal) is functional (producing mRNA) in an imprinted gene. Imprinted genes in the human genome have been identified and are sometimes associated with particular disorders. The two best-known conditions in human genetics that are the result of imprinted gene abnormalities are Prader-Willi (PWS) and Angelman (AS) syndromes. These two clinically distinct neurobehavioral disorders are the result of absent/deficient function of alleles of genes in the paternally (PWS) or maternally (AS) derived proximal segment of the long arm of chromosome 15 (15q11→q13). This chromosomal region is normally imprinted. At least four genes in this region were found to be expressed only by paternally derived alleles *(1)* and the deficiency of these and possibly other gene products in the region is believed to result in Prader-Willi syndrome. One gene, UBE3A, located within the PWS/AS region was found to be expressed in certain tissues only by its maternally derived allele. Mutations in this gene are associated with some cases of AS *(2)*.

Deletions of about 4Mb in 15q11→q13 were first observed in both PWS and AS. Such proximal 15q deletions result in PWS if paternally derived, and in AS if maternally derived. Approximately 70% of patients with either PWS or AS carry microdeletions. Most of the remaining PWS patients were later found to have maternal uniparental disomy (UPD) for chromosome 15. Approximately 3–5% of AS patients have paternal UPD 15 *(3)*. About 2–3% of both PWS and AS patients do not have 4Mb microdeletions or UPD, but are abnormally imprinted. As a result, genes located in the region are functioning as if they were only maternally derived (in cases of PWS) or paternally derived (in cases of AS). In about half of those patients small (1–2 kb) deletions near the 5' end of the gene for SNRPN, affecting a region with suggested functional importance in the regulation of imprinting, were identified *(4)*. Such patients are described as having a defect in the imprinting process (or imprinting center mutation).

The etiologic heterogeneity of PWS and AS makes it difficult to establish uniform diagnostic methods that would detect most positive cases. Fluorescence *in situ* hybridization (FISH) analysis to identify the large microdeletions on the proximal 15q is available, but identifies only 70% of Prader-Willi or Angelman syndrome cases. Since the proximal 15q chromosomal segment is subject to imprinting, parent-of-origin-specific DNA methylation patterns are normally present in this region, particularly in the CpG islands at the 5' portion of the SNRPN gene. The methylation affects cytosine at the CpG islands and occurs in the maternally derived chromosomal region. DNA, methylation abnormalities in the 5' region of SNRPN were reported in both PWS and AS. Namely, PWS patients have only methylated (maternally derived) DNA whereas most AS patients have only unmethylated (paternally derived) DNA *(5)*. Thus methodologies for the analysis of DNA methylation patterns in proximal 15q were developed and subsequently recommended for the initial PWS/AS screening *(3)*. Such tests can identify all patients with PWS and 75–85% of those with AS. The remaining AS patients do not have methylation abnormalities. About half of those patients were found to have mutations in the gene UBE3A *(2)*. Thus UBE3A mutation screening is recommended in methylation-negative cases of suspected AS.

The first methylation-specific methodology developed was Southern analysis using methylation-sensitive restriction enzymes. Polymerase chain reaction (PCR)-based methylation testing for the two disorders was recently developed in order to increase the rapidity of the testing and to minimize the amounts of material required *(6,7)*. The PCR-based methylation test is based on the ability of sodium bisulfite to convert cytosine residues to uracil under conditions whereby the methylated cytosine residues (5-methyl cytosines) remain essentially nonreactive. Following bisulfite modification, PCR amplification is carried out with strand-specific primers (since the two DNA strands are no longer complementary). In the resultant PCR product all uracil and thymine residues are amplified as thymine, and only the 5-methyl-cytosines are amplified as cytosines *(8)*. This PCR-based testing for PWS and AS is referred to as methylation-specific PCR (mPCR) *(6)*. In this protocol the methylation-specific changes in DNA sequence produced by bisulfite treatment are used to design PCR primers that would specifically amplify methylated (maternally derived) or unmethylated (paternally derived) sequences only. Since PWS patients are deficient for paternally derived DNA, amplification in PWS patients will occur only with the "methylated" primers, and the opposite will be the case with AS patients. Two sets of primers (paternally and maternally specific) were used in a multiplex amplification. While conducting mPCR testing, the paternal-specific primers were found to generate stronger amplification than the maternal-specific ones, so that using different amounts of primers was necessary *(6,9)* and adjustment of the primer amounts for every new primer batch was recommended *(9)*. We attempted to use this protocol by amplifying separately the maternally and paternally derived sequences. As such, false-positive cases due to a nonspecific inhibition of the PCR process were sometimes observed and they were difficult to identify because of the lack of internal controls.

Because of the aforementioned technical problems encountered using the mPCR approach, we have developed an alternative methylation test using methylation-specific digestion of the amplified, bisulfite-treated DNA *(7)*. In our protocol, referred to as Bisulfite Restriction Analysis (BRA), two rounds of nested amplification of the bisulfite-treated DNA are first carried out using two sets of primers that anneal to GC-free regions (**Fig. 1**). This is done in order to produce "methylation-independent" amplification. The nested PCR is required in order to produce a sufficient amount of amplified product so that the detection of the digested fragments is possible by simple ethidium bromide-

tctaga*ggcc ccctctcatt gcaacagtgc tgtggggccc* taggggtcca gtagccccct

cccccccaggt cattc**cg**gtg agggagggag ctgggacccc tgcactg**cg**g caaacaagca

cgcct*G**CG**'**C**g gc**cg*****c**agagg caggctg*G**CG**'**C**g*catgctca ggc**gg**ggatg tgtg**cg**aagc

ctgc**cg**ctgc tgcag**cg**agt ctgg**cg**caga gtggag*cggc cgccg*gagat gcctga**cg**ca

tctgtctgag gag**cg**gtcag tga**cgcg**atg gag**cg**ggcaa ggtcagctgt gc**cg**gtggct

tctctcaaga gacagcctgg ggag**cg**gcca cttttattca tcagatattc caagtttta

ggacttggag tactgaataa ac**gg**aatttg ggccctaaag tcctttgttc tggagaacca

gttc**cg**gaat

Fig. 1. The 5' region of the SNRPN gene amplified in our protocol is shown. This genomic region was previously reported to be fully methylated in cases of PWS and unmethylated in cases of AS *(5)*. The primers are designed according to the bisulfite-mediated changes in the DNA sequence (uracil substitutes cytosine). When cytosine is in the context of a CpG island it remains unchanged by the sodium bisulfite treatment, because it is methylated to form 5-methyl cytosine. The CpGs that have methylated cytosine residues are in bold. The sequences used to design the first set of primers are indicated in italic. The sequences for the nested primers are underlined. The CfoI digestion sites are shown in upper case. Since both sets of PCR primers are annealing to CG free sequences, the amplification is not methylation-dependent (it occurs for both methylated and unmethylated DNA). CfoI digestion will only occur with methylated (unmodified by the bisulfite treatment) DNA fragments. Thus in normal individuals, digestion will result in the presence of both an undigested (340 bp band) and digested (224, 22, and 94 bp) products. In individuals with PWS all DNA will be digested, whereas in AS no DNA will be digested. *See also* **Fig. 2**.

stained agarose gel analysis. Alternatively, this assay can be conducted with one-step PCR amplification. However specific labeling in order to visualize the fragments after digestion may be necessary. The amplified fragments are digested with CfoI, which recognizes the sequence GCGC. The bisulfite modification leads to the loss of the CfoI restriction site only in the paternally derived (unmethylated) DNA fragments (**Fig. 2**). Accordingly, the cleavage occurs only at the methylated (maternally derived) sites, thus allowing differentiation between maternally derived (digested) and paternally derived (undigested) alleles (**Figs. 2** and **3**). While our protocol takes longer to complete compared to the mPCR approach, it eliminates the risk for false positive results. As such, it is our method of choice for initial diagnostic work up of individuals with suspected PWS or AS. Because of the minimal amount of specimen required for the PCR-based methylation testing (100 ng or less of native genomic DNA is sufficient for testing); we were also able to conduct retrospective testing of archived fixed cell suspensions from individuals already evaluated and found to be negative by FISH. We have identified one new positive PW and one new AS case out of 20 specimens tested *(10)*. Such re-testing can help avoid obtaining additional blood specimens for methylation testing in FISH-negative individuals with suspected PWS/AS. In addition, sources of very small amounts of DNA such as blood specimens used in newborn screening may be used for methylation testing.

A Before bisulfite treatment

           Normal   PWS   AS

mat C^mGC^mG C^mGC^mG CGCG(or absent)
pat CGCG C^mGC^mG(or absent) CGCG

B After bisulfite treatment

           Normal   PWS   AS

mat C^mGC^mG C^mGC^mG UGUG(or absent)
pat UGUG C^mGC^mG(or absent) UGUG

Fig. 2. **(A)** A CfoI restriction site at the 5′ portion of SNRPN gene is presented. The methylated cytosine residues are indicated with C^m; mat, maternally derived sequence; pat, paternally derived sequence. The two CfoI restriction sites that are present in the SNRPN fragment studied (**Fig. 1**), are followed by a guanosine residue. Thus all cytosine residues within the restriction sites are in a CpG context. Accordingly, in the maternally derived fragments these residues are methylated to form 5-methyl cytosine. Normal individuals have one maternally derived (methylated), and one paternally derived (unmethylated) sequence. Individuals with PWS have only methylated sequence since they are deficient in paternally derived chromosomal material. Individuals with AS have only unmethylated sequence since they are deficient in maternally derived material. **(B)** Bisulfite modification converts the cytosine residues that are unmethylated into uracil. Thus in normal individuals one of the CfoI restriction sites (on the paternally derived sequence) is lost. In individuals with PWS no restriction site is lost. In individuals with AS both restriction sites are lost.

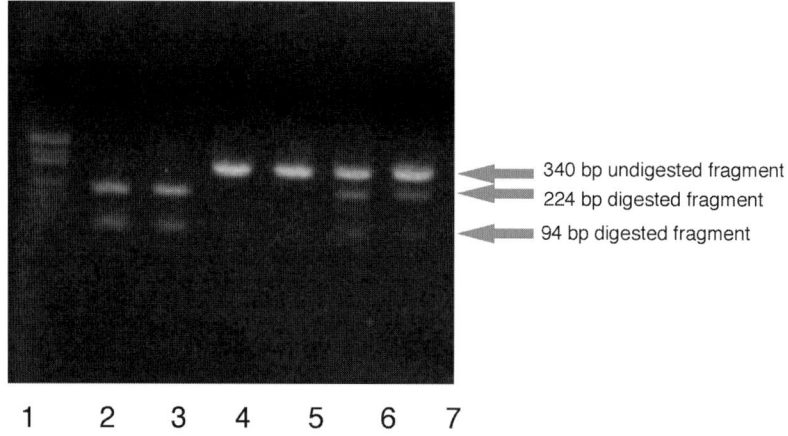

Fig. 3. BRA testing of patient and control specimens. Lane 1, DNA weight marker; lanes 2 and 3, PWS specimens showing full digestion (digested fragments of 224 bp and 94 bp are seen, undigested fragment 340 bp is absent); lanes 4 and 5, AS specimens with complete lack of digestion (only the undigested 340 bp fragment is seen); lanes 6 and 7, non-PWS non-AS specimens showing partial digestion (both digested 224 bp and 94 bp and undigested 340 bp fragments are seen).

2. Materials

2.1. DNA Extraction

1. Puregene DNA extraction kits (Gentra, Minneapolism MN, USA).
2. 100% isopropyl alcohol (2-propanol).
3. 70% ethanol.

2.2. Bisulfite Treatment

1. CpGenome DNA Modifiction Kit (Serologicals Co, Norcross, GA, USA).
2. Reagent I: For each sample add 571 µL of water to 227 mg of DNA Modification Reagent I (CpGenome DNA modification kit). Adjust the pH to 5.0 with approx 20 µL of 3 M NaOH, monitoring the pH using pH indicator paper.
3. Reagent II: Add 1 µL of β-mercaptoethanol to 20 mL of deionized water. Add 750 µL of this solution to 1.35 g of DNA Modification reagent II (CpGenome DNA modification kit) for each sample to be modified.
4. Reagent III is supplied ready to use in the CpGenome DNA modification kit.
5. Ethanol-NaOH solution: To prepare 1 mL of this solution, combine 900 µL of 100% Ethanol, 93.4 µL of H_2O, and 6.6 µL of 3 M NaOH.
6. TE buffer: 10 mM Tris-HCl, 0.1 mM EDTA, pH 7.5.

2.3. PCR

1. Primers (**Fig. 1**) used for the first PCR are:
   5'-GGTTTTTTTTTATTGTAATAGTGTTGTGGGG - 3' and
   5'-CTCCAAAACAAAAAACTTTAAAACCCAAATTC - 3'
   The primers used for the nested amplification are:
   5'- GGTTTTAGGGGTTTAGTAGTTTTTTTTTTTAGG -3' and
   5'-CAATACTCCAAATCCTAAAAACTTAAAATATC-3'
   The lyophilized primers received from the manufacturer are reconstituted as stock solutions to a concentration 1 µg/µL in Tris/EDTA solution. The working primer solutions are prepared from the stock solutions by diluting down to 250 ng/µL in dH_2O. From this solutions 4 µL of each primer are used for every reaction (for the 50 µL PCR format).
2. Polymerase: The polymerase used is AdvanTaq Plus (Clontech, Palo Alto, CA, USA) (a hot start enzyme).
3. PCR buffer: 10X AdvanTaq PCR buffer (Clontech).
4. dNTPs: Nucleotide mix is prepared from stock solutions of dATP, dCTP, dGTP, and dTTP (New England Biolabs) to a final concentration of 2.5 mM of each nucleotide.
5. Digestion: The digestion is done using restriction enzyme CfoI (Promega Inc.).
6. Electrophoresis: Electrophoresis is done on a 2.5% agarose minigel using the minigel system of Labnet Inc. *(11)*.
7. pUC18 (Sigma Chemical Co).

3. Methods

3.1. DNA Extraction

1. Genomic DNA is isolated using the Puregene DNA isolation kit of Gentra following the instructions of the manufacturer.

3.2. Bisulfite Treatment

1. 3 µg of DNA is bisulfite-treated using the CpGenome DNA modification kit following the manufacterer's instructions (*see* **Subheading 2.2.**).

2. In screwcap microcentrifuge tubes add the DNA and bring the volume to 100 µL with distilled H$_2$O. Add 7 µL of freshly made 3 M NaOH.
3. Incubate for 10 min at 37°C.
4. Add 550 µL of freshly prepared DNA modification Reagent I (*see* **Subheading 2.2., item 2**).
5. Incubate at 50°C for 16–20 h.
6. Resuspend DNA Modification Reagent III (*see* **Subheading 2.2., item 4**) by vortexing vigorously.
7. Add 5 µL of well-suspended DNA Modification Reagent III (*see* **Subheading 2.2., item 4**) to the DNA solutions.
8. Add 750 µL of DNA Modification Reagent II (*see* **Subheading 2.2., item 3**) and mix.
9. Incubate at room temperature for 5–10 min.
10. Spin for 10 s at 5,000g to pellet the DNA Modification Reagent III (*see* **Subheading 2.2., item 4**) with DNA. Discard supernatant.
11. Add 1.0 mL of 70% EtOH, vortex, centrifuge for 10 s at 5,000g and discard supernatant. Perform this step for a total of 3 times.
12. Centrifuge at 10,000g for 2–3 min, and remove the remaining supernatant with a pipet.
13. Add 50 µL of the freshly prepared 20 mM NaOH/90% EtOH solution (*see* **Subheading 2.2., item 5**).
14. Vortex and incubate at room temperature for 5 min.
15. Spin for 10 s at 5,000g. Add 1.0 mL of 90% EtOH and vortex. Spin again and remove the supernatant. Repeat this step one additional time.
16. Centrifuge the sample at high speed for 5 min.
17. Remove all the remaining supernatant and add 25 µL of TE buffer (**Subheading 2.2., item 6**) and vortex.
18. Incubate the sample for 15 min at 60°C.
19. Centrifuge at high speed for 2–3 min and transfer the sample (supernatant) to a new tube.

3.3. PCR

1. 2 µL of the bisulfite-treated DNA is amplified with the first set of primers.
2. Add 2 µL of the bisulfite treated DNA solution to a 0.2 mL PCR tube.
3. Add 48 µL PCR mix containing the following constituents as recommended by the manufacturer: 4 µL (1 µg) of each primer, 1 µL of AdvanTaq Plus polymerase, 5 µL 10X PCR buffer, 4 µL dNTP mix, and 30 µL dH$_2$O to make total of 48 µL.
4. The reaction is carried out with a two-step cycle: a denaturing step of 95°C for 30 s, and an annealing step of 65°C for 2 min, for the first five cycles, then 52°C for 2 min for the next 30 cycles, with a total of 35 cycles. After the first PCR program is over, 2 µL of the first reaction is used as a template for the nested PCR with the above conditions using the nested primers.

3.4. Electophoresis and Enzymatic Digestion

1. 5 µL of the PCR products are run on a 2.5% agarose gel to check for the presence of amplification products. A 340 bp fragment should be detected *(7)* (*see* also **Notes 1–4**).
2. 10 µL of amplification product are digested in the PCR buffer mix using 15 U (1 µL) of restriction enzyme CfoI. Digestion is carried out overnight at 37°C (minimum 3 h).
3. The digested specimens are run on a 2.5% agarose minigel for 15 min at 100 V. The DNA marker used for all runs is an MspI digest of pUC18.
4. For every run of amplification, four controls are included: a positive control for PWS (DNA from patient with known PWS), a positive control for AS (DNA from a patient with known AS), a positive control for a non-PWS, non-AS individual, and a negative PCR control (mix with no template). The three positive (PWS, AS and non-PWS-non-AS) PCR

controls after amplification are subjected to the next step of the testing protocol-CfoI digestion. They are thus used as controls for the digestion step.

5. Three patterns of digestion can be seen (**Figs. 2** and **3**):
   a. Complete digestion: only 224 bp, 22 bp, and 94 bp fragments that are the results of complete digestion of the PCR fragments are seen in cases of PWS since the genomic DNA in these patients is fully methylated and the restriction site is not modified by the bisulfite treatment.
   b. Partial digestion: both undigested 340 bp, and digested 224 bp, 22 bp and 94 bp fragments are seen when the DNA methylation pattern is normal (non-PWS, non AS patients).
   c. No digestion: only the 340 bp undigested fragment is seen in cases of AS since the restriction site is unmethylated and is thus completely modified with the bisulfite treatment. (*see* also **Notes 5** and **6**).

4. Notes

1. No amplification is observed including: Check PCR reaction components. Consider performing control PCR with the control DNA template and primers included in the AdvanTaq Plus PCR kit (Clontech Inc.). If the PCR kit is working most likely the problem consists of insufficient bisulfite treatment. Repeat bisulfite treatment with the following points of attention:
   a. Assure appropriate pH of Reagent I. In our experience 20 µL of 3 mM NaOH per specimen result in the desired pH and efficient bisulfite modification. However this parameter may vary.
   b. Discard all remaining supernatant at the last purification step after the bisulfite treatment. Remaining reagents after purification may inhibit amplification.
2. Amplification of most specimens but not all: Determine starting genomic DNA concentration for the specimens that were not successfully amplified. Consider electrophoretic analysis of the genomic DNA to check for degradation.
3. Weak amplification resulting in amounts insufficient for digestion or with nonspecific artifact bands: Consider changing the concentration of DNA template in the amplification mix. Either increasing or decreasing the amount of DNA template may result in more efficient amplification. In cases of artifact bands consider increasing the annealing temperature in the first 5 cycles to 68°C. Note that such modification is thermocycler-specific and enzyme-specific, and may lead to improvement of amplification only in certain types of thermocyclers/enzymes.
4. Positive amplification of blank controls: Most likely reason is contamination of the tubes used. Consider preparing new reagents, using new unopened packs of PCR-tubes. In such cases repeating the bisulfite treatment rather then only repeating the PCR-reactions is advisable.
5. Inefficient digestion (PWS controls are not fully digested): Consider using a fresh batch of restriction enzyme, longer incubation and repeating the amplification with 68°C annealing temperature to eliminate nonspecific amplification.
6. Excess digestion (partial digestion is observed for the AS specimens). This is most likely the result of contamination. Consider repeating of the bisulfite treatment using new tubes and reagents.

Acknowledgments

This work was supported in part by the New York State Office of Mental Retardation and Developmental Disabilities, and by a research grant no. 49701 from Maimonides Research Foundation (M.V.) We thank Dr. S. Nolin for her constructive criticism during the preparation of the manuscript.

References

1. Lee, S. and Wevrick, R. (2000) Identification of novel imprinted transcripts in the Prader-Willi syndrome and Angelman syndrome deletion region: further evidence for regional imprinting control. *Am. J. Hum. Genet.* **66(3),** 848–858.
2. Kishino, T., Lalande, M, Wagstaff, J. (1997) UBE3A/E6-AP mutations cause Angelman syndrome. *Nature Genet.* **16,** 16.
3. American Society of Human Genetics/American College of Medical Genetics Test and Technology Transfer Committee (1996): Diagnostic Testing for Prader-Willi and Angelman Syndromes: Report of the ASHG/ACMG Test and Technology Transfer Committee. *Am. J. Hum. Genet.* **58,** 1085–1088.
4. Ohta, T., Buiting, K., Kokkonen, H., McCandless, S., Heeger, S., Leisti, H., et al. (1999) Molecular mechanism of Angelman syndrome in two large families involves an imprinting mutation. *Am. J. Hum. Genet.* **64(2),** 385–396.
5. Zeschnigk, M., Schmitz, B., Dittrich, B., Buiting, K., Horsthemke, B., and Doerfler, W. (1997) Imprinted segments in the human genome: different DNA methylation patterns in the Prader-Willi/Angelman syndrome region as determined by the genomic sequencing method. *Hum. Mol. Genet.* **6,** 3875395.
6. Kubota, T., Das, S., Christian, S. L., Baylin, S. B., Herman, J. G., and Ledbetter, D. H. (1997) Methylation-specific PCR simplifies imprinting analysis. *Nature Genet.* **16,** 16.
7. Velinov, M., Gu, H., Genovese, M., Duncan, C., Brown, W. T., and Jenkins, E. (2000) The feasibility of PCR-based diagnosis of Prader-Willi and Angelman syndromes using restriction analysis after bisulfite modification of genomic DNA. *Mol. Genet. Metab.* **69,** 81–83.
8. Clark, S., Harrison, J., Paul, C. L., and Frommer, M. (1994) High sensitivity mapping of methylated cytosines, *Nucleic Acids Res.* **22,** 2990–2997.
9. Zeschnigk, M, Lich, C., Buiting, K., Doerfler, W., and Horsthemke, B. (1997) A single tube PCR test for the diagnosis of Angelman and Prader-Willi syndrome based on allelic methylation differences at the SNRPN locus. *Eur. J. Hum. Genet.* **5,** 94–98.
10. Velinov, M., Gu, H., Shah, K., Genovese, M., Duncan, C., Kupchik, G., and Jenkins, E. C. (2001) PCR-based methylation testing for Prader-Willi or Angelman syndromes using archived fixed-cell suspensions. *Genet. Test.* **5,** 153–155.
11. Kaczmarski, A. and Rosenblum M. (1999) Carrier detection and patient diagnosis of genetic mutations using a mini-gel electrophoresis system. *Am. Biotech. Lab.* **17,** 28.

V

Fluorescence *In Situ* Hybridization (FISH)

21

Fluorescence *In Situ* Hybridization (FISH) for Identifying the Genomic Rearrangements Associated with Three Myelinopathies

Charcot-Marie-Tooth Disease, Hereditary Neuropathy with Liability to Pressure Palsies, and Pelizaeus-Merzbacher Disease

Mansoor S. Mohammed and Lisa G. Shaffer

1. Introduction

1.1. Fluorescence In Situ Hybridization (FISH): A Brief Overview of its Application in Clinical Cytogenetics

The development of molecular probes by using DNA sequences of differing sizes, complexity, and specificity, coupled with technological innovations such as multicolor fluorochromes, computerized signal amplification, and image analysis, makes fluorescent *in situ* hybridization (FISH) a powerful investigative tool for use in clinical cytogenetics *(1–3)*. FISH is rapidly becoming routine in the clinical laboratory repertoire and, in many cases, has replaced high-resolution cytogenetic analyses (for a comprehensive overview of the applications of FISH in the cytogenetics laboratory the reader may refer to Shaffer (1995) *(4)*. Traditionally, routine cytogenetic analysis, with high-resolution banding levels of 650–850 bands per haploid karyotype, was limited to detecting deletions greater than 2–5 Mb in size. In contrast, by utilizing labeled DNA probes that are complementary to a desired gene or chromosomal locus, FISH analysis permits the detection of deletions significantly less than one Mb. In addition, FISH analysis has the distinct advantage of detecting not only cryptic deletions of a chromosomal locus but cryptic translocations *(5)* and as discussed below, cryptic duplications as well *(4)*.

In general, FISH analyses require two fundamental components: 1) target genomic DNA, usually in the form of metaphase and/or interphase nuclei prepared from a patient tissue (blood or other sample) that is immobilized on to a slide; and 2) probe DNA, usually in the form of single-stranded DNA, complementary to some target sequence within the immobilized genomic DNA and labeled so as to permit detection by fluorescent microscopy. Hybridization of the probe and target DNA is performed under conditions optimized for the specific hybridization of the probe DNA to its target genomic sequence (*see* **Subheading 3.6.**). Specifically, a FISH protocol designed for the clinical diagnosis of a chromosomal aberration simultaneously utilizes two probes (*see* **Fig. 1**).

From: *Methods in Molecular Biology, vol. 217: Neurogenetics: Methods and Protocols*
Edited by: N. T. Potter © Humana Press Inc., Totowa, NJ

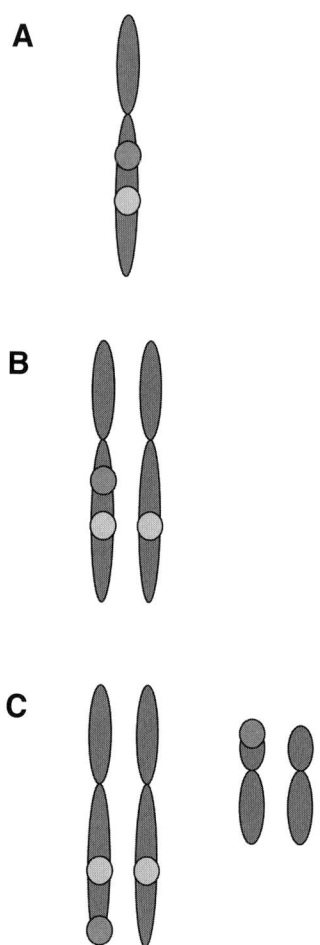

Fig. 1. Schematic representations to depict the typical hybridization patterns of test and control probes in the detection of microdeletions and cryptic translocations. (**A**) The components of a suitably designed probe set for the detection of microdeletions and translocations. The test probe (red) is designed to hybridize to a locus contained completely within the critical region. The control probe (green) is designed to hybridize to a locus that is not included in the critical region but located near to the test locus on the same chromosome. The test and control probes are differentially labeled and adequately spaced so that they are readily distinguishable on routine metaphase chromosomes (shown) or interphase nuclei (*see* **Fig. 2**). (**B**) Hybridization pattern resulting from a typical microdeletion. The signals from both the test (red) and control (green) probes are visible on the unaffected homolog. The unaffected homolog serves as an internal control for the FISH assay. The signal from the test probe is missing (red) on one homologue due to deletion. (**C**) Hybridization pattern resulting from a cryptic translocation. The signals for both the test (red) and control (green) probes are visible on the unaffected homologue. The signal from the test probe (red) is deleted from its original location on the affected homologue and is now located on a different (recipient) chromosome arm. The chromosomes would appear normal by routine cytogenetic analysis, with the cryptic translocation detected only by molecular cytogenetic methods. (See color plate 5 appearing in the insert following p. 82)

One probe is designed for use as the test probe and should hybridize to, and be completely contained within, the critical region of the chromosomal aberration. If a deleterious phenotype is associated with the chromosomal aberration, then ideally, the test probe should be designed to specifically hybridize to the gene implicated in the deleterious phenotype (i.e. the probe should contain the gene of interest). The other probe is designed for use as a control and should hybridize to a locus completely excluded from the critical region, but on the chromosome of interest. The test and control probes should be differentially labeled and adequately spaced so that they may be easily distinguishable from each other on a routine metaphase chromosome. Finally, the size and specificity of both the test and control probes should be optimized to ensure the most effective probe for FISH, producing a strong, and reliable hybridization signal.

As already noted, however, when analyzing high-resolution metaphase chromosomes, the inherent folding and condensation of the DNA limits the detection of chromosomal aberrations to greater than 2–5 Mb. Therefore, even when employing high-resolution metaphase chromosomes for FISH studies, the duplicated hybridization signals from a probe that hybridizes to a critical region involved in a microduplication (i.e., a duplication involving less than 2 Mb of DNA) are not resolvable. In such cases, interphase analysis is necessary in order to visualize the duplication (i.e., resolve the two hybridization signals) because the DNA in interphase nuclei is significantly less condensed and less folded (when compared to the DNA in metaphase chromosomes). Consequently, interphase analysis permits the resolution of hybridization signals spaced less than 2–5 Mb.

The interphase portion of the cell cycle includes gap phase 1, G1, which is characterized by protein synthesis and metabolic activity in preparation for the DNA replication phase or S phase. The S phase is then followed by the post-DNA replication phase, G2, during which DNA condensation occurs in preparation for the mitotic phase. During the S phase, the replication process of human chromosomes is not a linear end-to-end process but rather one in which replication is initiated at multiple specific sites along each chromosome, referred to as origins of replication (ori). The net result of the S phase is that each chromosome homologue is comprised of two identical strands of sister chromatids attached at their centromere regions (the reader may refer to any basic cell/molecular biology text for a review of the cell cycle). In general, homologous loci along each pair of chromosomes are expected to replicate synchronously during the S phase *(6)*. Therefore, if a test probe was hybridized to a random population of interphase cells, nuclei in G1 that harbor a duplication of the critical region would be expected to display a total of three signals; two signals corresponding to the duplicated chromosome, one for each critical region, and one signal for the normal homologue. Conversely, nuclei in G2 will display a total of six signals corresponding to two hybridization signals on each chromatid of the homologue with the duplication and a single hybridization signal on each chromatid of the normal homologue.

While homologous loci along each pair of chromosome homologues are expected to replicate synchronously during the S phase, there are exceptions, such as regions of the genome with inherently differential expression of the homologues or imprinted regions, where the transcriptionally active locus may replicate earlier than the transcriptionally silent locus *(7,8)* or asynchronously replication of the X chromosomes due to X-inactivation in females *(9)*. Considering these caveats of interphase nuclei, hybridization patterns to an asynchronously replicated locus may lead to the erroneous conclusion that the locus was duplicated. To avoid the possible complexity of hybridization results resulting from asynchronous replication of homologous loci, the ideal probe for use in detecting microduplications is one that hybridizes to a locus that exhibits synchronous replication between homologues. However, while designing a probe that hybridizes to synchronously replicated loci resolves some of the possible complexities involved in interphase analyses, it must also be considered that unless additional measures are specifically taken, a test sample of cells will contain cells at all stages of the cell cycle. Therefore, there is always the inherent variability in hybridization patterns associated with cells at the different stages of the cell cycle (as described earlier) and it requires some experience on the part of the person performing the interphase FISH analysis to distinguish between the dual signals resulting from replication of a locus vs those resulting from true duplication of that locus. Visually, the hybridization signals produced by replicated loci are often in much closer proximity when compared to the hybridization signals produced from duplicated loci. Of course, the extent of separation of the hybridization signals produced from duplicated loci will depend on both the size of the probe and the size of the duplicated region.

1.2. The Diagnostic Challenge: Gene Dosage Disturbances in Myelin Disorders of the Peripheral and Central Nervous Systems

In this chapter, we describe in detail our implementation of FISH technologies in the diagnosis of three myelinopathies: two neuropathies of the peripheral nervous system (PNS), Charcot-Marie-Tooth disease type 1A (CMT1A) and hereditary neuropathy with liability to pressure palsies (HNPP) and one leukodystrophy of the central nervous system (CNS), Pelizaeus-Merzbacher disease (PMD), in which either duplication or deletion of a large DNA sequence with concomitant gene-dosage alteration has been implicated.

These disorders present with specific diagnostic challenges, in particular, the genomic duplications. The duplications found in CMT1A and PMD are not resolvable on the metaphase chromosome. Therefore, the duplication analysis must be performed on interphase nuclei to allow for the separation of FISH signals so that the duplications can be visualized. Additionally, the laboratory performing the analysis must recognize the statistical distinction between replication of a locus (DNA synthesis during the cell cycle) and true, pathologic genomic duplication. However, these challenges can be met and a reliable FISH-based assay can be developed and implemented in the diagnostic laboratory.

1.2.1. CMT1A and HNPP

Inherited neuropathies are a heterogeneous spectrum of disorders with sometimes overlapping clinical features *(10)*. Essentially, every mode of Mendelian inheritance pattern (autosomal dominant, recessive, and X-linked), as well as sporadic cases, have been described for inherited neuropathies *(10,11)*. However, even within a particular mode of inheritance there is considerable variation in clinical severity among, not only unrelated individuals, but also affected family members including affected identical twins *(10)*. The heterogeneous nature of these neuropathies and the fact that they sometimes share clinical features with acquired conditions have often posed a diagnostic dilemma to the clinician. However, advances made in the last decade have significantly improved the ability to correctly diagnose these neuropathies. Consequently, patients who present with these conditions can now receive more appropriate genetic counseling and a more precise prognosis.

1.2.1.1. CLINICAL PRESENTATION AND MOLECULAR GENETIC ANALYSIS

Charcot and Marie *(12)* and, independently, Tooth *(13)*, described an unusual form of progressive muscular atrophy characterized by slow progression of symptoms with initial involvement of the feet and legs, often followed years later with involvement of the hands. Both studies highlighted the inherited nature of the disease well before Mendelian laws of inheritance were rediscovered and the so-called Charcot-Marie-Tooth disease eventually became recognized as the most common inherited disorder of the PNS, affecting one person in 2500 *(14)*. With this rate of incidence, CMT is also one of the most common genetic disorders. Tooth correctly classified the disorder as a neuropathy, and not a myelopathy, as was postulated by Charcot and Marie. Characteristically, the illness presents within the first two decades of life with early symptoms including complaints of frequent tripping and ankle-sprains and an equine-like gait resulting from footdrop. Progressive muscle atrophy of the distal muscles can lead to a stork-leg appearance in some patients. *Pes cavus* deformity, although usually not seen early in the course of the disease, eventually develops and seems to progress with age. Weakness of the intrinsic hand muscles usually occurs late in the course of the disease with severe cases resulting in claw-like deformities *(15)*.

Two major forms of CMT are delineated based on disease pathology and electrophysiology *(15,16)*. Type 1 CMT (CMT1) is characterized by a severely slowed motor nerve conduction velocity (NCV), usually less than 40 m/s *(17)*, and onion bulb structures evident upon sectioning of the peripheral nerve *(18)*. The clinical features of CMT1 appear to arise from myelinated fiber loss and denervation and it is referred to as the hypertrophic form *(18)*. In contrast, type 2 CMT (CMT2), retains an essentially normal NCV (otherwise only mildly reduced) with decreased amplitudes and onion bulb structures are seen only occasionally *(18)*. The clinical features of CMT2, for the most part, result from axonal dysfunction. Hence, CMT2 is referred to as the neuronal form. CMT1 can be further subcategorized, based on genetic linkage and the gene involved, into autosomal dominant CMT1A (*PMP22* in 17p12) *(19)* and CMT1B (*MPZ* in 1q21.2-q23) *(20)*, and X-linked dominant CMTX (*Cx32* in Xq13.1) *(21)*. Although likely to be less frequent, an autosomal recessive form of CMT has also been documented *(15)*. However, while multiple modes of inheritance have been documented for CMT, it is the autosomal dominant CMT1A that is the most common form of the disease. Furthermore, it is CMT1A and its unique molecular genetic relationship to another clinically distinct demyelinating neuropathy, HNPP, which has revealed a novel disease-causing mechanism in humans and which has led to the first Mendelian disorder to be diagnosed by FISH *(22)*.

Originally described in a family with recurrent peroneal neuropathy over multiple generations *(23)*, the clinical presentation of HNPP is usually much less severe than CMT with periodic episodes of numbness and muscular weakness that follow relatively minor compressions or trauma to the peripheral nerves *(24)*. Patients with HNPP sometimes show mildly slowed NCV. Peripheral nerve biopsies show segmental demyelina-

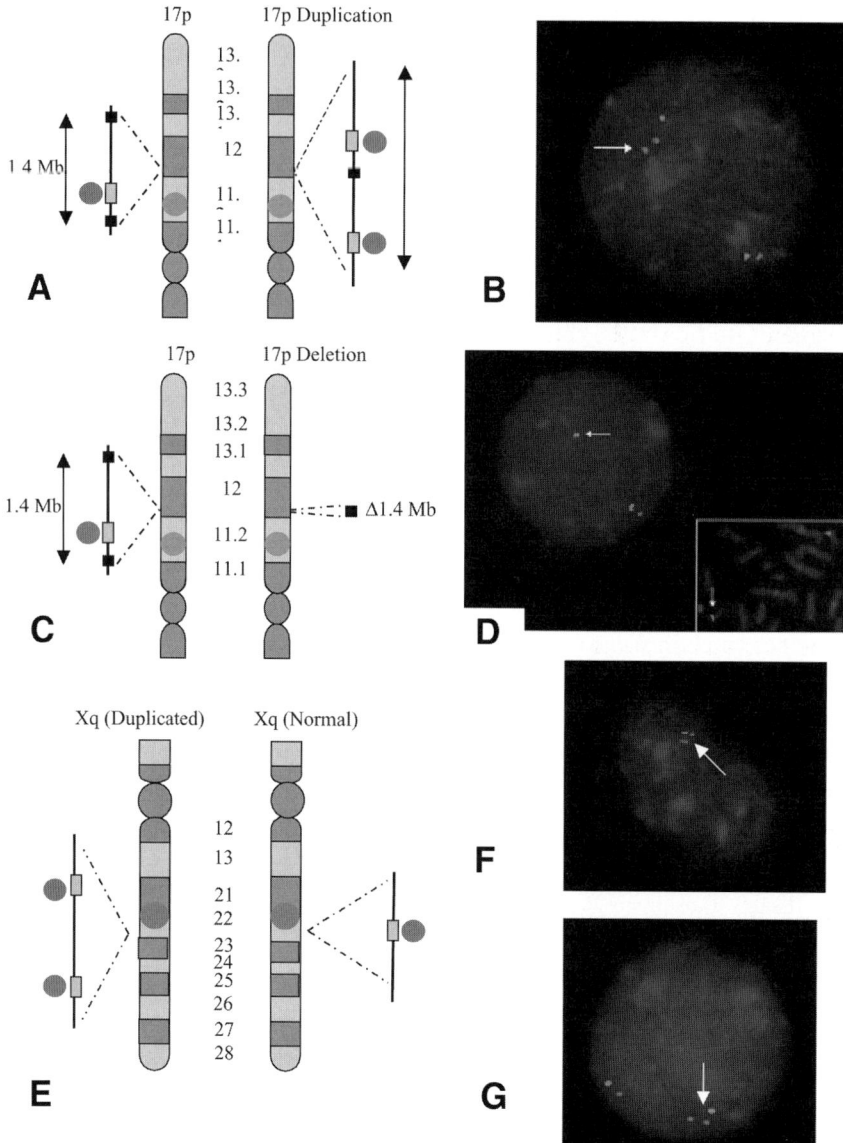

Fig. 2. Schematic representation of CMT1A duplication, HNPP deletion; and PMD duplication with accompanying FISH images from representative patient test samples. (**A**) Ideogram of G-banded 17p homologs from a typical CMT1A patient. In the normal homolog, the 1.5 Mb chromosomal region flanked by the CMT1A-REP (black box) is shown and includes a single copy of the *PMP22* locus (gray box) as detected by a digoxigenin/rhodamine labeled probe (red circle). A single copy of the *FLI* control locus is detected by a biotin/FITC labeled probe (green circle). The homolog with the cryptic duplication at the *PMP22* locus contains three copies of the CMT1A-REP, two copies of the *PMP22* gene and spans 3.0 Mb. A single copy of the FLI control locus is detected. (**B**) Typical results from of a FISH test on an interphase nucleus from peripheral blood of a CMT1A patient. The red signal represents hybridization of the test probe to the *PMP22* locus and the green signal hybridization of the control probe to the *FLI* locus. (**C**) Ideogram of G-banded 17p homologs from a typical HNPP patient. The normal homolog is essentially the same as that of the CMT1A patient, but

tion and remyelination with tomaculous or sausage-like focal thickenings of the myelin sheath *(25)*. While carpal tunnel syndrome and other entrapment neuropathies are frequent manifestations of HNPP, in severe cases, the clinical presentation can resemble that of CMT1A.

Molecular genetic analysis has demonstrated that DNA rearrangements have been implicated in the majority of patients with either CMT1A or HNPP: a 1.4Mb tandem DNA duplication in 17p12 *(22,26,27)* associated with CMT1A and the reciprocal 1.4Mb deletion in 17p12 *(28)* associated with HNPP. The 1.4Mb genomic region, which is either duplicated in CMT1A or deleted in HNPP, is flanked by a large 24kb direct repeat termed CMT1A-REP *(26,29)*. The identification of the CMT1A-REP in conjunction with genotype analysis of *de novo* patients *(30)*, was immediately suggestive of an unequal crossing-over mechanism that might explain the duplication and deletion observed in CMT1A and HNPP, respectively. Subsequently, molecular studies firmly established that the CMT1A duplication and the HNPP deletion were indeed products of a reciprocal recombination *(31)* involving a recombination hot spot *(32,33)* within the CMT1A-REP. Combined with the mapping of the peripheral myelin protein-22 gene *(PMP22)* to the duplicated region in17p12 *(34–36)*, a molecular understanding of the etiology and pathology of CMT1A and HNPP was revealed. The *PMP22* gene became the first and best example of a dosage-sensitive gene, which in the majority of CMT1A patients is present in three copies, while in the majority of HNPP patients, is present in one copy. The imbalance of the *PMP22* gene copy number ultimately results in a disturbance of the development or maintenance of myelin with the concomitant phenotype of patients with either of these neuropathies *(11,37)*.

1.2.2. CMT1A and HNPP

1.2.2.1. FISH-Based Strategy for the Diagnosis of CMT1A and HNPP

The chromosomal aberrations associated with CMT1A (the gain of a 1.4Mb genomic region in 17p12) and HNPP (the reciprocal loss of a 1.4 Mb genomic region in 17p12) (*see* **Fig. 2**) are both cryptic aberrations and are undetectable on routine G-banded chromosomes. However, knowledge of the gene implicated in the disease phenotype (e.g., *PMP22*), identification of a locus close to, but completely excluded from the micro

---

Fig. 2. *(continued)* in the homolog with the cryptic deletion, The *PMP22* locus has been deleted and only one copy of the CMT1A-REP remains. A single copy of the *FLI* locus is retained. (**D**) Typical results from both an interphase nucleus and metaphase chromosomes (insert) from a patient with HNPP. The white arrows in (B) and (D) point to the duplicated and deleted *PMP22* locus in the CMT1A and HNPP patients, respectively. (**E**) Ideogram of G-banded Xq homologs from a typical PMD patient with duplication of the *PLP* locus. A single copy of the *PLP* locus is detected by a digoxigenin/rhodamine labeled probe (red circle) and a single copy of the *BTK/GLA* locus (which is located about 2 Mb proximal to the PLP locus) is detected by a biotin/FITC probe (green circle). The homologue with the cryptic duplication contains two copies of the *PLP* locus within a genomic region which is variable in size and in most patients, one copy of the *BTK/GLA* locus. (**F**) Typical results from a FISH test on an interphase nucleus from peripheral blood of a PMD male patient. (**G**) FISH results on an interphase nucleus from a female carrier of a PMD duplication. The white arrows in (F) and (G) indicate the duplicated PLP loci. (See color plate 7 appearing in the insert following p. 82)

duplication/deletion (e.g., *FLI* in 17p11.2), and identification of cosmid contigs or PACs that are specifically mapped to either the *PMP22* locus or the *FLI* locus, are all the requisites needed (*see* **Fig. 1A**) for the design of an effective FISH strategy to detect the duplication and deletion observed in CMT1A and HNPP, respectively *(38)* (*see* **Fig. 2**). Therefore, an accurate and reliable interphase-based FISH strategy can be designed for the diagnosis of both CMT1A and HNPP and has been routinely implemented in a clinical cytogenetics laboratory (Kleberg Cytogenetics Laboratory, Baylor College of Medicine, Houston, TX, USA). As mentioned earlier, since the genomic region involved in the CMT1A duplication is not resolvable on routine metaphase chromosomes, hybridization analyses need to be performed on interphase nuclei to facilitate the visualization of the duplication *(22,38)* (*see* **Fig. 2B**). In the case of HNPP, since it is a loss of signal that is the diagnostic criteria, analyses can be performed on either interphase nuclei or metaphase chromosomes (*see* **Fig. 2C** and **D**). The interphase/metaphase-based FISH method developed for the diagnosis of CMT1A and HNPP *(38)* is presented in detail in **Subheading 3**.

1.2.3. PMD

1.2.3.1. CLINICAL PRESENTATION AND MOLECULAR GENETIC ANALYSIS

Pelizaeus-Merzbacher disease (PMD) belongs to a heterogeneous group of disorders called leukodystrophies. A leukodystrophy is a disruption of white matter of the brain. PMD is a disorder of myelin, which is divided into two groups, demyelinating and dysmyelinating diseases. PMD is the only characterized disease of dysmyelination *(39)*.

PMD is a rare and severe X-linked recessive leukodystrophy that has been recognized as a clinical entity for over a century *(39–41)*. Unlike the PNS neuropathies CMT1A and HNPP, PMD is rare and is characterized by oligodentrocyte dysmyelination in the CNS *(40–42)*. Presentation is early, with horizontal or rotary nystagmus that often disappears over time and psychomotor deterioration occurring before 2 y of age. Spastic paraplegia, dystonia and cerebellar ataxia are also often observed within the first years of life *(43)*. Magnetic resonance imaging (MRI) of the brain shows a delay of CNS myelination, which appears as diffuse hyperintensity in T2-weighted images of the entire white matter *(44)*. Abnormal delay or absence of waves in brainstem auditory evoked potentials is also helpful in the diagnosis of PMD.

Three subtypes of PMD have been recognized on the basis of disease severity *(45)*. The connatal type of PMD is the most severe and patients usually develop symptoms perinatally with subsequent demise within the first decade of life. A classical form is less severe and presents within the first year of life. Patients with classical PMD may survive into their second or third decade. Transitional PMD is intermediate in severity between the connatal and classical types. In addition, spastic paraplegia type 2 (SPG2) is allelic to PMD, which includes clinically distinct manifestations. SPG2 displays a wide spectrum of severity, with some patients showing features the overlap with PMD. However, as with the PNS neuropathies, PMD is highly pleiotrophic, making it difficult to make a diagnosis, and hence prognosis, based solely on clinical presentation *(45)*.

As with the association of the dosage sensitive gene *PMP22* to CMT1A and HNPP, gene dosage imbalances of the proteolipid protein gene (*PLP*) have been implicated as the etiological factor in the vast majority of PMD patients *(46–48)*. PLP maps to chromosome Xq22 and encodes, proteolipid protein (PLP), the predominant myelin protein in the CNS, constituting approx 50% of the total protein *(49)*. While both *PLP* deletion and duplication have been reported in PMD patients, duplications appear as the most frequent mutation *(46,47,50)* occurring in approximately 60–70% of patients. However, in addition to *PLP* duplications, more than 30 different mutations involving nucleotide substitutions or frameshifts have been identified in the *PLP* coding region of PMD patients *(39)*. A general correlation has been observed between the types of *PLP* aberrations and the various types of PMD *(50)*. In most cases, duplications of the *PLP* locus result in the classical form of PMD. Point and frameshift mutations often lead to a more severe presentation (connatal or transitional PMD) and also have been associated with SPG2. Deletion or null mutations of *PLP* result in a milder form of PMD or SPG2. Interestingly, unlike the duplications that occur in CMT1A patients, the breakpoints of the duplication in PMD patients exhibit a striking variation in location, resulting in at least six different sizes for the duplication *(50,51)*. Despite this variation, further molecular studies revealed that the mechanism for the recombination appears to be sister chromatid exchange in male meiosis, likely during spermatogenesis in the maternal grandfather *(50,51)*. Supporting this hypothesis, most mothers of the male patients with PMD are carriers. Thus, accurate diagnosis of carrier status in females is important for proper genetic counseling.

1.2.4. PMD

1.2.4.1. FISH-Based Strategy for the Diagnosis of PMD

Although the duplication breakpoints in PMD patients have exhibited considerable variation, the resulting chromosomal aberration is almost always cryptic and, hence, undetectable on routine G-banded chromosomes. However, as with the CMT1A duplication/HNPP deletion, the gene implicated in the disease phenotype (i.e., *PLP*) is known, as well as a locus close to, but almost always completely excluded from the micro duplication (i.e., *BTK/GLA* which is located about 2Mb centromeric to *PLP*). In addition, cosmid contigs or PACs that specifically map to these loci have been identified *(47)*. Therefore, an accurate and reliable interphase-based FISH strategy can be designed for the diagnosis of PMD *(47)* (*see* **Fig. 3**) and has been routinely implemented in a clinical cytogenetics laboratory (Kleberg Cytogenetics Laboratory). The interphase-based FISH method developed for the diagnosis of PMD is presented in detail **Subheading 3**.

2. Materials

2.1. Initiation of Blood Culture

1. Complete culture media for blood (*see* **Subheading 2.1., item 3**).
2. Peripheral blood (3.0–5.0 mL) collected in a sodium-heparin tube.
3. Complete culture media (RPMI) for blood (lymphocyte) culture.
    a. 500 mL RPMI 1640 medium with L-glutamine (Gibco, Cat. no.11875-093).
    b. 50 mL Fetal calf serum (Gibco, Cat. no. 26140-079).
    c. 5 mL L-Glutamine (Gibco, Cat. no. 25030-081).
    d. 2.5 mL Penicillin-Streptomycin (Gibco, Cat. no. 15140-122).
    e. 5 mL Phytohemagglutinin (Murex, Cat. no. HA15). Under a laminar flow hood, using aseptic technique, add components a–d. Filter the contents through a Nalgene 22-μm filter system into a sterile bottle. Add component e (5 mL Phytohemagglutinin) to the

filtered components. Label the bottle (e.g., Complete RPMI for bloods) including the preparation date and the expiration date (1 mo from date of preparation) and store at 4°C.
4. 15-mL sterile centrifuge tubes, polystyrene.
5. 50-mL serological pipets.
6. Slanted test tube racks.

2.2. Blood Culture Harvest

1. Colcemid (10 µg/mL). Store at 4°C
2. Potassium chloride (0.075 M) (hypotonic solution). Pre-warm to 37°C.
3. 3:1 methanol:acetic acid (referred to hereafter as fixative), freshly prepared.
4. 1-mL tuberculin syringe with hypodermic needle (26G).
5. 10-mL sterile glass serological pipet.
6. 9-in. glass Pasteur pipet.
7. Beaker with Clorox.

2.3. Blood Culture Slide Preparation

1. Lymphocyte cell pellet in fixative (*see* **Subheading 3.2.**).
2. Freshly prepared fixative (*see* **Subheading 2.2.**, **item 3**).
3. Glass microscope slides (Superfrost brand), precleaned and stored at 4°C in tap water.
4. Hotplate.
5. 5³⁄₄-in. Pasteur pipets.

2.4. Labeling Probes with Digoxigenin or Biotin by Nick-Translation

1. Unlabeled probe DNA (test probe DNA for CMT1A, HNPP, PMD, and control probe. DNA for chromosomes 17 and X) (*see* **Note 1**).
2. Deoxynucleotide Triphosphate set (Boehringer Mannheim, Cat. no. 1277049). Working solutions of the deoxynucleotide triphosphate set are made by diluting 2 µL of each dNTP (i.e., dATP, dCTP, dGTP, and dTTP) with 498 µL of sterile distilled water. These working solutions should be aliquoted and stored at –20°C until required (expiration date is the same as the purchased stock, usually 1 y from purchase).
3. Digoxigenin-11-dUTP (Boehringer Mannheim, Cat. no. 1093088).
4. Biotin-16-dUTP (Boehringer Mannheim, Cat. no. 1093070). 10 M digoxigenin-dUTP (dig-dUTP) and 20 nM biotion-dUTP (bio-dUTP) aliquots should be prepared by diluting 8 µL of 25 nM dig-dUTP or 50 nM bio-dUTP (the stock concentrations) in 12 µL of sterile distilled water. These aliquots should be stored at –20°C until required (expiration date is the same as the purchased stock, usually 1 y from purchase).
5. dNTP labeling mix. Using the working solution dilutions prepared in **items 2–4** prepare aliquots of the following (labeled as either 'dNTP dig-labeling mix' or 'dNTP bio-labeling mix'). For each aliquot, mix the following together in a 1.5-mL microcentrifuge tube:

    10 µL of 10 nM dig-dUTP or 20 nM bio-dUTP
    20 µL of dTTP
    30 µL of dATP
    30 µL of dCTP
    30 µL of dGTP

    Store aliquots at –20°C until required.
6. DNase I solution. From the manufacturer's stock of DNase I solution (Boehringer Mannheim, Cat. no. 104132) prepare a glycerol stock solution of 1 mg/mL in 50% glycerol and store at –20°C. Dilutions of this stock are made *fresh* when needed by diluting 1 µL in 249 µL of sterile distilled water (working stock).

7. DNA polymerase I (Boehringer Mannheim, Cat. no. 6427111).
8. 1Kb DNA ladder marker (Gibco-BRL, Cat. no. 15615-016).
9. Glycerol.
10. Sterile distilled water.
11. Nick-translation buffer. Prepare a solution containing 500 mM Tris base or HCl, pH 7.5, 10 mM DTT (dithioerythritol), 10 mM MgCl$_2$, in sterile water.
12. 5X TBE buffer. Bring to a final volume of 4 L (in sterile distilled water): 216 g Tris base, 110 g boric acid, 80 mL EDTA (pH 7.5–8.0).

2.5. Probe Precipitation

1. Labeled probe DNA.
2. Human Cot-1 DNA (1 mg/mL) (Gibco, Cat. no. 15279-D11).
3. Sonicated salmon sperm DNA (10 µg/µL)(Gibco, Cat. no. D9156).
4. 3 M sodium acetate, pH 5.2.
5. 100 % ethanol.
6. Hybridization solution. Mix the following reagents thoroughly: 6.5 mL Formamide (Fisher, Cat. no. F84-1), 1.0 mL 20X SSC (Ventura, Cat. no. S4073-10), 2.0 mL dextransulfate solution (Ventura, Cat. no.S4030), 0.5 mL water.

2.6. Denaturation and Hybridization of Blood Slides

1. 2X SSC, pH 7.0.
2. 70%, 90%, 95% ethanol at room temperature.
3. 70%, 90%, 95% ethanol at –20°.
4. Hybridization solution (*see* **Subheading 2.5.**, **item 6**).
5. Coverslips.
6. Rubber cement.
7. 20X SSC (Oncor, Cat. no. S4073-10).
8. Sterile distilled water.
9. Formamide (Fisher, Cat. no. F84-1).
10. 70% formamide denaturing solution. Mix thoroughly and aliquot into a coplin jar 4 mL 20X SSC, 8-mL sterile distilled water, 28 mL formamide.

2.7. Post-Hybridization Wash

1. 20X SSC (Ventura, Cat. no. S4073-10).
2. 2X SSC, pH 7.0.
3. Sterile distilled water.
4. Formamide (Fisher, Cat. no. F84-1).
5. 50% Formamide/2X SSC. Mix thoroughly: 4 mL 20X SSC, 16mL sterile distilled water, 20 mL formamide.
6. 1X PBD: Make 4 L of 4X SSC by mixing 800 mL of 20X SSC with 3200 mL of distilled water. Add 4 mL of Tween-20 and mix thoroughly. Store at 4°C for up to 1 y.

2.8. Detection of Probe

1. 1X PBD (*see* **Subheading 2.7.**, **item 6**).
2. FITC-avidin DCS (Vector, Cat. no. A2011).
3. Biotinylated Anti-avidin D (Vector, Cat. no. BA-0300).
4. Anti-Digoxigenin Monoclonal Antibody (R1) (Sigma, Cat. no. D8156).
5. Anti-Mouse IgG-Digoxigenin (R2) (Boehringer Mannheim, Cat. no. 1214624).
6. Anti-Digoxigenin-rhodamine FAB fragments (R3) (Boehringer Mannheim, Cat. no. 1207750).

7. TNT buffer: 100 m*M* Tris-HCl, pH 8.0, 150 m*M* NaCl, 0.05% Tween-20: For 500 mL TNT buffer: 50 mL of 1 *M* Tris, 15 mL of 5 *M* NaCl, and 250 µL of Tween-20.
8. TNB buffer: To 500 mL of TNT buffer (*see* **Subheading 2.8., item 7**), add 2.5 g of blocking reagent (Boehringer Mannheim, Cat. no. 1096176). Heat to boiling for several minutes to completely dissolve the blocking reagent.
9. DAPI with Vectasheild™: From a stock solution of DAPI (2.5 mg/mL in 2XSSC), prepare a final working solution of 0.2 µg/mL by diluting in Vectashield™ (Vector Laboratories, Cat. no. H-1000). It is helpful to make an intermediate concentration DAPI solution in Vectashield™ and then use this intermediate solution to make the final working dilution in Vectashield™.

3. Methods

3.1. Initiation of Blood Culture

Using aseptic techniques while working under a laminar flow hood and following OSHA universal guidelines, aliquot exactly 5 mL of complete blood medium with a serological pipet into a 15 mL sterile centrifuge tube (as needed). Inoculate with 0.4 mL of patient blood and mix thoroughly by capping the centrifuge tube tightly and inverting several times. Do not shake cultures vigorously to mix (*see* **Note 2**). Place thoroughly mixed samples in a slanted rack, and incubate at 37°C for 72 h.

3.2. Blood Culture Harvest (see Note 3)

1. Using a 1 mL tuberculin syringe with needle, add 0.05 mL of colcemid (10 µg/mL) to the cultured blood sample.
2. Mix thoroughly by capping the tube tightly and inverting the sample several times. If multiple patient cultures/tubes are being harvested, start the timer when colcemid is added to the first tube.
3. Incubate the sample at 37°C for 20 min.
4. With 1 min of incubation time left, transfer the tube to a table top centrifuge at the end of the 20 min, collect the cells by centrifugation at high speed ~430 RCF for 5 min.
5. Carefully remove as much supernatant as possible (*see* **Note 4**) with a Pasteur pipette, being careful not to disturb the top layer of the cell pellet.
6. Loosen the pellet by gently tapping the tube and then add 2 mL of prewarmed (37°C) hypotonic solution.
7. Add an additional 10 mL of hypotonic solution and incubate at room temperature for 20 min (*see* **Note 5**).
8. While the sample is incubating in hypotonic solution, prepare fresh fixative and slowly add approx 2 mL to the sample at the end of the hypotonic treatment.
9. Gently mix the sample by bubbling with a Pasteur pipet and pellet the cells by centrifugation at high speed for 7 min (*see* **Note 6**).
10. Remove and discard the supernatant with a Pasteur pipet without disturbing the pellet, and then gently tap the tube to loosen the pellet.
11. Repeat fixation by adding 5 mL of fixative and bubbling as before to thoroughly, but carefully, resuspend the cells.
12. Repeat this fixation and collection of cells twice more and either proceed to slide preparation directly (*see* **Subheading 3.3.**) or store pelleted cells at –20°C until required.

3.3. Blood Culture Slide Preparation (see Note 7)

1. Before placing fixed cell pellets onto slides, record the temperature and humidity of the preparation room and adjust conditions if necessary. Optimal temperature and humidity

ranges for preparing blood slides are between 76°C to 80°C and 65–70%, respectively.
2. If cell pellets were stored at –20°C prior to slide preparation, then re-pellet the sample by centrifugation at high speed ~430 RCF for 5 min, discard the supernatant, and resuspend the cells, as before, in freshly prepared fixative to achieve a slightly cloudy suspension.
3. Place wet, pre-cleaned slides on a hotplate preset to a temperature of about 64–65°C (*see* **Note 8**).
4. Wipe the frosted area and back of the slide with a Kimwipe, allowing a little of the water to drain off.
5. Apply a small amount of fresh fixative to the slide and allow it to drain.
6. Hold the slide at the frosted end, tilt the slide to a 45° angle and, using a Pasteur pipet, aspirate the cloudy cell suspension into the bottom ½ inch of the Pasteur pipet tip (this is enough for one slide).
7. Move the pipet tip along the slide just under the frosted area, allowing an even distribution of suspension to flow down the slide. Alternatively, dispense one drop of cell suspension onto the tilted slide from a height of approx 4 inches.
8. Place the slide at a 30° angle against a tube rack and allow to air dry in a non-drafty area (*see* **Note 9**).
9. Proceed directly to FISH analysis or store the slides –20°C prior to performing FISH analysis.

3.4. Labeling Probes with Digoxigenin or Biotin by Nick-Translation

1. Program a PCR machine to run at 15°C for 90 min, then 65°C for 10 min with an indefinite hold at 4°C.
2. Thaw the nick-translation buffer (*see* **Subheading 2.4.**, **item 10**) and appropriate aliquots of the dig-dNTP labeling and bio-dNTP labeling mix at room temperature.
3. Thaw the appropriate unlabeled probe DNA (*see* **Notes 1** and **10**) and ensure that the DNA is well mixed.
4. Prepare the following nick-translation reaction according to the amount of labeled probe desired:

	1 µg reaction	5 µg reaction	10 µg reaction
DNA (*see* **Note 11**)	X µL	X µL	X µL
10X Nick-translation buffer (*see* **Subheading 2.4.**, **item 11**)	4 µL	20 µL	40 µL
dNTP labeling mix (*see* **Note 12** and **Subheading 2.4.**, **item 5**)	10 µL	50 µL	100 µL
DNase I (working solution) (*see* **Subheading 2.4.**, **item 6**)	3.2 µL	16 µL	32 µL
DNA polymerase	2 µL	10 µL	20 µL
Sterile distilled water[a]	Y µL	Y µL	Y µL
Total volume	40 µL	200 µL	400 µL

[a]The amount of water added is dependent upon the amount of DNA solution added (*see* **Note 12**). Use an appropriate volume of water to reach the final volume of the reaction.

5. Run the labeling reaction through the pre-set PCR cycle.
6. Add 0.5 *M* EDTA to the reaction once the 65°C incubation begins (2 µL of 0.5 *M* EDTA to the 40 µL reaction, 5 µL to the 200 µL reaction and 10 µL to the 400 µL reaction).
7. Store the labeled probe at –20°C until required (*see* **Note 13**).

3.5. Probe Precipitation (see Note 14)

1. Add the following components in a 1.5-mL microcentrifuge tube: 400 µL of labeled probe DNA, 20 µL of human Cot-1 DNA, 35 µL of salmon sperm DNA, 40 µL of 3 M sodium acetate, 905 µL of 100% ethanol.
2. Thoroughly mix by brief vigorous vortexing.
3. Precipitate the DNA by incubating the tube overnight at –20°C.
4. After the overnight incubation, pellet the DNA on high speed for 10 min in a microcentrifuge.
5. Carefully decant the supernatant.
6. Wash the pellet with 100 µL of cold 70% ethanol and centrifuge (*see* **Subheading 3.5., item 4**).
7. Carefully pipet off the supernatant (*see* **Note 15**).
8. Dry the DNA pellet in a speed vacuum or allow to air dry, inverted on the bench-top for ½ to 1 h.
9. Resuspend the pellet in 250 µL of hybridization solution (*see* **Subheading 2.5., item 6**).
10. Incubate the resuspended pellet at 37°C for 1 to 16 h and then store at –20°C.

3.6. Denaturation and Hybridization of Blood Slides

1. Incubate slide(s) prepared from patient blood samples, containing interphase/metaphase nuclei (*see* **Subheading 3.3.**), in 2X SSC, pH 7.0 for 10 min at 37°C.
2. Dehydrate the slides by sequential transfer through an increasing alcohol gradient at room temperature (2 min each in 70%, 90%, and 95% ethanol).
3. After transferring the slides through the alcohol gradient, allow them to air dry at room temperature.
4. Denature the slides in pre-heated 70% formamide denaturing solution (*see* **Subheading 2.6., item 10**) for exactly 3 min at 70°C (*see* **Note 16**). Simultaneously, start the denaturation of the probes by incubating in the 70°C water-bath for 10 min.
5. After the slides have been denatured, immediately place into –20°C 70% ethanol for 2 min.
6. Transfer the slides through an alcohol gradient (2 min each in 80%, 90% and 100% ethanol kept at –20°C).
7. Air dry the slides, and once the probes have been denatured (*see* **Subheading 3.7.**), place 10 µL of each probe onto the slides
8. Carefully cover with a coverslip (avoiding trapping any air bubbles).
9. Seal with rubber cement.
10. Incubate at 37°C overnight in a humidified chamber.

3.7. Denaturation of the Probe

1. Aliquot an adequate volume of probe from the –20°C stock depending on the number of slides to be analyzed. 10 µL of probe is sufficient to cover half of a slide, which will typically contain enough interphase/metaphase nuclei for evaluation.
2. Denature the probes by incubating in the 70°C water-bath for 10 min.

3.8. Post-Hybridization Wash

1. After the overnight hybridization of the slides, gently remove the rubber cement (with a pair of forceps) and carefully pry the coverslip up at one corner (with a pair of forceps) paying particular attention to avoid rubbing the coverslip across the slide as it is being removed.
2. Wash slides in the 50% formamide/2X SSC post wash solution (*see* **Subheading 2.7., item 5**) for 15 min at 38°C.
3. Transfer to a 2X SSC solution for 8 min at 37°C.
4. Wash the slides in 1X PBD (*see* **Subheading 2.7., item 6**) for 1–2 min at room temperature and proceed to amplification and detection of probe signal.

3.9. Detection of Probe

1. Place 60 μL of FITC-avidin on the slide.
2. Place a piece of parafilm (cut to the size of the slide) on the slide. Gently press the parafilm down to remove any air bubbles that may appear.
3. Incubate the slide at 43°C for 15 min on a rack in a water bath.
4. Wash the slides in 1X PBD for 3 min at room temperature.
5. Prepare a mixture containing 1 μL of anti-digoxigenin monoclonal antibody (R1) and 250 μL of anti-avidin antibody.
6. Pipet 60 μL of this antibody mixture onto the slide and place piece of parafilm on the slide as before.
7. Re-incubate the slide at 43°C for 15 min on a rack in a water bath.
8. Wash in 1X PBD at room temperature for 3 min.
9. Prepare a 1–200 dilution of anti-mouse IgG-digoxigenin (R2) in TNB buffer and pipet 60 μL onto the slide.
10. Cover the slide with a piece of parafilm as before and incubate the slide at 43°C for 15 min on a rack in a water bath.
11. Wash the slide for 3 min in 1X PBD at room temperature.
12. Prepare a mixture containing 1 μL of anti-digoxigenin-rhodamine (FAB fragments) (R3) and 99 μL of FITC-avidin.
13. Pipet 60 μL of this mixture onto the slide, cover as before with parafilm and incubate the slide at 43°C for 15 min on a rack in a water bath.
14. Wash the slide at room temperature for 3 min with 1X PBD
15. Counterstain the slide by applying 18 μL of DAPI working solution (in Vectashield™) (*see* **Subheading 2.8., item 9**) to the slide and cover with a large coverslip.
16. View slides using a triple band pass filter that allows the simultaneous visualization of the DAPI (blue), rhodamine (red), and fluorescein (green).

4. Notes

1. We have successfully used BACs, cosmid contigs and most recently PACs as a source of probe DNA. For CMT1A and HNPP analysis we utilize a PAC test probe for the *PMP22* locus (RPCI-1-150M21) and a PAC control probe for the *FLI1* locus (RPCI-1-178F10). For PMD analysis we utilize a cosmid test probe for the *PLP* locus (LLOXNCO1 125A1) and a PAC control probe for the *BTK* locus (RPCI-1-39B6). Clone sequences can be retrieved from the Internet (*see* Website: http://www.ncbi.nlm.nih.gov).
2. It is of the utmost importance that patient samples are correctly labeled, from the very point of receiving the sample to the final slides for analysis. Patient samples should be labeled with patient and laboratory identifiers and, to avoid any mis-labeling, only one patient sample is worked with at a time.
3. The blood culture harvest is based on the principle that the formation of spindle fibers is essential for the cells to successfully complete mitosis. Thus, peripheral blood cultures are harvested using a spindle fiber inhibitor, colcemid, to stop dividing cells at metaphase, effectively increasing the overall number of metaphase cells in the harvest (referred to as the mitotic index). Following the metaphase block, the sample is treated with a hypotonic solution to swell the cells (to facilitate spreading of the metaphases once the samples are placed onto slides) and then with a fixative (methanol:acetic acid) to kill the cells, harden the membranes and preserve the cells.
4. It is important to remove as much supernatant as possible for optimization of the following step, which involves swelling the cells by the addition of hypotonic solution. If too

much residual supernatant remains, then the effective molarity of the hypotonic solution is altered and the efficiency of the hypotonic solution is compromised.

5. If multiple samples are being harvested then the supernatant should first be removed from all samples. Two mL of prewarmed hypotonic solution should then be quickly added to all samples, followed by the addition of a further 10 mL of hypotonic solution to each sample. The timer for 20 min should be started once the additional 10 mL of hypotonic solution is added to the first sample.

6. Several methods can be used to mix the samples once they are in fixative, including gentle vortexing, tapping the tube and bubbling the mixture. Since this step is crucial to mix the pellet yet avoid premature rupture of the cells, an optimal method should be used. We routinely and effectively employ bubbling as our choice for mixing of the samples.

7. This step is usually the most difficult to standardize, requiring optimized temperature and humidity settings to effectively accomplish the rupturing of the cell onto the slide so that the metaphase chromosomes or interphase nuclei are not trapped within a cytoplasmic envelope. The ranges given in the method **Subheading 3.3.** are precisely that, ranges. Actual optimal humidity and temperature parameters need to be established in each laboratory.

8. It is imperative that the slides used in FISH analyses are clean. Much of what can dirty a slide in routine handling, such as dust particles, flint, finger prints, and so on, either autofluoresce or nonspecifically bind to fluorescing particles with the consummate effect of nonspecific background fluorescence on the finished slide, which can significantly mar the ability to correctly analyze the slide.

9. As already noted in the introduction (*see* **Subheadings 1.2.2.** and **1.2.4.**) the main analysis for FISH on CMT1A, HNPP, and PMD is performed on interphase nuclei. However, as an internal quality control for each hybridization experiment, 1–2 of the metaphase chromosome spreads are examined, prior to analysis of the interphase nuclei, to ensure that proper and sufficient hybridization to the chromosome pair of interest has occurred. Therefore, slides to be used in the FISH analysis of CMT1A, HNPP, and PMD need not be particularly rich in metaphases. However, a small number of well spread metaphases should be visible on the slides (by viewing under a phase contrast microscope using low power) for FISH quality assessment and when necessary, confirmation of a deletion is needed for HNPP or PMD samples. If the metaphases are not sufficiently spread to facilitate differentiation of the chromosomes, then one or more of the following steps may be taken: 1) decrease the cell density by diluting the cell suspension in more fixative; 2) decrease the angle of the slide when applying the cell suspension; 3) leave more water on the slide when applying the cell suspension; 4) use colder water to wet the slides prior to preparation.

10. Particular attention should be paid to the correct labeling of tubes at this point. Be sure to label each tube to correctly match the particular probe that is being prepared and the dNTP labeling mix that is being used (e.g., dig-dUTP or bio-dUTP). Example, label tubes as dig-dUTP/CMT1A/HNPP test probe vs bio-dUTP/CMT1A/HNPP control probe. In addition, if probes for multiple FISH tests are to be prepared (e.g., for CMT1A/HNPP and PMD), it is advisable to perform these preparations at different times to avoid any erroneous labeling of tubes.

11. The volume of DNA needed is dependent upon the concentration of the probe stock. Example, if you are preparing 1 µL of your labeled CMT1A test probe and the concentration of the unlabeled probe DNA is 0.5 µg/µL, then use 2 µL of unlabeled probe DNA in your reaction. Empirically, this may be calculated as follows: (ensure that the concentration of the unlabeled probe DNA is in µg/µL) for a X µg/µL concentrated DNA solution then, 1 µg is contained in 1/X µL.

12. Be sure to use the appropriate dNTP labeling mix (i.e., either the dig-dNTP or bio-dNTP labeling mix). While there is no precedence for this, we find that it is helpful to standard-

ize which labeling mix is used for the control probes versus the test probes. Specifically, we always label our control probes with bio-dNTP and our test probes with dig-dNTP. Furthermore, we always detect our bio-dNTP labeled probes (control probes) with a fluorescein (green) based molecule and our dig-dNTP labeled probes (test probes) with a rhodamine (red) based molecule. Thus, at the point of analyzing our slides, we know that a green signal is always associated with the control probe whereas a red signal is always associated with the test probe.
13. While labeling the DNA probe, the nick-translation process also digests the DNA, producing labeled DNA fragments with a range of sizes. The longer the reaction proceeds, the shorter the overall size of the labeled fragments. Optimally, for FISH analyses, the labeled fragments should be 400–600 base pairs in length. Once the nick-translation has been completed, a sample can be assessed by agarose gel electrophoresis to determine if the size of the labeled DNA fragments is within the optimal range.
14. The purpose of this step is to concentrate an adequate amount of the labeled probe produced in **Subheading 3.4.** with human Cot-1 DNA and salmon sperm DNA into a hybridization solution, to facilitate the optimal hybridization of the probe to its homologous genomic DNA sequence. For example, the human Cot-1 DNA and salmon sperm DNA reduce nonspecific binding of the probe, particularly to repetitive sequences within the human genome. Whereas, the formamide and dextran sulphate components of the hybridization solution facilitate the partial denaturation of the target genomic DNA and the maintenance of a hydrophobic envirenment, respectively, to permit optimal hybridization of the probe.
15. As the DNA pellet may not always be obvious, the following is a helpful tip: Typical microcentrifuges use multi-well fix angle centrifuge buckets. Always place the microcentrifuge tube into the well of the centrifuge bucket that is directly in front of and in line with you. Place the microcentrifuge tube into the well, with the hinge to the outer side of the centrifuge bucket. The DNA pellet will form along the bottom, directly under the hinge. Remove the supernatant by carefully pipetting off from the bottom of the tube away from the pellet.
16. This is the most critical step in the preparation of the slides for hybridization. The timing and the temperature of the denaturing solution both need to be adhered to exactly. Shorter or longer incubation times can both lead to decreased hybridization efficiencies. Also, the temperature of the denaturing solution should be directly measured as opposed to taking the temperature of the water bath. Solutions contained within a coplin jar can be as much as 2°C below the temperature of the water bath itself.

Acknowledgments

We thank Drs. J. R. Lupski and K. Inuoe, Baylor College of Medicine for their critical reviews of the manuscript and helpful comments.

References

1. Escudero, T., Fuster, C., Coll, M. D., and Egozcue, J. (1998) Cytogenetic analysis using simultaneous and sequential fluorescence *in situ* hybridization. *Cancer Genet. Cytogenet.* **100**, 111–113.
2. Nath, J. and Johnson, K. L. (1998) Fluorescence *in situ* hybridization (FISH): DNA probe production and hybridization criteria. *Biotech. Histochem.* **73**, 6–22.
3. Raimondi, S. C. (2000) Fluorescence in situ hybridization: molecular probes for diagnosis of pediatric neoplastic diseases. *Cancer Invest.* **18**, 135–147.

4. Shaffer, L. G. (1997) Diagnosis of microdeletion syndromes by fluorescent *in situ* hybridization (FISH), in *Current Protocols in Human Genetics* (Dracopoli, N. C., Haines, J. L., Korf, B. R., et al., eds.) Supplement 14, 8.10.1.–8.10.14.
5. Kuwano, A., Ledbetter, S. A., Dobyns, W. B., Emanuel, B. S., and Ledbetter, D. H. (1991) Detection of deletions and cryptic translocations in Miller-Dieker syndrome by *in situ* hybridization. *Am. J. Hum. Genet.* **49,** 707–714.
6. Boggs, B. A. and Chinault, A. C. (1997) Analysis of DNA replication by fluorescence *in situ* hybridization. *Methods* **13,** 259–270.
7. Kitsberg, D., Selig, S., Brandeis, M., Simon, I., Keshet, I., Driscoll, D.J., et al. (1993) Allele-specific replication timing of imprinted gene regions. *Nature* **364,** 459–463.
8. Simon, I., Tenzen, T., Reubinoff, B. E., Hillman, D., McCarrey, J. R., and Cedar, H. (1999) Asynchronous replication of imprinted genes is established in the gametes and maintained during development. *Nature* **401,** 929–932.
9. Eden, S. and Cedar H. (1994) Role of DNA methylation in the regulation of transcription. *Curr. Opin. Genet. Dev.* **4,** 255–259.
10. Lupski, J. R. (1998) molecular genetics of peripheral neuropathies, in *Scientific American Molecular Neurology* (Martin, J. B., ed.), Scientific American, Inc., New York, NY, pp. 239–255.
11. Lupski, J. R. (1997) Charcot-Marie-Tooth disease: a gene-dosage effect. *Hospital Practice* **32,** 83–122.
12. Charcot, J-M and Marie P. (1886) Sur une forme particulaiere d'atrophie muscularie progressive souvent familiale debutante par les pied et les jambes et atteignant plus tard les mains. *Rev. Med.* **6,** 97–138.
13. Tooth, H. (1886) *The Peroneal Type of Progressive Muscular Atrophy*. HK Lewis; London, UK.
14. Skre, H. (1974) Genetic and clinical aspects of Charcot-Marie-Tooth disease. *Clin. Genet.* **6,** 98–118.
15. Lupski, J. R., Garcia, C. A., Parry, G., and Patel, P. I. (1991) Charcot-Marie-Tooth polyneuropathy syndrome: clinical electrophysiological and genetic aspects, in *Current Neurology* (Appel, S., ed.), Mosby-Yearbook Co., St. Louis, MO, USA, pp. 1–25.
16. Dyck, P. J. and Lambert, E. H. (1968) Lower motor and primary sensory neuron diseases with peroneal muscular atrophy, II: neurologic, genetic, and electrophysiologic findings in various neuronal degenerations. *Arch. Neurol.* **18,** 619–625.
17. Kaku, D. A., Parry, G. J., Malamut, R., Lupski, J. R., and Garcia, C. A. (1993) Nerve conduction studies in Charcot-Marie-Tooth polyneuropathy associated with a segmental duplication of chromosome 17. *Neurology* **43,** 1806–1808.
18. Lupski, J. R. and Garcia, C. A. (1992) Molecular genetics and neuropathology of Charcot-Marie-Tooth disease type 1A. *Brain Pathol.* **2,** 337–349.
19. Vance, J. M., Nicholson, G. A., Yamaoka, L. H., Stajich, J., Stewart, C. S., Speer, M. C., et al. (1989) Linkage of Charcot-Marie-Tooth neuropathy type 1a to chromosome 17. *Exp. Neurol.* **104,** 186–189.
20. Bird, T. D., Ott, J., and Giblett, E. R. (1982) Evidence for linkage of Charcot-Marie-Tooth neuropathy to the Duffy locus on chromosome 1. *Am. J. Hum. Genet.* **34,** 388–394.
21. Gal, A., Mucke, J., Theile, H., Wieacker, P. F., Ropers, H. H., and Wienker, T.F. (1985) X-linked dominant Charcot-Marie-Tooth disease: suggestion of linkage with a cloned DNA sequence from the proximal Xq. *Hum. Genet.* **70,** 38–42.
22. Lupski, J. R., de Oca-Luna, R. M., Slaugenhaupt, S., Pentao, L., Guzzetta, V., Trask, B. J., et al. (1991) DNA duplication associated with Charcot-Marie-Tooth disease type 1A. *Cell* **66,** 219–232.

23. De Jong, J. G. Y. (1947) Over Families met hereditaire dispositie tot het optreden van neuritiden gecorreleered met migraine. *Psychiatr. Neurol. Bull.* **50,** 60–76.
24. Staal, A., De Weerdt, C.J. and Went, L. N. (1965) Hereditary compression syndrome of peripheral nerves. *Neurology* **15,** 1008–1017.
25. Behse, F., Buchthal, F., Carlsen, F. and Knapplis, G. G. (1972) Hereditary neuropathy with liability to pressure palsies: electrophysiological and histopathological aspects. *Brain* **95,** 777–794.
26. Pentao, L., Wise, C. A., Chinault, A. C., Patel, P. I. and Lupski, J. R. (1992) Charcot-Marie-Tooth type 1A duplication appears to arise from recombination at repeat sequences flanking the 1.5 Mb monomer unit. *Nat. Genet.* **2,** 292–300.
27. Inoue, K., Dewar, K., Katsanis, N., Reiter, L.T., Lander, E.S., Devon K.L., et al. (2001) The 1.4-Mb CMT1A duplication/HNPP deletion genomic region reveals unique genome architectural features and provides insights into the recent evolution of new genes. *Genome Res.* **11,** 1018–1033.
28. Chance, P.F., Alderson, M.K., Leppig, K.A., Lensch, M.W., Matsunami, N., Smith, B., et al. (1993) DNA deletion associated with hereditary neuropathy with liability to pressure palsies. *Cell* **72,** 143–151.
29. Reiter, L. T., Murakami, T., Koeuth, T., Gibbs, R. A., Lupski, J. R. (1997) The human *COX10* gene is disrupted during homologous recombination between the 24 kb proximal and distal CMT1A-REPs. *Hum. Mol. Genet.* **6,** 1595–1603.
30. Raeymaekers, P., Timmerman, V., Nelis, E., de-Jonghe, P., Hoogendijk, J.E., Baas, F., et al. (1991) HMSN Collaborative Research Group: Duplication in chromosome 17p11.2 in Charcot-Marie-Tooth neuropathy type 1a (CMT 1a). *Neuromusc. Disord.* **1,** 93–97.
31. Chance, P. F., Abbas, N., Lensch, M. W., Pentao, L., Roa, B. B., Patel, P. I. and Lupski, J. R. (1994) Two autosomal dominant neuropathies result from reciprocal DNA duplication/deletion of a region on chromosome 17. *Hum. Mol. Genet.* **3,** 223–228.
32. Lopes, J., LeGuern, E., Gouider, R., Tardieu, S., Abbas, N., Birouk, N., et al. (1996) Recombination hot spot in a 3.2-kb region of the Charcot-Marie-Tooth type 1A repeat sequences: new tools for molecular diagnosis of hereditary neuropathy with liability to pressure palsies and of Charcot-Marie-Tooth type 1A. French CMT Collaborative Research Group. *Am. J. Hum. Genet.* **58,** 1223–1230.
33. Reiter, L. T., Murakami, T., Koeuth, T., Pentao, L., Muzny, D. M., Gibbs, R. A., and Lupski, J.R. (1996) A recombination hotspot responsible for two inherited peripheral neuropathies is located near a mariner transposon-like element. *Nat. Genet.* **12,** 288–297.
34. Matsunami, N., Smith, B., Ballard, L., Lensch, M. W., Robertson, M., Albertsen, H., et al. (1992) Peripheral myelin protein-22 gene maps in the duplication in chromosome 17p11.2 associated with Charcot-Marie-Tooth 1A. *Nat. Genet.* **1,** 176–179.
35. Timmerman, V., Nelis, E., Van, Hul. W., Nieuwenhuijsen, B.W., Chen, K.L., Wang, S., et al. (1992) The peripheral myelin protein gene PMP-22 is contained within the Charcot-Marie-Tooth disease type 1A duplication. *Nat. Genet.* **1,** 171–175.
36. Patel, P. I., Roa, B. B., Welcher, A. A., Schoener-Scott, R., Trask, B.J., Pentao, L., et al. (1992) The gene for the peripheral myelin protein PMP-22 is a candidate for Charcot-Marie-Tooth disease type 1A. *Nat. Genet.* **1,** 159–165.
37. Lupski, J. R. (1998) Charcot-Marie-Tooth disease: lessons in genetic mechanisms. *Mol. Med.* **4,** 3–11.
38. Shaffer, L. G., Kennedy, G. M., Spikes, A. S., and Lupski, J. R. (1997) Diagnosis of CMT1A and HNPP deletions by interphase FISH: implications for testing in the cytogenetics laboratory. *Am. J. Med. Genet.* **69,** 325–331.

39. Garbern, J., Cambi, F., Shy, M., and Kamholz, J. (1999) The molecular pathogenesis of Pelizaeus-Merzbacher disease. *Arch. Neurol.* **56,** 1210–1214.
40. Pelizaeus F. (1885) Uber eine eigenthumliche Form spastischer Lahmung mit Cerebralersheinungen auf hereditarer Grundlage (multiple Sklerose). *Arch. Psychiatr. Nervenkr.* **16,** 698–710.
41. Merzbacher, L. (1910) Eine eigenartige familiarhereditaire Erkrankungsform (Aplasia axialis extracorticalis congenita). *Z. Gesamte. Neurol. Pschiatr.* **3,** 1–138.
42. Seitelberger, F. (1995) Neuropathology and genetics of Pelizaeus-Merzbacher disease. *Brain Pathol.* **5,** 267–273.
43. Boulloche, J. and Aicardi, J. (1986) Pelizaeus-Merzbacher disease: Clinical and nosological study. *J. Child Neurol.* **1,** 233–239.
44. Kendall, B. E. (1993) Inborn errors and demyelination: MRI and the diagnosis of white matter disease. *J. Inherited Metab. Dis.* **16,** 771–786.
45. Hodes, M., Pratt, V., and Dlouhy, S. (1993) Genetics of Pelizaeus-Merzbacher disease. *Dev. Neurosci.* **15,** 383–394.
46. Sistermans, E. A., de Coo, R. F., De Wijs, I. J., Van Oost, B. A. (1998) Duplication of the proteolipid protein gene is a major cause of Pelizaeus-Merzbacher disease. *Neurology* **50,** 1749–1754.
47. Mimault, C., Giraud G., Courtois, V., Cailloux, F., Boire, J. Y., Dastugue, B., and Boespflug-Tanguy, O. (1999) Proteolipoprotein gene analysis in 82 patients with sporadic Pelizaeus-Merzbacher disease: duplications, the major cause of the disease, originate more frequently in male germ cells, but point mutations do not. The clinical European network on brain dysmyelinating disease. *Am. J. Hum. Genet.* **65,** 360–369.
48. Inoue, K., Osaka, H., Sugiyama, N., Kawanishi, C., Onishi, H, Nezu, A., et al. (1996) A duplicated PLP gene causing Pelizaeus-Merzbacher disease detected by comparative multiplex PCR. *Am. J. Hum. Genet.* **59,** 32–39.
49. Woodward, K. and Malcolm, S. (1999) Proteolipid protein gene: Pelizaeus-Merzbacher disease in humans and neurodegeneration in mice. *TIG* **15,** 125–128.
50. Inoue, K., Osaka, H., Imaizumi, K., Nezu, A., Takanashi, J., Arii, J., et al. (1999) Proteolipid protein gene duplications causing Pelizaeus-Merzbacher disease: molecular mechanism and phenotypic manifestations. *Ann. Neurol.* **45,** 624–632.
51. Woodward, K., Kendall, E., Vetrie, D., and Malcolm, S. (1998) Pelizaeus-Merzbacher disease: identification of Xq22 proteolipid-protein duplications and characterization of breakpoints by interphase FISH. *Am. J. Hum. Genet.* **63,** 207–217.

VI

IN VITRO EXPRESSION SYSTEMS AND STUDIES OF PROTEIN EXPRESSION AND FUNCTION

22

Drosophila Models of Polyglutamine Diseases

H. Y. Edwin Chan and Nancy M. Bonini

1. Introduction

The fruit fly, *Drosophila melanogaster*, has been used extensively as an experimental model organism since the beginning of the last century. More recently, the concept of large-scale genetic mutagenesis screens has been applied. In the first such screen, 15 loci spread throughout the genome were identified based on a common phenotype of disruption of embryonic segmentation *(1)*. This seminal work led to identification of orthologous genes that are important in human embryonic development. Mutations in these human genes have also been found to lead to congenital malformations. As one example, Waardenburg's syndrome is a human autosomal dominant disease that results in deafness and pigmentation defects in the eye, and is caused by mutations in the human homolog of the *Drosophila paired* gene, *PAX-3* *(2)*. Recently, comparative genomic analysis of the completed *Drosophila* genome has revealed that more than 60% of human disease genes are represented by *Drosophila* homologs *(3,4)*. This further fosters the use of *Drosophila* as a model organism to study processes relevant to human disease conditions.

The success of using *Drosophila* in human neurodegenerative disease modeling has largely been demonstrated during the last decade. *Swiss cheese* and *spongecake* are two mutants among others isolated from genetic screens that confer brain degeneration in *Drosophila* *(5,6)*. Subsequent molecular cloning of these genes revealed remarkable linkage between the degenerative phenotypes observed and brain degenerative diseases in humans. For example, in the *swiss cheese* (*sws*) mutant, progressive vacuolization is observed throughout the brain *(5)*. At the protein level, Sws shows 41% identity to the human neuropathy target esterase protein (NTE; *7*). NTE is a neuronal-specific serine proteinase, and covalent modification of NTE by organophosphorus esters leads to axonal degeneration *(8)*. Characterization of *sws* mutants therefore provides insight into the pathogenesis of organophosphate-induced axonopathy.

Here we discuss the approach of directly modeling a human neurodegenerative disease in *Drosophila* by directed expressed in the fly of the human disease genes. Spinobulbar muscular atrophy (SBMA) or Kennedy's disease was the first human disease shown to be caused by trinucleotide (CAG) expansion within the coding region of a gene, the androgen receptor gene *(9)*. Subsequently, at least seven more diseases have been added to the list and this class of diseases is now referred to as the polyglutamine diseases *(10)*. In this chapter, we discuss recent advances of using *Drosophila* as a model organism to: 1) decipher aspects of disease pathogenesis and 2) identify intervening pathways of disease progression.

2. Establishing Transgenic Fly Lines

To study the pathogenesis of polyglutamine-diseases, human polyglutamine disease genes have been introduced into the fly genome using routine *P*-element transposon-based transgenic techniques that have been well established *(11)*. In addition to gene over-expression, *P*-elements are widely adapted to various other applications that include genetic mapping *(12)*, gene trapping *(13)*, mutagenesis *(14)*, targeted gene knockout *(15,16)*, and RNAi analysis *(17)*.

There are two common approaches to ectopically express a gene in *Drosophila*. By one method, the protein coding sequence of the gene of interest, here a human disease gene, is cloned directly downstream of a tissue specific promoter element such as the *glass multiple reporter* element (*gmr*), which expresses predominantly in the photoreceptor neurons and pigment cells in the *Drosophila* eye *(18,19)*. This method allows constitutive expression of the disease protein in a stereotypic pattern in a nonessential tissue of the fly, the eye. However, if a more widespread expression of the disease protein is desired, such as expression throughout the nervous system (pan-neural expression), such a method may fail because global expression of a toxic disease protein might be detrimental to animal viability. To circumvent this general problem of transgene expression, other strategies of transgenic expression have been developed. Cloning the gene of interest downstream of a heat-inducible promoter allows the control of transgene expression through heat-shock treatment, usually at 37°C (the optimal growing temperature of *Drosophila* is 25°C). This allows fly lines that bear the toxic transgene to be maintained by growth at lower temperature, however this method suffers from a lack of spatial control over transgene expression. In order to achieve both spatial and temporal control of transgene expression, the two-component GAL4/UAS system was introduced into flies *(20)*. In yeast, the GAL4 protein is a transcriptional activator that binds to upstream activator sequence (UAS) of genes to direct transcription. GAL4-expressing transgenic lines can be produced either by random transposition of a GAL4-encoding vector (e.g., pGawB) in the fly genome or by cloning known regulatory sequences upstream in the GAL4 vector *(21)*. Genes to be expressed under the control of the GAL4 protein are cloned downstream of a UAS element in a UAS-containing vector, pUAST. When transgenic lines that bear a GAL4 driver are crossed to those bearing the UAS transgene, the transgene will be expressed in the progeny under the spatial and temporal control of the GAL4 driver *(20)*. Because of the usefulness of the *gmr* promoter, *gmr*-GAL4 lines were made so that it can be adapted to the GAL4/UAS system *(22)*. Additional practical considerations of the choice of expression systems are described in other articles *(21,23,24)*.

3. Features of Polyglutamine-Induced Phenotypes in Fly Models
3.1. Models

A variety of polyglutamine-encoding transgenic UAS lines have been generated and each of them has their own advantages and shortcomings. These UAS insert lines can be broadly divided into two groups: 1) disease model lines in which polyglutamine-containing protein of human disease genes are expressed *(25–27)*, and 2) polyglutamine model lines in which only a run of a polyglutamine-containing domain is expressed *(28,29)*. The first group is ideal for the studies of pathogenesis that might relate to specific human diseases, while the second group is useful to study general polyglutamine toxicity. These different UAS transgenic lines will allow subsequent comparison of the differences as well as common features of various aspects of polyglutamine-induced toxicity.

3.2. Tissue Specificity

A collection of GAL4 driver lines with different spatial and temporal expression patterns is maintained at the *Drosophila* stock center (*see* Website: http://flystocks.bio.indiana.edu/). Since polyglutamine-mediated neurodegeneration affects primarily neuronal tissues, one can easily examine the neurotoxic effect through the use of select neuronal specific GAL4 drivers, such as Ap^{VNC}-GAL4 (expresses in interneurons in the ventral nerve cord of the central nervous system [CNS]) *(30)*, *sev*-GAL4 (expresses in R1,-3, -4, -6, -7 photoreceptor cells) *(31)*; *gmr*-GAL4 (expresses in all cells of the eye) *(22)*; and *elav*-GAL4 (expresses in all neurons) *(32)*. Expression of expanded runs of polyglutamine as well as disease proteins in neuronal tissues either by such GAL4 drivers *(25,27,28)* or directly by a *gmr* promoter *(26)* form protein inclusions and cause degeneration within the developing and adult nervous system, as indicated by lethality or retinal photoreceptor loss. In contrast, no detectable cell degeneration is observed when a disease protein is expressed in non-neural tissues using *dpp*-GAL4 drivers, although nuclear inclusions still form *(25)*. The latter suggests selective tissue toxicity of polyglutamine protein.

3.3. Progressive Degeneration

Progressive loss of neurons has been a key feature of polyglutamine diseases. Progressive cell loss was observed when polyglutamine protein was expressed in adult photoreceptor neurons *(25,26)* and the nervous system *(25)*. Functionally, pan-neural expression of polyglutamine proteins leads to progressive degeneration *(33,34)* and lethality *(29,33,34)* in adults.

3.4. Protein Context

Developmental anomalies were observed when expression a mere run of polyglutamine was driven by the *dpp*-GAL4 driver, with this developmental phenotype being attenuated when additional non-polyglutamine sequences were added to the domain (comparable to the disease proteins) *(29)*. In addition, the severity of neuronal toxicity was also compromised by the same addition of nonpolyglutamine residues. These results emphasize the importance of protein context in polyglutamine toxicity, which is consistent with findings that proteolytic cleavage of full-length expanded polyglutamine disease protein may contribute to severity in pathogenesis *(35–37)*.

3.5. Toxic Conformation of Polyglutamine Proteins

It is generally accepted that repeat expansion of polyglutamine confers neurotoxicity to the normal protein containing a short polyglutamine repeat. However, in a *Drosophila* model for Spinocerebellar Ataxia Type I, increased levels of expression of the normal Ataxin-1 protein, like its expanded counterpart, caused similar disruption of adult eye structure *(27)*. The same effect was, later, also observed when high levels of normal Ataxin-1 protein were expressed in the Purkinje cell layer of a transgenic mouse model of the same disease. This raises the possibility that for some proteins, such as Ataxin-1, the protein itself may have inherent toxicity.

4. Genetic Approaches for Modifiers

The *Drosophila* eye has been used as a tissue of focus for genetic studies for several reasons: 1) its stereotypic and well-defined developmental events from larval development through adult eye formation have been well-described *(38)*; 2) at least 2,000 genes in the *Drosophila* genome are required for eye formation *(39)*; 3) the eye is dispensable for viability of *Drosophila* development which allows isolated analysis of genes that might be organismal lethal *(40)*; and 4) the orderly formed adult eye structure allows study of late-onset progressive degeneration *(25,26)*.

One advantage of genetic analysis is the ability to isolate mutants that are involved in virtually any given pathway of a biological phenomenon, such as the large-scale embryonic cuticular pattern screen *(1)*. As more and more genes have been identified and characterized, the understanding of complex interactions between genes that are involved in a given pathway has become invaluable, for example as with the *sevenless* receptor tyrosine kinase pathway that is involved in eye development *(41)*. Flies that are mutant for the *sevenless* (*sev*) gene fail to develop R7 photoreceptor cells *(42,43)*. Since receptor tyrosine kinases are essential for multiple developmental and signaling pathways, it is likely that other members of these pathways would be recessive lethal upon mutation, which would vastly limit recovery of mutants and their subsequent analyses. In order to circumvent this problem, dominant genetic screens were used to isolated mutants in second-site genes that would modify the *sev* mutant phenotype when a single copy of the corresponding gene was mutated *(44,45)*. In order to optimize the recovery of such dominant second-site modifiers, a "sensitized" genetic background can be used. Thus such a screen has been carried out in a temperature-sensitive *sev* mutant background using a temperature where just enough Sev activity is present to result in a normal eye *(44)*. Upon mutation of a single copy of a modifier gene, the phenotype at permissive temperature is now *sev*. Mutations in *Son of sevenless*, a guanyl-nucleotide exchange factor, were isolated from such a screen as were mutations in *Raf (44)*. This approach has been fruitful in identifying genes that function in the *sevenless* tyrosine kinase pathway.

As described previously, overexpression of expanded polyglutamine protein causes loss of pigment and retinal cells. Among the various pathways identified by genetic screens, the mechanism of action of the protein-folding pathway was the first to be defined and has been studied in detail *(33,34)*. Targeted-expression of the molecular chaperone Hsp70 *(33*; **Fig. 1**) and later dHdj1, a family member of the Hsp40 molecular chaperones *(28,34)*, using the GAL4/UAS system, were shown to suppress neural degeneration in the retina and in the nervous system. Not only does overexpression of chaperones suppress degeneration, but also targeted expression of dominant-negative mutants of the Hsp40 and Hsp70 families has been shown to enhance degeneration *(33,34*; **Fig. 2**). This indicates that the maintenance of cellular levels of chaperones is critical to disease pathogenesis, and modulation of chaperone activities may therefore be a means to intervene disease in progression.

Polyglutamine Diseases in Drosophila

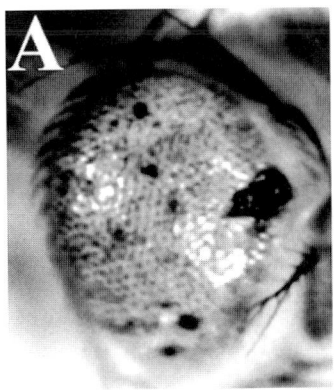 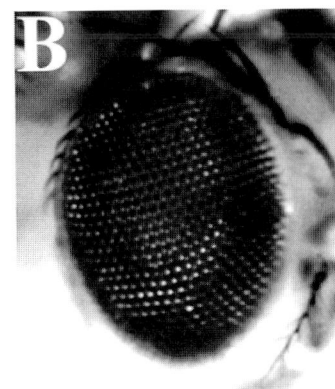

Fig. 1. Neurodegeneration induced by expression of expanded Machado-Joseph Disease (MJD) protein in the *Drosophila* retina, and its suppression by the molecular chaperone Hsp70. **(A)** Fly expressing the mutant disease form of the MJD protein with an expanded polyglutamine repeat (MJDtr-Q78), results in severe degeneration of the eye as indicated by the absence of red pigment in the eye. **(B)** Co-expression of the molecular chaperone Hsp70 with MJDtr-Q78 leads to dramatic suppression of degeneration, restoring eye structure back toward normal (*see* also **ref. 33**). (See color plate 8 appearing in the insert following p. 82)

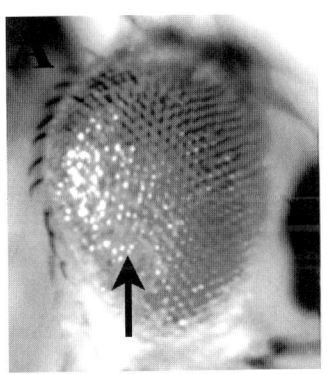 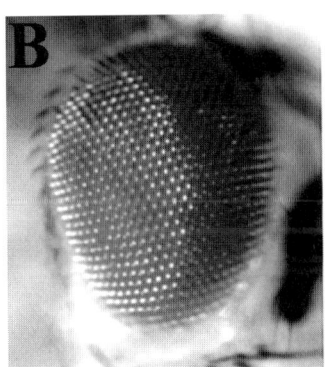 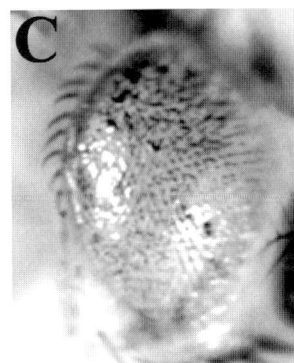

Fig. 2. Neurodegenerative phenotype caused by expression of pathogenic Machado-Joseph Disease (MJD) protein in the *Drosophila* retina and the modulatory effects of the molecular chaperone Hsp70. **(A)** Fly expressing the mutant disease form of the MJD protein with a moderately expanded polyglutamine repeat (MJDtr-Q61), exhibiting degeneration of the eye. Loss of pigmentation is noted by white patch in the eye (arrow). **(B)** Co-expression of the molecular chaperone Hsp70 with MJDtr-Q61 results in an eye that is phenotypically normal with the pigmentation restored. **(C)** Co-expression of a dominant negative mutant of the molecular chaperone Hsc70 with MJDtr-Q61 results in enhanced degeneration with severe loss of eye pigmentation. (*see* also **refs. 33** and **34**). (See color plate 9 appearing in the insert following p. 82)

The mechanistic relationship between dHdj1 and Hsp70 has also been elucidated using the genetic interaction technique. In the presence of the dominant negative Hsp70 mutant, dHdj1 can no longer suppress degeneration; however, Hsp70 still suppresses degeneration even when dominant-negative forms of dHdj1 are coexpressed *(34)*. This suggests that Hsp70 is dominant to dHdj1 in suppressing polyglutamine toxicity, which is consistent with biochemical data that dHdj1 is a co-chaperone for Hsp70 *(46)*.

Screening through collections of transposon insertion lines, several other second-site mutations have been identified as modifiers of the Ataxin-1 polyglutamine phenotype *(27)*. When expression of the Ataxin-1 protein is driven in the eye by *gmr*-GAL4 driver, an external disruption of the eye is observed at 25°C that is sensitive to both phenotypic enhancement and suppression *(27)*. A number of mild suppressors and enhancers were identified from both *P*-element and *EP*-element *Drosophila* stock collections *(27)*. Sequence information on Ataxin-1 modifiers suggests four cellular pathways might be involved in Ataxin-1 induced polyglutamine phenotypes, although mechanisms of modification have not been addressed *(27)*.

5. Immunocytochemical and Biochemical Approaches

In the last decade or so, developmental biologists have provided enormous amounts of information on the morphological events of *Drosophila* development from oogenesis to adult body structures, including the developing and adult nervous systems and the eyes *(38,47,48)*. Histochemical staining techniques were developed to decipher complex regulation of gene expression by antibody staining and *in situ* hybridization.

Antibodies that recognize recombinant polyglutamine proteins have been used to monitor the protein expression pattern in *Drosophila*. When expressed in the developing *Drosophila* eye tissues (the eye imaginal discs), progressive nuclear translocation and aggregation of expanded polyglutamine proteins are detected *(25,26)*. In adult photoreceptor neurons, nuclear translocation of the expanded polyglutamine was suggested to be dependent on the length of the polyglutamine expansion *(26)*. In addition, cellular proteins such as heat-inducible molecular chaperone Hsp70 *(33)* and polyglutamine-containing developmental protein EYA *(49)* have been found to co-localize with expanded polyglutamine protein in nuclear inclusions in eye tissues. Interestingly, when molecular chaperones are co-expressed with expanded polyglutamine proteins in neurons, no morphological alteration of the nuclear inclusions by immunocytochemistry was noted, despite suppression of neural degeneration *(28,33,34)*. These findings provide evidence that degeneration can be dissociated from the presence of nuclear inclusions.

Biochemical analysis has been applied to examine chaperone-mediated suppression of polyglutamine toxicity in vivo *(34)*. Coexpression of chaperones Hsp70 and dHdj1 increases the amount of monomeric disease protein upon detergent extraction (2% sodium dodecyl sulfate [SDS]) *(34)*. When dHdj2, a modest suppressor, is coexpressed with the polyglutamine protein, only a mild suppression is observed that parallels a limited increase in the amount of the monomeric disease protein seen by Western blot analysis *(34)*. These findings suggest that chaperones alter the solubility properties of polyglutamine protein, concomitant with the degree of suppression of toxicity in vivo. In addition to a change in solubility properties in detergent, chaperones also appear to have some ability to alter salt extractability (250 mM sodium chloride) of the disease polyglutamine protein *(34)*. These observations indicate that chaperones suppress polyglutamine toxicity by altering the biochemical properties of the toxic polyglutamine protein.

6. Bioinformatic and Proteomic Approaches

Length variation of trinucleotide repeats has been demonstrated in *Drosophila (50)*. In an early analysis, 50 *Drosophila* proteins out of a sample of 551 (9%) were shown to contain long polyglutamine runs *(51)*. Sequence analysis of the completed *Drosophila* genome revealed five orthologous polyglutamine disease genes, namely Huntingtin, TATA-binding protein, Ataxin-1, Ataxin-2, and Ataxin-6 *(3,4,52–54)*. It is worth noting that none of these disease genes are found in yeast and *C. elegans (4)*. Characterization of these orthologous disease proteins and their encoding genes, as well as their interacting proteins, together with the aforementioned molecular and classic genetic methods, will provide insightful information on the normal cellular functions of these disease proteins and conserved proteins with which they interact.

7. Concluding Remarks

Recently, peptide caspase inhibitors *(55)* and small molecules (for example, Congo red; *56*) have been shown to ameliorate neurodegeneration and inhibit aggregate formation of polyglutamine protein in vivo and in vitro respectively. This indicates that drug intervention is a viable strategy for disease treatment in patients. In vitro drug screens have been proven effective in identifying drugs. However, a relaying in vivo system would be of interest for rapid drug validation before testing in expensive, time-consuming mammalian models. Since *Drosophila* has a complex but extensively studied nervous system, a relatively short life-cycle, as well as established models for human polyglutamine disease, it would serve well the proposed role in the drug validation process. The combination of genetic, biochemical, bioinformatic, and pharmacological advantages of *Drosophila* will undoubtedly facilitate the understanding of molecular pathogenic mechanisms and the development of intervention methods and treatments for human polyglutamine disease.

8. Other Useful Technical/Practical Books

These books provide in depth and comprehensive reviews on the technical aspects of *Drosophila* methodologies.

***Drosophila* Protocols (2000)**
Edited by William Sullivan, Michael Ashburner, and R. Scott Hawley
Cold Spring Harbor Laboratory Press, USA
ISBN 0-87969-585-4

***Drosophila*: A Practical Approach, 2nd Ed. (1998)**
Edited by David B. Roberts
Oxford University Press, UK
ISBN 0-19-963660-5

Fly Pushing: The Theory and Practice of *Drosophila* Genetics (1997)
by Ralph J. Greenspan
Cold Spring Harbor Laboratory Press, USA
ISBN 0-87969-492-0

9. Useful Websites

FlyBase
see Website: http://flybase.bio.indiana.edu/

A database that carries molecular and genetic information about *Drosophila melanogaster* and other related species. FlyBase contains information on gene sequences, gene products, gene maps, and links to other useful *Drosophila* sites.

Drosophila Stock Centers

see Websites: http://flystocks.bio.indiana.edu/
http://stockcenter.arl.arizona.edu/
http://gen.bio.u-szeged.hu/stock/

Stock centers maintain collections of over 8,000 *Drosophila* stock lines for distribution among the *Drosophila* community.

The Interactive Fly

see Website: http://flybase.bio.indiana.edu/allied-data/lk/interactive-fly/aimain/1aahome.htm/

A regularly updated database that contains listings of developmentally-regulated genes in an alphabetical searchable format. Genes are also grouped by other criteria, such as expression patterns, to facilitate retrieval of information.

Bionet *Drosophila*

see Website: http://www.bio.net/hypermail/dros/

A newsgroup that allows rapid communications among the *Drosophila* community.

***Drosophila* Genome Projects**

see Websites: http://www.fruitfly.org/
http://edgp.ebi.ac.uk/

These sites provide information on genomic sequences, cDNA sequences, genome annotation data, and maintain a number of databases including Expressed Sequenced Tag (EST), cDNA, Single Nucleotide Polymorphisms (SNPs) maps, and *P*-element insert lines.

Homophila (Human Disease to *Drosophila* Gene Database)

see Website: http://homophila.sdsc.edu/

This site categorizes identified *Drosophila* orthologs of human disease genes, as defined in the Online Mendelain Inheritance in Man (OMIM) database.

FlyBrain: An Online Atlas and Database of the *Drosophila* Nervous System

see Website: http://flybrain.neurobio.arizona.edu/

This site provides an invaluable overview and in-depth description of developmental and anatomical features of the *Drosophila* nervous system.

Acknowledgments

H. Y. E. Chan was supported by the Wellcome Trust. Research in N. M. Bonini's laboratory is supported by grants from the Packard Foundation, the Huntington's Disease Society of America Coalition for the Cure, and the National Institutes of Health. N. M. Bonini is an Assistant Investigator of the Howard Hughes Medical Institute.

References

1. Nusslein-Volhard, C. and Wieschaus, E. (1980) Mutations affecting segment number and polarity in *Drosophila*. *Nature* **287,** 795–801.
2. Tassabehji, M., Read, A. P., Newton, V. E., Harris, R., Balling, R., Gruss, P. and Strachan, T. (1992) Waardenburg's snydrome patients have mutations in the human homologue of the *Pax-3* paired box gene. *Nature* **355,** 635–636.

3. Fortini, M. E., Skupski, M. P., Boguski, M. S. and Hariharan, I. K. (2000) A survey of human disease gene counterparts in the *Drosophila* genome. *J. Cell. Biol.* **150,** F23–F29.
4. Rubin, G. M., Yandell, M. D., Wortman, J. R., Gabor Miklos, G. L., Nelson, C. R., Hariharan, I. K., et al. (2000) Comparative genomics of the eukaryotes. *Science* **287,** 2204–2215.
5. Kretzschmar, D., Hasan, G., Sharma, S., Heisenberg, M., and Benzer, S. (1997) The *swiss cheese* mutant causes glial hyperwrapping and brain degeneration in *Drosophila*. *J. Neurosci.* **17,** 7425–7432.
6. Min, K. T. and Benzer, S. (1997) *Spongecake* and *eggroll*: two hereditary diseases in *Drosophila* resemble patterns of human brain degeneration. *Curr. Biol.* **7,** 885–888.
7. Lush, M. J., Li, Y., Read, D. J., Willis, A. C., and Glynn, P. (1998) Neuropathy target esterase and a homologous *Drosophila* neurodegeneration-associated mutant protein contain a novel domain conserved from bacteria to man. *Biochem. J.* **332,** 1–4.
8. Glynn, P., Holton, J. L., Nolan, C. C., Read, D. J., Brown, L., Hubbard, A., and Cavanagh, J. B. (1998) Neuropathy target esterase: immunolocalization of neuronal cell bodies and axons. *Neuroscience* **83,** 295–302.
9. La Spada, A. R., Wilson, E. M., Lubahn, D. B., Harding, A. E., and Fischbeck, K. H. (1991) Androgen receptor gene mutations in X-linked spinal and bulbar muscular atrophy. *Nature* **352,** 77–79.
10. Zoghbi, H. Y. and Orr, H. T. (2000) Glutamine repeats and neurodegeneration. *Annu. Rev. Neurosci.* **23,** 217–247.
11. Rubin, G. M. and Spradling, A. C. (1982) Genetic transformation of *Drosophila* with transposable element vectors. *Science* **218,** 348–353.
12. Chen, B., Chu, T., Harms, E., Gergen, J. P., and Strickland, S. (1998) Mapping of *Drosophila* mutations using site-specific male recombination. *Genetics* **149,** 157–163.
13. Lukacsovich, T., Asztalos, Z., Awano, W., Baba, K., Kondo, S., Niwa, S. and Yamamoto, D. (2001) Dual-tagging gene trap of novel genes in *Drosophila melanogaster*. *Genetics* **157,** 727–742.
14. Prokopenko, S. N., He, Y., Lu, Y. and Bellen, H. J. (2000) Mutations affecting the development of the peripheral nervous system in *Drosophila*: a molecular screen for novel proteins. *Genetics* **156,** 1691–1715.
15. Rong, Y. S. and Golic, K. G. (2000) Gene targeting by homologous recombination in *Drosophila*. *Science* **288,** 2013–2018.
16. Rong, Y. S. and Golic, K. G. (2001) A targeted gene knockout in *Drosophila*. *Genetics* **157,** 1307–1312.
17. Lam, G. and Thummel, C. S. (2000) Inducible expression of double-stranded RNA directs specific genetic interference in *Drosophila*. *Curr. Biol.* **10,** 957–963.
18. Hay, B. A., Maile, R., and Rubin, G. M. (1997) *P* element insertion-dependent gene activation in the *Drosophila* eye. *Proc. Natl. Acad. Sci. USA* **94,** 5195–5200.
19. Ellis, M. C., O'Neill, E. M., and Rubin, G. M. (1993) Expression of *Drosophila glass* protein and evidence for negative regulation of its activity in non-neuronal cells by another DNA-binding protein. *Development* **119,** 855–865.
20. Brand, A. H., and Perrimon, N. (1993) Targeted gene expression as a means of altering cell fates and generating dominant phenotypes. *Development* **118,** 401–415.
21. Brand, A. H., Manoukian, A. S., and Perrimon, N. (1994) Ectopic expression in *Drosophila*, in: *Drosophila melanogaster: Practical Uses in Cell and Molecular Biology* (Goldstein, L. S. B. and Fyrberg, E. A., eds.), Academic Press, San Diego, CA, USA, pp. 635–652.
22. Freeman, M. (1996) Reiterative use of the EGF receptor triggers differentiation of all cell types in the *Drosophila* eye. *Cell* **87,** 651–660.

23. Bonini, N. M. (2000) *Drosophila* as a genetic tool to define vertebrate pathway players, in: *Developmental Biology Protocols, Vol. II* (Tuan, R. S., and Lo, C. W., eds.), Humana Press, Totowa, NJ, USA, pp. 7–14.
24. Thomas, B. J. and Wassarman, D. A. (1999) A fly's eye view of biology. *Trends Genet.* **15**, 184–190.
25. Warrick, J. M., Paulson, H., Gray-Board, G. L., Bui, Q. T., Fischbeck, K., Pittman, R. N. and Bonini, N. M. (1998) Expanded polyglutamine protein forms nuclear inclusions and causes neural degeneration in *Drosophila*. *Cell* **93**, 939–949.
26. Jackson, G., Salecker, I., Dong, X., Yao, X., Arnheim, N., Faber, P., MacDonald, M. and Zipursky, S. (1998) Polyglutamine-expanded human Huntingtin transgenes induce degeneration of *Drosophila* photoreceptor neurons. *Neuron* **21**, 633–642.
27. Fernandez-Funez, P., Nino-Rosales, M. L., B., d. G., She, W., Luchak, J. M., Martinez, P., Turiegano, E., et al. (2000). Identification of genes that modify ataxin-1-induced neurodegeneration. *Nature* **408**, 101–106.
28. Kazemi-Esfarjani, P. and Benzer, S. (2000). Genetic suppression of polyglutamine toxicity in *Drosophila*. *Science* **287**, 1837–1840.
29. Marsh, J. L., Walker, H., Theisen, H., Zhu, Y., Fielder, T., Purcell, J. and Thompson, L. M. (2000) Expanded polyglutamine peptides alone are intrinsically cytotoxic and cause neurodegeneration in *Drosophila*. *Hum. Mol. Genet.* **9**, 13–25.
30. Rincon-Limas, D. E., Lu, C., Canal, I., Calleja, M., Rodriguez-Esteban, C., Izpisua-Belmonte, J. C. and Botas, J. (1999) Conservation of the expression and function of *aperous* orthologs in *Drosophila* and mammals. *Proc. Natl. Acad. Sci. USA* **96**, 2165–2170.
31. Grieder, N. C., Neelen, D., Burke, R., Basler, K., and Affolter, M. (1995) Schnurri is required for *Drosophila* Dpp signaling and encodes a zinc finger protein similar to the mammalian transcription factor PRDII-BF1. *Cell* **81**, 791–800.
32. Lin, D. M. and Goodman, C. S. (1994) Ectopic and increased expression of Fasciclin II alters motoneuron growth cone guidance. *Neuron* **13**, 507–523.
33. Warrick, J. M., Chan, H. Y. E., Gray-Board, G. L., Chai, Y., Paulson, H. L., and Bonini, N. M. (1999) Suppression of polyglutamine-mediated neurodegeneration in *Drosophila* by the molecular chaperone HSP70. *Nat. Genet.* **23**, 425–428.
34. Chan, H. Y. E., Warrick, J. M., Gray-Board, G. L., Paulson, H. L. and Bonini, N. M. (2000) Mechanisms of chaperone suppression of polyglutamine disease: selectivity, synergy, and modulation of protein solubility in *Drosophila*. *Hum. Mol. Genet.* **19**, 2811–2820.
35. Ellerby, L. M., Hackam, A. S., Propp, S. S., Ellerby, H. M., Rabizadeh, S., Cashman, N. R., et al. (1999) Kennedy's disease: caspase cleavage of the androgen receptor is a crucial event in cytotoxicity. *J. Neurochem.* **72**, 185–195.
36. Ellerby, L. M., Andrusiak, R. L., Wellington, C. L., Hackam, A. S., Propp, S. S., Wood, J. D., et al. (1999) Cleavage of atrophin-1 at caspase site aspartic acid 109 modulates cytotoxicity. *J. Biol. Chem.* **274**, 8730–8736.
37. Wellington, C. L., Singaraja, R., Ellerby, L., Savill, J., Roy, S., Leavitt, B., et al. (2000) Inhibiting caspase cleavage of huntingtin reduces toxicity and aggregate formation in neuronal and nonneuronal cells. *J. Biol. Chem.* **275**, 19,831–19,838.
38. Wolff, T. and Ready, D. F. (1993) Pattern formation in the *Drosophila* retina, in: *The Development of Drosophila melanogaster* (Bate, M. and Martinez Arias, A. eds.), Cold Spring Harbor Laboratory Press, Cold Spring Harbor, NY, USA, pp. 1277–1326.
39. Thaker, H. M., and Kankel, D. R. (1992) Mosaic analysis gives an estimate of the extent of genomic involvement in the development of the visual system in *Drosophila melanogaster*. *Genetics* **131**, 883–894.
40. Bergmann, A., Agapite, J., McCall, K. and Steller, H. (1998) The *Drosophila* gene *hid* is a direct molecular target of Ras-dependent survival signaling. *Cell* **95**, 331–341.
41. Simon, M. A. (2000) Receptor tyrosine kinases: specific outcomes from general signals. *Cell* **103**, 13–15.

42. Hafen, E., Basler, K., Edstroem, J. E., and Rubin, G. M. (1987) *Sevenless*, a cell-specific homeotic gene of *Drosophila*, encodes a putative transmembrane receptor with a tyrosine kinase domain. *Science* **236,** 55–63.
43. Banerjee, U., Renfranz, P. J., Hinton, D. R., Rabin, B. A., and Benzer, S. (1987) The *sevenless+* protein is expressed apically in cell membranes of developing *Drosophila* retina; it is not restricted to cell R7. *Cell* **51,** 151–158.
44. Simon, M. A., Bowtell, D. D., Dodson, G. S., Laverty, T. R., and Rubin, G. M. (1991) Ras1 and a putative guanine nucleotide exchange factor perform crucial steps in signaling by the *sevenless* protein tyrosine kinase. *Cell* **67,** 701–716.
45. Rogge, R. D., Karlovich, C. A., and Banerjee, U. (1991) Genetic dissection of a neurodevelopmental pathway: Son of sevenless functions downstream of the sevenless and EGF receptor tyrosine kinases. *Cell* **64,** 39–48.
46. Bukau, B., and Horwich, A. (1998) The Hsp70 and Hsp60 chapeone machines. *Cell* **92,** 351–366.
47. Goodman, C. S., and Doe, C. D. (1993) Embryonic develoment of the *Drosophila* central nervous system, in: *The development of Drosophila melanogaster* (Bate, M. and Martinez Arias, A., eds.), Cold Spring Harbor Laboratory Press, Cold Spring Harbor, NY, USA, pp. 1131–1206.
48. Truman, J. W., Taylor, B. J., and Awad, T. A. (1993) Formation of the adult nervous system, in: *The development of Drosophila melanogaster* (Bate, M. and Martinez Arias, A. eds.), Cold Spring Harbor Laboratory Press, Cold Spring Harbor, NY, USA, pp. 1245–1276.
49. Perez, M. K., Paulson, H. L., Pendse, S. J., Saionz, S. J., Bonini, N. M., and Pittman, R. N. (1998) Recruitment and the role of nuclear localization in polyglutamine-mediated aggregation. *J. Cell. Biol.* **143,** 1457–1470.
50. Michalakis, Y., and Veuille, M. (1996) Length variation of CAG/CAA trinucleotide repeats in natural populations of *Drosophila melanogaster* and its relation to the recombination rate. *Genetics* **143,** 1713–1725.
51. Karlin, S. and Burge, C. (1996) Trinucleotide repeats and long homopeptides in genes and proteins associated with nervous system disease and development. *Proc. Natl. Acad. Sci. USA* **93,** 1560–1565.
52. Li, Z., Karlovich, C. A., Fish, M. P., Scott, M. P., and Myers, R. M. (1999) A putative *Drosophila* homolog of the Huntington's disease gene. *Hum. Mol. Genet.* **8,** 1807–1815.
53. Hoey, T., Dynlacht, B. D., Peterson, M. G., Pugh, B. F., and Tjian, R. (1990) Isolation and characterization of the *Drosophila* gene encoding the TATA box binding protien, TFIID. *Cell* **61,** 1179–1186.
54. Reiter, L. T., Potocki, L., Chien, S., Gribskov, M., and Bier, E. (2001) A systematic analysis of human disease-associated gene sequences in *Drosophila melanogaster*. *Genome Res.* **11,** 1114–1125.
55. Ona, V. O., Li, M., Vonsattel, J. P., Andrews, L. J., Khan, S. Q., Chung, W. M., Frey, A. S., et al. (1999) Inhibition of caspase-1 slows disease progression in a mouse model of Huntington's disease. *Nature* **399,** 263–267.
56. Heiser, V., Scherizinger, E., Boeddrich, A., Nordhoff, E., Lurz, R., Schugardt, N., et al. (2000) Inhibition of huntingtin fibrillogenesis by specific antibodies and small molecules: implications for Huntington's disease therapy. *Proc. Natl. Acad. Sci. USA* **97,** 6739–6744.

23

A Comparative Gene Expression Analysis of Emery-Dreifuss Muscular Dystrophy Using a cDNA Microarray

Toshifumi Tsukahara and Kiichi Arahata

1. Introduction

The muscular dystrophies are a genetically heterogeneous group of disorders characterized by progressive wasting, weakness, and degeneration of the skeletal muscle. The types of muscular dystrophy have been classified according to clinical symptoms, disease progression, inheritance pattern, and genetic loci. Pathological changes, such as necrotic and regenerating fibers and inflammatory cells, present in biopsied muscle samples from muscular dystrophy cases, are caused by abnormalities in the muscle membrane. Defects in the plasma membrane or in extracellular matrix-associated proteins have been identified in muscular dystrophy cases, which leads to a fragile sarcolemma. In fact, most of the genetic defects responsible for muscular dystrophy pathology reside in genes that encode plasma membrane or extracellular matrix-associated proteins *(1–4)*. However, there are some muscular dystrophy genes that encode products not associated with the plasma membrane. In particular, deficiencies in two nuclear membrane associated proteins; emerin and lamin A/C, result in Emery-Dreifuss muscular dystrophy (EDMD) *(5–8)*. EDMD is an inherited muscular disorder that displays three characteristics: early contracture of the elbows, Achilles' tendons, and postcervical muscles; slowly progressive muscle wasting and weakness; and cardiomyopathy with severe conduction block. Most families with EDMD show an X-linked recessive inheritance, but a few autosomal dominant forms have been reported. Therefore, EDMD is a genetically heterogeneous disorder. The product of the gene responsible for the X-linked recessive EDMD is emerin, whereas that for the autosomal dominant form of EDMD is lamin A/C. However, the relationship between the abnormalities in these nuclear membranes and muscular dystrophy is unclear.

Recently, DNA microarray technology has been utilized to study gene expression and should be amenable for the study of pathologic disease processes *(9–11)*. **Figure 1** shows a diagram of our concept. Each molecular portrait represents a profile of gene expression in patients with EDMD, Duchenne muscular dystrophy (DMD), Fukuyama congenital muscular dystrophy (FCMD), and other muscular dystrophys. Common changes in the molecular portraits of all the muscular dystrophies may exist in the

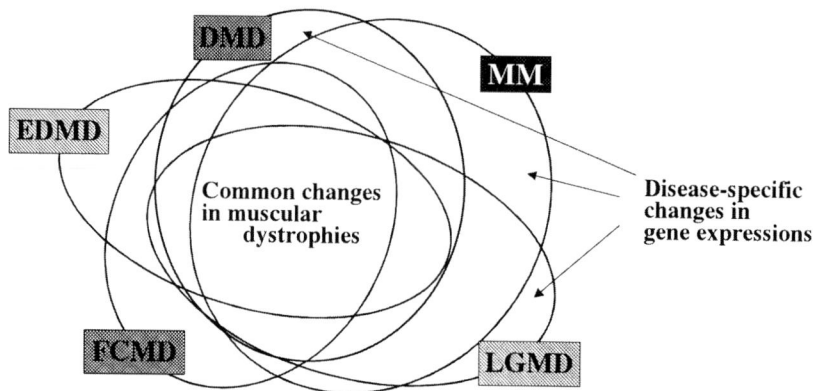

Fig. 1. Schematic description of "molecular portrait" and diseases. Each cluster shows the molecular portrait in each muscular dystrophy. The characteristic change of gene expression in each disease and the common change of the gene expression in general muscular dystrophies can be extracted by comparing these "molecular portrait" of each disease. DMD, Duchenne muscular dystrophy; FCMD, Fukuyama congenital muscular dystrophy; LGMD, Limb-girdle muscular dystrophy; MM, Miyoshi myopathy.

overlapping region. In addition, each specific change of the molecular portrait for muscular dystrophy is able to be extracted. These molecular portraits are obtained by a comprehensive gene expression analysis using muscle cDNA microarrays. A cDNA microarray is a device for analyzing the expression of thousands of genes simultaneously. Recently, the number of human genes was estimated at approx 40,000 *(12)*. Using a large-scale microarray, we are able to analyze over 1/10 of the entire set of human genes. Moreover, gene-clustering analysis may provide information to predict the function of an unknown gene. A clear advantage is that the microarray requires only a small sample for analysis. Only 0.5–1.0 µg of total RNA is sufficient for an assay. Therefore, we are able to monitor the gene expression profile from a biopsied sample using a cDNA microarray.

2. Materials

2.1. cDNA Microarray

2.1.1. Commercial Available cDNA Microarray

Pre made human cDNA microarrays can be obtained from Micromax (NEN, Boston, MA, USA), Atlas (Clontech, Palo Alto, CA, USA), ResGen (Invitrogen, Huntsville, AL, USA), LifeGrid (Incyte, Palo Alto, CA, USA), IntelliGene (Takara Biochemicals, Kyoto, Japan), GeneNavigator (Toyobo, Tokyo, Japan), and other resources. Custom-made arrays are also available from some of these suppliers. However, no suitable pre made cDNA microarrays are available for muscle research.

2.1.2. Preparation of cDNA Microarray

1. Hundreds of isolated plasmid-clones with identical linkers on both ends are necessary to contract the cDNA microarray. We cloned each cDNA fragment into pCR-Blunt vector (Invitrogen), carrying the M13 forward and reverse linkers.

2. Amplify each probe using M13 forward and reverse primers and each plasmid as a template in 100 µL reaction mixture (1X KOD plus reaction buffer (Toyobo), 0.2 mM dNTP mix, 1mM MgCl$_2$, and 5 U of KOD DNA polymerase (Toyobo).
3. Polymerase chain reaction (PCR) is performed using a 9700 thermalcycler, MicroAmp Optical 96-well reaction plates (Applied Biosystems, Foster City, CA, USA) and the following parameters: Initial denatruration at 98°C for 2 min, followed by 30 amplification cycles at 98°C for 15 s, 55°C for 15 s and 74°C for 30 s.
4. Purify amplified probes using a Biomek 2000 laboratory robot (Beckman Coulter, Fullerton, CA, USA) and Multiscreen PCR plates #MANU03050 (Millipore, Bedford, MA, USA).
5. The concentration of each purified probe was determined with PicoGreen dsDNA Quantitation kit (Molecular Probes, Eugene, OR, USA), then normalized as 0.1 µg/µL in distilled deionized water.
6. Transfer 30 µL (3 µg) each probe into 384-well plates.
7. Dry probes in a vacuum concentrator.
8. Add 6 µL 50% dimethyl sulfoxide (DMSO) into each well, dissolve probes, then briefly centrifuge the plate. The denaturing process is not necessary when the probes are dissolved in DMSO.
9. Spot the probes on GAPS II coated glass slides (Corning, Corning, NY, USA) using a SPBIO 2000 (Hitachi Software, Yokohama, Japan) microarray spotter.
10. Baking spotted glass slides at 80°C for 60 min.
11. Moisturize glass slides in a humid chamber at 80°C for 1–2 min.
12. Fix cDNAs to the glass slides by UV-cross linking at 120 mJ/cm^2.
13. Prepared microarrays should be kept in the dark at room temperature.
14. Multi-well plates, Falcon #35-3076 (BD Falcon, Franklin Lakes, NJ, USA), Beckman #140504 (Beckman Coulter) and Thermo-Fast 384 (AB gene, Epson, Surrey, UK)
15. TE buffer; 10 mM Tris-HCl, pH 8.0, 1 mM EDTA.
16. Pipct tips, Beckman #609044 and # 372654 (Beckman Coulte).
17. Incubator, humid box, and UV-cross linker.

2.2. RNA Isolation from Cultured Fibroblasts

1. In this study, fibroblasts from three cases of X-linked EDMD were examined. Cases 1 and 2 possess splice-site mutations; therefore, a smaller emerin gene product was detected in both cases *(13,14)*. Case 3 carries a large deletion between exon 1 and 6, resulting in a loss of a detectable emerin gene product *(15)*. Fibroblasts from a healthy volunteer were used as controls.
2. Fibroblasts were cultured in Dulbecco's modified Eagle's minimal essential medium (DMEM) supplemented with 10% fetal calf serum (FCS) from tissues derived from the normal control and the X-EDMD patients with written informed consent.
3. Total RNAs were isolated from the fibroblasts by acid phenol-guanidinium thiocyanate-chloroform extraction *(16)*, and purified with an SV total RNA isolation system (Promega, Madison, WI, USA).
4. Phosphate-buffered saline (PBS), sterile.
5. Cell scraper.
6. Mini grinder and disposable pestle to homogenize tissues.
7. Ethanol, 95%, RNase-free.
8. Microcentrifuge.
9. Micropipet (e.g., Gilson).
10. Incubator or heating block.

2.3. Recombinant Adenovirus

1. A recombinant adenovirus carrying the human emerin cDNA was constructed by the COS-TPC method using an Adenovirus Expression Vector kit (Takara Biochemicals). Expression of the human emerin is under the control of the CAG promoter, which is a modified chicken β-actin promoter with a CMV-IE enhancer. In addition, the polyadenylation signal of the rabbit β-globin gene was used.
2. A recombinant adenovirus carrying a LacZ expression unit also controlled by the CAG promoter (AxCALacZ) was used *(17)* (*see* **Note 1**).
3. Collagen-coated tissue-culture dish.
4. LB/ampicillin medium and LB/ampicillin plates.
5. DMEM containing 10% FCS.
6. *Eschericia coli*, recA-.
7. 293 cells.
8. λ packaging kit.
9. Transfection reagent.

2.4. cDNA Synthesis, Labeling, and Hybridization

1. Micromax™ TSA (tyramide signal amplification) kit (MPS521; NEN, Boston, MA, USA) was used for labeling and for detection in the microarray.
2. 5 M NH_4Ac, glycogen, isopropanol, 100% and 70% ethanol.
3. 0.5 M EDTA, pH 8.0 and 1 N NaOH.
4. Sterile TE buffer.
5. 20X Sodium Citrate-Sodium Chloride (SSC; 3 M Sodium Chloride, 0.3 M Sodium Citrate, pH 7.0) and distilled deionized water.
6. Hybridization camber (e.g., Corning, ArrayIt™), coverslip and incubator.
7. 10% Sodium Dodecyl Sulfate (SDS).
8. Dimethyl Sulfoxide (DMSO).
9. Bottle top filters, 0.22 µm and sterile bottles.
10. Conical tubes, 50 mL.
11. Gloves and forceps.
12. Refrigerated microcentrifuge.
13. Freezer.
14. Agitator.

2.5. TSA-Detection

1. TN Buffer: 0.1 M Tris-HCl, pH 7.5, 0.15 M NaCl. Pass buffer through a 0.22 µm filter.
2. TNT Buffer: 0.05% Tween-20 in TN Buffer. Pass buffer through a 0.22 µm filter.
3. TN Blocking Buffer (TNB): 0.5% blocking reagent in TN buffer. To prepare, heat to 60°C with stirring to completely dissolve the blocking reagent. Cool to room temperature. Do not filter.
4. TNB Goat Serum Buffer (TNB-G): 10% goat serum in TNB buffer.
5. Laser Scanner and image analysis software.

3. Methods

3.1. cDNA Synthesis and Labeling

1. Prepare two labeled cDNAs, a test and a reference, with FITC and Biotin, respectively.
2. Transfer one µg of total RNA in 13 µL DEPC-treated water to each tube.
3. Add 5 µL of a 10-fold-diluted control RNA solution (Micromax TSA kit) into the tube.

4. Add 1 µL each of Reaction Mix Concentrate (Micromax TSA kit) and FITC- or Biotin-labeled nucleotide.
5. Incubate the tubes at 65°C for 10 min, then cool the reaction mixture to room temperature for 5 min.
6. Incubate the tube to 42°C for 2–3 min.
7. Add 2.5 µL 10X RT Reaction Buffer (Micromax TSA kit), then mix well.
8. Add 2 µL AMV-RT/RNase Inhibitor Mix (Micromax TSA kit).
9. Leave the tube at 42°C for 60 min.
10. Cool the tube to 4°C for 5 min, add 2.5 µL each of 0.5 M EDTA, pH 8.0, and 1 N NaOH, then mix well.
11. Incubate 65°C for 30 min (do not exceed), then cool the tube to 4°C.
12. Add 2.7 µL 5 M NH$_4$Ac and 31 µL isopropanol, vortex and incubate at 4°C for 30 min.
13. Centrifuge for 20 min at 10,000g at 4°C, and carefully decant the supernatant.
14. Wash the pellet twice with 70% ethanol, and dry in a vacuum concentrator.
15. Resuspend the pellet in 20 µL Hybridization Buffer (Micromax TSA kit). Store at –20°C.

3.2. Hybridization

1. Combine equal amounts of cDNA from the FITC- and Biotin-labeled samples.
2. Adjust volume to 40 µL with Hybridization Buffer (Micromax TSA kit), and heat to 90°C for 2 min to denature.
3. Add combined samples to the cDNA microarray, and shield the microarray with a coverslip.
4. Incubate overnight at 65°C in a humid environment.

3.3. Washes and TSA Detection

1. All of the following treatments were carried out at room temperature.
2. Wash the microarray in 0.5X SSC, 0.01% SDS for 5 min, letting the coverslip fall off.
3. Wash in 0.06X SSC, 0.01% SDS for 5 min.
4. Wash in 0.06X SSC for 2 min.
5. Incubate in 600 µL of TNB-G for 10 min.
6. Wash slide with TNT buffer for 1 min.
7. Incubate in Anti-FITC-HRP conjugate solution (4 µL Anti-FITC-HRP conjugate; Micromax TSA kit, in 400 µL TNB-G) for 10 min.
8. Wash 3X in TNT buffer for 1 min each.
9. Incubate in Cy3 Solution (1/20 of Cyanine-3 Tyramide DMSO solution in 500 µL Amplification Diluent; Micromax TSA kit) for 10 min.
10. Wash 3X in TNT buffer for 5 min each.
11. Incubate in 300 µL of HRP Inactivation Solution (10 µL 3 M Na Acetate in 290 µL HRP Inactivation Reagent; Micromax TSA kit) for 10 min.
12. Wash 3X in TNT buffer for 1 min each.
13. Incubate in Streptavidin-HRP conjugate solution (4 µL Streptavidin-HRP conjugate; Micromax TSA kit, in 400 µL TNB-G) for 10 min.
14. Wash 3X in TNT buffer for 1 min each.
15. Incubate in Cy5 Solution (1/20 of Cyanine-5 Tyramide DMSO solution in 500 µL Amplification Diluent; Micromax TSA kit) for 10 min.
16. Wash 3X in TNT buffer for 5 min each.
17. Wash 1 min in 0.06X SSC.
18. Spin the slides to dry at room temperature.

3.4. Scanning and Data Processing

1. The microarray tagged with Cy3 and Cy5 were scanned, and fluorescence signals were detected by a microarray scanner (**Fig. 2**). Both confocal scanning devices and CCD cameras can be used. Prior to scanning, the scan area, laser power, PMT-gain, and focus should be set.
2. The intensities of Cy3- and Cy5-fluoresence from each spot should be determined and saved.
3. Scanned data may be obtained from a digital fluorescent image of a microarray using quantitative software. Quantitation is accomplished by superimposing a grid over the array image and measuring the average intensity of each spot on the microarray.
4. Quantitative gene expression and other data can be obtained from microarray analysis. Array data can be processed using analyzing software to extract and examining common features of genes and hierarchical clustering (*see* **Notes 2–9**).

3.5. Adenovirus Infection

1. The fibroblasts were infected with AxCA-Nemerin or AxCA-LacZ at a multiplicity of infection (MOI) 50.
2. The expression of exogenous emerin in EDMD fibroblasts was confirmed by immunohistochemistry and Western blot analysis.
3. After infecting the fibroblasts derived from EDMD patients with the adenoviral vectors AxCAN-Eme for emerin or AxCA-LacZ for control, RNAs were purified from the treated fibroblasts and analyzed using the cDNA microarray.

4. Notes

1. If the changes in gene expression were due to a deficiency of emerin, then emerin supplementation may reverse these changes. Therefore, we attempted to offset the change in gene expression by transferring the emerin gene into EDMD fibroblasts using a viral vector. After infecting fibroblasts derived from EDMD patients with the adenoviral vectors AxCAN-Eme for emerin or AxCA-LacZ for control, RNAs were purified from the treated fibroblasts and analyzed using the cDNA microarray.
2. To increase the reliability of the results, data were managed with two thresholds, 2.0 or 1.5. For example, genes that were expressed at levels two-fold greater than control levels, were regarded as increased genes with a threshold of 2.0. Genes that were expressed at levels $1/2$-fold less than control levels, were regarded as decreased genes with a threshold of 2.0.
3. In our experiments, the expression of 22 genes were decreased and 41 genes were increased in the EDMD fibroblasts when compared with control fibroblasts (**Table 1(A)** and **ref. 17**).
4. Our results from the infection experiment are summarized in **Table 1(B)**. As described in the table, 0, 10, and 12 genes corresponding to Groups A, B, and C, respectively, were decreased in EDMD fibroblasts when compared to control cells. In fibroblasts infected with AxCAN-Eme, the expression of 4 of 10 Group B genes and 5 of 12 Group C genes was increased. However, the expression of 13 of 22 Group A genes, 9 of 17 Group B genes, and 1 of 2 Group A genes were decreased by infection with AxCAN-Eme. Therefore, the expression of 32 of 63 genes, whose expression was altered with a deficiency of emerin, were recovered by infection of the EDMD fibroblasts with the viral vector carrying the emerin gene *(17)*.
5. Confirmation of the results from the microarray analysis is important. In our case, an RT-PCR analysis was performed. As described in **Fig. 3**, the expression levels of α2-spectrin, p63 transmembrane protein, lamin A/C, breast cancer LIV-1 genes were greater in the EDMD fibroblasts than in the control fibroblasts. Further, the expression of THRA1 and p35 cdk5 kinase genes were decreased in the EDMD fibroblasts than in the control fibroblasts.

cDNA Microarray Analysis of EDMD

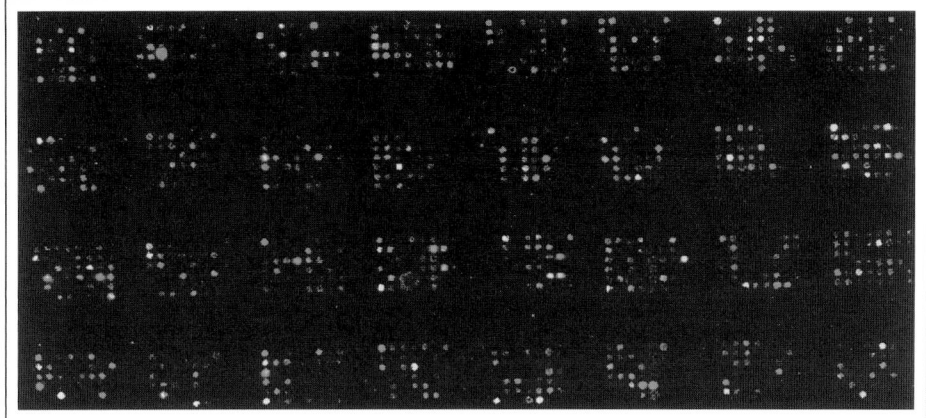

Fig. 2. Typical microarray image. One microgram of total RNAs from control fibroblasts and EDMD fibroblasts were labeled with biotin and FITC, respectively. Microarray hybridization and TSA detection was performed. The captured image was obtained by an overlay of both Cy3 and Cy5 images of the microarray. An in-house microarray on which 1536 different probes were spotted was used in this study. Our microarray showed low background and good resolution. (See color plate 6 appearing in the insert following p. 82)

Table 1
Altered Gene Expression as Detected by DNA Microarray

Ratio and Cases	Group A >2-fold in 3 cases	Group B >2-fold in 2 cases >1.5-fold in 1 case	Group C >2-fold in 1 case >1.5-fold in 2 cases	Total
Cont.>EDMD[a]	0 gene	10 genes	12 genes	22 genes
Cont.<EDMD	22 genes	17 genes	2 genes	41 genes
+Eme.>+LacZ[b]	0 gene	4 genes	5 genes	9 genes
+Eme.<+LacZ	13 genes	9 genes	1 genes	23 genes

[a]Gene expression was compared between the EDMD fibroblasts and the control fibroblasts by microarray analysis.

[b]We also compared the gene expression between EDMD fibroblasts infected with the adenoviral vectors AxCAN-Eme and AxCA-LacZ. To increase the reliability of the data, we divided the genes into three groups. Group A contains genes that had an altered, increased or decreased, expression with a threshold of >2.0 in all 3 cases of EDMD similar to the mean from 3 different fibroblasts. Genes that had an altered expression with a threshold of >2.0 in 2 cases and with a threshold of >1.5 in the remaining 1 case constitute group B. Group C consists of genes with an altered expression with a threshold of >2.0 in 1 case and with a threshold of >1.5 in the remaining 2 cases. The number of genes whose expression were altered in each group are represented in the table.

6. Importantly, the expression of the lamin A/C gene, which, when defective, results in autosomal dominant-EDMD (5,8), is increased in EDMD fibroblasts. Utrophin expression is upregulated to compensate for the decrease in dystrophin in the muscle of patients with Duchenne muscular dystrophy (18). The increased expression of lamin A/C gene may represent a similar phenomenon.

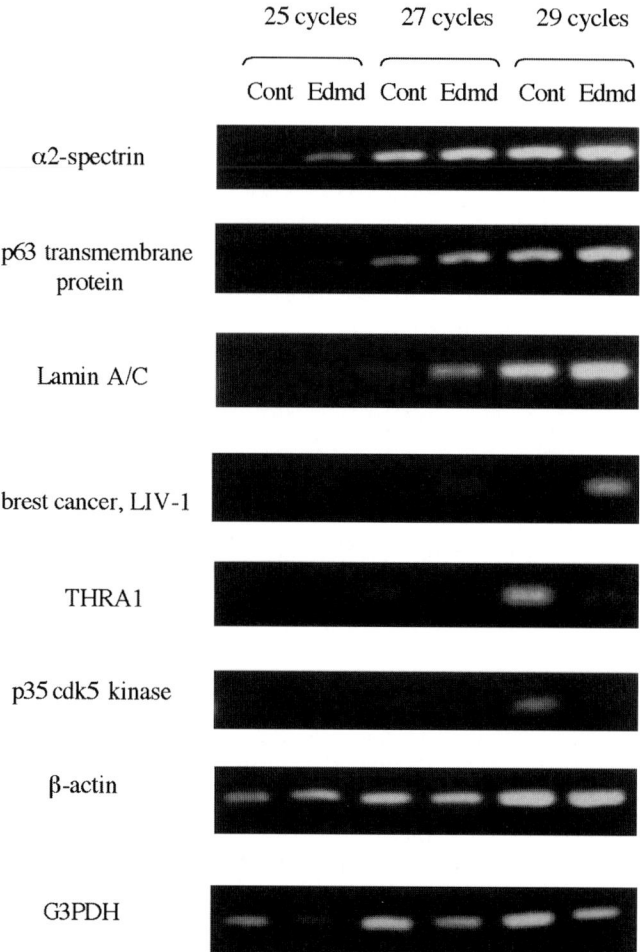

Fig. 3. Confirmation of altered gene expression by RT-PCR. To confirm the altered expression of genes detected by the microarray analysis, the relative amount of six genes, α2-spectrin, p63 transmembrane protein, lamin A/C, LIV-1, THRA1, and p35 cdk5 kinase were determined by RT-PCR. Total RNA from normal or EDMD fibroblasts was subjected to RT-PCR as described elsewhere *(15,17)*. PCR assays were performed under the following conditions: 1 cycle at 94°C for 2 min, followed by 25, 27, and 29 cycles at 94°C for 15 s, 55°C for 30 s and 68°C for 30 s. The PCR products were subjected to 2% agarose gel electrophoresis. G3PDH and β-actin mRNA were used as controls.

7. Our preliminary result suggests a correlation between disease similarity and gene expression. We submit that muscular disorders may be characterized by gene expression profiles using large-scale microarrays. However, some problems in microarray analysis exist. In many cases, sequences that are homologous may exist in several genes. Therefore, the homologous sequences in these genes may be hybridized by the same probe. Moreover, the specificities of the probes are unknown. Further, the different length of probes may result in differences in signal-strength. Therefore, the preparation of a suitable cDNA set for microarray is very important.

8. Currently, we are attempting to develop such a low-background human muscle cDNA microarray.
9. Our goal is to provide a "molecular portrait" of muscular dystrophy, and to clarify a common pathway to identify key genes in the disease. The cDNA microarray is a very efficient tool to analyze change in gene expression between two different samples. We are able to analyze gene expression in myogenesis and differentiation. In addition, we can compare gene expression between distal and proximal muscle or between specific muscles. (In distal myopathy, distal muscle, but not proximal muscle, is primarily affected. While reasons for this specificity are unclear, differences in gene expression between the regions may provide clues.) Using a comprehensive analysis of gene expression, we are able to clarify genetic and pathological features of diseases. Finding the key gene in the disease process may help to develop therapeutic approaches to the treatment of the disease.

Acknowledgments

This work was supported by a Research Grant-in-Aid for Scientific Research (B) and for Millennium Projects from the Ministry of Education, Culture, Sports, Science and Technology (T.T.) and from Core Research for Evolutional Science and Technology, Japan Science and Technology (K.A.) Authors thank Ms. Masako Fujita, Rumi Kurokawa, Akiyo Nishiyama, and Seiko Honma for their excellent technical assistance.

References

1. Chance, P. F., Ashizawa, T., Hoffman, E. P., and Crawford, T. O. (1998) Molecular basis of neuromuscular diseases. *Phys. Med. Rehabil. Clin. North Am.* **9**, 49–81.
2. Lim, L. E. and Campbell, K. P. (1998) The sarcoglycan complex in limb-girdle muscular dystrophy. *Curr. Opin. Neurol.* **11**, 443–452.
3. Hoffman, E. P. (1999) Muscular dystrophy: identification and use of genes for diagnostics and therapeutics. *Arch. Pathol. Lab. Med.* **123**, 1050–1052.
4. Zatz, M., Vainzof, M., and Passos-Bueno, M. R. (2000) Limb-girdle muscular dystrophy: one gene with different phenotypes, one phenotype with different genes. *Curr. Opin. Neurol.* **13**, 511–517.
5. Emery, A.E. (2000) Emery-Dreifuss muscular dystrophy: a 40 yr retrospective. *Neuromusc. Disord.* **10**, 228–232.
6. Bione, S., Maestrini, E., Rivella, S., Mancini, M., Regis, S., Romeo, G., and Toniolo, D. (1994) Identification of a novel X-linked gene responsible for Emery-Dreifuss muscular dystrophy. *Nat. Genet.* **8**, 323–327.
7. Nagano, A., Koga, R., Ogawa, M., Kurano, Y., Kawada, J., Okada, R., et al. (1996) Emerin deficiency at the nuclear membrane in patients with Emery-Dreifuss muscular dystrophy. *Nat. Genet.* **12**, 254–259.
8. Bonne, G., Di Barletta, M. R., Varnous, S., Becane, H. M., Hammouda, E. H., Merlini, L., et al. (1999) Mutations in the gene encoding lamin A/C cause autosomal dominant Emery-Dreifuss muscular dystrophy. *Nat. Genet.* **21**, 285–288.
9. Khan, J., Bittner, M. L., Chen, Y., Meltzer, P. S., and Trent, J. M. (1999) DNA microarray technology: the anticipated impact on the study of human disease. *Biochim. Biophys. Acta.* **1423**, M17–28.
10. Debouck, C., and Goodfellow, P. N. (1999) DNA microarrays in drug discovery and development. *Nat Genet.* **21(Suppl 1)**, 48–50.
11. Reeves, R. H. (2000) Recounting a genetic story. *Nature* **405**, 283–284.
12. Schena, M., ed. (1999) *DNA Microarrays: A Practical Approach*, Oxford University Press, UK.

13. Yorifuji, H., Tadano, Y., Tsuchiya, Y., Ogawa, M., Goto, K., Umetani, A. et al. (1997) Emerin, deficiency of which causes Emery-Dreifuss muscular dystrophy, is localized at the inner nuclear membrane. *Neurogenetics* **1,** 135–140.
14. Hasegawa, T., Kobayashi, K., Arahata, K., and Itoyama, Y. (1999) A novel splice-site mutation in the STA gene in a Japanese patient with Emery-Dreifuss muscular dystrophy. *Rinsho Shinkeigaku* **39,** 1138–1143. (In Japanese.)
15. Fujimoto, S., Ishikawa, T., Saito, M., Wada, Y., Wada, I., Arahata, K., and Nonaka. I. (1999) Early onset of X-linked Emery-Dreifuss muscular dystrophy in a boy with emerin gene deletion. *Neuropediatrics* **30,** 161–163.
16. Sambrook, J., and Rusell, D. W. (2001) *Molecular Cloning: A Laboratory Manual.* 3rd ed., Cold Spring Harbor Laboratory Press, Cold Spring Harbor, NY, USA.
17. Tsukahara, T., Tsujino, S., and Arahata, K. (2002) cDNA microarray of gene expression in fibroblasts of patients with X-linked Emery-Dreifuss muscular dystrophy. *Muscle Nerve* **25,** 896–901.
18. Mizuno, Y., Nonaka, I., Hirai, S., and Ozawa, E. (1993) Reciprocal expression of dystrophin and utrophin in muscles of Duchenne muscular dystrophy patients, female DMD-carriers and control subjects. *J. Neurol. Sci.* **119,** 43–52.

24

The COS-7 Cell In Vitro Paradigm to Study Myelin Proteolipid Protein 1 Gene Mutations

Alexander Gow

1. Introduction

Despite many shortcomings, a reductionist approach using cell culture paradigms to define basic principles underlying disease processes has considerable merit. One example of the utility of this approach is the expression of mutant forms of proteolipid protein 1 (PLP1) in transiently transfected COS-7 cells. In humans, the *PLP1* gene is located on the long arm of the X-chromosome and deletion, duplication, or coding region mutations in this gene cause the leukodystrophy, Pelizaeus-Merzbacher disease (PMD). Clinically, PMD is a heterogeneous disease that generally becomes apparent within the first year of life and is associated with hypomyelination in the central or peripheral nervous systems (CNS/PNS), breathing difficulties, poor motor coordination and paraparesis or paraplegia *(1–3)*. From simple beginnings using a transfection paradigm to express missense mutant gene products identified in PMD patients, we have developed an hypothesis to account for the cellular *(4–8)* and molecular pathogenesis of disease *(9)* and we have made use of several excellent mouse models of PMD to confirm our in vitro findings in vivo *(10)*.

The *PLP1* gene gives rise to two protein isoforms by alternative splicing of a cryptic splice-site in exon 3. Messenger RNAs (mRNAs) encompassing all of exon 3 (i.e. exon 3a and 3b) encode the 276 amino acid protein, PLP1. On the other hand, mRNAs in which exon 3b is absent encode the 242 amino acid protein, DM-20. The 35 amino acids encoded by exon 3b form part of a cytoplasmic loop near the center of PLP1 between transmembrane-spanning domains II and III. The presence of this polypeptide does not alter the topology of PLP1 compared to DM-20; both proteins exhibit 4 transmembrane domains with the amino and carboxyl-termini exposed to the cytoplasm *(11)*. Although the function of this PLP1-specific peptide is unknown, it arose recently in evolution with the emergence of terrestrial vertebrates *(12)*, confers properties on PLP1 that cannot be recapitulated by DM-20 *(13)* and clearly renders this protein more susceptible than DM-20 to the deleterious effects of missense mutations.

The initial phase of our research was to obtain cDNAs for wild-type and mutant forms of PLP1 and DM-20 and to examine the behavior of the encoded proteins in transfected COS-7 cells *(4–8)*. From wild-type PLP1 and DM-20 cDNA templates, we selected known mutations from the literature and generated the mutations for experimentation using site-directed mutagenesis. Thereafter, we sequenced and subcloned these cDNAs and constructed heterologous genes that force transfected cells to express extremely high levels of protein under transcriptional regulation of the human cytomegalovirus immediate-early promoter *(14)*.

A major motivation behind the choice of expression cassette was that during myelinogenesis in the CNS, 10% of the mRNA present in *PLP1* gene-expressing cells, called oligodendrocytes, encodes PLP1 and DM-20 *(15)*. This proportion exceeds that of other highly expressed genes, such as tubulin and actin, by two orders of magnitude. Accordingly, we wanted to recapitulate these levels as closely as possible in COS-7 cells. In this regard, the SV-40 T antigen expressed by these cells replicates the pCMV5 plasmid to high copy number from the SV-40 origin of replication site in the plasmid. Two other factors are also advantageous for high level expression: 1) the use of the human cytomegalovirus immediate-early promoter; and 2) transient transfections, which in general provide significant increases in expression compared to stably transfected cells. Finally, an additional benefit of transient transfection is the presence of nontransfected cells in all dishes, which serve as convenient internal negative controls for morphology, cell survival, and immunocytochemical labeling with antibodies.

Protein behavior in transfected fibroblasts was assessed by immunocytochemistry using a monoclonal antibody (MAb) raised against the carboxyl-terminus of PLP1/DM-20. As expected for polytopic membrane proteins, wild-type PLP1 and DM-20 are detected in all major compartments of the secretory pathway, the cell surface, and the endocytic pathway. In contrast, all single amino acid changes identified in PMD patients and animal models cause PLP1 to accumulate in the endoplasmic reticulum, which is consistent with a protein trafficking defect brought about by the inability of the mutant proteins to adopt stable three-dimensional conformations. Surprisingly, a portion of the same mutations introduced into DM-20 do not interrupt trafficking to the cell surface; these mutations are associated with mild forms of disease *(6)*.

The principal findings defined in the in vitro COS-7 cell transfection analyses were tested in vivo in mouse models of PMD *(10)*. Mutant PLP1 and DM-20 expressed in *jimpy^{msd}* and *jimpy^{rsh}* oligodendrocytes is confined to the perinuclear region of oligodendrocytes which likely reflects protein accumulation in the endoplasmic reticulum. In contrast, other myelin proteins such as myelin basic protein (MBP) are transported out to the myelin membranes that ensheath nearby axons. Thus, these data are consistent with earlier studies showing that PLP1 and DM-20 accumulate in the endoplasmic reticulum of *jimpy* mice *(16–18)*, and afford a novel interpretation of other published data for mild and severe PMD, that of a defect in protein trafficking *(9)*. Thus, the use of a simple in vitro paradigm to ask simple questions about PMD has led to new insights into the pathogenesis of this disease. Finally, the strong correlation between mild forms of disease and the trafficking of mutant DM-20 through the secretory pathway of transfected cells suggests that such an assay may be used to predict pathogenesis and severity of novel coding region mutations for the purposes of future genetic counseling. Towards this goal, we describe in detail the protocols we use to explore the disease mechanisms of PMD in COS-7 cells in culture.

2. Materials

2.1. Restriction Digest of Plasmids

1. Plasmid containing the cDNA for study. Store indefinitely at −20°C.
2. Plasmid, pUC19. Store indefinitely at −20°C.
3. Restriction enzyme, 10–20 U/µL, 10X restriction enzyme buffer, and 10 µg/mL bovine serum albumin (BSA) (New England Biolabs, MA, USA).
4. Gel loading dye, 6X: 4 mL of glycerol, 2.5 mL of 0.4 M EDTA, 3.5 mL distilled water, 0.25% bromophenol blue. Store indefinitely at room temperature (RT).

2.2. Agarose Gel Electrophoresis

1. Agarose, 0.5 g (Gibco-BRL).
2. Ethidium bromide (EtBr) in distilled water, 10 mg/mL stock. Store away from light for up to 1 yr at RT. Caution: EtBr is a powerful mutagen and should be handled with gloves.
3. Tris-acetate buffer, pH 8.3, 50X stock: 242 g Tris base, 57.1 mL glacial acetic acid, 100 mL of 0.5 M EDTA (pH 8.0). Store indefinitely at RT.
4. EtBr-TAE buffer, pH 8.3: dilute 50X stock to 1X with distilled water. Add 30 µL of EtBr stock per liter of buffer. Stable for 1 mo at RT.
5. DNA molecular weight markers, 0.5 µg/µL, 1 kb ladder (New England Biolabs).

2.3. Subcloning the cDNA

1. QIAquick gel extraction kit.
2. T4 DNA ligase, 4000 units/µL and 10X buffer (Stratagene).
3. XL-1 blue competent *Eschericia coli*, ≥ 10^6 transformants/µg, (Stratagene).
4. Luria broth (LB): 1% bacto-tryptone (Fisher), 0.5% bacto-yeast extract, 1% NaCl. Autoclave and store at RT for up to 6 mo.

2.4. Growth of Transformed Bacteria on Ampicillin Plates

1. LB (*see* **Subheading 2.3., item 4**).
2. Bacto-agar (Fisher).
3. Ampicillin in distilled water, 50 mg/mL stock. Filter through 0.45 µm Millipore membrane to sterilize. Store away from light up to 2 mo at 4°C.
4. Sterile 100 mm polystyrene petri dishes.
5. X-gal (Fisher) in N,N-dimethylformamide, 50 mg/mL stock. Store away from light up to 3 mo at −20°C. Discard if the solution turns noticeably yellow.
6. IPTG (Fisher) in distilled water, 0.1 M stock. Filter through 0.45 µm Millipore membrane to sterilize. Store indefinitely at −20°C.
7. Ethanol in a glass beaker. Caution: flammable, keep away from Bunsen flame.

2.5. Purifying Plasmids from Mini- and Maxi-Prep Cultures

1. Sterile tooth picks.
2. Ampicillin in distilled water, 50 mg/mL stock.
3. LB (*see* **Subheading 2.3., item 4**)
4. LB-ampicillin: to 500 mL of sterile LB at RT add 0.5 mL of ampicillin stock. Store away from light for up to 2 mo at 4°C.
5. Qiagen mini-prep kit, standard purity DNA.
6. Restriction enzyme (*see* **Subheading 2.1., item 3**).
7. Qiagen Endofree Plasmid Megakit.

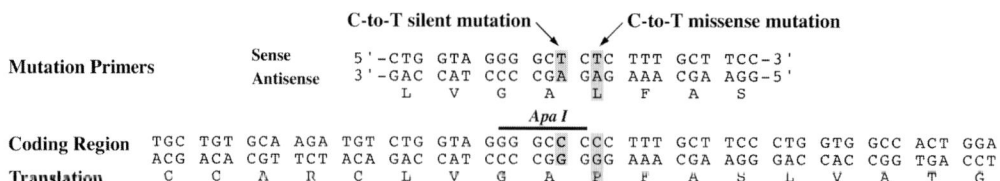

Fig. 1. Mutagenesis of the human *PLP1* coding region in exon 2. Codons 6–23 of the human *PLP1* cDNA are shown, which encode amino acids 5–22 of the mature PLP1 molecule. To generate the mutation identified in a connatal PMD patient, two PCR mutation primers were designed with single base changes (gray boxes) to effect the missense mutation, P14L, at codon 15 and to ablate an *Apa* I restriction enzyme site at codon 14. Note that the amino acid, alanine, encoded by codon 14 is unchanged by the single base change and that the new base triplet used, G-C-T, is one that commonly appears in mammalian cDNAs. Finally, the primers are of sufficient length so that 10–15 bases of perfect complementarity to the cDNA template flank the base changes on either side.

2.6. Site-Directed Mutagenesis to Generate the P14L Mutation in PLP1

1. Two overlapping and complementary oligodeoxynucleotide mutation primers. Store indefinitely at –20°C.
2. QuikChange mutagenesis kit (Stratagene).
3. LB (*see* **Subheading 2.3., item 4**)
4. LB-ampicillin (*see* **Subheading 2.5., item 4**).
5. Qiagen mini-prep kit.
6. *Apa* I restriction enzyme, 10X buffer, 10 µg/µL BSA (New England Biolabs) (*see* **Fig. 1**).

2.7. Subclone the cDNA into an Expression Cassette

1. pCMV5 expression plasmid.
2. Components listed in **Subheadings 2.1.–2.5.**

2.8. Maintaining COS-7 Cells in Culture

1. T75 flask of COS-7 cells between 10–100% confluence.
2. Sterile 9" Pasteur pipets.
3. Phosphate buffered saline (PBS) without Ca^{2+} or Mg^{2+} (BioWhittacker) (*see* **Note 1M**).
4. 0.25% trypsin /1 m*M* EDTA (Ginco-BRL). Store indefinitely in 10 mL aliquots at –20°C.
5. Dulbecco's Minimal Essential Medium (DMEM; Gibco-BRL).
6. 200 m*M* glutamine, 100X. (Ginco-BRL). Store indefinitely in 5 mL aliquots at –20°C.
7. Penicillin (10,000 U/mL)/streptomycin (10,000 U/mL), 100X. (Gibco-BRL). Store indefinitely in 5 mL aliquots at –20°C.
8. Fetal calf serum (Sigma). Heat-inactivate this reagent by incubating at 56°C for 30 min. Store indefinitely in 50 mL aliquots at –20°C.
9. Culture medium: 500 mL of DMEM, 50 mL of fetal calf serum (FCS), 5 mL of 100X penicillin/streptomycin, 5 mL of 100X glutamine added fresh every mo. Store medium at 4°C for up to 4 mo.
10. T75 culture flasks and individually wrapped sterile polystyrene pipets (Falcon).

2.9. Transfection of COS-7 Cells

1. Ethanol, 70%, in a spray bottle.
2. Culture medium (*see* **Subheading 2.8., item 9**).

3. PBS without Ca^{2+} or Mg^{2+}.
4. Confluent T75 culture flask of COS-7 cells.
5. T75 culture flasks, 100 mm and 60 mm culture dishes for mammalian cells, individually wrapped sterile polystyrene pipets, 35 mm NUNC culture dishes (Fisher) (*see* **Note 2**).
6. DMEM.
7. FUGENE 6 (Stratagene).
8. Expression plasmid DNA for transfection. Store indefinitely at 4°C.

2.10. Fixation and Immunocytochemistry

1. DMEM warmed to 37°C.
2. 0.1 M Sodium phosphate buffer, pH 7.4. Autoclave and store indefinitely at RT.
3. Paraformaldehyde powder (Sigma).
4. Stericup vacuum filter units, 100 mL capacity, 0.45-µm pore size (Millipore, MA, USA).
5. Tris-buffered saline (TBS), 10X stock: 250 mM Tris base, 1.5 M NaCl. Dilute with distilled water and adjust pH to 7.4 with conc. HCl in a fume hood. Add 50 mg sodium azide/100 mL and store indefinitely at RT. Cautions: HCl fumes are harmful to skin and lungs; sodium azide is extremely toxic, handle with gloves in a fume hood.
6. Saponin (Sigma) in distilled water, 10% stock. Add 50 mg sodium azide /100 mL and store indefinitely at RT. Caution: saponin is cardiotoxic, prepare in a fume hood.
7. Grease pen (e.g., Super HT PAP pen, Research Products International Corp.).
8. Gelatin (Type A, 300 Bloom, Sigma).
9. Bovine serum albumin (BSA), fraction V (Sigma).
10. Sodium azide (Sigma). Caution: extremely toxic, handle with gloves in a fume hood.
11. Normal goat serum (Sigma).
12. Combine reagents from **Subheading 2.10.**, **items 5,8–10** as follows to make 500 mL of TBSGBA solution. Add 0.5 g of gelatin (*see* **item 8**) to 450 mL of distilled water at RT and place in a water bath also at RT. Heat the solution to 45°C. Gently mix the gelatin crystals until they dissolve. Allow the solution to come to RT and add 5 g of BSA (*see* **item 9**). This component dissolves most easily if the solution is left unstirred for approx 30 min. With moderate stirring, add 50 mL of a 10X stock of TBS (*see* **item 5**). Add 250 mg of sodium azide (*see* **item 10**). Pass this TBSGBA solution through a 0.45-µm filter to remove small protein aggregates and store in 50 mL aliquots at RT for at least 1 y.
13. Blocking solution is made by combining the following reagents from **Subheading 2.10.**, **items 11** and **12**. To 10 mL of TBSGBA (*see* **item 12**), add 0.3 mL of normal goat serum (*see* **item 11**). This solution is stable at RT for 1 mo and should be centrifuged in a microfuge for 2 min immediately before use.
14. Humidified chambers (*see* **Note 3**).
15. DAPI (4,6, Diamidino-2-phenylindole, Sigma) in distilled water, 0.1 mg/mL, 100X stock.
16. 0.5 M Tris base in distilled water, pH 8.0.
17. DTG: dissolve 1 g of DABCO (Sigma) in 45 mL of glycerol at 45°C. Add 5 mL of 0.5 M Tris, pH 8.0 (*see* **item 16**) and aliquot for indefinite storage at –20°C. (*see* **Note 4**).
18. Soldering iron, 20–40 W, with a sharpened electrode to cut NUNC culture dishes.
19. Glass microscope slides, 22 mm square coverslips (No. 1, Fisher).
20. Superglue to attach NUNC dishes to glass slides.

3. Methods
3.1. Restriction Digest of Plasmids

1. Combine 1–2 µg of a plasmid containing the cDNA plus distilled water to 7 µL. Add in order: 1 µL of 10X restriction enzyme buffer, 1 µL of 1 µg/mL BSA, and 10–20 units of restriction enzyme (*see* **Note 5**).

2. Combine 0.5 µg of pUC19 with distilled water to 7 µL, then add the buffer, BSA, and enzyme.
3. Incubate samples for 1 h at the recommended temperature for the enzyme. Add 2 µL of 6X gel loading dye to each sample.

3.2. Agarose Gel Electrophoresis

1. Add 0.5 g of agarose to 50 mL of EtBr-TAE buffer in a 200 mL conical flask. Microwave 1 min to dissolve, cool for several min, and pour into 2 × 25 mL gel molds. Use a 5-well comb (preparative gel) in one gel and a 10-well comb (analytical) in the other (*see* **Note 6**).
2. Place gels at 4°C for 20 min to solidify.
3. Place the gels into the electrophoresis tank with sufficient EtBr-TAE buffer to submerse them by 0.5 cm.
4. Load digested DNAs into the preparative gel.
5. Combine 0.5 µg of DNA ladder, 1 µL of 6X loading dye and distilled water to 6 µL. Load onto gel.
6. Electrophorese at 80–100 V until the dye reaches the bottom of the gel or until the bands of interest are well separated from other bands on the gel.
7. Photograph the gel on a UV transilluminator. Caution: the UV illuminator can cause severe burns to exposed skin within 30 s. Wear a face mask and a laboratory coat.

3.3. Subcloning the cDNA

1. Excise the linearized pUC19 DNA and cDNA fragments from the gel using a razor blade and place slices into microfuge tubes.
2. Use the Qiagen gel extraction kit as directed to purify the DNAs.
3. Electrophorese 5 µL of each sample and 0.5 µg of DNA ladder on the analytical gel for 15 min to determine the DNA concentrations (*see* **Note 7**).
4. Combine the plasmid and cDNA, in a 1:3 molar ratio, into a microfuge tube and add distilled water to 8 µL. Add the same volume of either plasmid or cDNA into separate tubes and bring to 8 µL with distilled water. (*see* **Note 8**).
5. To each tube for ligation, add in order: 1 µL ligase buffer and 1 µL of ligase. Incubate for 1–2 h at RT or 16°C overnight (*see* **Note 9**).
6. Transform competent *E. coli* as recommended by the manufacturers.
7. After heat shocking the bacteria, add 10 competent cell-volumes of LB and incubate at 37°C for 1 h with gentle shaking (*see* **Note 10**).

3.4. Growth of Transformed Bacteria on LB-Agar Ampicillin Plates

1. Add 7.5 g bacto-agar per 500 mL of LB (LB-agar). Autoclave for 20 min, cool to 55°C in a water bath. Add 0.5 mL of ampicillin stock and pour 10–15 mL of LB-agar into sterile 100 mm polystyrene petri dishes. Leave covered plates to cool overnight at RT (*see* **Note 11**).
2. Spread 40 µL of X-gal onto 6 plates and allow to dry. Spread 20 µL of IPTG and allow to dry (*see* **Note 12**).
3. Spread 5, 15, and 80% of the *E. coli* transformed with plasmid + cDNA onto separate ampicillin plates. For plasmid-only and cDNA-only transformations, use amounts of bacteria equivalent to 15% of the plasmid + cDNA transformation. For the positive control, spread 10% of the pUC19-transformed E.coli (*see* **Note 13**).
4. Grow the bacteria overnight at 37°C or until colonies grow to 0.5–1 mm diameter.
5. Place plates at 4°C for 2 h (*see* **Note 14**).

3.5. Purifying Plasmids from Mini- and Maxi-Prep Cultures

1. Pick 8 white colonies from one of the plasmid + cDNA plates with a sterile tooth pick and place into 3 mL of LB-ampicillin (*see* **Note 15**).
2. Grow mini-prep cultures at 37°C overnight in a shaking incubator.
3. Purify the plasmid DNA from the cultures using a Qiagen mini-prep kit following the manufacturers recommendations (*see* **Note 16**).
4. Digest 0.5–1 μg of each plasmid with the restriction enzyme from **Subheading 3.1.** and electrophorese on an analytical gel to ensure that colonies have the cDNA insert.
5. Choose one correct colony to inoculate 500 mL of LB-ampicillin in a 2 L flask, and grow at 37°C overnight in a shaking incubator.
6. Purify the plasmid DNA using a Qiagen Megakit and following the manufacturer's recommendations.
7. Sequence the 5' untranslated region and coding region of the cDNA in both directions. Retain this stock for all mutagenesis experiments (*see* **Note 17**).

3.6. Site-Directed Mutagenesis to Generate the P14L Mutation in PLP1

1. The design of two overlapping and complementary oligodeoxynucleotide primers that harbor mutations at the center of each primer is shown in **Fig. 1**.
2. Use the mutation primers with the QuikChange mutagenesis kit as recommended by the manufacturers to generate a mutated plasmid using the Megakit DNA as template.
3. Choose 8 colonies and grow overnight at 37°C in separate tubes containing 3 mL of LB-ampicillin.
4. Use Qiagen mini-prep kit to purify plasmid DNA from overnight cultures.
5. Digest 0.5–1 μg of each mini-prep DNA with 10 U of *ApaI* and electrophorese on agarose gel (*see* **Note 18**).
6. Sequence the 5' untranslated and coding regions of two plasmids in both directions to ensure that the PCR did not introduce random point mutations (*see* **Note 17**).

3.7. Subclone the cDNA into an Expression Cassette

1. Subclone the mutagenized, sequenced cDNA into the polycloning site of the expression vector of choice (*see* **Subheadings 3.1.–3.5.**, and **Note 19**).

3.8. Maintaining COS-7 Cells in Culture (see Note 20)

1. Grow a flask of COS-7 cells to 80–100% confluence. These cells are easily maintained in T75 flasks at 37°C in 95% air, 5% CO_2. Replace culture medium every 3 d.
2. To passage the cells, remove the medium by suction using a Pasteur pipet and rinse the cells for a few s with PBS. Rinse the walls of the flask to remove traces of medium.
3. Add 5 mL of trypsin/EDTA to the cells and remove all but 0.5 mL by suction. Incubate for 3–5 min at 37°C (*see* **Note 21**).
4. Suspend the cells in 10 mL of culture medium (*see* **Note 22**).
5. Transfer 1 mL of cells to a fresh T75 flask containing 14 mL of fresh medium.
6. Grow the cells to confluence 37°C (*see* **Note 23**).

3.9. Transfection of COS-7 Cells

1. D 1 (afternoon). Begin with a T75 culture flask of COS-7 cells grown to confluence. Split the cells equally into 60 mm culture dishes. (*see* **Note 24**).
2. D 2 (morning). For each transfection, add 90 μL of DMEM (i.e., no serum) to a sterile microfuge tube. Add 6 μL of FUGENE 6 incubate for 5 min at RT (*see* **Note 25**).

3. Add the DMEM/FUGENE 6 solution, dropwise, to a sterile microfuge tube containing 4 μg of plasmid DNA. Tap the bottom of the tube gently and incubate for 15 min at RT.
4. Pipet the DNA solution, dropwise, across the COS-7 cells in the 60 mm dishes and culture overnight at 37°C. (*see* **Note 26**).
5. Day 3 (morning). Add 1.5 mL of culture medium to 35 mm NUNC culture dishes (*see* **Note 27**).
6. Wash the cells briefly with PBS, trypsinize for 5 min, and suspend in up to 4 mL of culture medium (*see* **Note 28**).
7. Add 0.5 mL of cell suspension to each 35 mm dish and culture the cells overnight at 37°C (*see* **Note 29**).

3.10. Fixation and Immunocytochemistry

1. Rinse cells twice for several s, by swirling, with 1 mL of DMEM. Add 1 mL of fixative to each dish and incubate for 30 min at RT (*see* **Note 30**).
2. Wash out excess fixative with 2 changes of TBS over 5 min.
3. Permeabilize cells for 30 min in 1 mL of 0.1% saponin in TBS. (*see* **Note 31**).
4. Suction away the buffer and clear the cells in a 2–3 mm track around the perimeter of each dish. (*see* **Note 32**).
5. Apply a PAP pen to the cleaned surface and quickly pipet 50 μL of a 0.1% saponin/TBSGBA block solution onto the cells and incubate for 15–30 min at RT (*see* **Note 33**).
6. Exchange the 0.1% saponin/block solution with 50 μL of primary antibody diluted in block solution and incubate overnight in a humidified chamber at RT.
7. Wash the cells 2 × 5 min with 1 mL of 0.1% saponin/TBS. Add 50 μL of secondary antibody diluted in 0.1% saponin/block solution for 1 h at RT. (*see* **Note 34**).
8. Wash the cells 2 × 5 min with 1 mL of 0.1% saponin/TBS. Remove buffer with suction and add 100 μL of DTG with a wide mouth tip.
9. Coverslip and stand for 30 min. Remove excess DTG and seal the coverslips with clear nail polish (*see* **Note 35**).
10. Cut away the wall of each NUNC dish using the soldering iron in a fume hood (*see* **Note 36**).
11. Mount each dish bottom onto a glass slide with drops of superglue at two points on the outer edge (*see* **Note 37**).
12. View cells using epifluorescence or confocal microscopes (*see* **Note 38**).

4. Notes

1. Ca^{2+} and Mg^{2+} reduce the effectiveness of cell detachment from culture dishes and cause detached cells to clump.
2. NUNC dishes have thin walls and low intrinsic fluorescence, which are important advantages for the immunocytochemistry protocol described herein. If necessary, coating these dishes with poly-L-lysine or poly-L-arginine (Sigma) will improve the adhesion of transfected cells to the dish but may increase the background staining somewhat. To coat dishes, dissolve either amino acid at 5 μg/mL in water and filter sterilize (0.45 μm filter). This solution is stable at 4°C for 1 yr. Dilute this stock 100-fold in PBS (50 μg/mL, stable at 4°C for 1 yr) and add 1 mL to each culture dish. After 3 h, rinse each dish with 3 × 2 mL of PBS. Coated dishes can be stored sterile for 1 mo at 4°C.
3. Place several wet paper towels into a 150 mm culture dish with a lid. Each chamber accommodates up to 6 × 35 mm dishes and will remain moist overnight at RT.
4. To maximize fluorescence signals from labeled cells, the pH of this solution is very important. DABCO is alkaline in solution, so the DTG should be approx pH 8.6 after

the addition of Tris. The pH can be easily measured after diluting the DTG 10-fold in distilled water.
5. Enzyme volume should be no more than 10% of the total.
6. Use protective gloves when handling the hot agarose solution, which can superheat and boil over, causing serious burns.
7. The known concentration of the ladder is used to estimate the concentrations of the samples by comparing band intensities under UV illumination.
8. Use 20–50 ng of plasmid DNA for the ligation. Increasing the molar ratio to more than 1:5 may cause 2 or more copies of the cDNA to be ligated into the pUC19 vector.
9. The 16°C incubation slows the enzyme activity and, theoretically, increases the ligation efficiency. However, for routine cloning the shorter ligation times yield sufficient numbers of bacterial colonies.
10. Do not use LB-ampicillin in this step because the bacteria need time to transcribe and translate the β-lactamase gene on the introduced plasmid. The 1 h incubation period can be eliminated to reduce processing time, which will reduce the number of colonies on the plates by two- or three-fold.
11. Ampicillin is heat labile and it is important to ensure the correct temperature of the LB-agar, which takes 1–2 h to cool to 55°C with occasional swirling of the bottle. After adding the ampicillin, mix thoroughly by swirling. When pouring the agar plates, large and small bubbles will collect on the liquid surface. Burst the bubbles by flaming the plates with a blue-yellow Bunsen flame for a few s. 500 mL of LB-agar makes up to 40 plates and can be stored for up to 1 mo at 4°C. Discard plates if fungal contamination is present.
12. To make a spreader, use a Bunsen burner to make a right-angle bend in the tapered end of a 9-in. Pasteur pipet. Sterilize with ethanol and briefly flame.
13. Spreading more than 100 µL of bacteria onto a plate should be avoided. Reduce the volume by microfuging the bacterial suspension for 20 s, withdrawing excess supernate to 100 µL and resuspending the soft pellet with a pipet.
14. The 4°C incubation increases the intensity of blue colonies. If the concentration of ampicillin in the plates is low, or if the transformed bacteria grow vigorously, true transformants will be ringed by small satellite colonies. Avoid picking the satellite colonies, which do not harbor plasmid DNA and will not grow in LB-ampicillin medium. In general, the longer the time of growth on the plates, the greater the number of satellite colonies. Thus, avoid growing transformants for longer times than is necessary to obtain 0.5–1 mm diameter colonies.
15. Colonies should be large, round, smooth and well-isolated.
16. The DNA concentration will be approx 0.1–0.2 µg/µL.
17. Although small, there is a finite risk of introducing point mutations into the cDNA during cloning and culturing the bacteria. Such mutations could confound all future work with the plasmid. Thus, the cDNA should be sequenced to ensure that random mutations are absent.
18. Include the PCR template DNA as a positive control. Correctly mutagenized plasmids will harbor one less *Apa* I site than the parent plasmid.
19. To achieve very high expression levels, pCMV5 is an ideal vector because it harbors an SV-40 origin of replication which the SV-40 T antigen expressed by COS-7 cells uses to replicate the plasmid to high copy number.
20. Use 70% ethanol to sterilize the surfaces of the culture hood, benches, Gilson pipets and outer surfaces of the culture-medium bottle. Allow the ethanol to evaporate.
21. Strike the side of the flask several times, which dislodges the cells when trypsinizing is complete.

22. Gently pipet cells up-and-down 5–10 times with a 10 mL pipet to break up cell clumps.
23. Growth time is generally 3–4 d if the cells are healthy. Slow growing cells may be due to a viral infection (e.g., mycoplasma), which will make the cells more difficult to transfect and will yield cells with poor morphology under the microscope. Discard the cultures and obtain a fresh flask of cells.
24. Cell number will approx double by d 2. 1 × T 75 flask at 100% confluence on d 1 yields approx 8 × 60 mm dishes at 50% confluence on d 2. Transfections also can be performed directly in individual 35-mm dishes rather than using a 60 mm dish intermediate. A disadvantage of this approach is that replicate dishes may have greater variability due to differences in transfection efficiencies. However, an advantage is that transfection times can be reduced to 15 h before fixation and staining which will yield higher apparent transfection efficiencies if the mutant protein being expressed is particularly toxic to the cells. We have used this approach to examine trafficking of a *PLP1* splice site mutation *(19)* that results in skipping exon 6.
25. The FUGENE 6 must be added directly into the DMEM, not down the wall of the tube.
26. Gently move the dishes several times from side-to-side to evenly distribute the DNA on the cells. Be careful not to move dishes in a circular motion, which will concentrate reagents in the centers of the dishes.
27. The number of dishes prepared depends on the number of DNAs for transfection and the number of replicate dishes for each DNA. Each 60 mm dish will provide sufficient cells for up to 8 × 35 mm dishes.
28. The levels of confluence in each 60 mm dish should approximate 100% if the transfected plasmid is not toxic to the cells. However, proportionately less medium should be used for subconfluent dishes (e.g., use 2 mL of medium for cells at 50% confluence).
29. Gently move the dishes from side-to-side to evenly distribute the cells. The desired level of confluence when the cells adhere is 20–30%. After overnight culturing, a confluence of 40–50% is expected, which provides sufficient transfected cells to examine and plenty of room for the cells to flatten out for good morphology.
30. Keep 35 mm dishes on a rocking platform (e.g., Bellco, NJ, USA) during staining. Warm DMEM and fixative minimize changes to cell morphology before the cells are fixed. Add solutions to the edge of each dish to minimize cell damage. Use a low vacuum to hasten solution changes.
31. Detergents such as Triton X-100 and other permeabilizing agents such as methanol can extract some proteins, such as PLP1, from the membranes of fixed cells, which reduces the staining intensity.
32. Attach a wide-mouth P200 or P1000 Gilson pipette tip to a vacuum line. It is most important that all cells and liquid be removed in preparation for applying the PAP pen. These pens are very expensive and are easily damaged by contact with debris and aqueous solutions.
33. This and subsequent steps use small volumes of antibody-containing solutions to label antigens in transfected cells spread over a wide area. Therefore, it is important to ensure that the culture dishes are level. If antibody solutions are in plentiful supply, larger volumes, e.g., 100 or 200 µL can be used.
34. In this step, the DNA-binding fluor, 1 µg/mL 4,6, Diamidino-2-phenylindole (DAPI, Sigma, MO, USA), can be included for visualizing the nuclei. This compound strongly labels nuclei and aids the location of cells that are unlabeled or weakly labeled by antibodies. Some antigens are expressed at sufficiently low levels to require signal amplification using a biotinylated secondary antibody and a streptavidin-conjugated fluorophore. Simply repeat this step with the necessary tertiary reagents.
35. Colored nail polish contains compounds that quench the fluorescence and can bleed into the DTG with time.

Myelin PLP1 Mutations

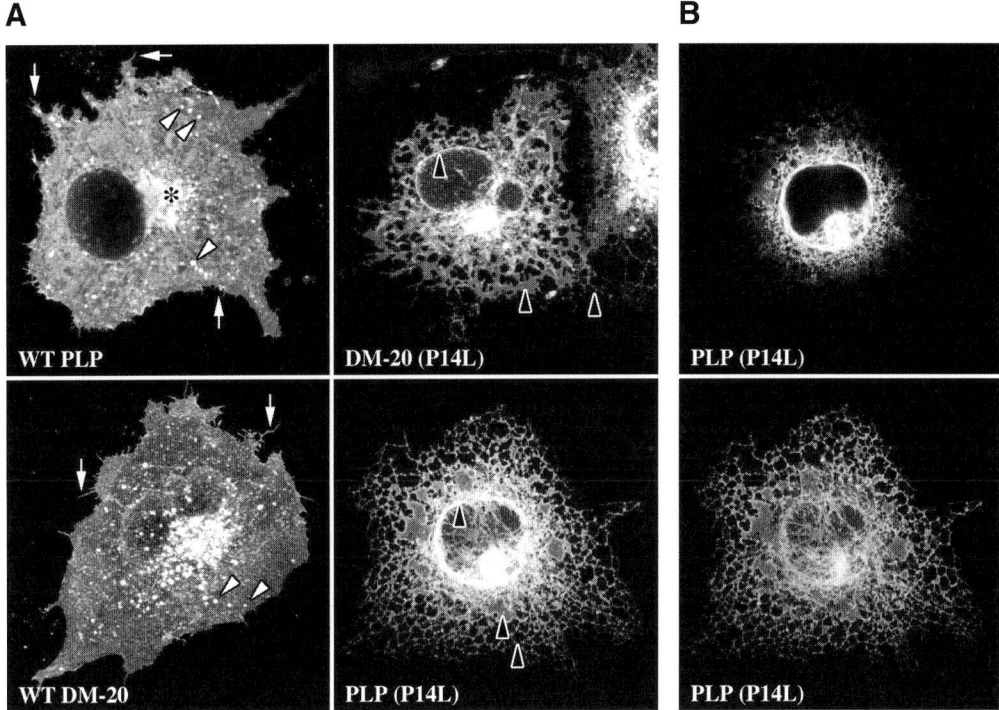

Fig. 2. Intracellular trafficking defect for PLP1 and DM-20 harboring the P14L mutation identified in a patient with severe PMD. **(A)** Extended-focus confocal series showing wild type PLP1 expressed in a transfected COS-7 cell. The protein is present in the Golgi (asterisk), on the cell surface (white arrows) and in lysosomes (white arrowheads). DM-20 expression in COS-7 cells is indistinguishable from PLP1. In contrast, mutant DM-20 (P14L) accumulates in the endoplasmic reticulum (black arrowheads) and does not reach the cell surface or lysosomes. The trafficking of PLP1 (P14L) is similar to that of DM-20. **(B)** Single optical slices through the middle of the nucleus (upper panel) and near the bottom of the nucleus (lower panel) from the extended focus series used in (A). These slices illustrate the intracellular detail gleaned from confocal microscopy.

36. This process is optional, but only takes approx 1 min per dish and increases the viewable field under the microscope by several-fold.
37. Dishes are stored at RT or 4°C for several wk without loss of signal. Strong, fluorescence signals remain visible for 6–12 mo.
38. Regular epifluorescence or confocal microscopes equipped with 20× or 40× lenses will usually provide sufficient resolution to document cell morphology and the general subcellular localization of heterologous proteins in transfected cells. However, 63× or 100× lenses may be necessary for high resolution imaging of organelles such as the endoplasmic reticulum and Golgi stacks. The shape of a COS-7 cell usually resembles that of a fried egg and many organelles of interest are located toward the bottom of the nucleus in the thick portion of the cell. Thus, confocal microscopy yields superior images to epifluorescence microscopy because optical sections obtained at the bottom of the cell are not contaminated with stray fluorescence from overlying structures (*see* **Fig. 2**).

Acknowledgments

I would like to thank Ms. Cherie Southwood for her critical review of the manuscript. This work was supported by a grant from the National Multiple Sclerosis Society, RG2891-B-2.

References

1. Garbern, J., Cambi, F., Shy, M., and Kamholz, J. (1999) The molecular pathogenesis of Pelizaeus-Merzbacher disease. *Arch. Neurol.* **56,** 1210–1214.
2. Hodes, M.E., Woodward, K., Spinner, N. B., Emanuel, B. S., Enrico-Simon, A., Kamholz, J., et al. (2000) Additional copies of the proteolipid protein gene causing Pelizaeus-Merzbacher disease arise by separate integration into the X chromosome. *Am. J. Hum. Genet.* **67,** 14–22.
3. Nave, K.-A. and Boespflug-Tanguy. O. (1996) X-linked developmental defects of myelination: from mouse mutants to human genetic diseases. *The Neuroscientist* **2,** 33–43.
4. Gow, A., Friedrich, V. L., and Lazzarini, R. A. (1994) Intracellular transport and sorting of the oligodendrocyte transmembrane proteolipid protein. *J. Neurosci. Res.* **37,** 563–573.
5. Gow, A., Friedrich, V. L., and Lazzarini, R. A. (1994) Many naturally occurring mutations of myelin proteolipid protein impair its intracellular transport. *J. Neurosci. Res.* **37,** 574–583.
6. Gow, A. and Lazzarini, R. A. (1996) A cellular mechanism governing the severity of Pelizaeus-Merzbacher disease. *Nat. Genet.* **13,** 422–428.
7. Tosic, M., Gow, A., Dolivo, M., Domanska-Janik, K., Lazzarini, R. A., and Matthieu, J. -M. (1996) Proteolipid/DM20 proteins bearing the paralytic tremor mutation in peripheral nerves and transfected COS-7 cells. *Neurochem. Res.* **21,** 423–430.
8. Tosic, M., Matthey, B., Gow, A., Lazzarini, R. A., and Matthieu, J. -M. (1997) Intracellular transport of the DM-20 bearing shaking pup (*shp*) mutation and its possible phenotypic consequences. *J. Neurosci. Res.* **50,** 844–852.
9. Southwood, C.M. and Gow, A. (2001) Molecular mechanisms of disease stemming from mutations in the proteolipid protein gene. *Micros. Res. Tech.* **52,** 700–708.
10. Gow, A., Southwood, C. M., and Lazzarini, R. A. (1998) Disrupted proteolipid protein trafficking results in oligodendrocyte apoptosis in an animal model of Pelizaeus-Merzbacher disease. *J. Cell Biol.* **140,** 925–934.
11. Gow, A., Gragerov, A., Gard, A., Colman, D. R., and Lazzarini, R. A. (1997) Conservation of topology, but not conformation, of the proteolipid proteins of the myelin sheath. *J. Neurosci.* **17,** 181–189.
12. Yoshida, M. and Colman, D. R. (1996) Parallel evolution and coexpression of the proteolipid proteins and protein zero in vertebrate myelin. *Neuron.* **16,** 1115–1126.
13. Stecca, B., Southwood, C. M., Gragerov, A., Kelley, K. A., Friedrich, V. L. J., and Gow, A. (2000) The evolution of lipophilin genes from invertebrates to tetrapods: DM-20 cannot replace PLP in CNS myelin. *J. Neurosci.* **20,** 4002–4010.
14. Lorence, M.C., Murry, B. A., Trant, J. M., and Mason, J. I. (1990) Human 3b-hydroxysteroid dehydrogenase/$\Delta^{5\text{-}4}$ isomerase from placenta: expression in nonsteroidogenic cells of a protein that catalyzes the dehydrogenation/isomerization of C21 and C19 steroids. *Endocrinology* **126,** 2493–2498.
15. Milner, R.J., Lai, C., Nave, K. A., Lenoir, D., Ogata, J., and Sutcliffe, G. (1985) Nucleotide sequences of two mRNAs for rat brain myelin proteolipid protein. *Cell* **42,** 931–939.
16. Nussbaum, J.L. and Roussel, G. (1983) Immunocytochemical demonstration of the transport of myelin proteins through the Golgi apparatus. *Cell Tissue Res..* **234,** 547–559.

17. Roussel, G., Meskovic, N. M., Trifilieff, E., Artault, J. -C., and Nussbaum, J. -L. (1987) Arrest of proteolipid transport through the Golgi apparatus in Jimpy brain. *J. Neurocytol.* **6,** 195–204.
18. Schwob, V.S., Clark, H. B., Agrawal,D., and Agrawal, H. C. (1985) Electron microscopic immunocytochemical localization of myelin proteolipid protein and myelin basic protein to oligodendrocytes in rat brain during myelination. *J. Neurochem.* **45,** 559–571.
19. Hobson, G.M., Davis, A. P., Stowell, N. C., Kolodny, E. H., Sistermans, E. A., de Coo, I. F., et al. (2000) Mutations in noncoding regions of the proteolipid protein gene in Pelizaeus-Merzbacher disease. *Neurology.* **55,** 1089–1096.

25

In Vitro Expression Systems for the Huntington Protein

Shi-Hua Li and Xiao-Jiang Li

1. Introduction

Huntington's disease (HD) is an autosomal dominant neurodegenerative disorder resulting from expansion (>37 units) of a polyglutamine tract in huntingtin, a 350 kDa protein of unknown function *(1)*. The N-terminal region of huntingtin contains the glutamine repeat, which is encoded by exon1 of the HD gene. Normal huntingtin is a cytoplasmic protein and is expressed ubiquitously, but N-terminal fragments of huntingtin with expanded polyglutamine tracts are able to accumulate in the nucleus and form aggregates. A large body of evidence has shown that N-terminal huntingtin with expanded glutamine repeats is toxic and can kill cells *(2–8)*. For example, transgenic mice (R6/2) expressing the HD exon1 protein with more than 115 glutamines develop neurological symptoms and neuronal intranuclear inclusions consisting of huntingtin aggregates *(2)*. Similar nuclear aggregates are found in patients with HD *(3–5)* and other polyglutamine diseases *(9)*.

Cell models that express mutant huntingtin are highly valuable for studying the mechanisms of huntingtin toxicity. The HD cellular models should recapitulate some pathological features of mutant huntingtin. These pathological features include huntingtin aggregate formation, intranuclear huntingtin accumulation, and decreased cell viability or increased vulnerability to apoptotic stimuli. Most of the reported cell models have used transient transfection in which the overexpression levels of transfected protein could lead to a rapid pathological process such as huntingtin aggregation and cell death. A stably transfected cell line that consistently expresses mutant huntingtin in the nucleus will provide a suitable approach to study whether intranuclear huntingtin affects cellular function at the transcriptional level.

We have reported studies using both transiently and stably transfected cells that express N-terminal huntingtin with an expanded polyglutamine tract. Using these cells, we have found that the formation of huntingtin aggregates is increased by expansion of the glutamine repeat and decreased with lengthening of the transfected protein. Similarly, more glutamines in the repeat and shorter huntingtin fragments lead to greater cellular toxicity. We further demonstrated that small huntingtin fragments with an expanded glutamine repeat are able to accumulate in the nuclei of cultured cells, affect gene expression, and trigger apoptotic process. We outline here our basic techniques for expressing mutant huntingtin in cultured cells and note variations performed by us.

2. Materials
2.1. DNA Constructs

1. Huntingtin exon1 DNA containing 20 or 150 CAG repeats was isolated from a lambda phage DNA library that contains the human HD gene and was provided by Dr. Gillian Bates at Guy's Hospital in London, UK. This huntingtin DNA was inserted into the pCIS expression vector that carries a CMV promoter *(10)* for expressing the HD exon1 protein. Nuclear localization sequences (PKKKRKV) can be tagged to the N-terminus of the HD exon1 protein to facilitate the localization of transiently transfected huntingtin in the nucleus.
2. pCDNA-3 vector (Invitrogen, Carlsbad, CA, USA).

2.2. Cell Lines

1. Human embryonic kidney (HEK) 293 cells (American Type Culture Collection, ATCC).
2. Rat pheochromocytoma PC12 cells (provided by Dr. James Lah at Emory University; **ref. *11***).

2.3. Cell Culture

1. Growth medium for PC12 cell culture: DMEM/F12 supplemented with 5% fetal bovine serum (FBS) and 10% horse serum, containing 100 µg/mL penicillin and 100 µg/mL streptomycin (all from Life Technologies, Inc., Gaithersburg, MD, USA).
2. Growth medium for HEK293 cell culture: DMEM/F12 supplemented with 15% FBS, 100 µg/mL penicillin, and 100 µg/mL streptomycin.
3. Trypsin-EDTA medium (1X) (Life Technologies) for splitting cells.
4. Transfection medium: serum-free DMEM containing 2–4 (µg/mL of plasmid DNA and 10 (µg/mL of lipofectAMINE (Life Technologies).
5. Cell storage medium: 10% DMSO and 20% FBS in DMEM.
6. Cell culture plates: Ten-cm dishes and 6-well plates (Corning Inc, Corning, NY, USA).
7. Two- or four-well chamber slides (Nunc Inc., Rochester, NY, USA).
8. Antibiotics: Geneticin (G418 sulfate), penicillin, and streptomycin (Life Technologies).

2.4. Immunofluorescence Staining

1. Fixation buffer: paraformaldehyde (Electron Microscopy Sciences, Fort Washington, PA, USA) is dissolved in phosphate-buffered saline (PBS, pH 7.4), which is preheated until boiling, at the final concentration of 4%. Keep 4% paraformaldehyde/PBS at 4°C for a couple of wk.
2. Triton buffer: 0.2% Triton X-100 in PBS.
3. Blocking buffer for immunostaining: 3% bovine serum albumin (BSA) in PBS and keep the buffer at –20°C until use.
4. Antibodies: EM48, a rabbit antibody that is against the first 256 amino acids of human huntingtin and was generated in our laboratory from previous studies *(5)*. Secondary antibodies: anti-rabbit IgG conjugated with either FITC or rhodamine (Jackson Immunoresearch Lab, West Grove, PA, USA) and used at 1:200 to 1:400 dilution for immunofluorescence staining. Donkey anti-rabbit antibody conjugated with peroxidase (Jackson Immunoresearch Lab) is used for Western blots.
5. Nuclear dye labeling: Hoechst dye (Molecular Probe, Eugene, OR, USA) used at 1 µg/mL.

2.5. Western Blots

1. SDS sample buffer: 5X sample buffer: 0.312 *M* Tris-HCl, pH, 6.8, 10% sodium dodecyl sulfate (SDS), 25% β-mercaptoethanol, 0.05% bromophenol blue.

2. SDS PAGE (polyacrylamide gel electrophoresis): 8–12% acrylamide gel in Tris/glycine and SDS buffer.
3. Transfer buffer: 12 mM Tris base and 96 mM glycine.
4. Nitrocellulose membrane (Amersham, Arlington Heights, IL, USA).
5. ECL kit: a chemiluminescence kit (Amersham).

3. Methods

3.1. Plasmid DNA Preparation

Huntingtin cDNA in pCIS vector is transformed into bacteria SURE cells (Strategene, La Jolla, CA, USA). QIANGEN (Valencia, CA, USA) DNA preparation kits are used to purify plasmid cDNA according to the instructions of the manufacture. Alternatively, CsCL gradient ultracentrifugation is used to prepare plasmid DNA for transfection (*see* **Note 1**). DNA is resuspended in sterilized water at about 1 µg/µL and kept at –20°C.

3.2. Transient Transfection

1. HEK293 or PC12 cells are incubated at 37°C in a humidified 5% CO_2 atmosphere and grown at densities ranging from 2–4×10^4 cells/cm^2. The culture media are replaced every 48–72 h and cells are split as needed.
2. To split cells, cultured cells are incubated with Trypsin-EDTA medium for 5–10 min until cells are floating.
3. Collect cells and medium into a 15-mL tube and spin down them in a centrifugator at 1100g for 5 min.
4. The collected cells are resuspended in fresh growth medium and placed into new dishes or plates. Cultured cells are split a day before transfection.
5. Cells for transfection are grown to 70–80% confluence in the 10-cm dishes or 6-well plates (*see* **Note 2**).
6. Plasmid DNA (2–4 µg/mL) are mixed with lipofecAMINE (10 µg/mL) in serum-free medium for 30 min to allow DNA-liposome complexes to form.
7. For transient transfection of HEK 293 cells in 6-well plates, the growth medium is removed and cells are incubated with 0.5 mL transfection medium per well for 4–6 h. For transfection of HEK293 cells in a 10-cm dish, we normally use 7 mL transfection medium (1 µg/mL of plasmid DNA and 10 µg/mL of lipofecAMINE).
8. The transfection medium is then removed and the cells are incubated with fresh growth medium for another 24–48 h before analysis.

3.3. Stable Transfection

1. For establishing stably transfected PC12 cells, subconfluent PC12 cells in 10-cm dishes are transfected with 7 µg huntingtin plasmid DNA and 10 µg/mL of lipofectAMINE in 7 mL serum-free DMEM/F12 medium. One µg pCDNA-3 containing the neomycin gene, which confers G418 resistance, is also included with huntingtin DNAs for transfection.
2. After 8–10 h incubation, the transfection medium is removed. Cells are washed once with serum-free DMEM and incubated with G418 selection growth medium (DMEM/F12 containing 5% fetal bovine serum [FBS], 10% horse serum, 100 µg/mL penicillin, 100 µg/mL streptomycin, and 500 µg/mL G418).
3. Every 2 d, the culture medium is replaced with fresh growth medium containing 500 µg/mL G418 to remove dead cells that are not resistant to G418 (*see* **Note 3**).

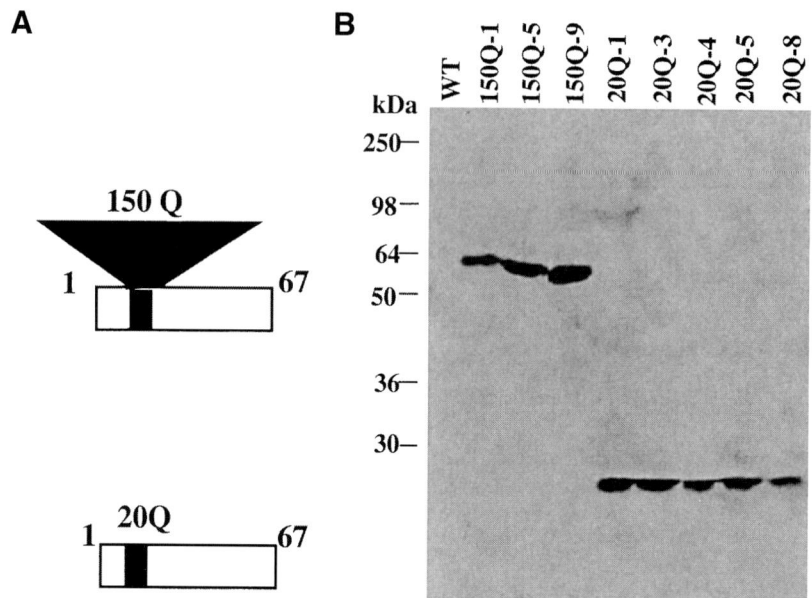

Fig. 1. Western blot analysis of stably transfected PC12 cells. (**A**) Schematic structure of truncated huntingtin expressed in PC12 cells. The transfected huntingtin is the huntingtin exon-1 protein with 150 (150Q) or 20 (20Q) glutamines and other 67 amino acid residues. (**B**) Western blots showing that three cell lines express 150Q and five cell lines express 20Q. Western blot was probed with antibody EM48. Note that the expanded polyglutamine of the 150Q protein greatly hinders its migration in the SDS gel (*see* also **ref. *13***).

3.4. Select Stably Transfected Cells

1. At 5–7 d after transfection, G418 resistant cells are split to separate cells and to allow individual colonies to form. The cells are maintained in G418-containing medium until they form large colonies.
2. Thirty to forty colonies are isolated with a pipet under an inverted microscope (*see* **Note 4**).
3. The cells of these colonies are placed in 24-well plates and separated by pipetting, each well containing one group of cells. These cells are maintained in the same G418-containing medium until each well contains enough cells to split and to analyze.
4. Normally, cells in one well of 6-well plates would be enough for Western blot analysis and cells in 24 or 12-well plates can be used for immunofluorescence examination.
5. We save stably transfected cells by freezing them in liquid nitrogen. To do so, cells are treated with Trypsin-EDTA medium and then collected by centrifugation. The cells are then resuspended in the freezing medium at $3–5 \times 10^6$ cells /mL. The cells are first placed in a $-80°C$ freezer for overnight or a couple of wk. For a long-term storage, the cells are transferred from the $-80°C$ freezer to a liquid nitrogen tank (*see* **Note 5**).

3.5. Western Blot Analysis of Transfected Proteins

Cultured cells are collected and solubilized in SDS sample buffer. Protein samples are then resolved by 10 or 12% SDS-PAGE. Blots are incubated with EM48 (1:1000). Immunoreactive bands are visualized using a chemiluminescence (ECL) kit (**Fig. 1** and *see* **Note 6**)

3.6. Immunofluorescence Examination of the Subcellular Localization of Transfected Proteins

Transfected cells grown in chamber slides are fixed in 4% paraformaldehyde in PBS for 15 min, permeabilized with 0.2% Triton X-100 in PBS for 30 min, blocked with 3% BSA in PBS for 1 h, and incubated with primary antibodies in 3% BSA/PBS overnight at 4°C. After several washes, the cells are incubated with secondary antibodies conjugated with either FITC or rhodamine at 4°C for 30 min. Hoechst dye (1 µg/mL) is used to label the nuclei. A Zeiss fluorescent microscope (Axioskop 2) and video system (Dage-MTI Inc., Michigan City, IN, USA) are used to capture images. The captured images are stored and processed using Adobe Photoshop software (*see* **Note 7**). Cytotoxicity associated with mutant huntingtin can be examined using TUNEL staining or other assays (*see* **Note 8**).

4. Notes

1. Mutant huntingtin cDNA contains CAG repeats that are unstable during propagation in bacteria. Thus, it is important to analyze a number of colonies to select one that contains a desired size of the CAG repeat. We normally prepare miniprep plasmid DNA and identify the size of the CAG repeat using DNA gel analysis. Bacteria containing the plasmid DNA with a desired size of the repeat are immediately grown to prepare a large amount of plasmid DNAs. DNA sequence analysis is then performed to confirm the size of the CAG repeat. We find that CsCl gradient ultracentrifugation method provides a higher quality of plasmid DNA for transfection than do other methods.
2. Cultured cells for transfection need to be healthy to achieve a high transfection rate. This is especially important for PC12 cells, as these cells have lower transfection efficiency than HEK293 cells and other types of cells. We split PC12 cells 24 h before transfection. However, if the cells do not look healthy or are not firmly attached to the plates or dishes, these cells need to grow for another day to be sure that they are healthy when transfected.
3. PC12 cells stably transfected with mutant huntingtin have a decrease in their viability. They are prone to metabolic insults and apoptotic stimuli. We normally change medium every 2 d and split them as needed to keep cells healthy.
4. Selection of stably transfected cells is time-consuming. Even an isolated single colony may still contain different populations of cells. Immunofluorescence staining of transfected cells will help determine the percentage of cells that express mutant huntingtin. Thus, several rounds of selection may be needed to ensure that a cell line containing homogenous transfected cells is isolated. Since each individual transfected cell may express transfected huntingtin at a different level, several cell lines need to be selected for comparison and studies. We find that Western blot analysis is more reliable to examine the expression level of transfected huntingtin, as this assay can detect the relative expression level of transfected proteins as compared with the total amount of proteins loaded in the gel. Also, Western blots can estimate the size of glutamine repeat in mutant huntingtin (**Fig. 1**).
5. Stably transfected cells can be frozen for a long-term storage. However, the recovery of these cells from thawing is slower than that of wild-type cells. The expression of transfected huntingtin in stably transfected cells needs to be periodically examined if the cell line has been amplified many times. We found that stably transfected PC12 cells could maintain a similar expression level of mutant huntingtin up to 50 passages.
6. Western blot analysis of huntingtin expression could determine whether the transfected huntingtin contains an expanded glutamine repeat or forms aggregates. The enlarged size

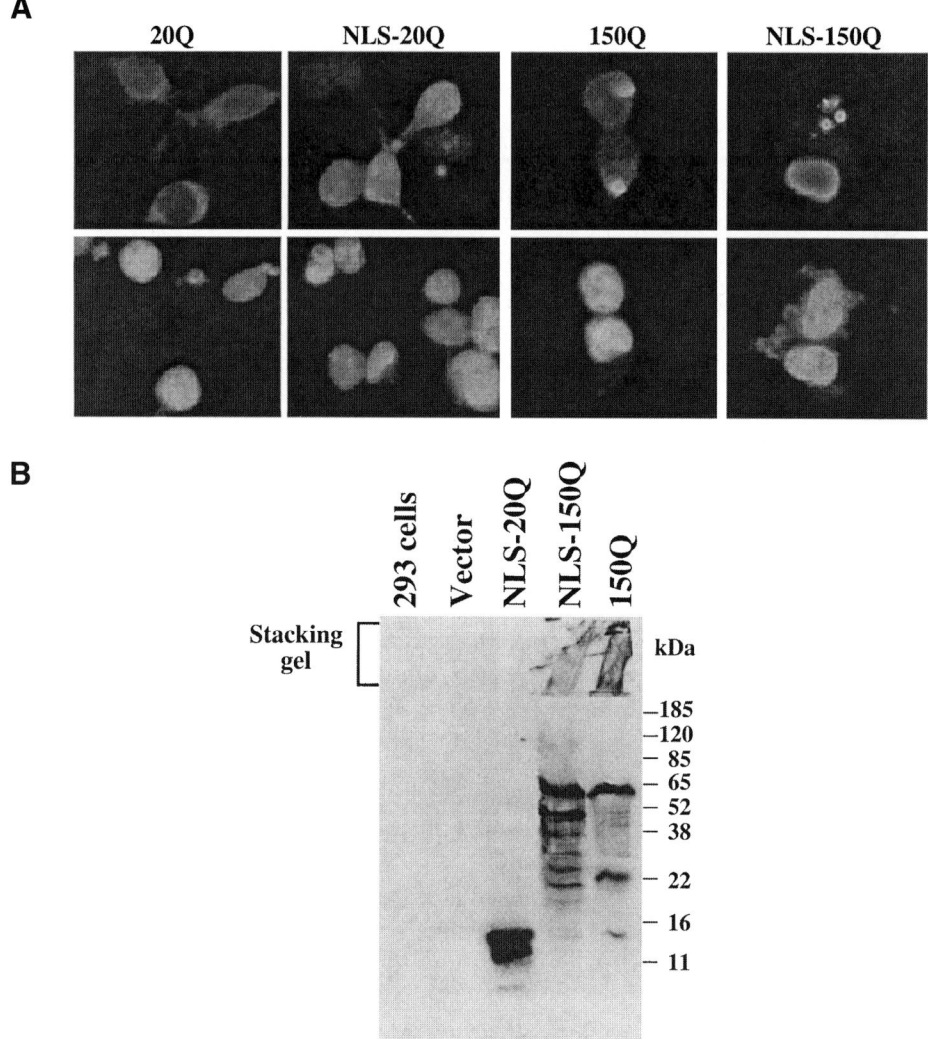

Fig. 2. Expression of mutant huntingtin in transiently transfected 293 cells. **(A)** EM48 immunofluorescent staining of transfected 293 cells (upper pannel). The HD exon1 proteins (20Q and 150Q), when overexpressed in 293 cells, are localized in the perinuclear region of the cytoplasm. Huntingtin proteins tagged with nuclear localization sequences (NLS-20Q and NLS-150Q), however, are localized in the nucleus. Hoechst staining was used to show the nucleus (lower panel). **(B)** EM48 western blot showing the expression of the NLS-20Q, NLS-150Q, and 150Q proteins. Aggregated huntingtin remains in the stacking gel (*see* also **ref. 12**).

of the glutamine repeat retards the migration of transfected huntingtin in SDS gel, leading to a higher molecular weight of band than that with a normal glutamine repeat. Thus, it is always helpful to include transfected huntingtin with a normal glutamine repeat as a control. Huntingtin aggregates are insoluble in SDS sample buffer and retain in the stacking gel. The presence of these huntingtin aggregates can be viewed as smear bands on the top of the gel. Due to its high expression level, transiently transfected mutant huntingtin often readily forms aggregates, which can be seen by Western blots (**Fig. 2**).

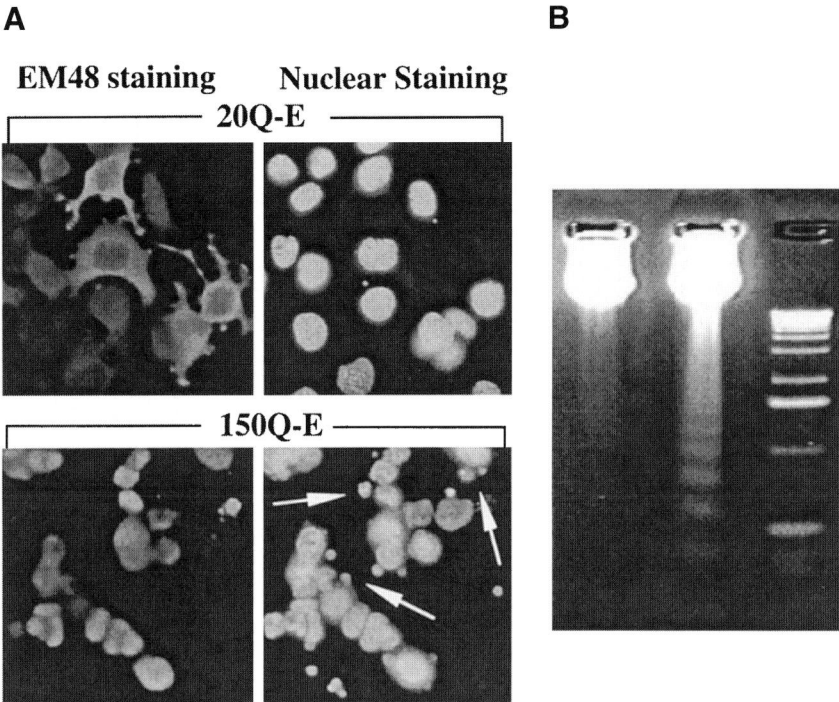

Fig. 3. Intranuclear mutant huntingtin causes apoptotic events in stably transfected PC12 cells. (**A**) PC12 cells stably expressing huntingtin exon1 protein with 150 (150Q-E) or 20 (20Q-E) glutamines were stained with anti-huntingtin antibody EM48 and Hoechst dye. Note that the 150Q-E protein is localized in the nucleus, whereas the 20Q-E is in the cytoplasm. Arrows indicate fragmented nuclear DNA. (**B**) Electrophoresis of genomic DNAs also shows DNA fragmentation in 150Q-E PC12 cells (*see* also **ref.** *12*).

7. Immunofluorescence staining provides a rapid and convenient assay to examine the subcellular distribution of transfected huntingtin in cultured cells. An inverted microscope with fluorescence will let one to readily examine the expression of huntingtin in cells cultured in dishes or plates. Aggregated huntingtin is normally present in the perinuclear area. To determine whether huntingtin is localized in the nuclei, it is necessary to stain cells with a nuclear dye. We normally use Hoechst to label the nuclei and capture photo images using different optical filters.
8. TUNEL assay kit (Promega, Madison, WI, USA) or Hoechst staining can be used for examining DNA fragmentation (**Fig. 3**). Alternatively, cellular toxicity caused by mutant huntingtin can be examined using colormetric and fluorometric assays for apoptosis. Cell viability can be determined by a modified 3-(4,5-dimethyl thiazol-2-yl)-2,5-diphenyl tetrazolium bromide (MTS) assay (Cell Titer 96, Promega). For these assays, a microplate reader is required.

Acknowledgments

This work was supported by National Institutes of Health Grants AG19206, NS41669. The Hereditary Disease Foundation Cure HD Initiative and The Huntington's Disease Society of America.

References

1. HD Collaborative Research Group (1993) A novel gene containing a trinucleotide repeat that is expanded and unstable on Huntington's disease chromosomes. *Cell*, **72,** 971–983.
2. Davies, S.W., Turmaine, M., Cozens, B.A., DiFiglia, M., Sharp, A. H., Ross, C. A., et al. (1997) Formation of neuronal intranuclear inclusions underlies the neurological dysfunction in mice transgenic for the HD mutation. *Cell* **90,** 537–548.
3. DiFiglia, M., Sapp, E., Chase, K. O., Davies, S. W., Bates, G. P., Vonsattel, J. P., and Aronin, N. (1997) Aggregation of huntingtin in neuronal intranuclear inclusions and dystrophic neurites in brain. *Science* **277,** 1990–1993.
4. Becher, M. W., Kotzuk, J. A., Pavlakis, G. N., Sharp, A. H., Davies, S. W., Bates, G. P., et al. (1998) Intranuclear neuronal inclusions in Huntington's disease and Dentatorubral and Pallidoluysian Arophy: Correlation between the density of inlcusions and IT15 CAG triplet repeat length. *Neurobiol. Dis.* **4,** 387–397.
5. Gutekunst, C.A., Li, S. H., Yi, H., Mulroy, J. S., Kuemmerle, S., Jones, R., et al. (1999) Nuclear and neuropil aggregates in Huntington's disease: relationship to neuropathology. *J. Neurosci.* **19,** 2522–2534.
6. Saudou, F., Finkbeiner, S., Devys, D. and Greenberg, M. E. (1998) Huntingtin acts in the nucleus to induce apoptosis but death does not correlate with the formation of intranuclear inclusions. *Cell* **95,** 55–66.
7. Peters, M. F., Nucifora, F. C., Jr., Kushi, J., Seaman, H. C., Cooper, J. K., Herring, W. J., et al. (1999) Nuclear targeting of mutant Huntingtin increases toxicity. *Mol. Cell. Neurosci.* **14,** 121–128.
8. Lunkes, A. and Mandel, J. L. (1998) A cellular model that recapitulates major pathogenic steps of Huntington's disease. *Hum. Mol. Genet.* **7,** 1355–1361.
9. Perutz, M.F., Johnson, T., Suzuki, M. and Finch, J.T. (1994) Glutamine repeats as polar zippers: their possible role in inherited neurodegenerative diseases. *Proc. Natl. Acad. Sci. USA* **91,** 5355–5358.
10. Li, S. H. and Li, X. J. (1998) Aggregation of N-terminal huntingtin is dependent on the length of its glutamine repeats. *Hum. Mol. Genet.* **7,** 777–782.
11. Lah, J. J. and Burry, R. W. (1993) Neuronotypic differentiation results in reduced levels and altered distribution of synaptophysin in PC12 cells. *J. Neurochem.* **60,** 503–512.
12. Li, S. -H., Lam. S., Cheng, A. L., and Li, X. -J. Intranuclear huntingtin increases the expression of caspase-1 and induces apoptosis. *Hum. Mol. Genet.* **9,** 2859–2867.
13. Li, S. H., Cheng, A. L., Li, H., and Li, X. J. (1999) Cellular defects and altered gene expression in PC12 cells stably expressing mutant huntingtin. *J. Neurosci.* **19,** 5159–5172.

26

Heterologous Expression of Ion Channels

Andrew R. Tapper and Alfred L. George, Jr.

1. Introduction

The use of recombinant DNA technology to clone, sequence, and express ion channels and transporters has powered an enormous acceleration in the understanding of structure-function relationships in these important proteins. Given that most ion channels reside in tissues that are largely inaccessible to direct recording techniques and the general paucity of continuous cell lines expressing defined populations of functional molecules, studying native channels is often difficult and impractical. However, the ability to introduce a recombinant complementary DNA (cDNA) selectively into cells normally devoid of highly expressed ion channels or transporters greatly facilitates the ability of scientists to study the function, subunit associations, regulation, and trafficking of these proteins. This approach has also enabled studies designed to investigate the role of ion channel mutations in inherited diseases.

The two most widely applied schemes for functional expression of ion channels utilize either *Xenopus laevis* oocytes *(1–3)* or transfected mammalian cell cultures (**Fig. 1**). Both approaches constitute powerful tools for the in vitro study of recombinant ion channels in a controlled experimental environment. By contrast to the study of native ion channels, these methods allow flexibility in defining subunit composition or stoichiometry and permit the characterization of engineered mutations or chimeric constructs. In general, the use of heterologous cells provides a reproducible system in which to express recombinant molecules in a setting where the background ionic currents are likely to be much less prominent than the introduced (i.e., overexpressed) channel. Each method has specific advantages and disadvantages (**Table 1**) that should be carefully considered prior to designing experiments. This chapter will summarize the basic methods involved in using these two strategies. Methods and procedures for performing detailed electrophysiological or biochemical analyses of heterologously expressed ion channels are available from several sources *(4,5)*.

Table 1
Advantages and Disadvantages of Heterologous Expression Systems

System	Advantages	Disadvantages
Xenopus oocytes (two-electrode voltage clamp)	Inexpensive High efficiency expression Simplicity of electrophysiological recording Permits varying ratios of subunits	Endogenous ion currents Can't control internal solution Poor voltage control Batch variability Amphibian cells
Transfected cells (whole cell patch clamp)	Can utilize mammalian cells Excellent voltage control Can control internal solution Stable cells lines possible	Variable expression levels Low transfection efficiency Difficult to control subunit ratios Endogenous ion currents

Heterologous Expression of Ion Channels

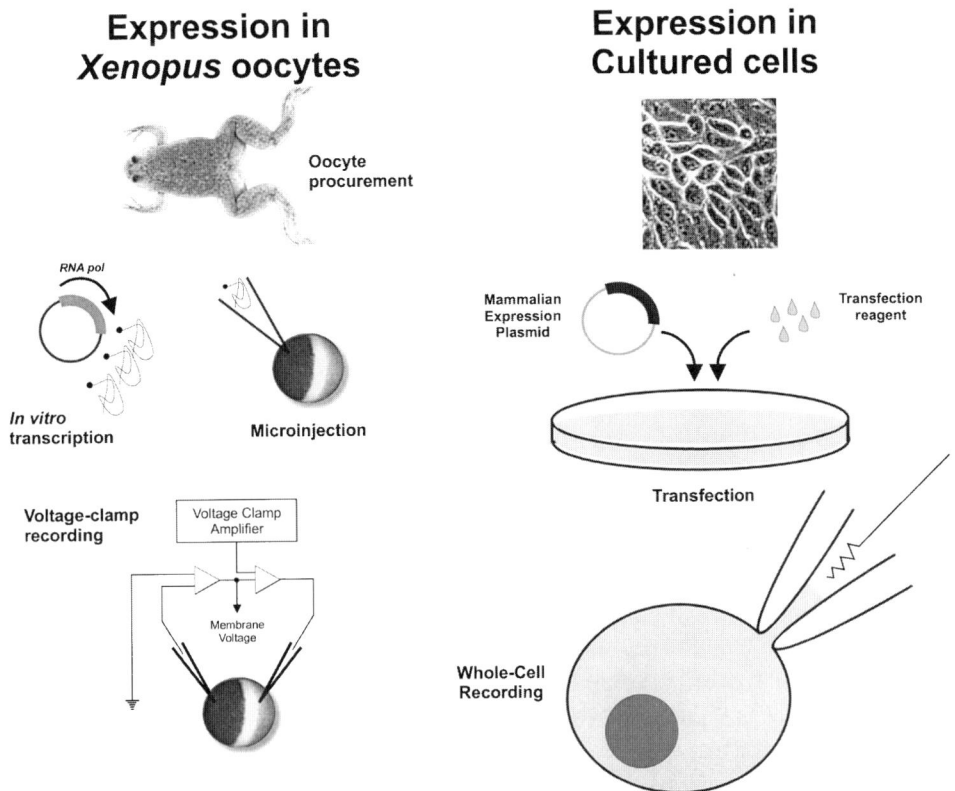

Fig. 1. Experimental paradigms for heterologous expression of ion channels.

2. Materials

2.1. Procurement of Xenopus Oocytes

1. Female adult *Xenopus laevis* frogs (Xenopus-Express, Homosassa, FL, USA, *see* Website: http://www.xenopus.com; or NASCO Science Supply, Fort Atkinson, WI, USA, *see* Website: http://www.enasco.com). Frogs must be kept in dechlorinated fresh water (one frog per gallon) at 18°C with a regulated light cycle (12 h light, 12 h dark). Animals should be fed twice weekly and have their tank water changed following meals (*see* **Note 1**).
2. Anesthetic stock solution: 3% benzocaine (ethyl-p-aminobenzoate) in ethanol. Dilute 1:100 with water for use with frogs.
3. Surgical instruments including small scissors, forceps, and needle holder.
4. Collagenase (type 1A from *clostridium histolyticum*, Sigma Chemical Co., cat. no. C-9891) dissolved in Ca^{2+} free ND-96 to final concentration of 2.5 mg/mL.
5. Ca^{2+} free ND-96: 96 mM NaCl, 2 mM KCl, 1 mM $MgCl_2$, 5 mM HEPES, adjust pH to 7.5 with NaOH.

2.2. In Vitro Transcription

1. RNase-free deionized water. Deionized water should be treated using 0.1% diethylpyrocarbonate (Sigma; 10% stock solution [v/v] prepared in absolute ethanol) overnight at room temperature followed by autoclaving (30 min liquid cycle).
2. 5X transcription optimized buffer (Promega Corp., Cat. no. P1181).

3. Ribonucleotide triphosphates: 10 m*M* rATP, UTP rCTP; and 2 m*M* rGTP.
4. m^{7}GpppG (5'-cap analog, Roche Molecular Biochemicals). Dissolve 5 A$_{260}$ units in 30 μL DEPC-treated water. Store –20°C.
5. SP6 polymerase (Promega Corp.).
6. RNase inhibitor (Roche Molecular Biochemicals).
7. Dithiothreitol (DTT), 0.1 *M*.
8. Buffer equilibrated phenol/chloroform/isoamyl alcohol (1:1:25, v:v:v).
9. Ethanol.
10. 7.5 *M* ammonium acetate.

2.3. Oocyte Microinjection

1. Low-power stereo dissecting microscope.
2. Manual coarse micromanipulator.
3. Microinjector (Picospritzer™ or Drummond Nanoject; *see* **Fig. 2**).
4. Pipet puller (Sutter Instrument Co., model P-97, or comparable model).
5. Glass capillary pipets (Drummond 10 μL pipets, Cat. no. 2-000-010).
6. ND-96: 96 m*M* NaCl, 2 m*M* KCl, 1.8 m*M* CaCl$_2$, 1 m*M* MgCl$_2$, 5 m*M* HEPES, adjust pH to 7.5 with NaOH. Add 275 mg/L pyruvic acid and 50 mg/L gentamicin.
7. Wide-bore Pasteur glass pipets for handling oocytes (modify a 9-in. Pasteur pipet by removing 4-in. of the tip and fire-polishing the end).
8. Oocyte injection chamber made by mounting polypropylene mesh (0.8 mm grid) onto the bottom of a small (30–60 mm) Petri dish.

2.4. Transient Transfection of Cultured Cells (Calcium Phosphate Method)

1. Cell lines: HEK-293, tsA201, CHO-K1, COS-7 and others (American Type Culture Collection, Rockville, MD, USA).
2. Media for tsA201 cells: Dulbecco's modified Eagle Medium (DMEM) with high glucose and NaHCO$_3$, 2 m*M* L-glutamine, 10% fetal bovine serum (FBS), penicillin (50 U/mL), and streptomycin (50 μg/mL), sterile-filtered.
3. 0.25% Trypsin, 1 m*M* EDTA solution (Gibco-BRL).
4. 2X HEBS solution: 274 m*M* NaCl, 40 m*M* HEPES, 12 m*M* dextrose, 10 m*M* KCl, 1.4 m*M* Na$_2$HPO$_4$, adjust final pH to 7.05, sterile-filtered.
5. 0.25 *M* CaCl$_2$ solution, sterile filtered.
6. Sheared salmon sperm DNA (1 mg/mL).

3. Methods

3.1. Oocyte Procurement

1. Anesthetize an adult female *Xenopus laevis* using 0.03% benzocaine. Immerse frogs in anesthetic solution for 3–5 min keeping the animal's nostrils above water. Monitor closely to determine the onset of anesthesia (decreased stimulated movement). Once anesthetized, place the frog (ventral side up) on a bed of crushed ice for approx 20 min prior to the start and throughout the surgical procedure to potentiate and maintain anesthesia.
2. Make a 1 cm incision in the lower abdominal wall near one leg to expose an ovary (**Fig. 3**).
3. Carefully remove a portion of one ovary (approx one-third) and cut open all lobes to expose oocytes.
4. Suture abdominal wall using 5–0 or 6–0 silk suture.

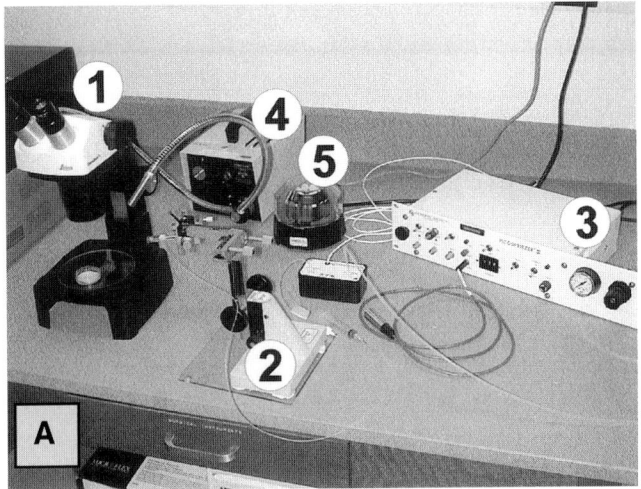

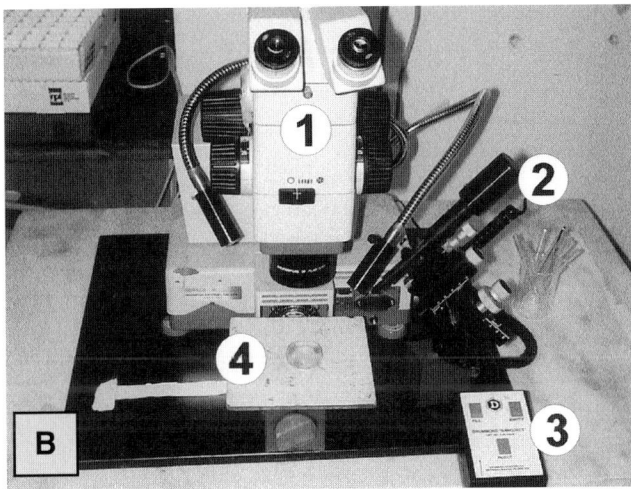

Fig. 2. *Xenopus* oocyte microinjection workstations. Two different microinjection workstations are shown. (**A**) Workstation based on use of a pressure driven microinjector (Picospritzer™ II, General Valve Corp., Fairfield, NJ, USA). Labeled components are as follows: 1) binocular dissecting microscope, 2) manual micromanipulator with pipette holder, 3) microinjector, 4) fiberoptic light source, and 5) small microcentrifuge. (**B**) Workstation based on use of piston-plunger type microinjector (Nanoject, Drummond Scientific Co., Broomall, PA, USA). Labeled components are as follows: 1) binocular dissecting microscope with fiberoptic light source, 2) microinjector mounted on a coarse micromanipulator with X-Y-Z axis controllers, 3) control switch for microinjector, and 4) microscope stage with oocyte injection chamber.

5. Allow animal to recover from anesthesia in shallow water inclined to keep nostrils above water.
6. Defolliculate oocytes using collagenase (2.5 mg/mL) in Ca^{2+} free ND-96 for 2 h with gentle agitation at room temperature. Wash 4 times in fresh ND-96 and select mature oocytes for microinjection. Oocytes may be injected on the same day as procurement or after overnight incubation (*see* **Notes 2** and **3**).

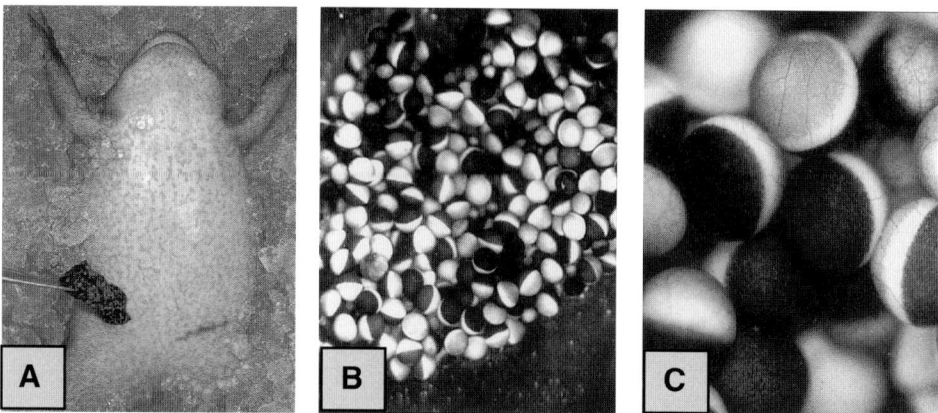

Fig. 3. Procurement of *Xenopus* oocytes. (**A**) Ventral surface of an anesthetized female *Xenopus laevis* with extraction of a partial ovary through an abdominal incision. (**B**) Oocytes immediately after removal from a frog and prior to defolliculation. Multiple stages of oocyte maturation are present (note the characteristic dark animal pole and light colored vegetal pole) (**C**). Higher magnification of oocytes prior to defolliculation (note the presence of small superficial blood vessels indicating the presence of follicular cell layers). Most of the oocytes in this photograph are stage VI indicated by the sharply demarcated animal and vegetal poles separated by a white equatorial band *(13)*.

3.2. In Vitro Transcription

1. Digest plasmid DNA (10–15 μg) in a large volume reaction (100–150 μL) using a restriction endonuclease that cleaves 3' to the cDNA insert. An enzyme that leaves a 5' overhang is optimal. A plasmid vector optimized for oocyte expression should be used (**Fig. 4**).
2. Extract twice with phenol/chloroform/isoamyl alcohol and once with chloroform (collect aqueous top layer after each extraction).
3. Ethanol precipitate using RNase-free 7.5 M ammonium acetate (one-half volume) and 3 volumes of 100% ethanol. Centrifuge 12,000g for 15 min at 4°C, wash once with 70% ethanol, decant supernatant completely and resuspend in DEPC-treated water (do not dry the DNA pellet) (*see* **Note 4**).
4. Mix together in a 1.5 mL microcentrifuge tube: 0.5–1.0 μg linear plasmid DNA (diluted to 15 μL total volume with DEPC-treated water), 10 μL transcription optimized buffer, 5 μL DTT, 2.5 μL ribonucleotides, 5 μL m^{7}GpppG, 2 μL RNase inhibitor, and 3 μL SP6 polymerase. Incubate 1–2 h at 37°C. Incubation times should be longer for large transcripts (*see* **Note 5**).
5. Ethanol precipitate mRNA by adding one half volume 7.5 M ammonium acetate, 3 volumes 100% ethanol and incubate at −80°C for 15 min or −20°C overnight. Collect mRNA by centrifugation for 20 min at 12,000g in a microfuge, wash once with 70% ethanol, and resuspend in DEPC water (20–30 μL depending on desired final concentration). Prepare small (2.5 μL) aliquots for storage at −80°C and avoid repeated freeze-thaw cycles.
6. Analyze and quantify mRNA by denaturing agarose gel electrophoresis (*see* **Note 6**).

3.3. Oocyte Microinjection

1. Thaw an aliquot of mRNA. Centrifuge 12,000g 2 min to pellet insoluble debris that can clog the injection pipet. Always handle mRNA using latex gloves.

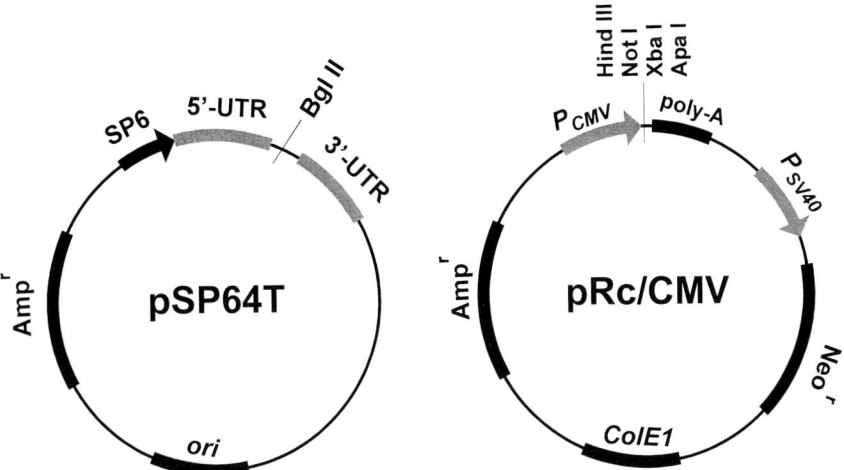

Fig. 4. Plasmid vectors for expression of ion channels in oocytes and cultured mammalian cells. The pSP64T plasmid *(18)* contains an SP6 RNA polymerase promoter and the 5' and 3' UTRs from *Xenopus laevis* β-globin flanking a unique Bgl II restriction endonuclease site. The β-globin sequences greatly enhance translation efficiency of heterologous mRNA transcripts in oocytes. The Bgl II site is used for inserting a cDNA sequence of interest. The pRc/CMV plasmid is optimized for expression in cultured mammalian cells. Four unique restriction endonuclease sites facilitate directional insertion of a cDNA sequence adjacent to the cytomegalovirus immediate-early promoter (P_{CMV}) and a polyadenylation signal derived from bovine growth hormone (poly-A). This plasmid also contains a neomycin resistance gene driven by the SV40 promoter for use in establishing cell lines stably expressing a cDNA. Other elements include an ampicillin resistance gene (Amp^r), and bacterial origin of replication (ori, ColE1).

2. Fabricate microinjection pipets to have tip diameter 10–15 microns. Pull glass capillaries as long as possible then break the tip with forceps that have been heated with a flame and allowed to cool (to remove trace RNA and RNase).
3. Select oocytes. Stage V and VI most typically used (*see* **Note 3**).
4. Gently inject with 20–40 nl of mRNA solution at 0.1–1.0 µg/µL (**Fig. 2**). Allow injection pipet to just pierce oocyte membrane and observe the oocyte "plump-up" when injected. Always change injection pipet between mRNA samples to avoid cross-contamination and plugging.
5. Incubate oocytes at room temperature or 19°C in ND-96 with antibiotics and pyruvate. Change incubation solution daily. (*see* **Note 2**).

3.4. Transient Transfection of Cultured Cells (Calcium Phosphate Method)

1. Grow cells using appropriate media, use standard cell culture technique and perform all procedures in a tissue culture hood.
2. Passage cells using trypsin-EDTA and seed several 100 mm sterile tissue culture grade dishes with 0.5–1.0 million cells. Grow until ~ 30% confluent (*see* **Notes 7** and **8**).
3. Prepare two 1.5 mL microcentrifuge tubes; place 0.5 mL of 250 m*M* $CaCl_2$ in one, and 0.5 mL of 2X HEBS in the s. Add 10 µg of plasmid DNA (*see* **Note 9**) and 10 µg salmon sperm DNA (carrier) to the $CaCl_2$ and then mix gently.

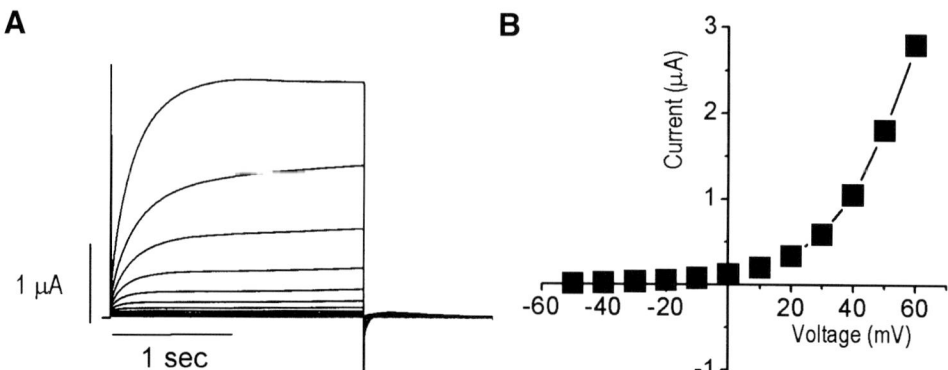

Fig. 5. Endogenous Cl⁻ currents recorded from a *Xenopus* oocyte. (**A**) Two-electrode voltage clamp recording made from an uninjected *Xenopus* oocyte in response to a series of 2 s test pulses (from −50 to + 60 mV from a holding potential of −120 mV). (**B**) Current-voltage relationship for the current shown in (A).

4. Add the DNA/$CaCl_2$ mixture slowly (drop by drop) to the 2X HEBS solution and gently tap the tube to mix (should get slightly cloudy). Incubate at room temperature for 20 min.
5. Change media (10 mL) on cells just before adding DNA. Dribble DNA mixture over cells, swirl to mix, and place in incubator (37°C, 5% CO_2) overnight. Change media the next day. Examine cells for ion channel expression 1–3 d later (*see* **Note 10**).

4. Notes

1. Only female *Xenopus* have oocytes and the gender of animals must be specified when ordering. Suppliers may inject frogs with human chorionic gonadotropin to induce new oocyte development. The quality of oocytes will vary from frog to frog. The variability will be manifest in the lifespan of oocytes, level of endogenous currents (especially Cl⁻ currents; *see* example in **Fig. 5**) *(6–10)* and level of non-specific "leak" current. Some seasonal variation in quality may also be evident but this can be minimized by keeping frogs in a temperature controlled environment with fixed light-dark cycle. It is critical to test uninjected or water injected oocytes from each batch to determine the level of endogenous currents. Repeat experiments using different oocytes batches whenever possible. Frogs may be fed beef liver, commercial trout chow, or NASCO frog brittle.
2. Incubation of oocytes with antibiotics (gentamicin, 100 μg/mL with or without tetracycline, 50 μg/mL) will prevent bacterial infection and lengthen the lifespan *(11)*. Damaged or infected oocytes will first become mottled then rupture. Promptly remove any unhealthy oocytes from the incubation solution or transfer healthy cells to fresh solution. Addition of pyruvate as an energy source will also extend oocyte lifespan in vitro *(12)*. Oocytes with residual follicular cells will adhere to the dish and not roll freely.
3. Typically, stage V and VI oocytes *(13)* are selected for microinjection. Stage VI oocytes are the most mature cells and are easily recognized by their size (1–1.2 mm diameter) and coloring (clearly demarcated animal and vegetal poles separated by a near-white equatorial band). Stage V oocytes lack the equatorial band and are slightly smaller. Stage IV oocytes closely resemble stage V cells but are smaller. The smaller oocytes may have the advantage of allowing better voltage-clamp control in electrophysiological experiments.

4. Plasmid DNA prepared for in vitro transcription may become insoluble if allowed to dry. After the final ethanol wash step, decant by inverting microtubes onto clean Kimwipes®, centrifuge briefly (5 s) and then manually remove remaining supernatant using a pipetter. Following this step, immediately redissolve DNA in DEPC-treated water at ~ 1 µg/µL. Check recovery of DNA and estimate its concentration by electrophoresing a 1 µL aliquot on a 1% agarose gel using an appropriate size standard of known quantity. Use latex gloves when handling DNA template or mRNA transcripts.

5. Commercial in vitro transcription systems are available that work well. The mMessage mMachine kit (Ambion Corp.) is especially popular for transcribing abundant quantities of capped mRNA for translation in oocytes. Protocols supplied with the reagent kit are suitable for most applications.

6. The products of in vitro transcription reactions will usually consist of complete and partial length transcripts. Incomplete transcription results from exhaustion of reaction components, time limitations, and template DNA secondary structure. Therefore, quantification of mRNA using spectrophotometric methods will not provide information only on the functional (i.e., full-length) transcript. Denaturing agarose gel electrophoresis using an RNA size standard of known quantity provides the best method for determining the yield of full-length reaction product. Determining the precise transcript size can also be problematic. Commercially available RNA size standards utilize uncapped, polyadenylated mRNAs and the apparent size of capped, nonpolyadenylated mRNA may not match predictions based on the length of the transcribed DNA template. In our experience, transcribed mRNAs may appear smaller than expected when analyzed this way.

7. We favor the use of tsA201 cells for most experiments because of their rapid growth, high transfection efficiency, high level of expression, and low levels of endogenous ionic currents. The tsA201 cell line was created by transforming native HEK-293 cells with the SV40 large T antigen. These cells grow very fast and it is important to have a consistent estimation of the degree of confluence for reproducible results with transfections. Using the calcium phosphate method, we generally observe 5–15% transfection efficiencies. After several passages (>20) the transfection efficiency may decrease significantly and it is wise to keep several vials of cells from an early passage stored in liquid nitrogen. Different lots of FBS may cause cells to become strongly or loosely adherent to the dish, and it is a good practice to screen different batches of sera if these problems occur.

8. Cultured cells may exhibit high levels of endogenous currents (especially outward Cl^- and K^+ currents). Using cells at the earliest passage will reduce the incidence of this problem.

9. When expressing recombinant ion channels for patch-clamp studies, it is often useful to co-transfect cells with channel cDNAs and a marker plasmid carrying the coding region of the leukocyte antigen CD-8 (pLeu2) *(14,15)* or enhanced green fluorescent protein (pEGFP) *(16,17)*. Typically we use a 3:1 molar ratio of channel cDNA to marker plasmid. Cells co-transfected with pLeu2 can be marked by brief incubation with Dynal micro-beads coated with anti-CD8 antibody. GFP transfected cells fluoresce green under UV illumination.

10. Several alternative methods may be used to transfect mammalian cells with plasmid DNA including lipid-mediated approaches and electroporation. We have had success with FuGENE 6 (Roche Molecular Biochemicals) in transfecting several cell types at high efficiency (20–60%). Not all cell lines transfect with equal efficiency and significant effort may be required to determine optimal ratios of DNA and transfection reagent.

Acknowledgments

The authors are grateful to the efforts of Craig Short for providing photographs of *Xenopus* frogs and oocytes, and help with developing protocols.

References

1. Shih, T. M., Smith, R. D., Toro, L., and Goldin, A. L. (1998) High-level expression and detection of ion channels in *Xenopus* oocytes. *Methods Enzymol.* **293,** 529–556.
2. Goldin, A. L. (1992) Maintenance of *Xenopus laevis* and oocyte injection. *Methods Enzymol.* **207,** 266–279.
3. Goldin, A. L. and Sumikawa, K. (1992) Preparation of RNA for injection into *Xenopus* oocytes. *Methods Enzymol.* **207,** 279–297.
4. Stuhmer, W. (1998) Electrophysiologic recordings from *Xenopus* oocytes. *Methods Enzymol.* **293,** 280–300.
5. Sherman-Gold, R. (ed.), (1993) *The Axon Guide.* Axon Instruments, Inc., Foster City, CA, USA.
6. Barish, M. E. (1983) A transient calcium-dependent chloride current in the immature *Xenopus* oocyte. *J. Physiol.* (London) **342,** 309–325.
7. Kowdley, G. C., Ackerman, S. J., John, E. J., Jones, L. R., and Moorman, J. R. (1994) Hyperpolarization-activated chloride currents in *Xenopus* oocytes. *J. Gen. Physiol.* **103,** 217–230.
8. Parker, I., and Miledi R. (1988) A calcium-independent chloride current activated by hyperpolarization in *Xenopus* oocytes. *Proc. R. Soc. Lond. B.* **233,** 191–199.
9. Landau, E. M., and Blitzer, R. D. (1994) Chloride current assay for phospholipase C in *Xenopus* oocytes. *Methods Enzymol.* **238,** 140–154.
10. Tokimasa, T. and North, R. A. (1996) Effects of barium, lanthanum, and gadolinium on endogenous chloride and potassium currents in *Xenopus* oocytes. *J. Physiol.* (London) **496,** 677–686.
11. Elsner, H. A., Honck, H. H., Willmann, F., Kreienkamp, H. J., and Iglauer, F. (2000) Poor quality of oocytes from *Xenopus laevis* used in laboratory experiments: prevention by use of antiseptic surgical technique and antibiotic supplementation. *Comp. Med.* **50,** 206–211.
12. Eppig, J. J. and Steckman, M. L. (1976) Comparison of exogenous energy sources for in vitro maintenance of follicle cell-free *Xenopus laevis* oocytes. *In Vitro* **12,** 173–179.
13. Dumont, J. N. (1972) Oogenesis in *Xenopus laevis* (Daudin). I. Stages of oocyte development in laboratory maintained animals. *J. Morphol.* **136,** 153–179
14. Margolskee, R. F., McHendry-Rinde, B., and Horn, R. (1993) Panning transfected cells for electrophysiological studies. *BioTechniques* **15,** 906–911.
15. Jurman, M. E., Boland, L. M., Liu, Y., and Yellen, G. (1994) Visual identification of individual transfected cells for electrophysiology using antibody-coated beads. *BioTechniques* **17,** 876–881.
16. Tsien, R. Y. (1998) The green fluorescent protein. *Annu. Rev. Biochem.* **67,** 509–544.
17. Zhang, G., Gurtu, V., and Kain, S. R. (1996) An enhanced green fluorescent protein allows sensitive detection of gene transfer in mammalian cells. *Biochem. Biophys. Res. Commun.* **227,** 707–711.
18. Krieg, P. A. and Melton, D. A. (1987) *In vitro* RNA synthesis with SP6 RNA polymerase. *Method Enzymol.* **155,** 397–415.

27

An Assay for Characterizing In Vitro the Kinetics of Polyglutamine Aggregation

Valerie Berthelier and Ronald Wetzel

1. Introduction

The expansion of trinucleotide CAG repeat sequences has been shown to be the underlying cause of eight human neurodegenerative disorders, including Huntington's disease (HD), spinal and bulbar muscular atrophy (SBMA), dentatorubral-pallidoluysian atrophy (DRPLA), Machado-Joseph disease (MJD), and the spinocerebellar ataxias (SCAs)(for review see **refs.** *[1–3]*). These diseases are progressive disorders characterized by motor and/or cognitive impairments and distinctive pathological patterns of neuronal degeneration within the central nervous system (CNS). Each is caused by CAG codon expansion within a unique gene producing a polyglutamine (polyGln) tract enlargement in the target protein *(4–6)*. All of these neurodegenerative disorders present a common feature: the accumulation of the polyGln repeat disease-related protein into neuronal intranuclear inclusions, which have become the neuropathological signature of polyGln disorders *(7–9)*. Several lines of reasoning suggest strongly that the expanded polyGln tract is itself responsible for the pathogenesis. First, the disease-related proteins do not present any homology in either size or amino acid composition outside of the expanded polyGln tract *(6,10–12)*. Second, the severity of the disease increases with the length of the polyGln repeat. Third, the length of the polyGln sequence that triggers increased risk for developing the pathology is very similar for almost of these diseases *(2,13,14)*. Finally, transgenic animals expressing protein fragments with an expanded polyGln repeat exhibit neurodegenerative phenotypes reminiscent of disease symptoms *(15–19)*. Recently, cellular experiments support the idea that polyGln aggregates are toxic for the cells due to their ability to recruit other critical cellular proteins, via their own polyGln components, into the growing aggregates *(20–22)*. The loss of protein activity due to this sequestration is toxic to the cell.

Given the potential role of polyGln aggregates and polyGln aggregate extension in the pathogenesis of expanded CAG repeat diseases, it is essential to characterize the fundamental aggregation behavior of polyGln sequences. To date, the only assay system for assessing polyGln aggregation behavior in vitro involves fusion proteins of polyGln-containing fragments of disease-related proteins expressed in *Escherichia coli (23)*.

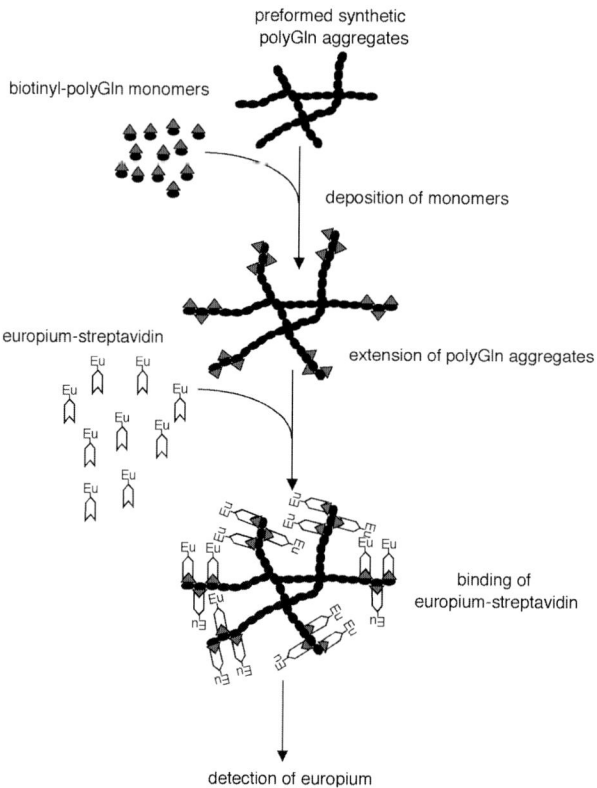

Fig. 1. Schematic representation of the extension of synthetic polyGln aggregates detected with the microtiter plate assay.

To be able to study the aggregation behavior of polyGln tract, we have established a highly sensitive, reproducible, and specific microtiter plate assay capable of monitoring aggregate-dependent deposition of polyGln peptides. This assay is similar in format to an Aβ deposition assay described by Maggio and colleagues (24). This previously described assay utilizes ^{125}I-labeled Aβ, and thus is capable of detecting very low levels of addition of monomeric Aβ to a pre-existing aggregate. In order to avoid using radioactivity, which raises safety issues and also introduces the inconvenience of requiring routine repeated syntheses of labeled peptide, we chose to use peptides tagged with biotin to follow the aggregation/deposition process. This microtiter plate assay for polyglutamine aggregation was made possible by the recent development of a protocol for solubilization of chemically synthesized polyGln peptides (25). This microtiter plate assay permits detailed studies on different aspects of aggregation kinetics (26) as well as detection and quantification of very low amounts of "extension-competent" aggregates (27). It is also a valuable tool for screening and characterizing anti-polyGln aggregation inhibitors.

Here, we describe in detail the microtiter plate assay formatted to follow the ability of preformed, synthetic polyGln aggregates to be extended via recruitment of polyGln monomers (**Fig. 1**). PolyGln aggregates, made from chemically synthesized peptides, are immobilized onto plastic and incubated various times with biotinylated-peptide.

A europium-streptavidin binding step, followed by time-resolved fluorescence detection of the europium, allows us to calculate the rate (fmoles/h) of incorporation of polyGln peptides into polyGln aggregates.

The kinetics of the extension of polyGln aggregates is carried out in reverse temporal order: the longest data time point is executed first and the shortest done last. This reverse kinetics approach is dictated by the use of the microplate.

2. Materials

2.1. Preparation of Biotinyl-polyGln Peptides and PolyGln Aggregates

1. PolyGln peptides and biotinyl-polyGln peptides (*see* **Note 1**) were obtained by custom syntheses from the Keck Biotechnology Center at Yale University (*see* Website: http://info.med.yale.edu/wmkeck/). They were purchased without purification (*see* **Note 2**). The vials containing the synthesized peptides should be stored at –80°C in a container desiccated with some anhydrous calcium sulfate.
2. A polyGln $K_2Q_{15}K_2$ peptide treated as described in **Subheading 3.1.** (**steps 1–6**) whose concentration is determined by amino acid composition analysis (Commonwealth Biotechnologies, *see* Website: http://www.cbi-biotech.com). This peptide serves to establish the concentration standard curve by high-performance liquid chromatography (HPLC).
3. Trifluoroacetic acid (TFA) (Pierce, Rockford, IL, USA). This solution should be used under a fume hood and handled wearing gloves, safety glasses, and a lab coat.
4. 1,1,1,3,3-hexafluoro-2-propanol (HFIP) (Sigma, St. Louis, MO, USA). This should also be handled using the same precautions as described in **item 3**.
5. Argon (*see* **Note 3**).
6. dH_2O adjusted to pH 3.0 by addition of TFA (H_2O/TFA-pH 3.0).
7. HPLC system fitted with a Zorbax SB-C3 column.

2.2. Extension Assay

1. A 96-well plate (EIA/RIA Plates, Costar, Atlanta, GA, USA).
2. A Pasteur pipet connected to a vacuum system.
3. Extension buffer: phosphate buffered saline (PBS) (Fisher Scientific, *see* Website: http://www.fishersci.com), 0.01% Tween 20, 0.05% sodium azide.
4. Blocking buffer: PBS, 0.1% gelatin, 0.01% Tween 20, 0.05% sodium azide.
5. Europium-streptavidin (Perkin Elmer, Boston, MA, USA).
6. Europium buffer: PBS, 0.5% BSA, 0.01% Tween 20, 0.05% sodium azide.
7. Enhancement solution (Perkin Elmer).
8. Victor 2 (Perkin Elmer) or other suitable time-resolved fluorescence microplate reader.

3. Methods

3.1. Solubilization of the $K_2Q_nK_2$ and Biotinyl-$K_2Q_nK_2$ Peptides (see Note 4)

1. Weigh out a small amount of peptide (*see* **Note 5**) and dissolve it in a mixture of 50% TFA and 50% HFIP. The peptide's concentration should be about 0.5 mg/mL. The solubilization of the peptides should be conducted in a glass 24-mL vial with a cap.
2. Vortex vigorously and incubate overnight at room temperature to ensure complete solubilization of the peptides.
3. The next day, blow off the solvent under a gentle stream of argon, using a Pasteur pipet at the end of the argon line.
4. Resolubilize the peptide by adding H_2O/TFA-pH 3.0 to a concentration of 0.5 mg/mL.

5. Ultracentrifuge the peptide at 4°C for 3 h at 50,000g (*see* **Note 6**).
6. Transfer the top 90% of the supernatant (*see* **Note 7**) into a 50 mL polypropylene tube (Falcon). Leave the tube at 4°C during the **steps 7** and **8**.
7. To 10 µL of the supernatant add 90 µL of a 0.1% aqueous solution of TFA. Inject 50 µL of the peptide onto the Zorbax SB-C3 column and apply a 0–50% (v/v) acetonitrile gradient with 0.05% TFA at a rate of 2% per min (*see* **Note 8**).
8. Integrate the peptide's elution peak (range of 11–18% acetonitrile, depending on polyGln length) by absorbance at 215 nm. The peak area obtained at A_{215} is compared with a standard curve previously established with the peptide $K_2Q_{15}K_2$ as described below in **Subheading 3.2.**

3.2. Preparation of an HPLC Standard Curve

1. To establish a concentration standard curve, first apply **steps 1–6** to the peptide $K_2Q_{15}K_2$. Determine the exact concentration of an appropriate dilution of this peptide stock solution by amino acid composition analysis.
2. Take the appropriate volumes of the calibrated $K_2Q_{15}K_2$ stock solution to get 100 µL of a final solution containing various amount of the peptide (for example 0.1, 0.2, 0.4, 0.8, and 1.6 µg). The dilutions should be done with a 0.1% aqueous solution of TFA.
3. Inject 50 µL of the various amounts of $K_2Q_{15}K_2$ peptide onto the Zorbax SB-C3 column and apply a 0–50% (v/v) acetonitrile gradient with 0.05% TFA at a rate of 2% per min.
4. Integrate the $K_2Q_{15}K_2$ A_{215} absorbance peak (14% acetonitrile). The area under the A_{215} peak in the HPLC profile obtained from a known weight of the $K_2Q_{15}K_2$ is used to generate a conversion factor (Peak area = f [µg of peptide]) to calculate the weight concentrations of other polyGln peptides.

3.3. Preparation of $K_2Q_nK_2$ Aggregates

1. Using the concentration for a pH 3.0 solution of polyGln peptide (**Subheading 3.1., step 6**) determined in **Subheading 3.1., step 8**, adjust the concentration to 10 µM with a H_2O/TFA-pH 3 solution and mix. Then, raise the pH to 7.4 by addition of a 1/9 volume of 10X PBS, with immediate mixing.
2. Cap the 50 mL Falcon tube, seal with parafilm, and incubate 24 h at 37°C.
3. Snap freeze the tube in liquid nitrogen, and incubate at −15 to −20°C for 48 h.
4. Thaw the tube, which now contains aggregated polyGln, and centrifuge 30 min at 20,800g at 4°C to collect the aggregates.
5. Resuspend the pellet to 10 µM in extension buffer and quickly aliquot into Eppendorf tubes, keeping the suspension well mixed during aliquoting. A convenient amount of aggregates per tube would be about 100 µL of a range concentration of 40–70 µg/mL.
6. Snap freeze in liquid nitrogen and store at −80°C.

3.4. Preparation of the Biotinyl-Q_n Peptides

1. Repeat exactly steps 1–8 described in **Subheading 3.1.**, but with the appropriate biotinyl-peptide.
2. After the determination of the exact concentration of the biotinyl-peptide, dilute it in extension buffer to a final concentration of 10 nM.
3. Aliquot, snap freeze in liquid nitrogen, and store at −80°C (*see* **Note 9**).

3.5. Extension Assay

1. We described here a kinetic extension of 5 h including 8 time points (5, 15, 30, 45, 60, 120, 240, and 300 min); each time point is done in triplicate. Dilute the aggregates into

2.5 mL of extension buffer to get the desired amount of aggregate/well. A convenient amount is 80 ng/well.
2. In the 96-well plate, distribute 100 µL of extension buffer in all of the rows of columns 1, 2, and 3 (*see* **Note 10**). In columns 4, 5, and 6 distribute 100 µL of the aggregate suspension made in **step 1**. Make sure the suspension is well mixed during the aliquoting process. Allow the plate to dry uncovered 17 h in a 37°C incubator.
3. Wash the plate 3 times with 200 µL of extension buffer.
4. Add 100 µL/well of the blocking buffer, seal the plate, and incubate 1 h at 37°C.
5. During this time, thaw the biotinyl-peptide at room temperature. When the peptide is thawed but still cold place it at 4°C.
6. After the 1 h incubation with the blocking buffer, discard the blocking buffer and repeat **step 3** (*see* **Note 11**).
7. Add 100 µL of extension buffer in each well of columns 1–6 except for row H.
8. In the row H, distribute 100 µL of the biotinyl-peptide in the wells. Immediately put the sealed plate at 37°C and initiate timing. This step (row H samples) corresponds to the 5 h incubation.
9. Fifty-five min later, empty the row G by aspirating with a pipet connected to the vacuum system. Then, fill row G with 100 µL of the biotinyl-peptide, reseal, and return the plate to 37°C. Return the biotinyl-peptide stock to 4°C. This step is the 4-h incubation of the aggregates with the peptide.
10. Repeat **step 9** for the rows F, E, D, C, B, A at times 125, 65, 50, 35, 20, and 10 min respectively, returning the plate to 37°C after each addition. From the time 50 min, the peptide stock solution is left at room temperature.
11. Five min before the end of the incubation step with the peptide, make a fresh europium-streptavidin solution at 100 ng/mL in europium buffer (a 1/1000 dilution from the purchased stock solution).
12. When the kinetic reaction is complete, stop the extension reaction by washing the plate 3 times with 200 µL of extension buffer.
13. Add 100 µL/well of the europium-streptavidin solution made in **step 11**, and incubate the sealed plate 1 h at room temperature in the dark.
14. Wash the plate 3 times with extension buffer.
15. Add 100 µL/well of Enhancement Solution and incubate the plate 5 min at room temperature in the dark (*see* **Note 12**).
16. Place the plate in the Victor 2 microtiter plate reader, or other suitable time-resolved fluorescence microplate reader, using the programmed parameters for counting europium.
17. Analyze the data (*see* **Note 13**).

4. Notes

1. In addition to the polyGln tract comprising between 5 and 50 Gln residues, the peptides contain a flanking pairs of Lys residues. This design allows the peptides to be charged in the neutral pH region and then enhances their general solubility. The biotinyl-polyGln peptides have the same sequence as the polyGln peptides, except for the addition of a biotin molecule on their N-terminal residue.
2. Solid-phase synthesis of long polyGln sequences can generate significant levels of deletion peptides. For example, analysis by mass spectrometry revealed that the synthetic peptide synthesized as Q_{45} was predominantly a peptide 42 glutamines in length. To get precise, reproducible, and interpretable kinetic data it may be important to purify longer peptides.
3. Nitrogen can be used instead of argon. Nevertheless, we recommend the use of argon, which is a more pure and heavier gas than nitrogen.

4. This procedure must be stringently followed. To ensure reproducibility of results all the steps have to be executed consecutively without any interruption during the process.
5. A convenient amount is about 1 mg of peptide. It should give approx a final stock of 20 mL of aggregates at 50 μg/mL.
6. This step is to ensure that the peptide solution does not contain any residual aggregates even after the TFA/HFIP treatment.
7. Care must be taken not to disturb the pellet, which may not be visible but which can still compromise the results. For this reason, we strongly recommend leaving at least 10% of the supernatant.
8. The HPLC analysis has to be conducted immediately after the centrifugation step to minimize the time the disaggregated peptide stands in concentrated solution.
9. We noticed that 10 nM solutions of disaggregated biotinyl-peptides stored at −80°C develop over time small amounts of aggregated polyGln that give very high backgrounds in the extension assay (*see* **Fig. 2**). The stability of biotinyl-peptides against storage-related aggregation was then tested using the microplate assay (*see* **Subheading 3.5.**). In the absence of deposited aggregates in the wells, the signal given by biotinyl-peptides alone does not increase significantly after 5 h incubation under standard conditions. The magnitude of this signal background (corresponding to 0.2 fmoles biotinyl-peptide) is unchanged for biotinyl peptide stock solutions up to 1 mo at −80°C. After 1 mo, however, backgrounds determined in this manner tend to increase. This increase becomes increasingly significant (up to 5 times higher than the normal background) when biotinyl-peptides are stored at high concentrations (500 nM). For this reason we prepare fresh stocks of biotinyl-peptides at low concentrations (less than 60 nM) every month. Recently, we observed that higher concentration stock solutions of biotinyl-peptides (up to 1 μM) are stable for 2 mo if they are prepared and stored at −80°C in extension buffer with 5% dimethyl sulfoxide (DMSO).
10. After the 17 h drying step, we determined that at least 96% of the aggregates are immobilized to the microplate wells *(27)*.
11. It is possible to stop at this point. The 96-well plate can be filled with 200 μL per well of PBS with 0.05% sodium azide, sealed and stored at 4°C for a period of 5 d.
12. We observed that biotinyl-peptides recruited into polyGln aggregates do not dissociate appreciably even after 5 h at 37°C. This result suggests that there is no significant loss of biotinyl-peptide from aggregates during the incubation with the europium-streptavidin.
13. The addition of enhancement solution releases the europium into solution. Europium counts detected by the microtiter plate reader are converted to fmoles using a standard curve established using a calibrated europium solution obtained from Perkin Elmer. Fmoles of europium are converted into fmoles of deposited biotinyl-peptide using the manufacturer determination of the number of Eu^{3+} ions per streptavidin molecule (7 in the reagent we used). The background signal is subtracted from the extension signal. **Figure 3** shows a representative result of the extension kinetics. The data was fit to two linear portions by manually grouping data points into two sets, each of which was then fit by linear regression. The two-phase kinetic behavior observed in the microplate assay is very similar to that obtained for Aβ deposition by Maggio and collaborators, which they have interpreted in terms of a "dock-and-lock mechanism" *(24)*. According to this mechanism, the most recently added molecules of Aβ are relatively loosely bound and can still dissociate from the fibrils. A slow rearrangement, on the fibril surface, of this loosely bound Aβ to a more tightly bound conformation is required in order to create new binding sites for additional molecules of Aβ The similarity in reaction rate profiles between Aβ and polyGln deposition kinetics reinforces the idea that polyGln aggregates may have an amyloid-like aggregation pathway and substructure *(23,28)*.

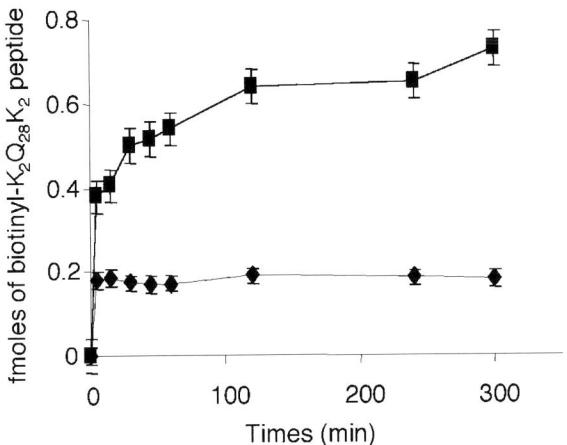

Fig. 2. Influence of storage conditions of the biotinyl-peptide on the signal. 10 nM of the biotinyl-Q_{28} peptide stored at –80°C for either 1 mo (◆) or 2 mo (■) were incubated in wells blocked with 0.3% gelatin but not containing deposited preformed polyGln aggregates. Presumably, aggregates formed on storage of biotinyl-peptide deposit on the plastic to give this high background.

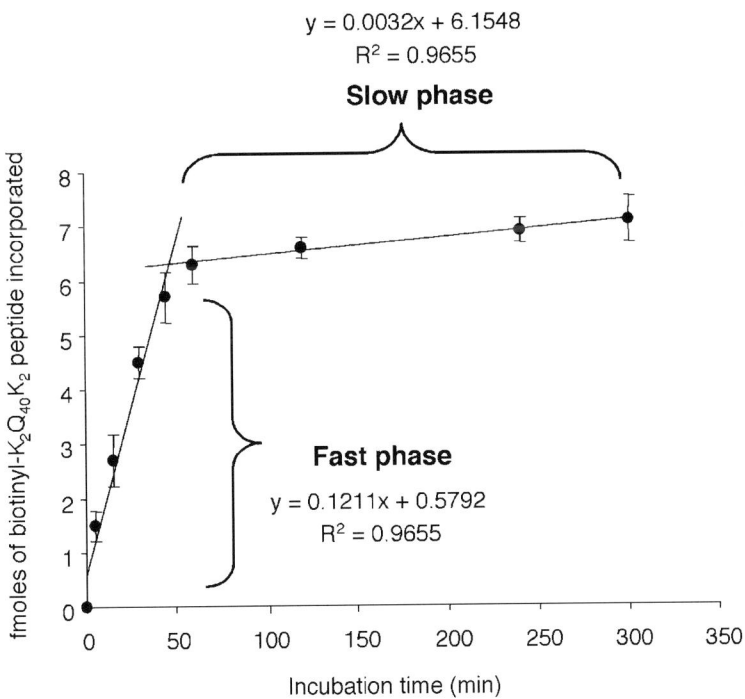

Fig. 3. Kinetic diagram of the extension of polyGln aggregate by the biotinyl-peptide. A 96-well plate was coated with 80 ng/well of polyGln $K_2Q_{40}K_2$ aggregates and incubated with 10 nM of the biotinyl-$K_2Q_{40}K_2$ peptide. The extension of polyGln aggregate is characterized by two phase kinetics: a fast phase, corresponding to the docking of the polyGln peptide, and a slower second phase, which describes a locking step of the polyGln peptide required before another polyGln peptide can bind *(25)*.

Acknowledgments

This work was supported by a Lieberman Award (RW) and a Cure Huntington's Disease Initiative grant (RW), both from the Hereditary Disease Foundation, and by the Lindsay Young Alzheimer's Disease Research Gift Fund (RW).

References

1. Cummings, C. J. and Zoghbi, H. Y. (2000) Fourteen and counting: unraveling trinucleotide repeat diseases. *Hum. Mol. Genet.* **9,** 909–916.
2. Evert, B. O., Wullner, U., and Klockgether, T. (2000) Cell death in polyglutamine diseases. *Cell Tissue Res.* **301,** 189–204.
3. Zoghbi, H. Y. and Orr, H. T. (2000) Glutamine repeats and neurodegeneration. *Annu. Rev. Neurosci.* **23,** 217–247.
4. Hackam, A. S., Singaraja, R., Wellington, C. L., Metzler, M., McCutcheon, K., Zhang, T., et al. (1998) The influence of huntingtin protein size on nuclear localization and cellular toxicity *J. Cell Biol.* **141,** 1097–1105.
5. MacDonald, M. E. and Gusella, J. F. (1996) Huntington's disease: translating a CAG repeat into a pathogenic mechanism. *Curr. Opin. Neurobiol.* **6,** 638–643.
6. Paulson, H. L., Perez, M. K., Trottier, Y., Trojanowski, J. Q., Subramony, S. H., Das, S. S., et al. (1997) Intranuclear inclusions of expanded polyglutamine protein in spinocerebellar ataxia type 3. *Neuron* **19,** 333–344.
7. Davies, S. W., Turmaine, M., Cozens, B. A., Raza, A. S., Mahal, A., Mangiarini, L., and Bates, G. P. (1999) From neuronal inclusions to neurodegeneration: neuropathological investigation of a transgenic mouse model of Huntington's disease *Philos. Trans. R. Soc. Lond. B. Biol. Sci.* **354,** 981–989.
8. DiFiglia, M., Sapp, E., Chase, K. O., Davies, S. W., Bates, G. P., Vonsattel, J. P., and Aronin, N. (1997) Aggregation of huntingtin in neuronal intranuclear inclusions and dystrophic neurites in brain. *Science* **277,** 1990–1993.
9. Yamada, M., Tsuji, S., and Takahashi, H. (2000) Pathology of CAG repeat diseases. *Neuropathology* **20,** 319–325.
10. Becker, M., Martin, E., Schneikert, J., Krug, H. F., and Cato, A. C. (2000) Cytoplasmic localization and the choice of ligand determine aggregate formation by androgen receptor with amplified polyglutamine stretch. *J. Cell. Biol.* **149,** 255–262.
11. Holmberg, M., Duyckaerts, C., Durr, A., Cancel, G., Gourfinkel-An, I., Damier, P., et al. (1998) Spinocerebellar ataxia type 7 (SCA7): a neurodegenerative disorder with neuronal intranuclear inclusions. *Hum. Mol. Genet.* **7,** 913–918.
12. Li, S. H. and Li, X. J. (1998) Aggregation of N-terminal huntingtin is dependent on the length of its glutamine repeats. *Hum. Mol. Genet.* **7,** 777–782.
13. Harper, P. S. and Newcombe, R. G. (1992) Age at onset and life table risks in genetic counselling for Huntington's disease. *J. Med. Genet.* **29,** 239–242.
14. Schelhaas, H. J., Ippel, P. F., Hageman, G., Sinke, R. J., van der Laan, E. N., and Beemer, F. A. (2001) Clinical and genetic analysis of a four-generation family with a distinct autosomal dominant cerebellar ataxia. *J. Neurol.* **248,** 113–120.
15. Abel, A., Walcott, J., Woods, J., Duda, J., and Merry, D. E. (2001) Expression of expanded repeat androgen receptor produces neurologic disease in transgenic mice. *Hum. Mol. Genet.* **10,** 107–116.
16. Ikeda, H., Yamaguchi, M., Sugai, S., Aze, Y., Narumiya, S., and Kakizuka, A. (1996) Expanded polyglutamine in the Machado-Joseph disease protein induces cell death in vitro and in vivo. *Nat. Genet.* **13,** 196–202.

17. Li, H., Li, S. H., Cheng, A. L., Mangiarini, L., Bates, G. P., and Li, X. J. (1999) Ultrastructural localization and progressive formation of neuropil aggregates in Huntington's disease transgenic mice. *Hum. Mol. Genet.* **8,** 1227–1236.
18. Mangiarini, L., Sathasivam, K., Seller, M., Cozens, B., Harper, A., Hetherington, C., et al. (1996) Exon 1 of the HD gene with an expanded CAG repeat is sufficient to cause a progressive neurological phenotype in transgenic mice. *Cell* **87,** 493–506.
19. Satyal, S. H., Schmidt, E., Kitagawa, K., Sondheimer, N., Lindquist, S., Kramer, J. M., and Morimoto, R. I. (2000) Polyglutamine aggregates alter protein folding homeostasis in Caenorhabditis elegans. *Proc. Natl. Acad. Sci. USA* **97,** 5750–5755.
20. McCampbell, A., Taylor, J. P., Taye, A. A., Robitschek, J., Li, M., Walcott, J., et al. (2000) CREB-binding protein sequestration by expanded polyglutamine. *Hum. Mol. Genet.* **9,** 2197–2202.
21. Nucifora, F. C., Jr., Sasaki, M., Peters, M. F., Huang, H., Cooper, J. K., Yamada, M., et al. (2001) Interference by huntingtin and atrophin-1 with cbp-mediated transcription leading to cellular toxicity. *Science* **291,** 2423–2428.
22. Steffan, J. S., Kazantsev, A., Spasic-Boskovic, O., Greenwald, M., Zhu, Y. Z., Gohler, H., et al. (2000) The Huntington's disease protein interacts with p53 and CREB-binding protein and represses transcription. *Proc. Natl. Acad. Sci. USA* **97,** 6763–6768.
23. Scherzinger, E., Sittler, A., Schweiger, K., Heiser, V., Lurz, R., Hasenbank, R., et al. (1999) Self-assembly of polyglutamine-containing huntingtin fragments into amyloid-like fibrils: implications for Huntington's disease pathology. *Proc. Natl. Acad. Sci. USA* **96,** 4604–4609.
24. Esler, W. P., Stimson, E. R., Jennings, J. M., Vinters, H. V., Ghilardi, J. R., Lee, J. P., et al. (2000) Alzheimer's disease amyloid propagation by a template-dependent dock-lock mechanism. *Biochemistry* **39,** 6288–6295.
25. Chen, S. and Wetzel, R. (2001) Solubilization and disaggregation of polyglutamine peptides. *Protein Sci.* **10,** 887–891.
26. Chen, S., Berthelier, V., Wang, W., and Wetzel, R. (2001) Polyglutamine aggregation behavior in vitro supports a recruitment mechanism of cytotoxicity. *J. Mol. Biol.* **311,** 173–182.
27. Berthelier, V., Hamilton, J. B., Chen, S., and Wetzel, R. (2001) A microtiter plate assay for polyglutamine aggregate extension. *Anal. Biochem.* **295,** 227–236.
28. McGowan, D. P., van Roon-Mom, W., Holloway, H., Bates, G. P., Mangiarini, L., Cooper, G. J., et al. (2000) Amyloid-like inclusions in Huntington's disease. *Neuroscience* **100,** 677–680.

28

Characterization of Prion Proteins

Wenquan Zou, Monica Colucci, Pierluigi Gambetti, and Shu G. Chen

1. Introduction

Prion disease represents a group of transmissible neurodegenerative disorders that include scrapie in sheep and goats, bovine spongiform encephalopathy (BSE), and Creutzfeldt-Jakob disease (CJD), Gerstmann-Sträussler-Scheinker syndrome, fatal familial insomnia, and kuru in humans. The disease is characterized clinically by ataxia, dementia, and myoclonus as well as pathologically by spongiosis, astrocytic gliosis, and neuronal loss (reviewed in **refs.** *1* and *2*). CJD occurs mostly (about 85% of all cases) as sporadic cases (unknown etiology), with the rest of the cases being familial (owing to inheritance of mutations in the prion protein gene) or iatrogenic (due to accidental transmission during medical procedure). Recently, a new variant form of CJD (vCJD) has emerged in the United Kingdom that may originate from the BSE epidemic *(3)*.

According to the "protein only" hypothesis *(1,4)*, prion disease is caused by abnormalities in the structure or conformation of the prion protein (PrP). PrP include two isoforms, the normal cellular form (PrP^C), and the pathogenic isoform (PrP^{Sc}) present in patients with prion disease. PrP^C is a 209-residue glycoprotein with a single disulfide bond, two N-glycosylation sites, and a glycosyl phosphatidylinositol anchor. While it is expressed in a number of tissues, the highest amount of PrP^C is found in the central nervous system (CNS). Its function is unknown but may be involved in the copper metabolism and signal transduction *(5,6)*. The conversion of PrP^C to PrP^{Sc} is believed to be the key molecular event in the pathogenesis of prion disease, which involves a conformational transition of α-helical structure to β-sheet structure *(1,4)*. Although PrP^C and PrP^{Sc} share almost identical primary structure *(7)*, the two isoforms differ in conformation and physicochemical properties. PrP^C is soluble in aqueous solutions and can be readily degraded by proteinase K (PK), whereas PrP^{Sc} is insoluble and shows partial resistance to PK under the same digestion conditions *(8–10)*. These unique properties of PrP^{Sc} underlie much of the current methodology for its characterization and the diagnosis of prion disease.

Our previous work has provided evidence for the existence of two subtypes of PrP^{Sc} differing in the size of the PK-resistant core fragments and the extent of glycosylation *(11,12)*. Treatment of human PrP^{Sc} with PK generates three core fragments representing PrP with no, one or two N-linked glycans. We have previously identified two distinct conformers of PrP^{Sc} that are associated with different phenotypes of human prion disease *(11–14)*. Type 1 PrP^{Sc} is characterized by three PK-resistant fragments, which upon deglycosylation migrate at ~21 kDa. In contrast, type 2 PrP^{Sc} yields a ~19 kDa protein under the identical experimental conditions. The difference in size of the PK-resistant fragments results from differential PK cleavage within the N terminal regions of PrP^{Sc} conformers, reflecting different conformations between type 1 and type 2 PrP^{Sc}. To analyze the origin of the PrP^{Sc} heterogeneity, we have determined the precise PK cleavage sites of type 1 and type 2 PrP^{Sc} by the N-terminal sequencing and mass spectrometry *(15)*. We demonstrated two primary cleavage sites at residue 82 and residue 97 for type 1 and type 2 PrP^{Sc}, respectively, and numerous secondary cleavages distributed along the region spanning residues 74–102. Characterization of different subtypes of PrP^{Sc} has been shown to provide a more reliable diagnosis for a variety of disease phenotypes *(12–14)*.

In this chapter we outline several widely used techniques for the characterization of PrP^{Sc} and diagnosis of prion diseases. We describe, in detail, the experimental procedures, reagents and equipments required, reagent preparation, and interpretation of results. These protocols are mainly used for the detection of PrP^{Sc} in brain tissue of clinically suspected cases of human prion disease. The protocols may also be applied to animal prion disease such as BSE, chronic wasting disease, and transmissible mink encephalopathy. In addition, we also make a brief introduction to some promising assays reported in the more recent literature.

1.1. Western Blotting

Definitive diagnosis of prion disease is usually accomplished by the detection of the PK-resistant PrP^{Sc} in brain tissues *(16,17)*. The most convenient method for this purpose is Western blot analysis using an antibody against PrP after digestion of brain homogenate with PK. Several antibodies against PrP including the widely used 3F4 antibody *(18)* are now commercially available. The immunoassay has shown an excellent sensitivity, ease in interpretation, and accuracy in definitive diagnosis of CJD with small brain biopsy specimens. *(19,20)*. The detection of either type 1 or type 2 PrP^{Sc}, in combination with the PrP genotype (mutation and polymorphism) and histological lesion profiles, provides an accurate diagnosis and classification of different variants of human prion disease *(21)*.

1.2. Purification of Human PrP^{Sc} for Structural Characterization

For detailed structural studies of multiple PrP^{Sc} subtypes in different disease phenotypes, it is necessary to obtain sufficient amounts of purified PrP^{Sc} from diseased brains. Due to its relative insolubility in aqueous solutions and resistance to protease digestion, PrP^{Sc} can be readily separated from PrP^{C} and other proteins using cycles of detergent extraction, ultracentrifugation, and PK digestion *(15,22,23)*. Various chemical and spectroscopic methods can be used to study biochemical and conformational properties of purified PrP^{Sc}. These include: 1) analysis of primary structure and glycosylation of PrP^{Sc} using enzymatic digestion, protein sequencing, and mass spectrometry *(7,24)*; 2) analysis of the conformational difference between PrP^{Sc} subtypes through the mapping of cleavage sites by PK using protein sequencing and mass spectrometry *(14,15,25)*; and 3) analysis of secondary structure of PrP^{Sc} using circular dichroism and

Fourier transform infrared spectroscopy *(26–29)*. Taken together, these investigations have provided strong evidence for the existence of distinct PrP^{Sc} conformations associated with different disease phenotypes.

1.3. Immunohistochemistry

For diagnosis purposes, Western blotting is of sufficient sensitivity for the detection of PrP^{Sc} in the brain. However, this method provides the lowest anatomical resolution. Immunohistochemistry is capable of providing information on the PrP^{Sc} distribution in various brain structures, and it does not require fresh tissue. Its sensitivity, however, is quite low compared to the Western blotting. Special processing steps are also required to reveal optimal PrP^{Sc} immunoreactivity in fixed tissues. Treatments of tissue sections with formic acid and hydrolytic autoclaving have been used to reduce the background immunoreactivity of PrP^{C} *(30)*. Recently, this immunohistochemical method has also been used to detect PrP^{Sc} immunoreactivity in lymphoreticular tissues such as appendix and tonsil from patients with vCJD *(31,32)*.

1.4. Recent Developments

Several methods based on the enzyme-linked immunosorbent assay (ELISA) using various antibodies against PrP have been recently developed *(33)*. The ELISA format has the advantage of greater sensitivity, better quantitation, and potential for high throughput and automation. In addition, fluorescence-correlated spectroscopy has been used to detect PrP aggregates in cerebrospinal fluid *(34)*. Signal amplification of PK-resistant PrP by cyclic sonication has been reported to increase the detection limit *(35)*.

1.5. Biosafety Considerations

PrP^{Sc} is an essential component of the transmissible pathogen causing prion disease *(1)*. Kuru and iatrogenic CJD are the result of accidental transmission of the disease between human beings. Occurrence of vCJD in the UK has raised the concern that prion disease may also be transmitted from animals to humans through consumption of contaminated foodstuffs *(3)*. Normal social or routine clinical contact with affected patients does not present a risk to health care workers and relatives since human prion disease is not communicable or contagious. However, the potential risk arising from the handling brain tissues of patients with prion disease cannot be ruled out. In addition, it is possible that tissue samples may contain other pathogens. Therefore, strict precautions are necessary to minimize the potential exposure to such pathogens.

In general, brain tissues should be handled in a Class II Biological Safety Cabinet in a specialized Biosafety Level 2 or 3 facility according to the current guidelines *(36,37)*. Skilled research personnel performing the experiments must wear protective clothing including protective gowns, gloves, and face shields (or goggles). All protective clothing should be discarded for incineration. The disposable materials such as gloves, aprons, sleeves, tips, and tubes are discarded for incineration. The reusable equipment should be pretreated with 2 N NaOH for 1 h, followed by autoclaving at 134°C for 2 h or longer *(36–38)*. Contaminated liquids should be treated with 2 N NaOH (final concentration) followed by autoclaving.

2. Materials

2.1. Western Blotting

1. Tissue samples (*see* **Note 1**).
2. 2X lysis buffer: 20 mM Tris-HCl, pH 7.5, 20 mM EDTA, 0.5% Nonidet P-40, 0.5% Na deoxycholate, and 200 mM NaCl. Store at −20°C. Dilute to 1X lysis buffer prior to use.
3. Disposable pellet pestles and matched tubes (Fisher).
4. 0.55 mg/mL Proteinase K (PK) in distilled water. Store in 50 µL aliquots at −80°C.
5. 100 mM Pefabloc SC (Roche). Store at −20°C.
6. 2X SDS sample buffer: 125 mM Tris-HCl, pH 6.8, 4 mM EDTA, 6% SDS, 20% glycerol, 8% β-mercaptolethanol, and a trace of dye (either bromophenol or pyronine Y).
7. 10X Electrode Stock: 576.4 g glycine, 121.14 g Tris. Make up to 4 L with distilled water. Store at 4°C for several mo.
8. 1X Electrode buffer: 400 mL 10X Electrode stock, 40 mL 10% SDS. Make up to 4 L with distilled water.
9. 1X Transfer buffer: 400 mL 10X Electrode stock, 800 mL Methanol. Make up to 4 L with distilled water.
10. Tris-buffered saline (TBS): 0.9% (w/v) NaCl, 100 mM Tris-HCl, pH 7.6.
11. TBS-T: 0.1% (v/v) Tween-20 in TBS.
12. Blocking buffer: 5% non-fat dried milk in TBS-T.
13. Anti-PrP primary antibodies such as 3F4 (Chemicon), working dilution 1:50,000 in blocking buffer.
14. Horsenadish peroxidase (HRP)-conjugated anti-mouse secondary antibody such as HRP-conjugated sheep anti-mouse IgG (Amersham), working dilution 1:3,000 in blocking buffer.
15. ECL plus Western blotting detection reagents (Amersham).
16. Kodak scientific imaging film X-OMAT AR, and Kodak film cassette.
17. Premade 12% SDS-polyacrymide gels and electrophoresis apparatus (Novex).
18. Apparatus for electroblotting (Bio-Rad).

2.2. Purification of Human PrP^{Sc}

1. Tissue grinder (Fisher).
2. 2X TEND solution: 20 mM Tris-HCl, pH 8.0, 2 mM EDTA, 1.5% NaCl, and 2 mM DTT.
3. TNSS solution (10 mL): 5 mL of 2X TEND, 1.0 g NaCl, and 0.5 mL 20% Sarkosyl. Make up to 10 mL with distilled water.
4. 20% (w/v) N-lauroylsarcosine sodium salt (Sarkosyl).
5. Ultracentrifuge (Beckman).
6. Tris-buffered saline (TBS): 0.9% (w/v) NaCl, 100 mM Tris-HCl, pH 7.6.
7. 10 mg/mL RNase A (Sigma). Store at −20°C.
8. 15 mg/mL DNase I (Sigma). Store at −20°C.
9. 20 mM Tris, pH 7.6.
10. 200 mM dithiothreitol (DTT).

2.3. Immunohistochemistry

1. 98% formic acid.
2. 10% buffered Formalin.
3. 60%, 80%, 95%, 100% ethyl alcohol.
4. 100% Methanol.
5. Hydrogen peroxide 30% (H_2O_2).
6. TBS.
7. Normal goat serum (NGS) (Vector Laboratories).

Characterization of Prion Proteins

8. 3F4 monoclonal antibody (MAb) to PrP (Chemicon).
9. Goat anti-mouse IgG (1:50 in 1% NGS).
10. Mouse peroxidase anti-peroxidase (MPAP, 1:250 in 1%NGS).
11. Dimethylaminoazobenzene (DAB).
12. HCl (60 mM).
13. Hematoxylin.
14. Safetysolve (IMEB Inc.).
15. Microwave Pressure Cooker (NordicWare, Tender Cooker 2 $\frac{1}{2}$ quart).
16. Microwave (Kenmore, 2.0 cu ft, 1100 watts).
17. Microwavable Staining dishes (IMEB Inc.).

3. Methods

3.1. Western Blotting

3.1.1. Sample Preparation (see **Note 2**).

1. Transfer 40–80 mg of brain tissue to a 1.5-mL plastic tube.
2. Add 9 volumes of 1X lysis buffer.
3. Homogenize tissue using a disposable pellet pestle with matched tube (Fisher).
4. Treat 20 µL of brain homogenate from a test case with 2 µL of 0.55 mg/mL proteinase K (final concentration of 50 µg/mL) at 37°C for 1 h. Terminate digestion reaction by adding 1 µL of 100 mM Pefablock, 23 µL of 2X SDS sample buffer, and boiling for 10 min in a dry bath.
5. For untreated control sample: Take 10 µL brain homogenate from a control case. Add 13 µL 1X lysis buffer, and 23 µL 2X SDS sample buffer. Boiling for 10 min in a dry bath.
6. Store unused brain homogenates in a secure freezer at –80°C.

3.1.2. Electrophoresis, Protein Blotting, and Immunodetection

1. Load 30 µL boiled samples onto 12% Tris-glycine gel. Run electrophoresis at 120 V for about 1.5–2.0 h until the dye reaches the bottom of the gel. Following electrophoresis, take out the gel.
2. Electrotransfer proteins from the gel onto the Immobilon P membrane (Millipore) using Bio-Rad at electroblottig apparatus at 70 V for 2 h at 4°C.
3. Block membrane with 10% non-fat dry milk in TBS-T for 1 h at 37°C
4. Incubate the membrane with the primary antibody 3F4 (diluted 1:50,000) for 2 h at room temperature.
5. Wash the membrane with T-TBS 4 times, 12 min each.
6. Incubate the secondary antibody, diluted 1:4,000 in the antibody dilution buffer, for 1 h at room temperature.
7. Wash the membrane with T-TBS 4 times, 12 min each.
8. Add 50 µL ECL solution B to 2 mL of the ECL solution A, mix well, then put the mixture on the membrane and incubate for 5 min.
9. Expose the membrane to the Kodak Scientific Imaging Film. Develop the films in the Kodak Film Processor
10. Examine the film for the presence of PK-resistant PrP^{Sc} bands (**Fig. 1**) and interpret the results accordingly (*see* **Note 3**).

3.2. Purification of PrP^{Sc} for Structural Characterization

1. All procedures should be performed in a Biosafety Level 2 or 3 facility (*see* **Subheading 1.5.**).
2. Add 9.0 mL of 2 X TEND solution to ~5 g of brain tissue (gray matter).

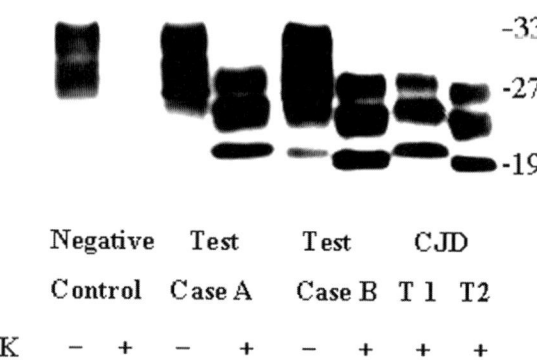

Fig. 1. Detection of PrP^{Sc} subtypes in suspected cases of human prion disease. Brain homogenates were prepared from a non-CJD case (negative control), two cases of suspected CJD (test case A and test case B), one case of CJD with type 1 PrP^{Sc} (CJD T1), and one case of CJD with type 2 PrP^{Sc} (CJD T2). Samples were either untreated (PK−) or treated with PK (PK+), then denatured in SDS sample buffer and run on SDS-PAGE gel (12%). Detection of PrP bands were made on Western blots using the anti-PrP monoclonal antibody 3F4. The positions of molecular weight markers are indicated on the right (in kDa). In the PK-untreated samples (lanes 1, 3 and 5), PrP bands are broad with a molecular weight between 25–35 kDa. After treatment with PK, no bands are detected in the negative control (lane 2), while three bands of PK resistant PrP^{Sc} are present in test case A (lane 4), test case B (lane 6), and CJD cases with either type 1 PrP^{Sc} (lane 7) or type 2 PrP^{Sc} (lane 8). The lowest band of type 1 PrP^{Sc} (lane 7) migrates at ~21 kDa while that of type 2 PrP^{Sc} (lane 8) migrates at ~19 kDa. Judging from the positions of the lowest PrP^{Sc} band in test case A (lane 4, ~21 kDa) and test case B (lane 6, ~19 kDa), it is clear that test case A contains type 1 PrP^{Sc} while test case B is positive for type 2 PrP^{Sc}.

3. Homogenize in a tissue grinder with 15 up and down strokes on ice.
4. Transfer the homogenate into a 50-mL culture tube.
5. Add 9.5 mL of 20% Sarkosyl to the homogenate slowly. Freeze (on dry ice) and thaw (by sonication) the mixture 3–4 times to reduce viscosity. Centrifuge at 20,000g and 4°C for 25 min.
6. Discard the pellet (P1) and centrifuge the supernatant (S1) at 200,000g and 4°C for 120 min.
7. Resuspend the pellet (P2) in 1 mL of TNSS by sonication. Centrifuge at 200,000g and 18°C for 120 min.
8. Resuspend P3 in 750 μL TBS, sonicate at 4°C, then add 7 μL of 0.5 M CaCl$_2$ and 0.5 M MgCl$_2$, 7 μL of 10 mg/mL RNase A and 1 μL 14.6 mg/mL DNase I, incubate the mixture at 4°C overnight.
9. Add the following reagents: 100 μL of 300 mM EDTA, 560 μL 20 mM Tris, pH 7.6; 8 μL of 200 mM DTT and 0.15 g NaCl. Centrifuge at 200,000g and 18°C for 120 min.
10. Resuspend the pellet (P4) in 375 μL of TNSS, and incubate with proteinase K at 50 μg/mL and 37°C for 1 h. Then add 4 μL of 100 mM Pefablock to terminate the digestion. Add 375 μL of TNSS and centrifuge at 200,000g and 18°C for 120 min.

11. Resuspend the pellet (P5) in 750 µL of TNSS and centrifuge again at 200,000g at 18°C for 120 min.
12. Resuspend pellet (P6) in a small volume of an appropriate buffer. Save this preparation at −80°C until used for further structural studies (*see* **Note 4**).

3.3. Immunohistochemistry

1. Cut paraffin-embedded tissue sections at 8–10 µm on charged slides. Dry overnight at 60°C.
2. Deparaffinize sections through two 5 min changes of Safetysolve, then 3 min changes in 100%, 95%, 80% and finally 60% ethyl alcohol.
3. Immerse in Methanol/8% H_2O_2 solution for 20 min and rinse in H_2O.
4. Microwave in 250 mL of 60 m*M* solution of HCl in a microwavable slide dish using a microwave pressure cooker that has 600 mL of distilled water added (*see* **Note 5**).
5. Rinse in H_2O and treat in 98% formic acid for 10 min if not previously treated and rinse thoroughly in H_2O.
6. Immerse in 2 changes of TBS solution at 15-min intervals.
7. Incubate with 10% NGS for 10 min.
8. Blot section and incubate with 3F4 (1:200 in 1% NGS) overnight at 4°C.
9. Repeat **step 6**.
10. Repeat **step 7**.
11. Blot section and incubate with goat anti-mouse IgG (1:50 in 1% NGS) for 60 min.
12. Repeat **step 6**.
13. Repeat **step 7**.
14. Blot section and incubate with MPAP (1:250 in 1% NGS) for 1 h.
15. Repeat **step 6**.
16. Develop in DAB solution for 1–3 min or until staining is visualized.
17. Stain in hematoxylin (1 dip), wash, 2 dips in ammonium hydroxide, wash and dehydrate sections through 70%, 95%, 100% ethyl alcohol for 3 min each, then two 5 min changes in Safetysolve.
18. Mount the section with a coverslip. Examine the section under a microscope (*see* **Note 6**).

4. Notes

1. Brain tissues are obtained at autopsy or biopsy from patients with prion disease and are kept frozen at −80°C until use. The tissue for PrP^{Sc} detection by Western blotting should not be treated and especially not be fixed in formalin or other fixatives. Tissue from more than one brain region should be sampled whenever available. Tissue samples for processing should contain mostly the gray matter after dissecting off white matter.
2. Brain samples must be processed in the biosafety hoods up to boiling samples in SDS sample buffer that will denature PrP^{Sc} and greatly diminish the associated infectivity.
3. In our experience, the possibility of false positive is remote and can be minimized by including the non-CJD cases in the assay. However, there is a possibility of false negative if the biopsy from suspected CJD cases is taken from an unaffected brain region, or if the tissue is fixed prior to Western blotting.
4. The purified PrP^{Sc} preparations may be further characterized using a variety of biochemical and biophysical methods as outlined in **Subheading 1.2.**
5. One example is given here for performing hydrolytic autoclaving in a microwave pressure cooker (NordicWare) using a Kenmore microwave. First, place slides in slide dish containing 60 m*M* HCl, which are then put in the pressure cooker. Place the cooker in a microwave (Kenmore) at 100% power until the pressure regulator starts to rock back and forth. Reduce the microwave power at 30% and microwave for 15 min. Remove the cooker

from the microwave. Once cooker's pressure has vented, remove cover and let slides remain in the cooker for 20 min to cool. Under these conditions, the hydrolytic treatment presumably destroys immunoreactivity of PrP^C while preserving that of PrP^{Sc} *(39,40)*.

6. PrP^{Sc} deposition visualized by DAB may appear with various patterns. These patterns include: synaptic, perivacuolar, plaque-like deposits, kuru plaques, or laminar depending on the subtypes of sporadic CJD *(30,39)*. Other patterns may be observed in familial prion disease and in acquired prion disease such as vCJD *(40)*.

Acknowledgments

This work was supported in part by National Institutes of Health, and Center for Disease Control and Prevention. The authors wish to thank Taocong Jin, Diane Kofsky, Manuela Pastore, Phyllis Scalzo, and Zhiliang Xie for technical assistance.

References

1. Prusiner, S. B. (1991) Molecular biology of prion diseases. *Science* **252**, 1515–1522.
2. Prusiner, S. B. (1997) Prion diseases and the BSE crisis. *Science* **278**, 245–251.
3. Hill, A. F., Desbruslais, M., Joiner, S., Sidle, K. C. L., Gowland, I., Collinge, J., et al. (1997) The same prion strain causes vCJD and BSE. *Nature* **389**, 448–450.
4. Cohen, F. E. and Prusiner, S. B. (1998) Pathologic conformations of prion proteins. *Annu. Rev. Biochem.* **67**, 793–819.
5. Brown, D. R., Qin, K., Herms, J. W., Madlung, A., Manson, J., Strome, R., Fraser, P. E., et al. (1997) The cellular prion protein binds copper in vivo. *Nature* **390**, 684–687.
6. Mouillet-Richard, S., Ermonval, M., Chbassier, C., Laplanche, J. L., Lehmann, S., Launay, J. M., and Kellermann, O. (2000) Signal transduction through prion protein. *Science* **289**, 1925–1928.
7. Stahl, N., Baldwin, M. A., Teplow, D. B., Hood, L., Gibson, B. W., Burlingame, A. L., and Prusiner, S. B. (1993) Structural studies of the scrapie prion protein using mass spectrometry and amino acid sequencing. *Biochemistry* **32**, 1991–2002
8. Meyer, R. K., McKinley, M. P., Bowman, K. A., Braunfeld, M. B., Barry, R. A., and Prusiner, S. B. (1986). Separation and properties of cellular and scrapie prion proteins. *Proc. Natl. Acad. Sci. USA* **83**, 2310–2314.
9. Bassen, R. A. and Marsh, R. F. (1994) Distinct PrP properties suggest the molecular basis of strain variation in transmissible mink encephalopathy. *J. Virol.* **68**, 7859–7868.
10. Chen, S. G., Teplow, D. B., Parchi, P., Teller, J. K., Gambetti, P., and Autilio-Gambetti, L. (1995) Truncated forms of the human prion protein in normal brain and in prion diseases. *J. Biol. Chem.* **270**, 19173–19180.
11. Monari, L., Chen, S. G., Brown, P., Parchi, P., Petersen, R. B., Mikol, J., Gray, F., et al. (1994) Fatal familial insomnia and familial Creutzfeldt-Jakob disease: different prion proteins determined by a DNA polymorphism. *Proc. Natl. Acad. Sci. USA* **91**, 2839–2842.
12. Parchi, P., Castellani, R., Capellari, S., Ghetti, B., Young, K., Chen, S. G., et al. (1996) Molecular basis of phenotypic variability in sporadic Creutzfeldt-Jakob disease. *Ann. Neurol.* **39**, 767–778.
13. Parchi, P., Capellari, S., Chen, S. G., Petersen, R. B., Gambetti, P., Kopp, N., et al. (1997) Typing prion isoforms. *Nature* **386**, 232–234.
14. Parchi, P., Chen, S. G., Brown, P., Zou, W., Capellari, S., Budka, H., et al. (1998) Different patterns of truncated prion protein fragments correlate with distinct phenotypes in P102L Gerstmann-Sträussler-Scheinker disease. *Proc. Natl. Acad. Sci. USA* **95**, 8322–8327.

15. Parchi, P., Zou, W., Wang, W., Brown, P., Capellari, S., Ghetti, B., et al. (2000) Genetic influence on the structural variations of the abnormal prion protein. *Proc. Natl. Acad. Sci. USA* **97,** 10,168–10,172.
16. McKinley, M. P., Bolton, D. C., and Prusiner, S. B. (1983) A protease-resistant protein is a structural component of the scrapie prion. *Cell* **35,** 57–62.
17. Brown, P., Coker-Vann, M., Pomeroy, K., Franko, M., Asher, D. M., Gibbs, C. J., Jr., and Gajdusek, D. C. (1986) Diagnosis of Creutzfeldt-Jakob disease by Western blot identification of marker protein in human brain tissue. *N. Engl. J. Med.* **314,** 547–551.
18. Kascsak, R. J., Rubenstein, R., Merz, P. A., Tonna-DeMasi, M., Fersko, R., Carp, R. I., et al. (1987) Mouse polyclonal and monoclonal antibody to scrapie-associated fibril proteins. *J. Virol.* **61,** 3688–3693.
19. Castellani, R., Parchi, P., Stahl, J., Capellari, S., Cohen, M., and Gambetti, P. (1996) Early pathologic and biochemical changes in Creutzfeldt-Jakob disease: study of brain biopsies. *Neurology* **46,** 1690–1693.
20. Castellani, R. J., Parchi, P., Madoff, L., Gambetti, P., and McKeever, P. (1997) Biopsy diagnosis of Creutzfeldt-Jakob disease by western blot: a case report. *Hum. Pathol.* **28,** 623–626.
21. Parchi, P., Giese, A., Capellari, S., Brown, P., Schulz-Schaeffer, W., Windl, O., et al. (1999) Classification of sporadic Creutzfeldt-Jakob disease based on molecular and phenotypic analysis of 300 subjects. *Ann. Neurol.* **46,** 224–233.
22. Bolton, D. C., Bendheim, P. E., Marmorstein, A. D., and Potempska, A. (1987) Isolation and structural studies of the intact scrapie agent protein. *Arch. Biochem. Biophys.* **258,** 579–590.
23. Chen, S. G., Parchi, P., Brown, P., Capellari, S., Zou, W., Cochran, E. J., et al. (1997) Allelic origin of the abnormal prion protein isoform in familial prion diseases. *Nat. Med.* **3,** 1009–1015.
24. Stimson, E., Hope, J., Chong, A., and Burlingame, A. L. (1999). Site-specific characterization of the N-linked glycans of murine prion protein by high-performance liquid chromatography/electrospray mass spectrometry and exoglycosidase digestions. *Biochemistry* **38,** 4885–4895.
25. Chen, S. G., Zou, W., Parchi, P., and Gambetti, P. (2000) PrP^{Sc} typing by N-terminal sequencing and mass spectrometry. *Arch. Virol.* **16(Suppl),** 209–216.
26. Caughey, B. W., Dong, A., Bhat, K. S., Ernst, D., Hayes, S. F., and Caughey, W. S. (1991) Secondary structure analysis of the scrapie-associated protein PrP 27-30 in water by infrared spectroscopy. *Biochemistry* **30,** 7672–7680.
27. Pan, K. M., Baldwin, M., Nguyen, J., Gasset, M., Serban, A., Groth, D., et al. (1993) Conversion of α-helices into β-sheets features in the formation of he scrapie prion proteins. *Proc. Natl. Acad. Sci. USA* **90,** 10,962–10,966.
28. Safar, J., Roller, P. P., Gajdusek, D. C., and Gibbs, C. J., Jr. (1993) Thermal stability and conformational transitions of scrapie amyloid (prion) protein correlate with infectivity. *Protein Sci.* **2,** 2206–2216.
29. Caughey, B., Raymond, G. J., and Bessen, R. A. (1998) Strain-dependent differences in beta-sheet conformations of abnormal prion protein. *J. Biol. Chem.* **273,** 32,230–32,235.
30. Kitamoto, T., Shin, R. W., Doh-ura, K., Tomokane, N., Miyazono, M., Muramoto, T., and Tateishi, J. (1992) Abnormal isoform of prion proteins accumulates in the synaptic structures of the central nervous system in patients with Creutzfeldt-Jakob disease. *Am. J. Pathol.* **140,** 1285–1294.
31. Hilton, D. A., Fathers, E., Edwards, P., Ironside, J. W., and Zajicek, J. (1998). Prion immunoreactivity in appendix before clinical onset of variant Creutzfeldt-Jakob disease. *Lancet* **352,** 703–704.

32. Hill, A. F., Butterworth, R. J., Joiner, S., Jackson, G., Rossor, M. N., Thomas, D. J., et al. (1999). Investigation of variant Creutzfeldt-Jakob disease and other human prion diseases with tonsil biopsy samples. *Lancet* **353,** 183–189.
33. European Commission (1999) The evaluation of tests for the diagnosis of transmissible spongiform encephalopathy in bovines. (*see* Website: http://europa.eu.int/comm/food/fs/bse12 en.html)
34. Bieschke, J., Giese, A., Schulz-Schaeffer, W., Zerr, I., Poser, S., Eigen, M., and Kretzschmar, H. (2000). Ultrasensitive detection of pathological prion protein aggregates by dual-color scanning for intensely fluorescent targets. *Proc. Natl. Acad. Sci. USA* **97,** 5468–5473.
35. Saborio, G. P., Permanne, B., and Soto, C. (2001). Sensitive detection of pathological prion protein by cyclic amplification of protein misfolding. *Nature* **411,** 810–813.
36. Center for Disease Control and Prevention (1999). Biosafety in Microbiological and Biomedical Laboratories (BMBL), 4th Ed., Section VII-D: Prions. (*see* Website: http://www.cdc.gov/od/ohs/biosfty/bmbl4/bmbl4s7d.htm)
37. Baron, H., Safar, J., Groth, D., DeArmond, S. J., and Prusiner, S. B. (1999) Biosafety issues in prion diseases, in *Prion Biology and Diseases*. (Prusiner, S. B., ed.), Cold Spring Harbor Laboratory Press, Cold Spring Harbor, NY, USA, pp. 743–777.
38. Tateishi, J., Tashima, T., and Kitamoto, T. (1991) Practical methods for chemical inactivation of CJD pathogen. *Microbiol. Immunol.* **35,** 163–166.
39. Budka, H., Aguzzi, A., Brown, P., Brucher, J. M., Bugiani, O., Gullotta, F., et al. (1995) Neuropathological diagnostic criteria for Creutzfeldt-Jakob disease (CJD) and other human spongiform encephalopathies (prion diseases). *Brain Pathol.* **5,** 459–466.
40. Ironside, J. W., Head, M. W., Bell, J. E., McCardle, L., and Will, R. G. (2000) Laboratory diagnosis of variant Creutzfeldt-Jakob disease. *Histopathology* **37,** 1–9.

29

Detection of *NF1* Mutations Utilizing the Protein Truncation Test (PTT)

Meena Upadhyaya, Michael Osborn, and David N. Cooper

1. Introduction

The protein truncation test (PTT) *(1–3)*, also known as the in vitro coupled transcription/translation synthesis assay, was designed as a tool to detect mutations that lead to premature translational termination. It was originally developed to screen for mutations in the dystrophin (*DMD*) gene causing Duchenne/Becker muscular dystrophies. The ability of the PTT to detect mutations at the protein level offers various advantages over other screening methods such as single-strand conformational polymorphism analysis (SSCP), heteroduplex analysis (HA), and denaturing gradient gel electrophoresis (DGGE) that reveal polymorphisms and rare variants that may not be disease causing. The PTT has been applied to the mutation screening of a number of large and complex genes including *BRCA1* *(4)*, *BRCA2* *(5)*, *ATM* *(6)*, *TSC2* *(7)* and *NF2* *(8)*. For *BRCA1* and *BRCA2*, more than 3kb of coding sequence can be screened in a single PCR reaction combined with coupled in vitro transcription/translation reaction and run on a single lane on an sodium dodecyl sulfate polyacrylamide gel electrophoresis (SDS-PAGE) gel *(9)*.

PTT is performed in five stages (*see* **Fig. 1**):

1. Isolation of cellular RNA.
2. Conversion of RNA to cDNA by reverse transcriptase polymerase chain reaction (RT-PCR).
3. PCR amplification of gene-specific cDNA fragments.
4. Use of PCR-amplified product as a template for the in vitro synthesis of RNA which is subsequently translated into protein.
5. The synthesised protein is then resolved by electrophoresis on an SDS-PAGE gel. Shorter protein products resulting from truncated transcript can be distinguished from the full-length products of normal protein.

Neurofibromatosis type 1 (NF1) is an autosomal dominant disorder with an incidence of 1 in 3500. The *NF1* gene is located at 17q11.2, spans 350kb of genomic DNA, contains 60 exons, and encodes a 12 kb mRNA *(10,11)*. The *NF1* gene product, neurofibromin, comprises 2818 amino acids and has an estimated molecular mass of 327 kDa *(12)*. The central region of neurofibromin (encoded by exons 21–27a) possesses marked homology to Ras-GTPase activation proteins (GAPs). The identification of mutations in

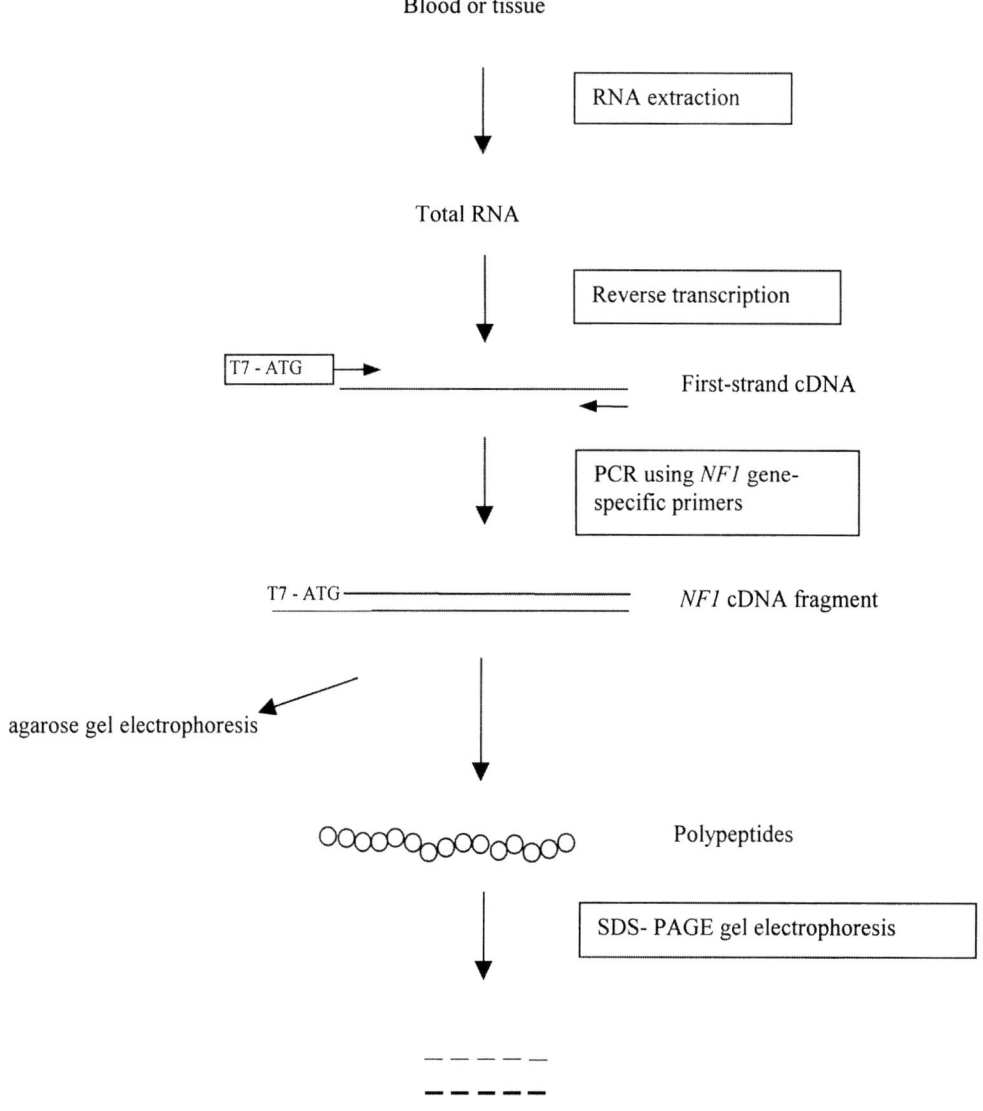

Fig. 1. Schematic diagram for PTT protocol.

the *NF1* gene causing type 1 neurofibromatosis (NF1) has presented a considerable challenge owing to the large size of the gene, the lack of significant mutational clustering, and the diversity of the underlying pathological lesions. In addition, the presence in the genome of numerous *NF1* pseudogenes has rendered the PCR-based analysis of genomic DNA somewhat problematic *(13,14)*.

The *NF1* gene is considered a classical tumor suppressor in that the first hit is inherited whereas the second hit is somatic *(15)*. The mutation rate for the *NF1* gene (approx 1×10^{-4}/gamete/generation) is some 10-fold higher than observed for many other disease genes and about half of all NF1 individuals appear to represent cases of new mutation *(16)*. To date, over 440 *NF1* gene lesions have been logged in either the Human Gene Mutation Database (*see* Website: http://www.hgmd.org) or the National Neurofibromatosis Foundation (NNFF) Mutation Database (*see* Website: http://www.clam.com/nf/nf1gene). At least 70% of reported mutations are predicted to lead to the synthesis of a truncated protein *(16)*. The high percentage of truncating mutations renders the *NF1* gene amenable to reverse transcription-polymerase chain reaction (RT-PCR) combined with the protein truncation test (PTT).

The known germline mutations of the *NF1* gene range in size from large genomic deletions that remove the entire gene and its flanking regions (>1.5 Mb) *(17,18)* to single base-pair substitutions. Thus no single mutation screening technique can be expected to detect all *NF1* gene lesions. SSCP and heteroduplex analysis (HA) have often been used to screen for microlesions in the *NF1* gene *(19,20)*. However, more sensitive detection techniques such as DGGE, direct genomic sequencing (DGS), or denaturing high-performance liquid chromatography (DHPLC) are now being introduced *(21,22)*.

The use of the RNA-based PTT for screening the *NF1* gene has yielded mutation detection rates of between 50 and 85% *(20,21,23–25)*. In addition to PTT, two methods have been developed to screen the entire *NF1* coding region: long RT-PCR *(26)* and SSCP/HA *(27)*. However, in their retrospective analysis, these authors *(26)* found that their method yielded positive results in only 30% of *NF1* mutations. In the case of diseases characterised by a high frequency of truncating mutations, the use of PTT is recommended (*see* **Fig. 1**).

PTT is most powerful as a technique when RNA is used. For example, one of the key genes involved in familial colorectal cancer (*APC*) has a large exon 15 which contains the majority of detected truncating mutations. Since most laboratories only possess stored DNA, the genomic DNA-based PTT is preferable. RNA-based PTT requires one to contact patients for collection of fresh blood samples.

The RNA-based analysis also allows the detection of splicing abnormalities. The high proportion of *NF1* splicing aberrations emphasises the potential importance of intronic disease-causing mutations and the application of RNA-based screening methods to confirm their effects *(21,27)*. In an earlier study, we were able using RNA-based PTT to identify an in-frame insertion caused by splicing error in an NF1 patient by PTT *(25)* (**Fig. 2**, R151). In this patient, cloning and sequencing revealed a 171bp insertion between exons 30 and 31 that corresponded to nucleotides 4047 to 4217 of intron 30. This extended new exon is spliced in-frame and encodes an additional 57 amino acids.

The major advantage of PTT over conventional screening techniques is that 1) it generally discriminates automatically between pathogenic and non-pathogenic sequence changes. This is because the mutation is detected at the protein level, and therefore only truncating mutations or large insertions/deletions are usually identified; 2) the length of the truncated protein product on the gel pinpoints the likely position of the mutation thereby facilitating its confirmation by direct sequencing; 3) the entire coding region of the *NF1* gene can be screened using between 5 and 10 overlapping fragments rather than by examining all 60 exons individually. Obviously, this reduces the workload and expedites the mutational analysis.

Use of the technique is not however without its disadvantages. Missense mutations and small in-frame insertions or deletions are normally undetectable in terms of the relative size of the encoded protein. Obviously, therefore, a whole category of biologically

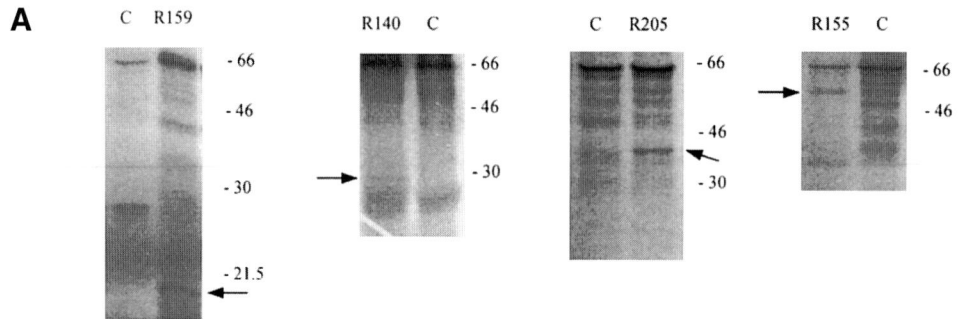
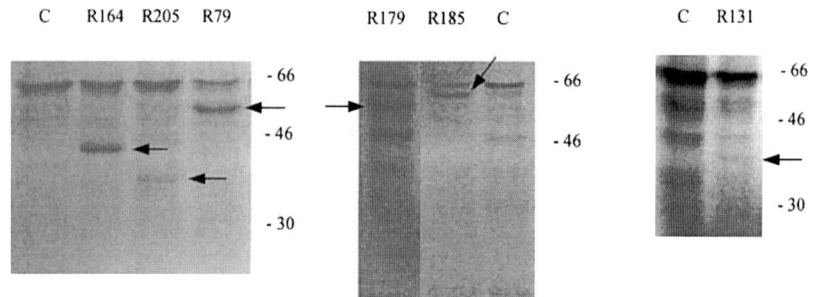
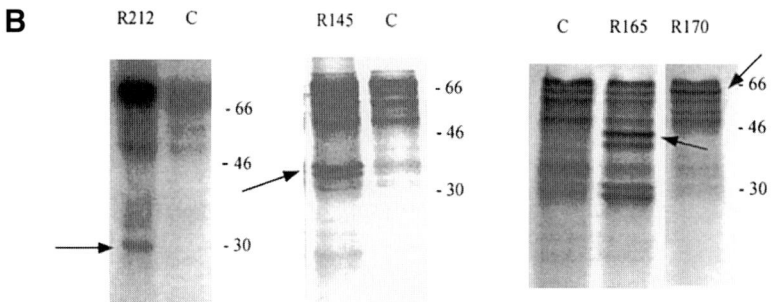
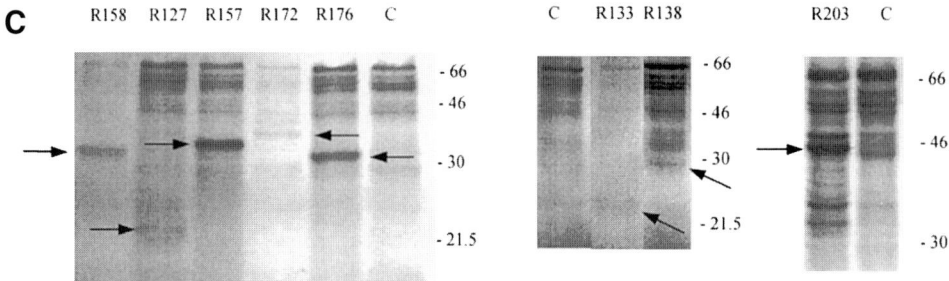

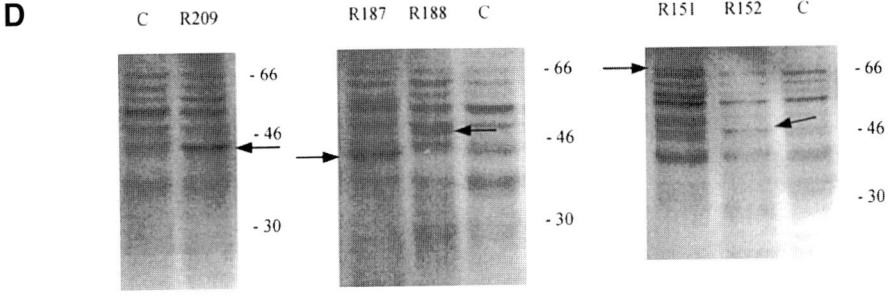

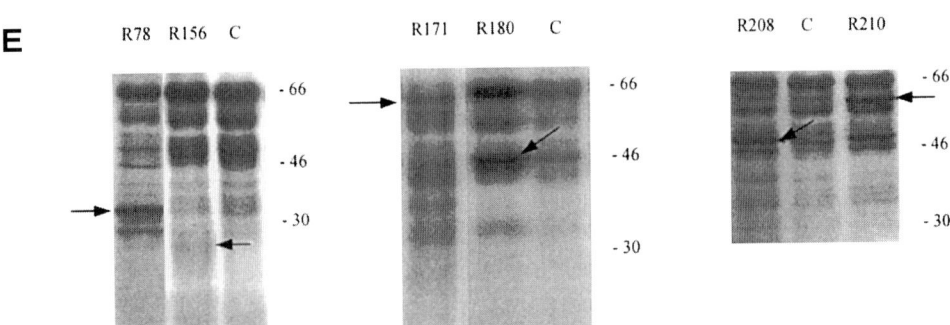

Fig. 2. Composite of protein gels showing positive PTT results from segments 1–5 (A–E). An arrow indicates the position of the abnormal protein product. The corresponding patient number is given above the lane. C denotes a normal control lane. The position of protein molecular weight standards is indicated (in kDa) to the right of each photograph. PTT results from segment 3 (**C**), segment 4 (**D**), and segment 5 (**E**).

meaningful mutations will not be detected. PTT would thus be an inappropriate choice of screening technique for genes manifesting a high proportion of nontruncating lesions, or if missense mutations are sought for structure/function studies. The occasional missense mutation will, however, be detected if it also activates a cryptic splice site. One example of this is provided by the *NF1* gene coding sequence which appears to convert tyrosine to cysteine at nucleotide 1466 *(25)*. However, RT-PCR analysis of *NF1* cDNA from the patient revealed a deletion of 62 nucleotides *(25)* (**Fig 2**, R79). The A to G substitution introduces a new splice donor site whose use is predicted to lead to the loss of the 3' end of exon 10b and a truncated protein product. Other types of mutations that one might expect to be invariably missed include mutations in the noncoding regions and large deletions encompassing overlapping PTT fragments.

PTT will obviously not detect a mutation if the mutant allele is not PCR amplified as a consequence of 1) the destruction of one or more primer binding sites by polymorphism or by the mutation itself, 2) the presence of a large insertion that dramatically increases the size of the PCR product, or 3) instability of the mutant transcript. Truncating mutations may also be missed if the size of the truncated protein product falls outside the resolving ability of the gel system. This can occur if a nonsense mutation or frameshift change is located immediately adjacent to either end of one of the screening segments. Early (N-terminal) mutations may also generate products which have either too little or no label incorporated or they may be difficult to resolve on a gel. Late (C-terminal) mutations may result in a size difference that is unresolved near the top of the gel. However, it should be noted that the deleterious effect of C-terminal truncating mutations may sometimes be questionable as found for example with the Lys 3326X mutation in the *BRCA2* gene *(28)*.

As with any RNA-based test, the major drawback of the PTT is the frequent direct association of truncating mutations with a low level of mRNA expression from the mutant allele. Nonsense and frameshift mutations are commonly associated with reduced mutant mRNA levels as a consequence of nonsense-mediated mRNA decay (NMRD), which often reduces the transcript level to 10–30% of the normal amount *(29)*. The occurrence of NMRD leading to "allelic exclusion" has now been reported for many genes, including the β-globin *(30)*, protein S *(PROS1; 31)* and cystic fibrosis (*CFTR*; *32*) genes.

NMRD compromises most RNA-based mutation detection methods but can be circumvented either by treating cultured cells with cycloheximide, which blocks protein synthesis *(9)* or by puromycin treatment *(33)*. However, establishment of Epstein-Barr Virus (EBV) lymphoblastoid cell cultures is both expensive and time consuming. Short term cultures of phytohaemagglutinin- stimulation of lymphocytes prior to RNA extraction has been shown to diminish aberrant splicing *(20)*. The relative sensitivity of PTT/ direct cycle sequencing was compared in 13 EBV lymphoblastoid cell cultures treated with and without puromycin. PTT was able to detect all the truncating mutations in both culture types. However, in only seven EBV cultures not treated with puromycin was the mutant transcript clearly identified by direct sequencing. In the remaining six cultures, the expression of mutant transcript was markedly reduced with the ratio of mutant to wild-type peak height in sequencing chromatograms being <0.35. In the cultures treated with puromycin, the ratio between mutant and wild-type peak heights in sequencing chromatograms varied between 0.8 and 1.00 *(20)*.

Allelic exclusion may only permit detection of the wild-type allele, thereby producing a false-negative result. However, a previous report has demonstrated that allelic exclusion to be present in the *NF1* gene occured in only 10% of mutations for which this phenomenon is a potential problem *(25)*. In this study, RNA was analyszed from 15 NF1 patients with known truncating mutations previously found by screeing genomic DNA; lack of expression of the mutant allele was noted in only two of these patient samples.

Various other studies have also screened the *NF1* gene using the PTT. Thus Heim et al. *(23)* identified mutations in 14 of 21 (67%) individuals studied whereas Side et al. *(34)* found mutations in 8 of 18 (44%) children with NF1 and an associated malignant myeloid disorder. The study by Park and Pivnick *(24)* identified mutations in >70% of their NF1 patients (11/15) using modified primer sets that amplified the *NF1* gene in 10 overlapping segments. We only found *NF1* gene mutations in 52% of NF1 patients screened *(25)*, whereas in the most recent study *(20)*, mutations were detected in up to 80% of NF1 patients, using five overlapping fragments spanning the entire coding region of the *NF1* gene.

Table 1
PCR Primers for Amplifying the *NF1* cDNA in Five Overlapping Segments[a]

Segment	Nucleotides	Exons	Size	Primer sequence[b]
1	1–1868	1–12b	1868	1F: ATG GCC GCG CAC AGG CCG GTG AAA T 1R: TG ACA GG ATC TGC CTG CTT A
2	1468–3583	10b–21	2115	2F : ATG GTG AAA CTA ATT CAT GCA GAT 2R :T GTC AAA TTC TGT GCC TTG
3	3217–5256	19b–29	2039	3F: ATG GAA GCA GTA GTT TCA CTT 3R: TAG GAC TTT TGT TCG CTC TGC TGA
4	4998–6987	28–38	1989	4F :G GAG TAC ACC AAG TAT CAT GAG 4R :TAT ACG GAG ACT ATC TAA ATG CAG
5	6574–8404	35–49	1830	5F :ATG GAG GCA TGC ATG AGA ATT C 5R : T CTG CAC TTG GCT TGC GGA T

[a]Annealing temperatures fragment 1–5: 61°C, 59°C, 61°C, 62°C, 62°C.
[b]The 5' end of the upstream primer for each segment comprised sequence for T7 promoter and translation initiation site that allow efficient eukaryotic transcription and translation in vitro: GGATCC TAATACGACTCACTATAGGGAGACCACCATG-3'. Adapted with permission from **ref. (23)**.

2. Materials

Stored at room temperature unless otherwise stated.

2.1. RNA Extraction

1. Lysis buffer: 150 mM ammonium chloride, 10 mM potassium hydrogen carbonate, 0.1 mM EDTA, pH 8.0. Store at 4°C.
2. RNAzol B (Biogenesis Ltd, Poole, UK). TOXIC. Store at 4°C.
3. Chloroform: HARMFUL.
4. Isopropanol.
5. 70% ethanol.

2.2. Reverse Transcription

1. Random hexamer primers, 100 μM. Store at –20°C.
2. RNase inhibitor (RNAguard, Amersham Pharmacia Biotech, Bucks, UK). Store at –20°C.
3. Superscript II reverse transcriptase, 200 U/μL (provided with 5X first strand buffer and 0.1 M DTT,) (Life Technologies Ltd., Paisley, Scotland). Store at –20°C.
4. dNTP mix: 1.25 mM each dNTP (Amersham Pharmacia Biotech). Store at –20°C.
5. Ribonuclease H, 1 U/μL (Life Technologies Ltd.). Store at –20°C.

2.3. cDNA Amplification

1. Expand High Fidelity PCR system (Roche Diagnostics Ltd., Lewes, UK). Store at –20°C.
2. PTT primers (**Table 1**), 5 μM.

2.4. In Vitro Transcription/Translation

1. TNT T7 coupled reticulocyte lysate system (Promega, Southampton, UK). Aliquot rabbit reticulocyte lysate into a convenient volume, e.g. 100 μL. This prevents multiple freeze/thaw cycles. Store at –70°C.
2. RNase inhibitor (*see* **Subheading 2.2.**, **item 2**).
3. ^{35}S-methionine (Amersham Pharmacia Biotech). Store at –20°C.

2.5. SDS-Polyacrylamide Gel Electrophoresis

1. 30% acrylamide/bis solution, 29:1 (Biorad Laboratories Ltd., Herts, UK). TOXIC. Store at 4°C.
2. 10% ammonium persulphate. Make up fresh every 2 wk. Store at 4°C.
3. TEMED.
4. Rainbow coloured protein molecular weight marker (Amersham Pharmacia Biotech). Store at −20°C.
5. 4X separating buffer: 1.5 M Tris-HCl, 0.4% sodium dodecyl sulfate (SDS), pH 8.8. Store at 4°C.
6. 4X stacking buffer: 0.5 M Tris-HCl, 0.4% SDS, pH 6.8. Store at 4°C.
7. Running buffer: 192 mM glycine (14.4 g/L), 25 mM Tris, 0.1% SDS.
8. Loading buffer: 20% glycerol, 2% SDS, 1X stacking buffer, 0.025 mg/mL bromophenol blue. Prior to use add 5% β-mercaptoethanol.
9. Fixing solution: 7% glacial acetic acid.
10. Whatman 3 MM paper.

3. Methods

3.1. RNA Extraction

1. Dilute blood sample (*see* **Note 1**) with 3 volumes cold lysis buffer, place on ice for 5 min.
2. Centrifuge at 200g for 10 min at 4°C, discard supernatent.
3. Resuspend pellet, repeat **steps 1** and **2**.
4. Homogenize pellet with 800 µL RNAzol B, transfer to a 1.5 mL eppendorf tube (*see* **Note 2**) and place on ice for 5 min.
5. Add 80 µL chloroform, mix by inversion and centrifuge at 16,000g at 4°C.
6. Collect aqueous phase (top layer) into a new tube and add an equal volume of isopropanol. Incubate on ice for 45 min or overnight at 4°C.
7. Centrifuge at 16,000g for 15 min at 4°C.
8. Wash pellet in 1 mL 70% ethanol. Centrifuge at 16,000g for 5 min.
9. Dissolve RNA in 20 µL sterile H_2O. An incubation at 55°C may be required for complete resuspension.
10. Determine RNA concentration by measuring absorbance at 260 nm (A_{260} = 1 = 40 µg/mL).
11. Store RNA samples at −70° C.

3.2. Reverse Transcription

1. Combine 1–5 µg total RNA with sterile H_2O to a final volume of 9.5 µL (*see* **Note 2**). Add 1 µL random hexamers and 0.5 µL RNase inhibitor. Incubate at 25°C for 10 min.
2. Add 4 µL 5X first-strand buffer, 2 µL DTT, 2 µL dNTP mix, and heat at 42°C for 5 min.
3. Add 1 µL Superscript II reverse transcriptase and incubate at 42°C for 60 min.
4. Inactivate the enzymes by heating to 70°C for 15 min.
5. Add 0.5 µL ribonuclease H, incubate at 37°C for 30 min.
6. Store at −20°C.

3.3. cDNA Amplification

1. In a thin-walled PCR tube, mix 6 µL dNTPs, 1.5 µL forward and reverse PTT primers (*see* **Note 3** and **4**), 2 µL buffer (1.5mM $MgCl_2$), 0.5 µL cDNA, and 10.5 µL H_2O.
2. Overlay with mineral oil and heat to 94°C.
3. Add second mix, containing 0.5 µL buffer, 0.25 µL enzyme (*see* **Note 5**), and 4.25 µL H_2O.

4. Perform the following PCR conditions:

1 cycle	94 °C for 2 min
10 cycles	94 °C for 15s, A for 30s, 72 °C for 90s
30 cycles	94 °C for 15s, A for 30s, 72 °C for 90s + 20s per cycle
1 cycle	72 °C for 10 min

A = annealing temperature (see **Table 1**)

5. Store samples at −20 °C (see **Notes 6** and **7**).

3.4. In Vitro Transcription/Translation

1. Make up a master mix containing: 0.4 μL reaction buffer, 0.2 μL T7 RNA polymerase, 0.2 μL amino acid mix (minus methionine), 0.2 μL RNase inhibitor, 0.5 μL ^{35}S-methionine (see **Note 8**) and 1.5 μL H2O.
2. Aliquot 3 μL into each reaction tube, add 5 μL rabbit reticulocyte lysate and 2 μL cDNA amplification product.
3. Incubate at 30°C for 90 min.
4. Store at −20°C.

3.5. SDS-PAGE

1. Clean gel plates using soap solution, rinse with H_2O and wipe over with ethanol.
2. Assemble glass plates in BioRad pouring apparatus, ensuring the spacers are flush with the base of the plates.
3. Separating gel mix (see **Note 9**) is made by combining 3.1 mL 30% acrylamide/bis solution, 1.9 mL 4X separating buffer, 2.4 mL of H_2O, 100 μL ammonium persulphate and 5 μL TEMED.
4. Pour between the plates, leaving a gap of 2 cm at the top.
5. Cover the surface with a layer of 0.1% SDS and leave to polymerize for 45 min.
6. Remove the SDS layer and wash the surface of the gel with 4X stacking buffer.
7. The stacking gel contains 500 μL 30% acrylamide/bis solution, 750 μL 4X stacking buffer, 1.7 mL H_2O, 30 μL ammonium persulphate, and 5 μL TEMED.
8. Pour onto the surface of the separating gel, filling the space to the top of the plate. Insert the comb carefully, at an angle, before leveling to avoid bubbles at the base of the wells. Leave to polymerise for 45 min.
9. Remove the comb and wash the wells with running buffer.
10. Dilute in vitro transcription/translation product 1 in 5 with loading buffer and denature at 99°C for 3 min. Place on ice.
11. Mix 5 μL Rainbow colored protein molecular weight marker with equal volume of loading buffer containing 10% β-mercaptoethanol.
12. Load samples and marker and perform electrophoresis at constant current of 15 mA in the stacking gel and 30 mA in the separating gel. Stop when the smallest molecular weight marker (14.3 kDa) reaches the bottom of the gel.
13. Separate the plates and soak the gel in fixing solution for 30 min.
14. Transfer the gel to Whatman paper, cover with cling film, and dry on a vacuum gel dryer.
15. Expose to X-ray film overnight (see **Notes 10** and **11**).

4. Notes

1. Blood samples should be collected in EDTA tubes and stored at 4 °C or in freezer as soon as possible. Delays in transit of a blood sample can cause decreased viability of lymphocytes and reduced fidelity in splice site selection *(35)*. Cold shock to blood samples has

been demonstrated to lead to the inclusion of a 31 bp cryptic exon between exons 4a and 4b of the NF1 mRNA *(27)*.
2. To avoid contamination with RNases, all RNA work should be performed using gloves, RNase-free eppendorf tubes and pipet tips.
3. The forward PTT primer possesses a bacteriophage T7 RNA polymerase promoter sequence along with a translation start codon (ATG) in-frame to the gene-specific sequence. The downstream primer is specific to the relevant coding region. The primer sequences used in our laboratory (*see* **Table 1**) are taken from Heim et al. *(23)*, and amplify the total coding sequence of the *NF1* gene in five overlapping fragments.
4. Park and Pivnick *(24)* used ten smaller overlapping fragments which may increase the sensitivity of the technique. Also *see* **Note 9**.
5. Expand High Fidelity PCR system utilises an enzyme mix containing thermostable *Taq* and *Pwo* DNA polymerases resulting in a high yield and low error rate. Two separate mixes are set up to facilitate a 'hot' start, which prevents the enzyme mix from partially degrading the primers and template.
6. We always check that the reverse transcription and PCR reactions have worked by performing agarose gel electrophoresis of the PCR product.
7. After detecting a mutation within a cDNA fragment by PTT, it is not always easy to identify the change in genomic DNA. In general, the formation of a shorter protein product is due to the presence of a new stop codon or by an alteration in the reading frame. In these cases, the size of the protein product should provide the approximate location of the sequence change (kDa × 9 × 3 = approximate position in bp). However, mutations causing the loss of an exon can result in the maintenance of the reading frame. In the case of the *NF1* gene, 24 exons (excluding those known to be alternatively spliced) fit into this category. The resulting shortened protein in these cases would not as it might appear indicative of a mutation toward the 3' end of the screened cDNA, but rather of a change anywhere within the fragment. Therefore the detection of aberrant exon splicing, exon skipping or the deletion of genomic DNA encompassing an exon may be useful in characterizing the pathological lesion. Separation of the cDNA amplification products on a 1.5% agarose gel can detect these exon deletion events and aid the interpretation of the PTT result.
8. A nonradioactive alternative utilises a reticulocyte lysate that contains charged lysine-tRNA labeled with biotin (Roche Diagnostics Ltd.). The resulting biotin-labeled protein is then electrophoretically separated by SDS-polyacrylamide gel electrophoresis (PAGE), blotted onto a PVDF membrane and subsequently detected with the BM chemiluminescence Western blotting kit (Biotin/Streptavidin). This methodology has been successfully applied to mutation detection in the *APC* gene *(36)*.
9. We normally use a 12.5% acrylamide separating gel. However, we feel that the maximum resolution of the five overlapping *NF1* fragments is probably achieved by running the samples on two protein gels containing different amounts of acrylamide. An 8% acrylamide gel would optimize separation of the larger protein fragments (45–66 kDa), whereas the 15% gel would separate the small products (6.5–14.4 kDa).
10. Photographs of PTT autoradiograms showing positive results for each of the five *NF1* fragments are shown in **Fig. 2** *(25)*.
11. The presence of a background of smaller translation products can complicate the detection of a genuine truncated product. The background bands are usually derived from spurious transcripts, initiated mostly from an internal translational initiation site. This problem can be overcome by modifying the forward PTT primer to include a sequence encoding a small known protein sequence, for example N-terminal C-*myc* tag *(37)*. By using specific antibodies against this protein tag, only the correctly initiated protein is detected using

Western blotting or immunoprecipitation. However, in most cases, examination of the relative intensity of bands within a sample lane compared to the equivalent bands for an adjacent sample can identify the presence of a truncated product (e.g., sample R209 in **Fig. 2D**).

References

1. Roest, P. A. M., Roberts, R. G., Sugino, S., van Ommen, G. J. B., and den Dunnen, J. T. (1993) Protein truncation test (PTT) for rapid detection of translation-terminating mutations. *Hum. Mol. Genet.* **2**, 1719–1721.
2. van der Luijt, Meera, K. P., Vasen, H., Van Leeuwen, C., Tops, C., Roest, P. A. M., et al. (1994) Rapid detection of translation-terminating mutations at the adenomatous polyposis coli (*APC*) gene by direct protein truncation test. *Genomics* **20**, 1–4.
3. den Dunnen, J. T. and van Ommen, G. J. B. (1999) The protein truncation test: a review. *Hum. Mutat.* **14**, 95–102.
4. Hogervorst, F. B. L., Cornelis, R. S., Bout, M., Van Vliet, M., Oosterwijk, J. C., Olmer, R., et al. (1995) Rapid detection of *BRCA1* mutations by the protein truncation test. *Nat. Genet.* **10**, 208–212.
5. Lancaster, J. M., Wooster, R., Mangion, J., Phelan, C. M., Cochran, C., Gumbs, C., et al. (1996) *BRCA2* mutations in primary breast and ovarian cancers. *Nat. Genet.* **13**, 238–240.
6. Telater, M., Wang, Z., Udar, N., Liang, T., Bernatowska-Matuszkiewicz, E., Lavin, M., et al. (1996) Ataxia-telangiectasia: mutations in ATM cDNA detected by protein truncation screening. *Am. J. Hum. Genet.* **59**, 40–44.
7. van Bakel, I., Sepp, T., Ward, S., Yates, H. R. W., and Green, A. J. (1997) Mutations in the TSC2 gene: analysis of the complete coding sequence using the protein truncation test (PTT). *Hum. Mol. Genet.* **6**, 1409–1414.
8. Parry, D. M., MacCollin, M. M., Kaiser Kupfer, M. I., Pulaski, K., Nicholson, H. S., Bplesta, M., et al. (1996) Germline mutations in neurofibromatosis 2 gene: correlations with disease severity and retinal abnormalities. *Am. J. Hum. Genet.* **59**, 529–539.
9. Garvin, A. M. (1998) A complete protein truncation test for *BRCA1* and *BRCA2*. *Euro. J. Hum. Genet.* **6**, 226–234.
10. Wallace, M. R., Marchuk, D. A., Andersen, L. B., Letcher, R., Odeh, H. M., Fountain, J. W., et al. (1990) Type 1 neurofibromatosis gene: identification of a large transcript disrupted in three *NF1* patients. *Science* **249**, 181–186.
11. Viskochil, D., Buchberg, A. M., Xu, G., Cawthorn, R. M., Stevens, J., Wolff, R. K., et al. (1990) Deletions and a translocation interrupt a cloned gene at the neurofibromatosis type 1 locus. *Cell* **62**, 187–192.
12. Viskochil, D. (1998) Gene structure and expression, in: *Neurofibromatosis Type 1; From Genotype to Phenotype*, (Upadhyaya, M., and Cooper, D. N. eds.), BIOS Scientific, Oxford, pp. 39–53.
13. Purandare, S. M., Breidenbach, H. H., Ling, L., Zhu, X. L., Sawada, S., Neil, S. M, et al. (1995) Identification of neurofibromatosis (*NF1*) homologous loci by direct sequencing fluorescence *in-situ* hybridization, and PCR amplification of somatic cell hybrids. *Genomics* **30**, 476–485.
14. Cummings, L. M., Trent, J. M., and Marchuk, D.A. (1996) Identification and mapping of type 1 neurofibromatosis (*NF1*) homologous loci. *Cytogenet. Cell Genet.* **73**, 334–340.
15. Side, L. E. and Shannon, K. M. (1998) The *NF1* gene as a tumor suppressor, in: *Neurofibromatosis Type 1; From Genotype to Phenotype*, (Upadhyaya, M. and Cooper, D. N., eds.), BIOS Scientific, Oxford, UK, pp. 133–146.

16. Upadhyaya, M. and Cooper, D. N. (1998) The mutational spectrum in neurofibromatosis type 1 and its underlying mechanisms, in: *Neurofibromatosis Type 1; From Genotype to Phenotype,* (Upadhyaya, M. and Cooper, D. N., eds.), BIOS Scientific, Oxford, UK, pp. 65–82.
17. Dorschner, M. O., Sybert, V. P., Weaver, M., Pletcher, B. A., and Stephens, K. (2000) NF1 microdeletion breakpoints are clustered at flanking repetitive sequences. *Hum. Mol. Genet.* **9,** 35–46.
18. Lopez-Correa, C., Brems, H., Lazaro, C., Marynen, P., Legius, E. (2000) Unequal meiotic crossover: a frequent cause of *NF1* microdeletions. *Am. J. Hum. Genet.* **66,** 1969–1974.
19. Upadhyaya, M., Ruggierim M,, Maynard, J., Osborn, M., Hartog, C., Mudd, S., et al. (1998) Gross deletions of the neurofibromatosis type 1 (*NF1*) gene are predominantly of maternal origin and commonly associated with learning disability, dysmorphic features and developmental delay. *Hum. Genet.* **102,** 591–597.
20. Messiaen, L. M., Callens, T., Mortier, G., Beysen, D., Vandenbroucke, E., van Roy, N., et al. (2000) Exhaustive mutation analysis of the *NF1* gene allows identification of 95% of mutations and reveals a high frequency of unusual splicing defects. *Hum. Mutat.* **15,** 41–555.
21. Fahsold, R., Hoffmeyer, S., Mischung, C., Gille, C., Ehlers, C., Kucukceylan, N., et al. (2000) Minor lesion mutational spectrum of the entire *NF1* gene does not explain its high mutability but points to a functional domain upstream of the GAP-related domain. *Am. J. Hum. Genet.* **66,** 790–818.
22. Han, S., Cooper, D. N., and Upadhyaya, M. (2001) Evaluation of denaturing high performance liquid chromatography (DHPLC) for the mutational analysis of the neurofibromatosis type 1 (*NF1*) gene. *Hum. Genet.* **109,** 487–497.
23. Heim, R. A., Kam-Morgan, L. N. W., Binnie, C. G., Corns, D. D., Cayouette, M. C., Faber, R. A., et al. (1995) Distribution of 13 truncating mutations in the neurofibromatosis 1 gene. *Hum. Mol. Genet.* **4,** 975–981.
24. Park, V. M., and Pivnick, E. K. (1998) Neurofibromatosis type 1 (NF1): a protein truncation assay yielding identification of mutations in 73% of patients. *J. Med. Gene.t* **35,** 813–820.
25. Osborn, M. J., and Upadhyaya, M. (1999) Evaluation of the protein truncation test and mutation detection in the *NF1* gene: mutation analysis of 15 known and 40 unknown mutations. *Hum. Genet.* **105,** 327–332.
26. Martinez, J. M., Breidenbach, H. H., and Cawthon, R. (1996) Long RT-PCR of the entire 8.5kb NF1 open reading frame and mutation detection on agarose gels. *Genome Res.* **6,** 58–66.
27. Ars, E., Serra, E., de la Luna, S., Estivill, X., and Lazaro, C. (2000) Cold shock induces the insertion of a cryptic exon in the neurofibromatosis type 1 (*NF1*) mRNA. *Nucleic Acids. Res.* **28,** 1307–1312.
28. Mazoyer, S., Dunning, A. M., Serova, O., Dearden, J., Puget, N., Healey, C. S., et al. (1996) A polymorphic stop codon in *BRCA2*. *Nat. Genet.* **14,** 253–254.
29. Maquat, L. E. (1995) When cells stop making sense: effects of nonsense codons on RNA metabolism in vertebrate cells. *RNA* **1,** 453–465.
30. Baserga, S. .J, Benz, E. J., Jr. (1992) β-globin nonsense mutation: deficient accumulation of mRNA occurs despite normal cytoplasmic stability. *Proc. Natl. Acad. Sci. USA* **89,** 2935–2939.
31. Formstone, C. J., Wacey, A. I., Berg, L. P., Rahman, S., Bevan, D., Rowley, M., et al.(1995) Detection and characterisation of 7 novel protein-S (PROS) gene lesions—evaluation of reverse transcript-polymerase chain -reaction as a mutation screening strategy. *Blood* **86,** 2632–2641.

32. Will, K., Dörk, T., Stuhrmann, M., von der Hardt, H., Ellemunter, H., Tummler, B., and Schmidtke, J. (1995) Transcript analysis of CFTR nonsense mutations in lymphocytes and nasal epithelial cells from cystic fibrosis patients. *Hum. Mutat.* **5,** 210–220.
33. Andreutti-Zaugg, C., Scott, R., and Iggo, R. (1997) Inhibition of nonsense-mediated messenger RNA decay in clinical samples facilitates detection of human *MSH2* mutations with an in vivo fusion protein assay and conventional techniques. *Cancer. Res.* **57,** 3288–3293.
34. Side, L. E., Taylor, B., Cayouette, M., Connor, E., Thompson, P., Luce, M., and Shannon, K. M. (1997) Homozygous inactivation of *NF1* in the bone marrows of children with neurofibromatosis type 1 and malignant myeloid disorders. *New. Engl. J. Med.* **336,** 1713–1720.
35. Wimmer, K., Eckart, M., Rehder, H., and Fonatsch, C. (2000) Illegitimate splicing of the *NF1* gene in healthy individuals mimics mutation-induced splicing alterations in NF1 patients. *Hum. Genet.* **106,** 311–313.
36. Kirchgesser, M., Albers, A., Vossen, R., den Dunnen, J., van Ommen, G. J., Gebert, J., et al. (1998) Optimised non-radioactive protein truncation test for mutation analysis of the adenomatous polyposis coli (*APC*) gene. *Clin. Chem. Lab. Med.* **36,** 567–570.
37. Rowan, A. J. and Bodmer, W. E. (1997) Introduction of a myc reporter tag to improve the quality of mutation detection using the protein truncation test. *Hum. Mutat.* **9,** 172–176.

30

Application of the Protein Truncation Test (PTT) for the Detection of Tuberosis Sclerosis Complex Type 1 and 2 (TSC1 and TSC2) Mutations

Karin Mayer

1. Introduction

Tuberous sclerosis complex (TSC) is an autosomal dominant disorder characterized by the development of widespread hamartomas affecting most organs including the brain, heart, skin, kidney, lung, and eyes. Involvement of the brain is associated with some of the most severe clinical problems of TSC, including epilepsy, intellectual disability, and abnormal behavioral phenotypes. Inactivating mutations have been identified in two disease-causing genes, *TSC1* on chromosome 9q34 (MIM *191100) *(1)* and *TSC2* on chromosome 16p13 (MIM *191092) *(2)*. TSC affects up to 1 in 6000 individuals with two-thirds representing sporadic cases, reflecting a high spontaneous mutation rate. Both TSC genes cover about 50 kb and 40 kb of genomic DNA, respectively. The *TSC1* gene contains 23 exons with 21 exons coding for a 8.6 kb transcript which is translated into the protein hamartin. The TSC2 gene product tuberin is encoded by an mRNA of 5.5 kb consisting of 41 exons.

The C-terminal region of tuberin has homology to *rap1GAP* and has been shown to interact with *rabaptin5*. Moreover this region exhibits GTPase activating properties for the Ras-related proteins Rap1 and Rab5 *(3,4)*. Its implication in neuronal differentiation is discussed *(5)*. Hamartin contains a stretch of hydrophobic amino acids as well as two putative coiled coil domains. Its ability to interact with ezrin/ radixin/ moesin (ERM) proteins and the capability to activate the GTPase *Rho* implicates a role in cellular adhesion *(6)*. Physical interaction of tuberin and hamartin via their coiled coil domains and intracellular colocalization of both proteins in vivo support a common function *(7,8)* and explains the observed indistinguishable clinical phenotype in TSC1 and TSC2 patients. Recent findings obtained with the *Drosophila* homologues of TSC1 and TSC2 suggest a common function of both proteins in the insulin signaling pathway *(9)* with a phenotype characterized by increased cell size, cell number, and organ size resulting from inactivating mutations in either of the two genes. LOH of *TSC1* and *TSC2* loci in TSC-associated hamartomas as well as in sporadic tumors of non-TSC patients support the tumor-suppressor function of both genes *(10–12)*.

Up to now more than 127 different *TSC1* and 348 different *TSC2* mutations have been reported, of which less than 25% are recurrent and which are summarized in The Human Gene Mutation Database, Cardiff (*see* **Subheading 5.2.**), and the TSC Varation Database (*see* **Subheading 5.3.**). Large deletions and rearrangements as well as missense mutations occur almost exclusively in TSC2 and account for approx 20% for each of the two mutation types. Subtle mutations in both genes composed of nonsense mutations, small deletions and insertions, and splice site changes, which all lead to a premature translational stop comprise 98% of all TSC1 mutations and about 60% of all TSC2 mutations. The fact that these apparently inactivating mutations are scattered over the entire coding region in both genes requires a comprehensive screening procedure represented by the protein truncation test (PTT).

Application of the PTT in TSC has first been described for screening of the TSC2 gene in 1997 *(13)* and has been extended to the TSC1 gene in 1999 *(14)*. We reported the first RNA-based comprehensive mutation screening procedure for both TSC genes applying the PTT in 1999 *(15)*. All these initial descriptions are based on using RNA extracted from EBV-transformed lymphocytes as starting material. The establishment and maintenance of lymphoblastoid cell lines requires a specialized cell culture facility and is a time-consuming procedure. Therefore we tried to establish a short term culture of T lymphocytes stimulated with phytohaemagglutinin (PHA) as an alternative to EBV transformation in order to achieve biologically active and undegraded RNA as starting material in contrast to RNA extracted directly from blood lymphocytes, which is susceptible to nonsense mRNA decay *(16)*. Subsequent cDNA synthesis was performed in parallel with oligo (dT) and random hexanucleotides to enhance the amplification of different transcript lengths. Amplification of the entire coding region of both TSC genes was performed in partially overlapping fragments; four for TSC1 and six for TSC2, respectively. Overlapping sequences of at least 400 bp in length (**Table 1**) have proved most suitable for the detection of mutations located less than 150 bp apart from the fragment ends (**Fig. 1**). Careful analysis of RT-PCR products prior to and in addition to the PTT analysis has proved to be useful for the interpretation of splicing aberrations due to intronic mutations (**Fig. 2B**). In vitro transcription/ translation of reverse transcriptase polymerase chain reaction (RT-PCR) products was carried out in a TNT-coupled reticulocyte system in the presence of ^{35}S-methionone. Radiolabeled proteins were separeted through SDS-PAGE in 10% or 12% denaturing gels, respectively, dependent on the size of the wild-type protein.

In cases of aberrant migrating protein bands the approximate position of the putative mutation was estimated by comparing the molecular weights of the wild type and the shortened polypeptide with a molecular-weight marker. In a first attempt to identify the underlying mutation, amplification products generated from cDNA and genomic DNA containing the putative mutation site were subjected to direct sequencing. Cases with suspected splicing aberrations recognized by size differences of transcripts after agarose gel electrophoresis of RT-PCR products were investigated more intensively to interpret the consequences of an underlying mutation: if RT-PCR fragments representing different transcript lengths could be separated by agarose gel electrophoresis they were cut out and either directly subjected to sequencing or cloned into a T-cloning vector prior to sequence analysis of individual clones.

Table 1
Primer Pairs Used for Amplification of RT-PCR Fragments Applied to PTT

PCR fragment	Forward primer[a]	Reverse primer[b]	T_M (°C)	TSC specific fragment length (bp)	Overlapping region (bp)	Polypeptide size (kD)
TSC1						
exon 3–12	<u>A</u>TGGCCCAACAAGCAAATGTCGG	CGTAGTCATCCGAATGACAGAGTG	63	1219		44.7
exon 10–15	G<u>T</u>GTGCTACTTCTACCCCTTACTC	ACTTGTTCAGCTCCTTGCTGTGC	63	1085	306	40.0
exon 15–20	CTTTCTGAGATCACCACAGCAGAG	GGTATCTGAGTGCTTGTTCTGCAG	63	1164	544	42.7
exon 18–23	GAGAAGGACATCCAGATGTGGAAG	TGCATTCACACCTCCTGTTCTGTG	63	1329	390	46.6
TSC2						
exon 1–11	<u>CCACC</u>ATGGCCAAACCAACAAGC	CCACCAGTTCAAAGTATCTCTCCTG	64/63	1239		45.3
exon 9–20	<u>T</u>GCTGAGAGGAGCCGTGTTTTTG	CCGCTCCAGTGTCTTTGGGCC	63/65	1376	362	50.5
exon 17–26	GCCTTTGACTTCCTGTTGCTGCTG	CCGTGAAGTTGGAGAAGACGTATC	64/63	1279	408	46.8
exon 23–34	AATCAGTACATCGTGTGTCTGGCC	CATTGGGCAGCAGGATTGGCTTG	63	1924	475	67.3
exon 28–37	<u>T</u>GCATGTGAGACAGACCAAGGAG	CCTTGA<u>T</u>GGTGCCAAGCTTGAAG	64/63	1692	1269	64.5
exon 33–41	G<u>T</u>CGTCCTCAGTCTCCAGCCAG	TCACTGACAGGCAATACCGTCCAA	66/63	1510	413	54.0
T7 promoter sequence	GGATCCTAATACGACTCACTATAGGAACAGACCACCATG					
M13 reverse sequence	CAGGAAACAGCTATGACC					

[a] Each TSC specific forward primer was supplemented with a T7 promoter and translation initiation sequence (the partially overlapping region between T7 and gene-specific sequences is underlined).
[b] Each TSC specific reverse primer was supplemented with a M13 reverse sequence for sequencing with fluorescence labeled primers.

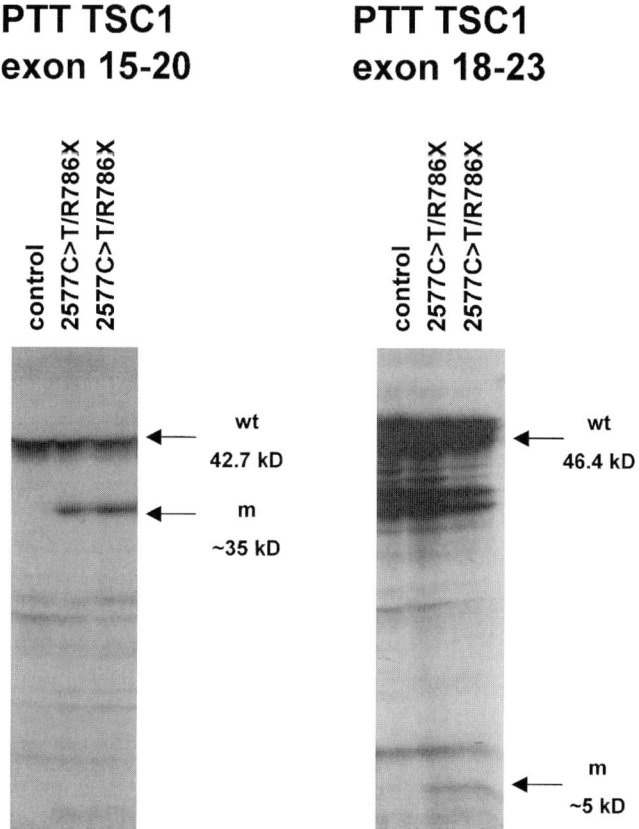

Fig. 1. Detection of a truncated polypeptide owing to a nonsense mutation in TSC1 exon 18 in both overlapping fragments after PTT of (**A**) exon 15–20 and (**B**) exon 18–23. Due to the observation of a shortened protein band in both cases, the premature stop has to be positioned in the overlapping region. The causative nonsense mutation 2577 C>T is located 900 bp downstream of the 5' end of the exon 15–20 fragment which is translated into a 35 kD polypeptide (**A**). Although the distance of the mutation to the 5' end of the exon 18–23 fragment is only 127 bp, the resulting polypeptide of about 5 kD can be recognized in a 12% PAA gel (**B**), (*see* **Notes 11** and **12**).

In the remaining cases in which the aforementioned procedures did not unravel the disease-causing mutation RT-PCR products resulting in aberrant migrating polypeptides were cloned into a T-vector. Individual clones were subsequently subjected to a second PTT reaction to separate wild-type and mutated alleles of a patients' transcripts *(17)* and then chosen for sequence analysis.

Each mutation identified on the transcript level was verified by direct sequencing of a PCR product generated from genomic DNA and every putative splicing mutation recognized on the genomic level was confirmed by the detection of aberrantly spliced transcripts.

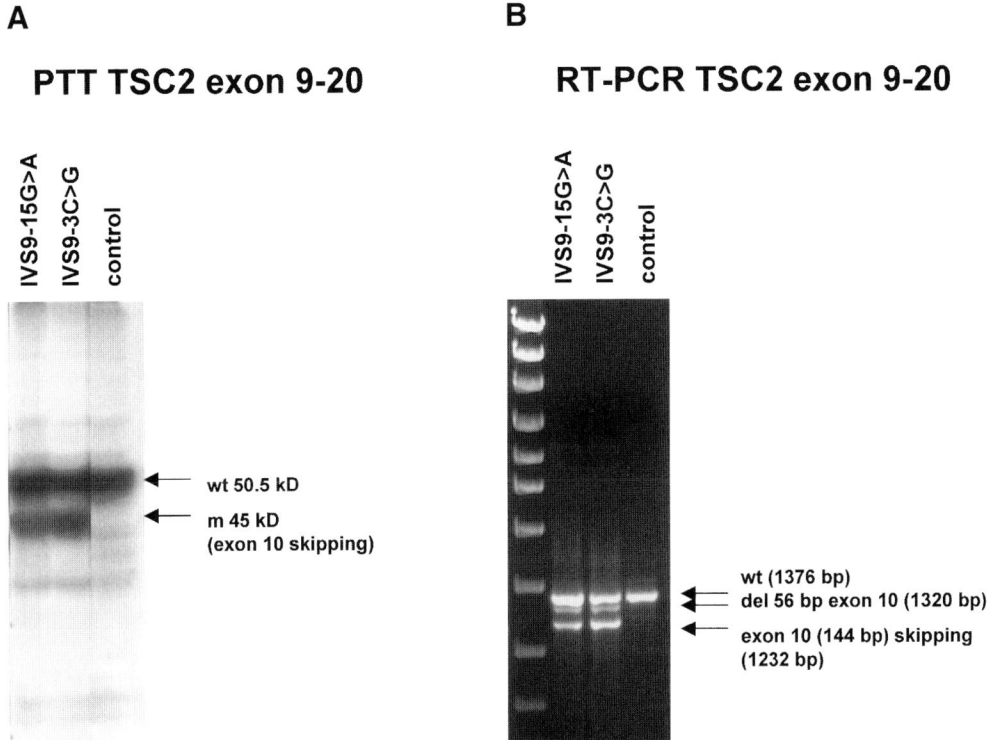

Fig. 2. Confirmation of aberrant splicing recognized by a shortened polypeptide band in PTT by agarose gel electrophoresis of different transcripts. (**A**) PTT of RT-PCR fragments spanning TSC2 exon 9–20 from two patients carrying different splicing mutations in intron 9. The shortened band corresponds to a polypeptide with an in frame deletion of exon 10. A third band expected from a transcript lacking the first 56 bp of exon 10 would terminate after 40 amino acids, which is too short to be visible in a 10% PAA gel used in this experiment (*see* **Notes 11** and **12**). (**B**) The existence of aberrantly spliced transcripts in both patients could be proven by separation of three RT-PCR in an 1% agarose gel (*see* **Note 8**). Products correspond to 1376 bp wild-type transcript, and two shorter bands of 1320 bp lacking 56 bp exon 10 and 1232 bp lacking the entire exon 10.

Applying this comprehensive RNA-based technique in 70 TSC patients that have been prescreened for large TSC2 deletions we identified small truncating mutations in 13 TSC1 and 28 TSC2 cases that correspond to an efficiency of 55% without considering possible missense mutations in TSC2, which cannot be detected with this approach. A high proportion of identified genetic alterations were splice defects, some inducing complex splicing aberrations that mainly affect the *TSC2* gene and therefore extending the known diversity of small mutations in TSC. The PTT offers the advantage to test several exons in one step compared to DNA based exon-scanning techniques. Moreover, the inclusion of intronic sequences and the demonstration of the effects of identified mutations on the transcript levels emphasize the significance of an RNA-based screening method as PTT for a disease with a wide variety and complexity of disease-causing mutations like TSC.

2. Materials

2.1. EBV Transformation of B Lymphocytes

1. RPMI 1640 media with L-glutamin (Seromed Biochrom, Berlin, Germany).
2. Fetal calf serum (FCS) (Seromed Biochrom) (inactivate at 56°C for 30 min before use).
3. Penicillin/Streptomycin (10.000 IE/10,000 µg/mL) (Seromed Biochrom).
4. Cyclosporin A 100 µg/µL.
5. Epstein Barr virus EB95-8.
6. Biocoll 1.077 g/mL (Seromed Biochrom).
7. Puromycin dihydrochloride (Sigma, St. Louis, MO, USA) 50 mg/mL in sterile water.
8. Phosphate-buffered saline (PBS) (Seromed Biochrom).
9. Nunclon TC tubes flat (Nunc).

2.2. PHA Stimulated Short Time Cultures of T Lymphocytes

1. RPMI 1640 media with L-glutamin (Seromed Biochrom).
2. Fetal calf serum (FCS) (Seromed Biochrom).
3. Phytohaemagglutinin (PHA) in H_2O (Seromed Biochrom).
4. Puromycin dihydrochloride (Sigma) 50 mg/mL in sterile water.
5. Nunclon TC tubes flat (Nunc).

2.3. Extraction of Genomic DNA

1. Blood lysis: 15.5 mM NH_4Cl, 1 mM $KHCO_3$, 0.1 mM Na_2EDTA, pH 7.4
2. SE buffer: 75 mM NaCl, 24 mM Na_2EDTA.
3. 20% SDS.
4. Pronase from *streptomyces griseus* (20 mg/mL).
5. 5 M NaCl.
6. Isopropanol.
7. Low TE: 10 mM Tris, 0.1 mM EDTA.

2.4. RNA Extraction

1. Tri reagent (Sigma).
2. Chloroform.
3. Isopropanol.
4. Ethanol (absolute and 70%).

2.5. cDNA Synthesis

1. Superscript II RNase H⁻ reverse transcriptase (Life Technologies, Bethesda, MD, USA).
2. 5X First-strand buffer: 250 mM Tris-HCl, pH 8.3, 375 mM KCl, 15 mM MgCl,
3. Oligo (dT) $(d(pT)_{15})$ (500 ng/µL) (Roche , Mannheim, Germany).
4. Random hexanucleotide mix 100 ng/µL (62.5 A_{260} /mL) (Roche).
5. dNTP mix: 20 mM of each dNTP (MBI Fermentas, Lithuania).
6. Ribonuclease inhibitor (40 U/µL) (MBI Fermentas, Lithuania).
7. Ribonuclease H (5 U/µL) (MBI Fermentas)
8. 0.1 M DTT.

2.6. PCR

1. Taq DNA polymerase, recombinant, 5 U/µL (Life Technologies).
2. 10X PCR buffer: 200 mM Tris-HCl, pH 8.0, 500 mM KCl,
3. dNTP mix: 20mM of each dNTP (MBI Fermentas),

4. 50 m*M* MgCl$_2$.
5. Primers for the amplification of RT-PCR fragments subjected to PTT, 10µ*M* (**Table 1**)
6. Primers for the amplification of single exons from genomic DNA, 10 µ*M* (published at "The TSC Project" website, *see* **Subheading 5.3.**).

2.7. Agarose Gel Electrophoresis

1. Agarose (Life Technologies).
2. 10XTris borate (TBE) buffer: 1 *M* Tris, 0.83 *M* boric acid, 0.01 *M* Na$_2$EDTA.
3. Ethidium bromide 1% (Sigma).
4. 6X gel loading buffer: 0.2% bromphenol blue, 0.2% xylene cyanol, 60% glycerol, 60 m*M* EDTA.

2.8. In Vitro Transcription/ Translation

1. TNT-coupled reticulocyte lysate system (Promega, Madison, WI, USA): including TNT rabbit reticulocyte lysate (thaw only once, aliquot, and store at –70°C for further use), TNT reaction buffer, TNT T7 RNA Polymerase, amino acid mixture minus methionine
2. Ribonuclease inhibitor (40 U/µL) (MBI Fermentas).
3. [^{35}S] methionine (1000 Ci/mmol at 10 mCi/mL) (consider appropriate precautions for dealing with radioactive material).

2.9. SDS-PAGE

1. Rotiphorese Gel 40 (19:1): 38% Acrylamid, 2% Bisacrylamid (Roth, Karlsruhe, Germany) (PAA is toxic; avoid inhalation and contact with skin).
2. Separating gel 4X buffer: 18.7 g Tris, 2 mL 20% SDS ad. 100 mL, pH 8.0, with HCl.
3. Stacking gel 4X buffer: 6.06 g Tris, 2 mL 20% SDS ad. 100 mL, pH 6.8, with HCl.
4. Ammoniumperoxodisulfate (APS) (Sigma) 10% in H$_2$O.
5. N,N,N'N'-Tetramethylethylendiamine (TEMED) (Sigma).
6. 10X SDS running buffer: 30 g Tris, 144 g glycine, 50 mL 20% SDS ad. 1 L.
7. SDS sample buffer: 2.0 mL glycerol, 1.0 mL 20% SDS, 0.25 mg bromphenol blue, 2.5 mL stacking gel 4X buffer, 0.5 mL β-mercaptoethanol (add the β-mercaptoethanol to the sample buffer immediately before use).
8. SDS-7B molecular weight marker (Sigma).

2.10. Autoradiography

1. Fixing solution: 25% ethanol, 7% acetic acid.
2. 1 *M* sodium salicylate (Sigma) (avoid inhalation and contact with skin).
3. BioMax MR film (Eastman Kodak, Rochester, NY, USA).

2.11. Cloning of RT-PCR Products

1. Gel extraction kit (Qiagen, Hilden, Germany).
2. TA-cloning kit (Invitrogen, San Diego, CA, USA) including T-vector pCR2.1, T4 DNA Ligase (4 Weiss U/µL; 10X ligation buffer (60 m*M* Tris-HCl, pH 7.5, 60 m*M* MgCl, 50 m*M* NaCl, 1 mg/mL BSA, 70 m*M* β-mercaptoethanol, 1 m*M* ATP, 20 m*M* DTT, 10 m*M* spermidine) ; 100m*M* DTT; *E. coli* INVαF-competent cells; SOC media.
3. LB media: 1% tryptone, 0.5% yeast extract, 1% NaCl.
4. Agar-agar (Merk, Darmstadt, Germany).
5. Ampicillin 50 mg/mL (Sigma).
6. X-gal (5-Bromo-4-chloro-3-indolyl-β-D-galactopyranoside) 40 mg/mL in dimethylformamide (Sigma).

2.12. Sequence Analysis

1. PCR purification kit (Qiagen).
2. Sequitherm EXCEL II cycle sequencing kit (EpiCentre, Madison, WI, USA) including 3.5X sequencing buffer, DNA polymerase, ddNTP mixes for A,C,G,T; Dimethylsulfoxid (DMSO); IRD-700 labeled M13 universal sequencing primers corresponding to the 5' end of the reverse primer used for RT-PCR (Note). Sequagel complete gel solutions (National Diagnostics, Atlanta, GA, USA).
3. ABI Prism Big Dye Terminator cycle sequencing ready reaction kit (PE Biosystems, Foster City, CA, USA) including A-Dye terminator labeled with dichloro[R6G], C-Dye terminator labeled with dichloro[ROX], G-Dye terminator labeled with dichloro[R110], T-Dye terminator labeled with dichloro[TAMRA], deoxynucleoside triphosphates (dATP, dCTP, dITP, dUTP), Ampli*Taq* DNA polymerase FS, $MgCl_2$, Tris-HCl pH 9.0. DyeEx Spin Kit (Qiagen)

3. Methods

3.1. EBV Transformation of B Lymphocytes

1. Dilute 1–5 mL heparinized blood with an equal amount of RPMI media.
2. To separate lymphocytes overlay three parts of biocoll with four parts of diluted blood and centrifuge with 300*g* at room temperature for 40 min without brake.
3. Pull out the B lymphocytes inclosed in the fishy fraction and dilute it to 11 mL with RPMI.
4. Centrifuge with 170*g* for 10 min and wash the B lymphocytes with RPMI by repeating this step two times.
5. Resuspend the cell pellet in 0.5 mL RPMI media supplemented with 20% FCS, 0.3% penicillin/streptomycin and 20 µL cyclosporin A and add 0.5 mL EB95–8 virus.
6. Cultivate the cells in Nunclon tubes at 37°C with 5% CO_2 for 1 wk without changing the media.
7. Change RPMI media supplemented with FCS and penicillin/streptomycin weekly and subdivide the cells into culture flasks if necessary.
8. Incubate the B lymphocytes for 12 h with 200 µg puromycin/ mL media prior to harvest to prevent nonsense mRNA decay (*see* **Note 1**).
9. Harvest the cells by centrifugation at 500*g* and 4°C for 10 min and wash the pellet once with icecold PBS.

3.2. PHA Stimulated Short Time Cultures of T Lymphocytes

1. Combine 4 mL RPMI media supplemented with 10% FCS and 0.3% penicillin/streptomycin with 0.3% PHA and add six drops of heparinized blood (*see* **Note 2**).
2. Cultivate the blood cells in Nunclon tubes for 72 h at 37°C without CO_2 and shake them gently every 12 h.
3. Incubate the lymphocytes for 6 h with 200 µg puromycin/ mL media prior to harvest to prevent nonsense mRNA decay (*see* **Note 1**).
4. For harvesting the lymphocytes add three volumes of icecold blood lysis and shake gently on ice for 30 min. Centrifuge at 500*g* and 4°C for 10 min, remove the supernatant and resuspend the pellet in 1 mL TRI reagent. Proceed to RNA extraction or store the cells at –70°C for further use.

3.3. Extraction of Genomic DNA

1. Lyse the erythrocytes by combining 10 mL whole peripheral blood with 30 mL ice cold blood lysis and incubate for 2 h at 4°C with recurrent mixing.

2. Centrifuge at 2000 rpm and 4°C for 15 min and resuspend the pellet containing the leucocytes in 10 mL ice-cold blood lysis. Centrifuge again at 500g and 4°C for 15 min and wash the leucocytes by repeating the last two steps until the pellet is clear.
3. Resuspend the leucocytes in 5 mL SE buffer and incubate with 125 µL 20% SDS and 100 µL Pronase over night at room temperature to release the DNA.
4. Add 1.5 mL 6 M NaCl and vortex 15 s. Centrifuge at 8000g and room temperature for 15 min to precipitate proteins.
5. Transfer the supernatant into a fresh tube and precipitate the DNA by mixing with an equal amount of isopropanol.
6. Remove the high molecular genomic DNA with a glass heel, wash it once in 70% ethanol, and dissolve it in 4000 µL low TE.
7. Typical yields of DNA isolated from 10 mL blood are 300–800 µg.
8. Determine the DNA concentration at 260nm in a UV spectrophotometer (1 A_{260} = 50 ng/µL ds DNA).
9. Store concentrated DNA and 100 ng/µL working dilutions at 4°C.

3.4. RNA Extraction

1. Lyse the pellet obtained from $5-10 \times 10^6$ EBV transformed B lymphocytes, leukocytes from 10 mL heparinized blood (*see* **Subheading 3.2.**), or a PHA culture of T lymphocytes in 1 mL TRI reagent by repeated pipeting. Stand the samples for 5 min at room temperature to completely dissolve nucleoprotein complexes.
2. Add 0.2 mL chloroform per mL of TRI reagent, shake vigorously for 15 s and stand for 5 min at room temperature. Centrifuge at 12000g for 15 min at 4°C.
3. Transfer the colorless upper aqueous phase to a fresh tube (*see* **Note 3**) and add 0.5 mL isopropanol per mL TRI reagent used in step 1 and mix. Stand the samples for 5 min at room temperature. Centrifuge at 12000g for 10 min at 4°C.
4. Remove the supernatant and wash the RNA pellet with 1 mL 75% ethanol, vortex and centrifuge again at 12000g for 10 min at 4°C.
5. Briefly dry the RNA pellet by air drying or under vacuum in a Speed-Vac concentrator. Dissolve the RNA in 20 µL sterile RNase free water and incubate at 65°C for 10 min for complete resuspension. Store RNA in H_2O at –70°C (*see* **Note 4**).
6. Typical yields obtained from $5-10 \times 10^6$ EBV transformed B lymphocytes are 10–20 µg RNA, from peripheral blood leukocytes about 5 µg, and from short term PHA cultures 1–5 µg RNA (**Fig. 1**).
7. RNA quantity and quality was always checked by agarose gel electrophoresis (*see* **Note 5**).

3.5. cDNA Synthesis

1. Dilute 1–5 µg of total RNA with H_2O to a final volume of 10.25 µL. Denature RNA at 90°C for 5 min and cool on ice.
2. Set up the following 20 µL reactions with: a) 0.5 µg oligo (dT) and b) 0.1 µg random hexanucleotides (*see* **Note 6**): 10.25 µL RNA, 4 µL 5X first-strand buffer, 1 µL primer, 0.1 M DTT, 0.5 µL RNase inhibitor, 1 µL dNTP mix, 1.25 µL Superscript II.
3. Incubate at 37°C for 1 h.
4. Terminate the reaction by heating at 94°C for 3 min.
5. Degrade remaining RNA in DNA/ RNA hybrids by incubation with 0.5 µL RNaseH for 20 min at 37°C.
6. Store cDNA at –20°C.

3.6. PCR

3.6.1. RT-PCR

1. Set up a 30 µL reaction containing the following components: 1 µL cDNA of the 20 µL reaction, 3 µL 10X PCR buffer, 0.3 µL dNTP mix, 0.9 µL 50mM MgCl$_2$, 0.6 µL of T7 promoter sequence coupled forward primer and M13 coupled reverse primer, 0.18 µL *Taq* polymerase, 23.42 µL H$_2$O.
2. PCR is carried out in a PTC 200 Thermocycler (MJ, Rockville, MA, USA) using the following touch down conditions (*see* **Note 7**): initial denaturation: 95°C for 1 min; stage I (6 cycles): denaturation at 95°C for 1 min (decreasing to 30 s minus 5 s per cycle), annealing at 66°C to 63°C (annealing temperature was adapted to T$_M$ (**Table 1**) for 1 min, elongation at 68°C for 2 min; stage II (21 cycles): denaturation at 95°C for 30 s (decreasing to 10 s minus 1 s per cycle), annealing at 63°C for 1 min, elongation at 68°C for 2 min (increasing 3 s per cycle); stage III (11 cycles): denaturation at 95°C for 10 s, annealing at 63°C for 1 min, elongation at 68°C for 3 min; final extension: 68°C for 7 min.
3. Store RT-PCR products at 4°C.

3.6.2. PCR of Genomic DNA

1. Set up a 30 µL reaction containing the following components: 1 µL genomic DNA (100 ng), 3 µL 10X PCR buffer, 0.3 µL dNTP mix, 0.9 µL 50mM MgCl$_2$, 0.6 µL of intronic forward and reverse primer, 0.18 µL *Taq* polymerase, 23.42 µL H$_2$O.
2. PCR is carried out in a PTC 200 Thermocycler (MJ, Rockville, MA) using the following conditions: initial denaturation: 95°C for 1 min; stage I (6 cycles): denaturation at 95°C for 1 min (decreasing to 30 s minus 5 s per cycle), annealing at 60°C to 50°C (annealing temperature was adapted to T$_M$ (published at "The TSC Project" website, *see* **Subheading 5.3.**) for 30 s, elongation at 68°C for 45 s; stage II (30 cycles): denaturation at 95°C for 30 s, annealing according to T$_M$ for 30 s, elongation at 68°C for 45 s; final extension: 68°C for 4 min.

3.7. Agarose Gel Electrophoresis of DNA Fragments

1. Prepare 1% and 2% agarose gels for the separation of RT-PCR (**Fig. 2B**) and genomic PCR products, respectively (*see* **Note 8**) by dissolving 1 g and 2 g agarose per 100 mL 1X TBE buffer by heating in a microwave oven, cool to 55°C, add 2.5 µL 10% ethidium bromide, and pour into gel tray.
2. An aliquot of the PCR product (10 µL for RT-PCR, 5 µL for genomic PCR) is mixed with 2 µL and 1 µL 6X gel loading buffer, respectively, and loaded into gel lanes.
3. Gel electrophoresis is carried out at 100 V constant voltage for 1–2 h depending on fragment sizes.
4. PCR products are visualized by UV illumination and photographed using a video camera (MWG Biotech, Ebersberg, Germany).

3.8. In Vitro Transcription/Translation

1. 100–200 ng of RT–PCR products were directly subjected to transcription and translation in a 12.5 µL reaction volume (*see* **Note 9**). Prepare a mix composed of: 4 µL RT-PCR product diluted in H$_2$O, 0.5 µL TNT reaction buffer, 0.25 µL amino acid mixture minus methionine, 0.25 µL TNT T7 RNA Polymerase, 0.25 µL Ribonuclease inhibitor, and add it to 6.25 µL pre-cooled reticulocyte lysate aliquots (*see* **Subheading 2.8.**). Finally add 1 µL [^{35}S] methionine (*see* **Note 10**).
2. Incubate at 30°C for 1 h.
3. Proceed immediately with SDS-PAGE or store the reaction at −20°C.

3.9. SDS-PAGE

1. Prepare a 10 or 12% PAA vertical separating gel depending on the size of the expected proteins (see **Note 11**). Mix the following components sufficient for a 1.0 mm thick, 11 × 11cm vertical gel (30 mL) and fill it between the two glass plates avoiding air bubbles: 14.85 mL H_2O, 7.5 mL 40% PAA, 15 mL 4X separating gel buffer, 150 µL APS, 7.5 µL TEMED.
2. Overlay the separating gel with H_2O prior to polymerization.
3. After polymerization pour off the H_2O and rinse the surface with 20% SDS.
4. Prepare a 4% stacking gel by mixing the following components (7.5 mL) and fill it into the remaining space between the two glass plates and insert the comb: 4.6 mL H_2O, 0.94 mL 40% PAA, 1.88 mL 4X stacking gel buffer, 75 µL APS, 7.5 µL TEMED.
5. After polymerization remove the comb, mount the gel apparatus, fill the chamber with 1X SDS running buffer, and rinse the wells with 1X SDS running buffer.
6. 12.5 µL in vitro transcription/ translation reactions are mixed with an equal volume of SDS loading buffer and denatured at 95°C before loading.
7. Electrophoresis is carried out starting with 200 V and constant 25 mA with permanent cooling at 10°C for 2.5 h until the blue dye has reached the end of the separating gel.

3.10. Autoradiography

1. Proteins are fixed by immersing the gel in ethanol/ acetic acid for 30 min, followed by a 20 min rinse with H_2O.
2. Signals are amplified by incubating the gel in 1 M sodium salicylate for 30 min.
3. Gels are vacuum dried for 1 h and exposed with BioMax MR X-ray film for 1–3 d at –70°C.

3.11. Cloning of RT-PCR Products

1. Either "full length" RT-PCR products resulting in aberrant migrating proteins (**Fig. 3**) or shorter RT-PCR fragments—which are suspected to containing the mutation—obtained with different combinations of primers are separated on 1–2% agarose gels dependent on size (see **Note 12**).
2. Bands were exercised with a scalpel and the DNA was extracted using gel extraction kit: gel slices are weighed to calculate the necessary buffer volumes (100 mg ~ 100 µL); gel slices are dissolved with 3 volumes of buffer QG at 50°C for 10 min with repeated vortexing. Mix the dissolved gel slice with 1 volume isopropanol and load it onto the column. Centrifuge at 10,000g for 1 min to bind the DNA to the membrane. Wash the column once with 0.5 mL buffer QG to remove traces of agarose and re-centrifuge. Wash the DNA on the column with 0.75 mL buffer PE and re-centrifuge. Remove traces of ethanol contained in buffer PE by centrifuging an additional 1 min. Elute the DNA with 20 µL 10 m*M* Tris-HCl, pH 8.5, pipet onto the membrane, and re-centrifuge.
3. The amount of purified PCR product is checked by agarose gel electrophoresis.
4. 10 µL ligation reactions were set up as follows: 6 µL (5–20 ng, equimolar to the vector) fresh purified PCR product diluted in H_2O, 1 µL 10X ligation buffer, 1 µL (25 ng) pCR2.1 vector, 1 µL T4 DNA ligase, and incubated overnight at 14°C.
5. For bacterial transformation 50 µL competent cells are gently mixed with 2 µL β-ME. 2 µL ligation reaction was added by pipetting and stirring gently. Incubate on ice for 30 min, heat shock at 42°C for 30 s, and place on ice for 2 min.
6. Add 250 µL SOC media to bacterial cells and incubate at 37°C in a shaker with 225 rpm for 1 h before plating different dilutions on agar plates.
7. Prepare agar plates by mixing equal volumes of 2X LB media with H_2O containing 3 g agar, both autoclaved and cooled down to 55°C. Add 100 µL ampicillin and 25 µL X-gal per 100 mL and pour 25 mL per plate.

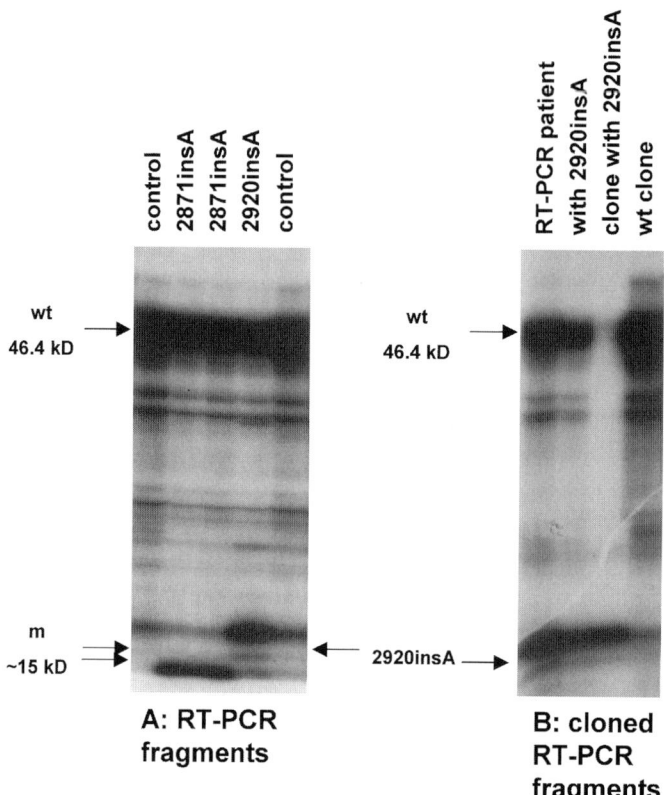

Fig. 3. Identification of a truncating mutation in TSC1 exon 21 by separating the patient's alleles after cloning the corresponding RT-PCR products spanning exon 18–23 and identification of the mutated transcripts in second PTT reaction. **(A)** PTT of RT-PCR products of exon 18–23 amplified from patients carrying different insertions in exon 21. Although the causative mutations are 50 bp apart, translation stops in both cases at the same nucleotide at position 2929 in exon 21, which leads to a polypeptide of about 15 kD. The migration differences are due to the different C termini of both truncated polypeptides (*see* **Note 12**). **(B)** The RT-PCR fragment of the patient carrying the 2929insA mutation was cloned in pCR2.1 and individual clones were tested in a second PTT reaction to separate wild-type and mutated transcripts, leading to different protein sizes (*see* **Note 13**).

8. Pellet bacterial cells by spinning at 12,000*g* for 30 s, pour off the supernatant and resuspend the pellet to a final volume of 100 μL. Spread aliquots of 10 μL and 90 μL bacterial cell suspension per plate and incubate overnight at 37°C.
9. Pick several white colonies per plate and isolate it by spreading on a fresh plate followed by incubation at 37°C.
10. Analyze transformants by amplifcation of the inserts with the same TSC specific primers used for the first amplification. Place a single colony into the prepared PCR mix.
11. PCR products representing parts of specific transcripts were either directly subjected to sequencing or tested in a sond PTT analysis (*see* **Note 13**; **ref. *17***).

3.12. Sequence Analysis

1. PCR products generated from cDNA, genomic DNA or bacterial clones were purified using PCR purification kit: add 5 volumes buffer PB to 1 volume PCR product, mix, and load it onto the column. Centrifuge at 10,000g for 1 min to bind the DNA to the membrane. Wash the DNA on the column with 0.75 mL buffer PE and re-centrifuge. Remove traces of ethanol contained in buffer PE by centrifuging an additional 1 min. Elute the DNA with 20 µL 10 mM Tris-HCl pH, 8.5 pipetted onto the membrane and re-centrifuge.
2. The amount of purified PCR product is checked by agarose gel electrophoresis.
3. Purified PCR products were subjected to cycle sequencing reactions:
   a. With IRD-700 labeled M13 universal sequencing primers binding to the target sequence at the end of the chimera gene specific primer (see **Table 1**; **Note 14**) using Sequitherm EXCEL II kit. Combine the following premix: 3.6 µL 3.5X sequencing buffer, 1 µL (1 pmol) IRD-700 labeled M13 sequencing primer, 50 fmol DNA template diluted to 3.225 µL with H$_2$O, 0.175 µL DMSO, 0.5 µL DNA polymerase. Prepare 1 µL of A, C, G, and T termination mix placed on ice and add 2 µL of premix. Perform 30 cycles using the following conditions: initial denaturation at 95°C for 5 min; stage I (10 cycles): denaturation at 95°C for 30 s, annealing at 57°C for 15 s, elongation at 70°C for 1 min; stage II (10 cycles): denaturation at 95°C for 30 s, annealing at 60°C for 15 s, elongation at 70°C for 1 min; stage III (10 cycles): denaturation at 95°C for 30 s, annealing at 63°C for 15 s, elongation at 70°C for 1 min. Add 2 µL stop loading buffer and proceed with electrophoresis or store at –20 °C. For gel electrophoresis denature at 70°C for 2 min and place on ice. Load 1–2 µL onto a Sequagel sequencing gel containing 6% PAA (length 41 cm, thickness 0.25 mm). Electrophoresis is carried out with 1X TBE running buffer at 1500 V, 35 mA, 31.5 W, 50°C using a LI-COR automated sequencer (MWG Biotech, Ebersberg, Germany).
   b. With Ampli*Taq* FS BigDye Terminator sequencing kit using both amplification primers in separate reactions (see **Note 15**): 2.0 µL premix, DNA template (10–50 ng) diluted to 7 µL with H$_2$O, 1 µL (10 pmol) gene specific primer. Perform 25 cycles with denaturation at 96°C for 10 s, annealing at 55°C for 10 s, elongation at 60°C for 2 min. To remove unincorporated BigDye terminators we use Dye Ex Spin Kit: resuspend the resin contained in the spin column and centrifuge at 750g for 3 min; pipet the 10 µL sequencing reaction directly onto the center of the slanted gel bed surface and centrifuge at 750g for 3 min; the eluate contains the purified DNA. 4–6 µL of purified sequencing reactions were diluted with 20 µL H$_2$O, denatured at 90°C for 2 min and analyzed on an ABI 310 automated capillary sequencer.

4. Notes

1. Translational inhibitors like cycloheximide and puromycin have been shown to prevent nonsense-mediated mRNA decay and have been used successfully in RNA based mutation screening stratgies with several genes like COL1A1, COL6A1, MLH1, HNPCC, APC, and BRCA1 *(18)*.
2. PHA-stimulated short-term culture has been shown to reduce illegitimate splicing observed in aged blood in the NF1 gene *(19)*. Moreover, short-term culturing additionally offers the possibility to prevent NMD by addition of puromycin compared to direct RNA isolation from peripheral blood without treatment.
3. TRI reagent offers the possibility to simultaneously extract RNA, DNA, and protein. After adding chloroform to the homogenized sample and centrifugation the mixture separates into three phases with the aqueous phase containing the RNA, the interphase containing DNA,

RNA extraction

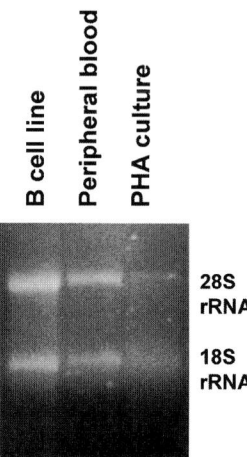

Fig. 4. Extraction of whole RNA from different sources. RNA was extracted from 2×10^5 transformed B lymphocytes, from 10 mL heparinized blood, and from a PHA short term culture of T lymphocytes with TRI reagent (*see* **Note 2**). 1.5 µL of a total of 20 µL isolated RNA was separated by electrophoresis in an 1.2% agarose gel containing 6.3% formaldehyde (*see* **Note 5**). Concentration of the B cell RNA was measured by UV absorption at 260nm (1 $A2_{60} = 40$ µg/mL) and adjusted to 1 µg/µL as a reference. Quality of RNA was checked according to the visibility of the two bands representing 28S rRNA and 18S rRNA.

and an organic lower phase containing proteins. To isolate high-quality RNA, it is important not to transfer any of the other phases before precipitation.

4. Alternatively RNA in H_2O can be precipitated with 0.1 volumes 3 *M* sodium acetate and 2.5 volumes absolute ethanol and stored at –70°C without losing quality after repeated thawing compared to RNA stored in aqueous solution.

5. For electrophoresis of whole RNA in a 1.2% agarose gel containing 6.3% formaldehyde dissolve 0.6 g agarose in 36.5 mL H_2O and 5 mL 10X MOPS (3-Morpholino-propanesulfonic acid), heat in a microwave oven, cool down to 55°C, and add 8.5 mL 37% formaldehyde. 1.5 µL of the RNA sample is mixed with 1.5 µL H_2O, 7.5µL formamide, 1.5 µL 10X MOPS, 3.0 µL formaldehyde, and 0.5 µL ethidiumbromide (0.2%). After denaturation at 65°C for 5 min the sample is mixed with 1.5 µL gel loading dye, loaded into gel lanes and separated at constant 100V for 1 h with 1X MOPS as running buffer. 28S and 18S rRNA bands are visualized by UV illumination. An example is given in **Fig. 4**.

6. cDNA synthesis is performed in parallel with oligo(dT) and random hexanucleotides to improve the amplification of RT-PCR fragments spread over the entire gene regions and as an internal control to exclude the detection of illegitimate transcripts.

7. Touch down PCR conditions have additionally been optimized by decreasing denaturation and increasing elongation time.

8. For separation of RT-PCR fragments longer than 1000 bp, we used 1% agarose gels with a long separation time to achieve a high resolution (**Fig. 2B**). For shorter subdivided RT-PCR fragments and genomic exons ranging from 200 bp to 1000 bp, we used 2% agarose gels.

9. Reaction volumes for in vitro transcription/ translation can be reduced to one fourth or even one-eighth compared to the manufacturer's protocol.
10. As an alternative to [^{35}S] methionine, y other radiolabeled amino acids such as [^{3}H] leucine, [^{14}C] leucine, or [^{35}S] cysteine recommended by the manufacturer dependent on the amino acid composition of your protein of interest. We do not have satisfactory experience with non-radioactive PTT.
11. If the expected size of the wild-type polypeptide extends 50 kD, we use 10% PAA; otherwise we use 12% PAA gels (compare **Fig. 2A** with **Figs. 1** and **3**).
12. Depending on the size of the truncated protein band compared to the wild-type band, it is possible to estimate the position of the premature stop codon and the position of the causative mutation, especially if a truncated band is visible in both of two adjacent overlapping fragments (**Fig. 1**). Sometimes proteins show different migration patterns dependent on the amino acid composition of the C terminus even if they end with the same premature stop codon (**Fig. 3A**). Other exceptions are complex splicing mutations in which the truncated band is misleading (**Fig. 2A**) and it is necessary to investigate the RT-PCR fragments directly (**Fig. 2B**) or after cloning.
13. A reliable method to separate the normal and the mutated allele in a heterozygous patient carrying a truncating mutation has been described in *(17)*. The cloned RT-PCR fragments are tested in a second PTT reaction in parallel to the directly amplified RT-PCR fragment (**Fig. 3B**), and the so-identified clone harboring the mutation can be directly sequenced.
14. For sequencing directly amplified RT-PCR fragments we used fluorescence-labeled primers and the Sequitherm protocol in combination with a LiCOR sequencer. The more quantitative nature offers the possibility to detect transcripts present in a heterozygous patient even in low amounts and the protocol has proved to be useful for sequencing fragments up to 1000 bp.
15. For sequencing cloned RT-PCR products, and genomic PCR products we used the BigDye terminator protocol in combination with an ABI 310 sequencer.

5. Electronic Database Information

Accession numbers and URLs for data in this article are as follows:

1. Online Mendelian Inheritance in Man:
   see Website: http://www.ncbi.nlm.nih.gov/omim
   For TSC1 [MIM 191100] and TSC2 [MIM 191092].
2. The Human Gene Mutation Data Base, Cardiff (HGMD):
   see Website: http://www.uwcm.acuk/uwcm/mg
   For TSC1 120735 and for TSC2 120466.
3. TSC Variation Database:
   see Website: http://www.expmed.bwh.harvard.edu/projects/tsc_database

Acknowledgments

Special thanks to Mrs Silke Schirdewahn for expert technical assistance. This study was supported in part by grants from the Tuberoese Sklerose Deutschland e.V. and from the Bayerisches Ministerium fuer Wissenschaft, Forschung und Kunst.

References

1. Van Slegtenhorst, M., De Hoogt, R., Hermans, C., Nellist, M., Janssen, B., Verhoef, S., et al. (1997) Identification of the tuberous sclerosis gene TSC1 on chromosome 9q34. *Science* **277,** 805–808.

2. The European Chromosome 16 Tuberous Sclerosis Consortium (1993) Identification and characterization of the tuberous sclerosis gene on chromosome 16. *Cell* **75**, 1305–1315.
3. Wienecke, R., König, A., and DeClue, J. E. (1995) Identification of tuberin, the tuberous sclerosis-2 product. *J. Biol. Chem.* **270**, 16,409–16,414.
4. Xiao, G. H., Shoarinejad, F., Jin, F., Golemis, E.A., and Yeung, R.S. (1997) The tuberous sclerosis 2 gene product, tuberin, functions as a Rab5GTPase activating protein (GAP) in modulating endocytosis. *J. Biol. Chem.* **272**, 6097–6100.
5. Soucek, T., Hölzl, G., Bernaschek, G., and Hengstschläger, M. (1998) A role of the tuberous sclerosis gene-2 product during neuronal differentiation. *Oncogene* **16**, 2197–2204.
6. Lamb, R. F., Roy, C., Diefenbach, T. J., Vinters, H. V., Johnson, M. W., Jay, D. G., and Hall, A. (2000) The TSC1 tumour suppressor hamartin regulates cell adhesion through ERM proteins and the GTPase Rho. *Nat. Cell. Biol.* **2**, 281–287.
7. Van Slegtenhorst, M., Nellist, M., Nagelkerken, B., Cheadle, J., Snell, R., van den Ouweland, A., et al. (1998) Interaction between hamartin and tuberin, the TSC1 and TSC2 gene products. *Hum. Mol. Genet.* **7**, 1053–1057.
8. Plank, T. L., Yeung, R. S., and Petri Henske, E. (1998) Hamartin, the product of the tuberous sclerosis 1 (TSC1) gene, interacts with tuberin and appears to be localized to cytoplasmic vesicels. *Cancer Res.* **58**, 4766–4770.
9. Potter, C. J., Huang, H., and Xu, T. (2001) Drosophila TSC1 functions with TSC2 to antagonize insulin signaling in regulating cell growth, cell proliferation, and organ size. *Cell* **105**, 357–368.
10. Green, A. J., Johnson, P. H., and Yates, J. R. (1994) The tuberous sclerosis gene on chromosome 9q34 acts as a growth suppressor. *Hum. Mol. Genet.* **3**, 1833–1834.
11. Henske, E. P., Scheithauer, B. W., Short, M. P., Van Slegtenhorst, M., Welsh, C. T., and Kwiatkowski, D, J. (1996) Allelic loss is frequent in tuberous sclerosis kidney lesions but rare in brain lesions. *Am. J. Hum. Genet.* **59**, 400–406.
12. Carbonara, C., Longa, L., Grosso, E., Mazzucco, G., Borrone, C., Garre, M. L., et al. (1996) Apparent preferential loss of heterozygosity at TSC2 over TSC1 chromosomal region in tuberous sclerosis hamartomas. *Genes Chrom. Cancer* **15**, 18–25.
13. Van Bakel, I., Sepp, T., Ward, S., Yates, J. R., and Green, A, J. (1997) Mutations in the TSC2 gene: analysis of the complete coding sequence using the protein truncation test. *Hum. Mol. Genet.* **6**, 1409–1414.
14. Benit, P., Kara-Mostefa, A., Hadj-Rabia, S., Munnich, A., and Bonnefont, J. P. (1999) Protein truncation test for screening hamartin gene mutations and report of new disease-causing mutations. *Hum Mutat.* **14**, 428–432.
15. Mayer, K., Ballhausen, W., and Rott, H. D. (1999) Mutation screening of the entire coding regions of the TSC1 and the TSC2 gene with the protein truncation test (PTT) identifies frequent splicing defects. *Hum Mutat.* **14**, 401–411.
16. Jacobson, A. and Peltz, S. W. (1996) Interrelationships of the pathways of mRNA decay and translation in eukaryotic cells. *Ann. Rev. Biochem.* **63**, 693–739.
17. Binnie, C. G., Kam-Morgan, L. N., Cayouette, M. C., Marra, G., Boland, C. R., and Luce, M. (1997) Rapid identification of RT-PCR clones containing translation-terminating mutations. *Mutat Res.* **388**, 21–26.
18. Bateman, J. E., Freddi, S., Lamandé, S. R., Byers, P., Nasioulas, S., Douglas, J., et al. (1999) Reliable and sensitive detection of premature termination mutations using a protein truncation test designed to overcome problems with nonsense-mediated mRNA instability. *Hum. Mutat.* **13**, 311–317.
19. Wimmer, K., Eckart, M., Rehder, H., Fonatsch, C. (2000) Illegitimate splicing of the NF1 gene in healthy individuals mimics mutation-induced splicing alterations in NF1 patients. *Hum. Genet.* **106**, 311–313.

31

Development and Characterization of Antibodies that Immunoprecipitate the FMR1 Protein

Stephanie Ceman, Fuping Zhang, Tamika Johnson, and Stephen T. Warren

1. Introduction

The ability to generate antibodies that recognize a given protein relies on that protein's "antigenicity", i.e., its ability to appear foreign to the host immune system. Thus, proteins that are highly conserved among species are often poor antigens when used for immunization *(1)*. This lack of immunoreactivity is not owing to the inherent nature of the protein; rather, it is because of the mechanism of tolerance, which exists to minimize autoimmune disease. Immunological tolerance occurs in an organism to prevent the development of a destructive immune response to self-proteins. It operates by eliminating or inactivating immune cells that bear receptors that recognize and respond to autologous protein antigens *(2)*. Thus, it is very difficult to generate an immune response in mice to either murine (i.e., self) antigens or to highly conserved proteins *(1)*.

One way to circumvent tolerance is to immunize mice in which the gene encoding that protein has been inactivated or "knocked-out" *(3)*. Such animals will no longer be tolerized to that protein because it is not expressed and is therefore not involved in lymphoid education. Such proteins, when used as an immunogen, will be viewed as foreign by the host immune system. Thus, when knock-out animals are immunized, a vigorous immune response will be elicited to that protein *(4–6)*.

Our interest in this problem arose from initial, unsuccessful efforts to generate high-affinity antibodies to the Fragile X Mental Retardation Protein (FMRP). The absence of FMRP in individuals results in mild to moderate mental retardation; thus, FMRP is thought to play a role in learning and memory (reviewed in **ref. 7**). Fragile X syndrome is one of the most common forms of inherited mental retardation with an estimated incidence of 1/4000 males and 1/8000 females (reviwed in **ref. 8**). Individuals with fragile X syndrome have IQs in the range of 20–60, mildly abnormal facial features of a prominent jaw and large ears, and macroorchidism in post-pubescent males. Many patients also display subtle connective tissue abnormalities, hyperactivity, and attention deficit disorder, and autistic-like behavior (reviewed in *8*).

FMRP is highly conserved among species: murine FMRP (Fmrp) is 97% identical by amino acid sequence to human FMRP *(9)*; chicken FMRP and *Xenopus Laevis* FMRP are 92% and 86% identical, respectively, to human FMRP *(10,11)*. This high conservation of

FMRP among species suggests that it plays an important function in many different organisms, but also confounds the development of high-affinity antibodies.

Our goal was to generate monoclonal antibodies (MAbs) that had high enough affinity to facilitate immunoprecipitation of both mouse and human FMRP. Such antibodies would be used for biochemical characterization of FMRP, as well as for the development of an enzyme-linked immunosorbent assay (ELISA) to be used for patient screening. Both polyclonal and monoclonal anti-FMRP antibodies had been developed in the past (12); however, these antibodies have only been useful for intracellular staining and for probing Western blots, they were not able to capture the native protein from tissue. Toward this end, we have now developed two immunoprecipitating antibodies, 7G1-1 and 6G2 using the approach outlined in **Fig. 1**.

1.1. Immunization

1.1.1. Bleeds

Before undertaking any work in animals, the investigators need to have their experimental protocol approved by the animal care committee at their specific institution. The sera from the mice will be analyzed before and during the immunization protocol. It is desirable to obtain a serum sample before immunization, which will be referred to as the "pre-immune" bleed to determine the baseline antibody activity or background levels in the screen.

1.1.2. Immunization with the Antigen

To elicit antibodies, the antigen first must be introduced in the presence of an adjuvant that elicits a vigorous immune response (reviewed in *[13]*). Because of its efficacy and low toxicity, we chose to use TiterMax® and have essentially followed the manufacturer's instructions. It is important to have milligram quantities of purified protein for both the immunizations and screening.

1.2. Screening by ELISA for the Presence of Antibodies

It is critical that a reliable screen for antibody-detection is developed before the immunizations begin. ELISAs are often used because large numbers of samples can be screened quickly. The basic principle is shown in **Fig. 2A**. The wells of a plate are coated with the immunizing protein antigen. The serum or supernatant to be tested is added to the coated plate. The interaction between the coating protein and antibody is visualized by adding an enzyme-conjugated second antibody that recognizes mouse immunoglobulin. The assay is then developed with a colorimetric substrate, an example of which is shown in **Fig. 2B**.

1.3. Fusion

Splenocytes, the B cells producing antibodies, are only viable for approx 7 d in culture; thus, they are "immortalized" by fusion to a myeloma that has two important characteristics: 1) it is deficient for an enzyme in the purine salvage pathway, HPRT; and 2) it is unable to produce antibody (reviewed in *[13]*). After fusion, the unfused tumor cells are killed in "HAT" medium, which is toxic for the HPRT-deficient cells. The only cells able to survive after 7 d will be the hybridomas, because the B cells provide HPRT and the tumors cells provide the genes for continuous cell division in tissue culture.

FMR1 Protein

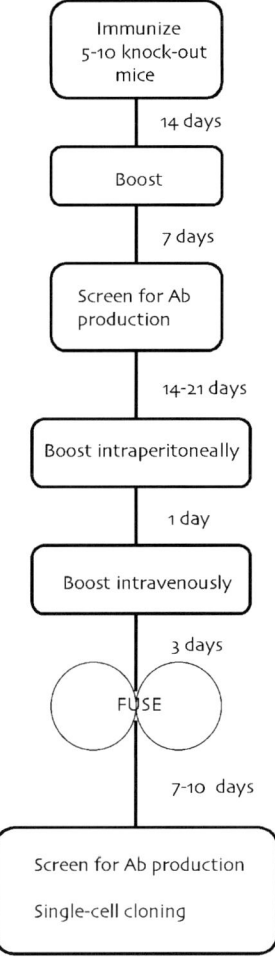

Fig. 1. Time line for monoclonal antibody (MAb) production. Ideally, the procedure can take approx 9–10 wk. The mice are immunized on d 0 and d 7 and the sera are screened for antibody production on d 14. Sero-positive mice are boosted again on d 28 intraperitonealy and on d 29 intravenously. On d 32, the splenocytes are harvested and fused to form hybridomas, which are screened 7–10 d later. Positive wells must be subcloned by limiting dilution, which will take an additional 3–4 wk.

1.4. Single-Cell Cloning

Each well of the plated-out fusion will contain more than one hybridoma, many of which may not be making any antibody. These "nonproducing" cells tend to grow faster than the antibody-producing cells, thus, it is essential that wells that are positive in the screen be subcloned immediately so that each well contains only one hybridoma. Poisson statistics indicate that if < 22% of the wells have growing cells (which is the proportion of positive wells one would expect if .3 cell per well were seeded with a 100% cloning efficiency), then 88% of these wells have only one clone *(14)*. Because hybridomas rarely have 100% cloning efficiency, i.e., the ability to grow-up from a single cell, more cells need to be seeded in each well to derive a reasonable number of growing clones.

A

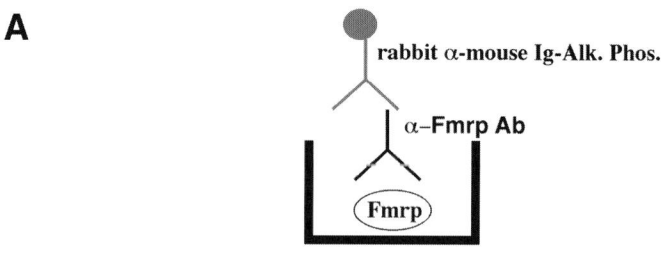

B

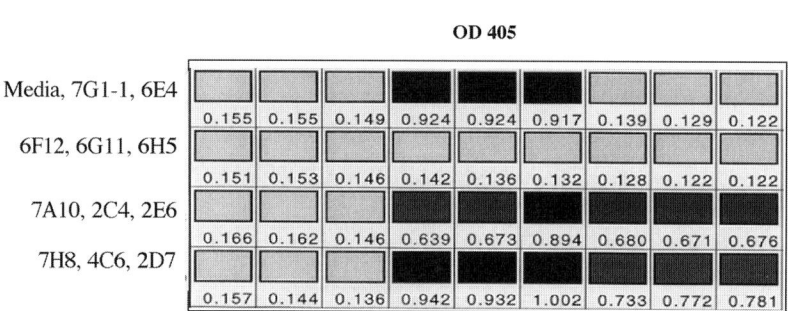

C

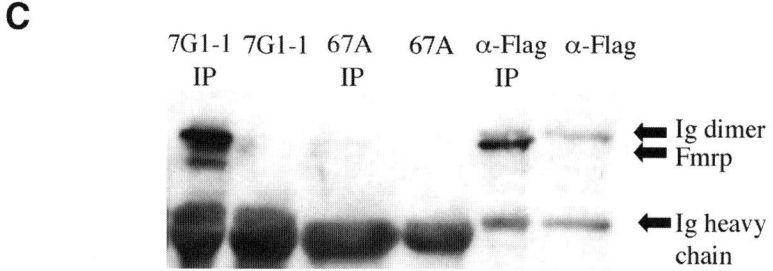

Fig. 2. Screen for antibody production. (**A**) Schematic of a capture ELISA where the protein of interest (FMRP) is bound to the well. The sample containing the antibody of interest (α-FMRP antibody) is depicted bound to FMRP. This interaction is visualized by addition of an antibody that recognizes murine antibodies, which is conjugated to an enzyme (alkaline phosphatase ●) able to form a colorimetric product, which is read at an absorbance of 405. (**B**) Example of an ELISA. Each hybridoma supernatant is indicated on the left and was tested in triplicate. The addition of hybridoma culture media alone is shown in the upper left-hand triplicate. The degree of shading correlates with intensity of the interaction between FMRP and the hybridoma supernatant, which was measured at an of absorbance 405. We found that the most strongly reactive supernatants in the ELISA correlated with the ability to immunoprecipitate. (**C**) An immunoprecipitation experiment with the candidate hybridoma supernatants 7G1-1 and 67A, and the commerically available anti-Flag antibody (Sigma), which recognizes an epitope tag engineered into the Fmr1 gene *(17)* immunoprecipitates Flag-FMRP and was used as a positive control. Lanes 1, 3, and 5 show the immunoprecipitation (IP) using the three antibodies. Lanes 2, 4, and 6 are the antibodies alone. The heavy chain of the antibodies is indicated on the right (Ig heavy chain) as is the position of FMRP and the undenatured antibody (Ig dimer). The presence of FMRP is visualized with the 1C3 antibody *(12)*, which recognizes Fmrp in a Western blot.

1.5. Immunoprecipitation

To test whether the antibodies are able to immunoprecipitate FMRP, the antibodies are first captured from the supernatant by rotating with agarose beads coupled to protein G, a recombinant bacterial protein that binds the Fc portion of most immunoglobulin molecules (reviewed in *[15]*). It is then determined whether FMRP can be captured from cellular lysates by the antibody-coupled beads in a process referred to as immunoprecipitation. The success of this experiment is determined by Western blotting, which requires a second antibody to FMRP *(12)*. An example of this experiment is shown in **Fig. 2C**.

2. Materials

Special equipment needed for this work include a laminar flow hood to create a sterile work environment for the cell culture and a 5% CO_2, humidified, 37°C incubator. For the ELISAs, it is useful to have the Spectra MAX 340 plate reader from Molecular Devices or comparable equipment and an attached computer with the SOFT MAX software.

2.1. Bleeds

1. Extract of wintergreen (local pharmacy).
2. Blood collection tubes (Microtainer Brand Serum Separator Tubes, Becton Dickinson).

2.2. Immunization

1. TiterMax adjuvant, 18 gauge double-hub emulsification needle, all-plastic syringes (CytRx Corp., Norcross, GA, USA).
2. PBS: 137 mM NaCl, 2.7 mM KCl, 0.6 mM Na_2HPO_4, 1.5 mM KH_2PO_4.

2.3. ELISA

1. 96 well Nunc-immunosorb ELISA plates (Fisher, Pittsburgh, PA, USA).
2. 10% milk solution in PBS made with powdered milk purchased from a grocery store. We use Carnation brand powdered milk.
3. Tween-20 (Fisher).
4. Rabbit anti-mouse antibody, conjugated to alkaline phosphatase (Zymed, San Francisco, CA, USA).
5. Alkaline phosphatase substrate (BioRad, Hercules, CA, USA).

2.4. Fusion

1. 4 push pins.
2. 150 mm and 100 mm petri dishes.
3. Teri towels.
4. 3cc syringes, 18 1/2 gauge needles and 27 1/2 gauge needles.
5. Styrofoam block (15-mL conical tube holder).
6. DMEM, sodium pyruvate, penicillin/streptomycin (Gibco-BRL, Rockville, MD, USA) Grow all cell lines in DMEM, 10% fetal calf serum (FCS), 1 mM Na pyruvate, 10 mM HEPES, and 100 U/mL of penicillin/streptomycin.
7. DMEM with 10% FCS, 8-azaguanine (Sigma, St. Louis, MO, USA).
8. HAT medium: DMEM with 10% FCS, 0.01 mM hypoxanthine (Sigma), 0.4 µM aminopterin (Sigma), 16 µM thymidine (Sigma), and 3 µM glycine.
9. Multi-channel pipettor (Flow Laboratories, McLean, VA, USA).
10. PEG-4000 (ATCC, Gaithersburg, MD, USA).

11. Fusion partner cell line, SP2/0Ag14 (ATCC).
12. Three sets of sterilized scissors and forceps.

2.5. Single-Cell Cloning

1. 96-well, flat-bottom plates.
2. HT medium: DMEM with 10% FCS, 0.01 μM hypoxanthine (Sigma), and 16 μM thymidine (Sigma).

2.6. Immunoprecipitation

1. Protein G agarose (Roche, Indianapolis, IN, USA).

3. Methods
3.1. Immunization
3.1.1. Bleeds

1. Obtain 5–10 knock–out mice *(16)*, approx 2 mo in age (*see* **Note 1**).
2. The underside of the tail is vaso-dialated by applying extract of wintergreen topically or by heating with a lamp.
3. Nick the underside of the tail with a razor blade. Have a blood collection tube ready to catch the drops of blood (50–100 µL). Tape a piece of gauze over the cut to stop the bleeding, which is removed after 1 d.
4. Sera is prepared from the blood as described *(13)*. First incubate the blood at 37°C for 1 h with occasional flicking to dislodge the blood clot. The tube is then transferred to 4°C for 2–16 h and spun at 10,000g for 10 min at 4°C. The serum is removed from the pellet and spun a second time for 10 min and stored at –20°C.

3.1.2. Immunization with the Antigen

1. Warm the adjuvant (TiterMax®) to room temperature and then vortex for 30 s.
2. Dissolve the immunizing protein in PBS.
3. Prepare an emulsion of adjuvant and antigen by first loading a syringe, which is all-plastic, through an 18 gauge needle with 0.5 mL of adjuvant.
4. Load a second syringe with half of the total antigen volume: in this case, 0.25 mL.
5. Remove the needles and connect the two syringes by an 18-gauge double-hub emulsifying needle.
6. Push the antigen into the adjuvant first, mixing the adjuvant with the antigen by forcing the solution back and forth through the needle for approx 1 min. It is important to push the antigen into the adjuvant first so that the aqueous phase enters the oil phase.
7. After 1 min, take the syringes apart and load the empty syringe with the remaining 0.25 mL of antigen. Reconnect and emulsify for another min. It will become difficult to continue mixing when the antigen-adjuvant is properly prepared, as it has the consistency of whipped cream (*see* **Note 2**).
8. Load the antigen/adjuvant mixture into a 28-gauge hypodermic needle and deliver 25 µL into two sites subcutaneously at the base of the tail (*see* **Note 3**).
9. The yield of antigen/adjuvant will be approx 30%, thus, this protocol gives enough to immunize 6 mice.
10. Fourteen d later, repeat **steps 1–10**.
11. Twenty-one d later, collect serum from each animal and assay for the presence of antibodies to the antigen of interest. Use the same assay that will be used to screen the hybridomas by ELISA.

12. Three to four wk after the last boost but 5 d before harvesting the spleen, inject the animal previously identified as making antibody intraperitonealy with 20–100 μg of soluble antigen (no adjuvant) (*see* **Note 4**). On the second d, immunize the animals intravenously with soluble antigen (*see* **Note 5**). On the fifth d, harvest the splenocytes for fusion.

3.2. Screening by ELISA for the Presence of Antibodies

1. Coat a 96-well immunosorb plate with 50 μL of 2.5–10 μg/mL of soluble protein antigen in PBS overnight at 4°C. Dispense the protein with a multi-channel pipet.
2. The next d, remove the antigen and block the plate for 30 min at room temperature (RT) with 10% milk in PBS.
3. Wash the plate 2× with 150 μL/well of PBS /0.2% Tween-20 and 1% milk,
4. Add 50 μL of the samples to be tested for the presence of antibodies to the wells. For antisera, initially set up 1/5 dilutions; for hybridoma supernatants test them undiluted. Incubate at room temperature for 2 h.
5. Wash the plate 2× with 150 μL/well of PBS/0.2% Tween-20 and 1% milk.
6. Add 50 μL of a rabbit a mouse antibody conjugated to alkaline phosphatase at a dilution of 1/1000. Incubate this reaction in the dark RT for 1–2 h.
7. Wash the plate 3× for 3 min each with 150 μL/well of PBS /0.2% Tween-20 and 1% milk.
8. Add 100 μL/ well of the alkaline phosphatase substrate.
9. Incubate at RT in the dark until a bright yellow color develops (2–60 min) (*see* **Note 6**).
10. Stop the reaction by adding 100 μL/ well of 0.4 *M* NaOH. Read the plate at 405–420 nm using the plate reader.

3.3. Fusion

1. The mouse myeloma, SP2/0Ag14, is used as a fusion partner. To insure that it is still HPRT-, culture it for a few wk in 20 μg/mL of 8 azo-guanine (*see* **Note 7**).
2. To prepare the spleen cells, the mouse is sacrificed by either cervical dislocation or asphyxiation with CO_2 outside of the tissue-culture area. Place the mouse in a 150-mm dish and douse with 70% ethanol. Place the mouse on its back on the styrofoam block, which has been covered with a teri towel. Use the push pins to anchor all four paws such that the abdomen of the mouse is clearly exposed. Place the mouse in the laminar flow hood.
3. Use the first set of sterile instruments to snip the skin near the throat (not the underlying musculature) and peel the skin down toward the tail, off of the abdomen. Wipe away any hair with a sterile 70% ethanol swab. Use the second set of sterile instruments to open the body cavity and locate the stomach on the left side of the mouse. The spleen is a dark red long oval organ underneath the stomach. Use the third set of instruments to remove the spleen to a sterile 100 mm petri dish. Dissect away the fat.
4. Add 5 mL of DMEM with no serum (wash medium) to the petri dish containing the spleen. Cut off one end of the spleen. Fill a 3-cc syringe with an 18 $^1/_2$-gauge needle and then change to a 27 $^1/_2$-gauge needle for the perfusions. While holding the spleen with the forceps, perfuse the spleen with media, washing the splenocytes out the open end.
5. Repeat until the splenic capsule appears clear of splenocytes.
6. Finally, bend the 27 $^1/_2$-gauge needle on the bottom of the petri dish so that it is at a 90° angle to the syringe and use it to massage any remaining splenocytes from the spleen.
7. Transfer the splenocytes to a 50-mL conical tube, increase the volume to 50 mL with wash medium (DMEM without serum or additives) and count. Pellet the splenocytes at 400*g* for 5–10 min (*see* **Note 8**).

8. Resuspend the splenocytes in wash medium and pellet. Simultaneously, melt the PEG-4000 by placing it in a 50°C water bath. Once it is melted, place it at 37°C and add an equal volume of DMEM, serum-free. Place this 50% solution at 37°C.
9. Resuspend the splenocytes and the fusion partner, SP2/0Ag14, which has been washed once in wash media, at 10^8 cells/mL in wash medium.
10. Mix 10^8 SP2/0Ag14 with 2×10^8 splenocytes. Spin at 400g for 5 min and remove the media. Wash again with wash media and pellet the cells. Remove the media.
11. Disrupt the cell pellet by tapping the tube.
12. Over a period of 1 min and with gentle mixing using the pipet tip, add 1 mL of a 50% PEG 4000 solution to the cells.
13. Continue stirring for an additional min.
14. Fill a 10-mL pipet with wash medium and add 1 mL to the cell suspension over the next min, while continuing to stir with the end of the 10 mL pipet. Add the remaining 9 mL over the next 2 min with stirring. Centrifuge the cells at 400g for 5 min.
15. Remove the supernatant and resuspend the cells in 12 mL of DMEM with 20% FCS, pen/strep and HAT with glycine as described in the Methods section. Dispense 100 µL of cells using a multi-channel pipetor into the wells of ten, 96-well, flat-bottom plates. Place at 37°C in a CO_2 incubator.
16. Clones should be visible by eye between d 7 and 10 (*see* **Note 9**).
17. Begin screening by ELISA between d 7 and 14 after fusion.

3.4. Single-Cell Cloning

1. Once a positive well is identified, it should be expanded into a single well of a 24-well dish. At this time the hybridomas can also be weaned from the HAT media by growing them in media containing only hypoxanthine and thymidine (HT medium).
2. Once the 24-well plate is confluent, count the number of cells using a hemacytometer.
3. Prepare 10 mL of cells at 50 cells/mL and 10 mL of cells at 5 cells/mL. Then seed a 96-well plate at 200 µL/well, thus, the top half of the plate will contain 10 cells/ well and the bottom half will contain 1 cell/well.
4. After 7–10 d, count the number of wells containing colonies—note whether they appear to be single colonies or multiple colonies in one well. Use the screening assay (ELISA) to check the supernatants of the wells for the presence of antibody.
5. Reclone the positive hybridoma clones by seeding two new 96-well plates at 0.3 cells/well. Repeat the screening assay and expand the positive clones (*see* **Note 10**).

3.5. Immunoprecipitation

1. Couple 50 µL of 50% Protein G agarose solution to 1 mL of hybridoma supernatant by rotating this solution for ≥ 0.5 h at 4°C.
2. Prepare a cellular lysate by washing in PBS approx 5×10^6 cells which express the protein of interest. Alternatively, approx 300 µg of tissue extract can be used (*see* **Note 11**).
3. After a lysate is prepared, the antibody-coupled beads are spun down by briefly centrifuging 3–5 s at 6000g. The supernatant is removed and the lysate is added.
4. Rotate mixture at 4°C for ≥ 2 h.
5. Pellet the antigen-antibody coupled sepharose and wash with 1 mL of lysis buffer.
6. Discard the supernatant and repeat the wash.
7. Resuspend the antigen-antibody coupled sepharose in SDS sample buffer, boil and resolve the proteins on an SDS-PAGE gel and perform standard Western blotting procedures.

4. Notes

1. Clearly identify each animal by ear punch, for example, animal #1 will have no punch, #2 will have a single punch in the left ear, and so on.
2. To test whether the adjuvant is sufficiently emulsified, it can be transferred to a 27-gauge needle and expressed onto the surface of water. If it spreads and mixes with the water, it is not sufficiently mixed. Reconnect the syringe to the double-hub needle and mix for another min. If the emulsion is performed correctly, the mixture will remain separate from the water and maintain its shape.
3. Subcutaneously is between the skin and the underlying musculature. It is most accessible by tenting the skin first by pinching the skin with your fingers.
4. Intraperitonealy is in the abdomen. Scruff the mouse and flip it over so its belly is up. Inject off of the midline.
5. Intravenously is often done through the tail vein. This is a difficult procedure and best performed by a trained animal technician.
6. Place the plate in a drawer on a piece of white paper and check it every 2 min.
7. SP2/0Ag14 cells divide more than once every 24 h and must not be overgrown when used for fusions. They should be in logarithmic growth phase for use.
8. Cells are usually pelleted in a clinical centrifuge which may not have a 'g-force' conversion. Use 1000–1500 rpms.
9. Carefully hold the plate up to the light, looking up through the bottom of the wells. Take care not to tip the plate, as media on the lid is a vector for contamination to enter the wells.
10. Many investigators add feeder cells (i.e., peritoneal wash-out cells, or splenocytes) to produce conditioned media that seems to enhance hybridoma growth and cloning. However, the addition of cells can sometimes be a source of contamination, thus, as a source of conditioned media, we have harvested the supernatants from cultured splenocytes, filter-sterilized it, and used it to supplement the single-cell cloning experiments. The following is a protocol for obtaining peritoneal wash-out cells. Peritoneal cells are obtained from normal mice by flushing the peritoneal cavities with 3–4 mL of $0.34\ M$ sucrose that has been filter-sterilized. The mouse is sacrificed, dipped in 70% ethanol, and laid on its back on toweling in a laminar flow hood. A snip is made in the skin between the hind legs and the skin is peeled forward. A pair of forceps is used to hold up the thin musculature over the abdomen and the sucrose can then be injected and withdrawn. Be careful not to puncture the gut. The peritoneal washes are pooled and centrifuged and plated at approx 2×10^4 cells/well.
11. The amount of lysate will vary depending on how abundant the protein is and where in the cell it is located, for example, nuclear proteins usually require 3 times as much lysate.

Acknowledgments

We would like to thank Julie Mowrey for her thoughtful comments on this manuscript.

References

1. Goding, J. W. (1983) *Monoclonal Antibodies: Principles and Practice.* In: Production and application of monoclonal antibodies in cell biology, biochemistry, and immunology. Academic Press, London, UK, pp. 5–55.
2. Burnet, F. M. (1959) *The Clonal Selection Theory of Acquired Immunity.* Vanderbilt University Press, Nashville, TN, USA.
3. Thomas, K. R. and Capecchi, M. R. (1987) Site-directed mutagenesis by gene targeting in mouse embryo-derived stem cells. *Cell.* **51,** 503–512.

4. Claesson, M. H., Endel, B., Ulrik, J., Pedersen, L. O., Skov, S., and Buus, S. (1994). Antibodies directed against monomorphic and evolutionary conserved self epitopes may be generated in 'knock-out' mice. Development of monoclonal antibodies directed against monomorphic MHC class I determinants. *Scand. J. Immunol.* **40(2),** 257–264.
5. Castrop, J., Verbeek, S., Hofhuis, F., and Clevers, H. (1995) Circumvention of tolerance for the nuclear T cell protein TCF-1 by immunization of TCF-1 knock-out mice. *Immunobiology* **193,** 281–287.
6. Declerck, P.J., Carmeliet, P., Verstreken, M., De Cock, F., and Collen, D. (1995). Generation of monoclonal antibodies against autologous proteins in gene-inactivated mice. *J. Biol. Chem.* **270,** 8397–8400.
7. Warren, S.T., and Nelson, D. L. (1994) Advances in molecular analysis of fragile X syndrome. *JAMA* **271,** 536–542.
8. Jin, P. and Warren, S. T. (2000) Understanding the molecular basis of fragile X syndrome. *Hum. Mol. Gen.* **9,** 901–908.
9. Ashley, C. T., Sutcliffe, J. S., Kunst, C. B., Leiner, H. A., Eichler, E. E., Nelson, D. L., and Warren, S. T. (1993) Human and murine *FMR-1*: alternative splicing and translational initiation downstream of the CGG repeat. *Nat. Genet.* **4,** 244–251.
10. Price, D. K., Zhang, F., Ashley, C. T., and Warren, S. T. (1996) The chicken FMR1 gene is highly conserved with a CCT 5'-untranslated repeat and encodes an RNA-binding protein. *Genomics.* **31,** 3–12.
11. Siomi, M., Siomi, H., Sauer, W. H., Srinivasan, S., Nussbaum, R. L., and Dreyfuss, G. (1995) *FXR1*, an autosomal homolog of the fragile X mental retardation gene. *EMBO J.* **14,** 2401–2408.
12. Devys, D., Lutz, Y., Rouyer, N., Bellocq, J. -P., and Mandel, J. -L. (1993) The *FMR-1* protein is cytoplasmic, most abundant in neurons, and appears normal in carriers of the fragile X premutation. *Nat. Genet.* **4,** 335–340.
13. Harlow, E. and Lane, D. (1988) *Antibodies: A laboratory manual.* Cold Spring Harbor Laboratory Press, Cold Spring Harbor, NY, USA.
14. Coligan, J. E., Kruisbeek, A. M., Margulies, D. M., Shevach, E. M., and Strober, W. (1999) *Current Protocols in Immunology.* John Wiley and Sons, Inc., New York, NY, USA.
15. Roitt, I. (1988) *Essential Immunology,* 6th ed., Blackwell Scientific Publication, Oxford, UK.
16. Anonymous. (1994). Fmr1 knockout mice: a model to study fragile X mental retardation. *Cell.* **78(1),** 23–33.
17. Ceman, S., Brown, V., and Warren., S. T. (1999) Isolation of an FMRP-associated messenger ribonucleoprotein particle and identification of nucleolin and the fragile X-related proteins as components of the complex. *Mol. Cell. Biol.* **19(12),** 7925–7932.

32

Immunological Methods for the Analysis of Protein Expression in Neuromuscular Diseases

Mariz Vainzof, Maria Rita Passos-Bueno, Mayana Zatz

1. Introduction

Protein studies are of utmost importance for enhancing our understanding of genotype:phenotype correlations, as well as for diagnostic purposes. This is particularly true for the study of neuromuscular diseases, where defects in protein expression directly contribute to the ethiology of disease.

Different approaches have been used for studying proteins, including assays for specific biological activities and methods for the detection and localization of the whole or part of a protein. The development of sensitive techniques to allow the measurement of very small amounts of proteins are very important, and the use of antibodies that react specifically with entire proteins or specific epitopes became the preferred used methodologies *(1)*.

Studies of protein on unfixed frozen sections represent a close approximation to the situation in vivo, because they are in an almost native form. Immunohistochemical analysis of proteins can provide us with information about their localization in tissues or cell structures, and the presence or absence of specific epitopes. In contrast, assessment of tissue proteins by Western blot analysis, allows one to study denatured proteins. Through this methodology, it is possible to assess the presence of a specific protein (band visible or not), its size (through migration distance) and approximate amount (density of the band) *(2)*.

1.1. The Muscular Dystrophies

Duchenne (DMD) and Becker (BMD) muscular dystrophies are allelic conditions caused by mutations in the dystrophin gene, at Xp21 *(3–5)*. The limb-girdle muscular dystrophies (LGMDs) include a heterogeneous group of progressive disorders mainly affecting the pelvic and shoulder girdle musculature. The inheritance may be autosomal dominant (LGMD1) or recessive (LGMD2). Clinically, LGMD ranges from severe forms with onset in the first decade and rapid progression, to milder forms with later onset and slower progression *(6)*. During the last decade, at least 15 LGMD genes, six autosomal dominant (AD) and nine autosomal recessive (AR), have been mapped. The AD forms are relatively rare and probably represent less than 10% of all LGMD *(6,7)*.

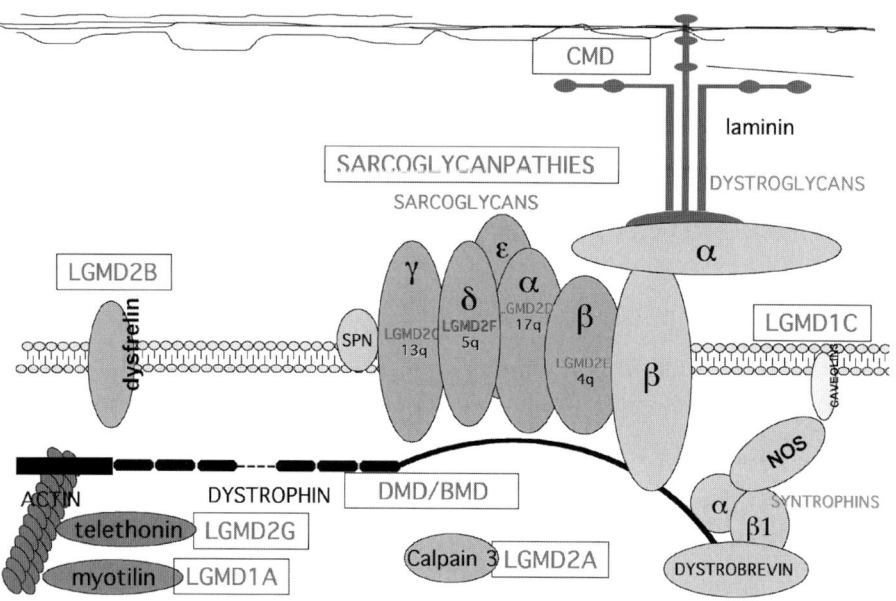

Fig. 1. Schematic representation of proteins from the dystrophin-glycoprotein complex, sarcolemmal, sarcomeric, and cytosolic proteins involved in the different forms of neuromuscular disorders. (See color plate 10 appearing in the insert following p. 82)

The six AD-LGMD forms are: LGMD1A at 5q22, coding for the protein myotilin *(8,9)*, LGMD1B at 1q11, coding for lamin A/C *(10)*, LGMD1C at 3p25 coding for caveolin-3 *(11–12)*, LGMD1D at 6q23 *(13)*, LGMD1E at 7q *(14)* and LGMD1F at 5q31 *(15)*.

Up to now, nine AR forms have been mapped and with the exception of LGMD2H at 9q31-33 *(16)* and LGMD2I, at 19q13.3 *(17)*, all the others have had their protein products identified. Four of them, mapped at 17q21, 4q12, 13q12 and 5q33, encode respectively for α-sarcoglycan (α-SG), β-SG, γ-SG and δ-SG, which are glycoproteins of the sarcoglycan sub-complex of the dystrophin-glycoprotein complex (DGC) *(18,19)*. Mutations in these genes cause, respectively, LGMD2C, 2D, 2E and 2F and constitute a distinct subgroup of LGMDs, the sarcoglycanopathies *(20–33)*. The three other identified forms are LGMD2A, at 15q15.1, coding for calpain 3 *(34–35)*, LGMD2B, at 2p31, coding for dysferlin *(36–37)* and LGMD2G, at 17q11-12, coding for the sarcomeric telethonin *(33,38)* (**Fig. 1**).

Protein studies through the analyses of the expression of the proteins of the DGC and of the sarcomere, are of utmost importance for the diagnosis and elucidation of the physiopathology of these diseases. This chapter will focus on immunological methods for the analysis of protein expression in neuromuscular diseases.

2. Materials

2.1. Antibodies for Neuromuscular Disorders

Most antibodies against muscle proteins are now commercially available. Novacastra (Newcastle, UK) is offering the antibodies developed by Louise V.B. Anderson, from

Newcastle (UK). Dystrophin (3 different domains), the 4 sarcoglycans, β-dystroglycan, calpain 3, dysferlin, α2-laminin, and emerin.

Additionally, Chemicon, Gibco (USA) has antibodies for the different domains of merosin, while Transduction Laboratories (USA), antibodies for caveolin 3. Sigma also commercializes some antibodies for dystrophin.

Secondary antibodies: for immunohistochemistry, a secondary antibody against the immunoglobulin G (IgGs) of the species in which the primary antibody was raised (anti-mouse, rabbit, sheep, goat) is used, conjugated to fluorochromes, such as fluorescein, rodamin, texas, CY3, and so on (*see* **Note 1**). For double reactions, it is important to use fluorochromes that do not have overlapping wave lengths. For Western blotting, the secondary antibodies can be conjugated to enzymes, such as peroxidase, or alkaline phosphatase. All are commercially available.

2.2. Tissue Sample

Immunological studies of proteins can be done on a wide range of different tissues and cells. In neuromuscular disorders they are done in muscle samples usually obtained through open biceps biopsies, under local anesthesia. Commonly, 3–5 small fragments of 5 × 5 mm are incised. Three fragments, maintaining the orientation of the fibers, are mounted on a cork, using OCT-tissue Tek as crio-protector, and frozen in liquid nitrogen. These fragments are used for morphologic and *in situ* studies on sections. The other 2–3 fragments are quickly frozen in liquid nitrogen, and stored in an eppendorf tube, for protein extract preparations.

The fragments can be stored both in liquid nitrogen or at –70°C until use. Muscle biopsies stored up to 12 yr in our lab, have been maintained in good conditions.

A routine histological and histochemical analysis is done in all muscle samples *(39)*.

2.3. Apparatus

The BioRad mini Protean II 16 cm system, with 1.5 mm spacers and 15-wells combs, have been used for electrophoresis, while the mini-Transblot Western blotting system from BioRad or other suppliers have been used for blotting.

For electrophoresis, as well as for electro-blotting, any power supply with 250V/ 500mA is suitable.

Small additional facilities, such as a boiling water bath, microcentrifuge, gel dryer, and rocking table, are also necessary.

For densitometric analysis, several densitometric softwares are commercially available. An Imaging Densitometer with software and SCSI card for use with a PC computer is supplied by BioRad.

2.4. Solutions and Buffers

It is highly recommended all chemicals and reagents to be as pure as possible. Only distilled deionized water should be used for preparing all reagents. Additional requirements are liquid nitrogen and Ponceau S (from Sigma). For nitrocellulose, we use the hybond C-extra filter (Amersham, USA).

1. Phosphate buffered saline (PBS)-S, pH 7.3: 137 mM NaCl, 2.7 mM KCl, 4.3 mM Na$_2$HPO$_4$, 1.4 mM KH$_2$PO$_4$
2. Prepare fresh daily: 5% horse serum (Sigma) 0.05% Triton X-100.

3. Solubilization solution (stock solution): 1% sodium dodecyl sulfate (SDS) (*see* **Note 2**), 10 m*M* EDTA, 0.1 *M* Tris, pH 8.0. Filter and store at room temperature
4. 1*M* DTT: Aliquot (1 mL eppendorf and store at −20°C)
5. Prepare fresh: 5 mL Solution 3 + 0.05 mL solution 4.
6. Sample buffer (2X): 0.130 *M* Tris, 20% glycerol, correct pH to 6.8 (HCl), 4% SDS, 2% β-mercaptoetanol, 1 mg bromophenol blue; aliquot and freeze (−20°C).
7. Separating gel buffer (4X): 1.5 *M* Tris, 0.4% SDS, pH 8.7 (correct with HCl). Filter and store at room temperature.
8. Acrylamide stock (30%:0,8% cross linker) (*see* **Note 3**): 30% acrylamide, 0.8% bis-acrilamide. Filter and store in a dark bottle in a refrigerator.
9. Ammonium persulfate 10%: prepare fresh daily.
10. Stacking gel buffer (4X): 0.5 *M* Tris, 0.4% SDS, pH 6.8. Filter and store at room temperature.
11. Electrophoresis buffer (5X): 0.125 *M* Tris, 0.96 *M* glycine, 0.5% SDS, pH 8.3.
12. Staining solution: 0.05% Coomassie blue, 25% mL isopropanol, 10% acetic acid. Filter and store in a dark bottle at room temperature.
13. Destaining solution: 7.5% acetic acid, 5% methanol.
14. Transfer buffer (10X): 25 m*M* Tris, 250 m*M* glycine, 0.24% SDS.
15. Prepare freshely: 1X transfer buffer, 20% methanol.
16. Ponceau S (Sigma) 20%. It can be used several times.
17. TBS (10X): 150 m*M* NaCl, 10m*M* Tris, pH 7.5, filter, and autoclave.
18. Washing solution (TBS-Tween- TTBS): 1X TBS, 0.05% Tween 20.
19. Blocking solution: 5% non fat milk in TBS, filter, add 1% human (or horse) serum; aliquot (5 mL) and freeze (−20°C).
20. Antibody buffer: Dilute the buffer-milk 5% to 1%, 1% of human serum (or horse serum), 0.05% Tween 20; aliquot (5 mL), and freeze (−20°C).
21. Alkaline Phosphatase (AP) buffer: 1*M* Tris, 1*M* NaCl, 1*M* MgCl$_2$·6H$_2$O, pH 9.5.
22. NBT-Nitro-blue-Tetrazolium : 33 mg/mL in DMF (N,N-Dimethylformamide) 70%.
23. BCIP- 5-Bromo-4-Chloro-3-indolyl Phosphate: 16.5 mg/mL DMF 100%.
24. Reaction solution: 5 mL AP buffer + 0.5 mL NBT + 0.5 mL BCIP.

3. Methods

3.1. Immunohistochemistry

We use previously described routine methodologies *(40, 41)*, with some small adaptations, as follows:

1. Muscle sections of 5–6 mm thick are cut in a cryostat microtome and mounted in slides coated with polylisine (*see* **Note 4**).
2. The sections can be stored and maintained at −70°C, wrapped in cling-film, until use.
3. Before use, slides are defrosted and allowed to air dry at room temperature for 1 h.
4. The sections are incubated with PBS-S for 10 min (*see* **Note 5**).
5. The PBS-S is removed, by inclining the slide and surrounding the section with a Kleenex paper.

3.1.1. Incubation with the First Antibody

Dilution of primary antibodies: it depends on the specificity of each antibody (*see* **Note 6**) (monoclonal antibodies [MAbs] can be diluted between 1/5 to 1/20 and polyclonal antibody [PAb], between 1/100 to 1/2000) (*see* **Note 7**).

1. The dilution of the antibody is done in PBS-S in the previously tested concentration.
2. The antibody is quickly centrifuged (3 min, 18,000*g*)

3. 20–30 μl of the antibody is applied to sections, which are covered with a cover-slide and incubated for at least 1 h (*see* **Note 8**).
4. After incubation, the antibody is removed and the sections are washed 3–4 times for 10 min with PBS-S.
5. Incubation with the second antibody, usually diluted 1/200, is done for 1 h.
6. The sections are washed again 3–4 times for 10 min with PBS-S
7. The sections are mounted with Vectashield mounting medium for fluorescence (Vector).
8. The slides can be stored at 4°C for several wk.
9. The analysis is done in a microscope with epi-fluorescence, using specific filters for each fluorochrome.

3.1.2. Double Immunofluorescence Analysis

This methodology is adapted from the single immunofluorescence (IF) and has been used routinely in our laboratory, when two different antibodies, made from different animals, such as an N-terminal dystrophin antibody made in rabbit, and a C-terminal made in mouse, are available *(42)*. This procedure allows to detect the presence and possible co-localization of two proteins, or two different epitopes of a specific protein in exactly the same section.

1. Two primary antibodies are mixed, maintaining their original concentrations tested before use (*see* **Note 9**).
2. The sections are incubated with the primary antibodies, during the period of time previously tested for single reactions.
3. The sections are washed 3 × 10 min with PBS-S.
4. The sections are incubated with a mixture of the 2 secondary antibodies, as for example, anti-rabbit conjugated to fluorescein and anti-mouse conjugated to texas red or CY3 (1 μL of each in 100 μL of PBS-S), for 1 h.
5. After 3 × 10 min washes, the sections are mounted in vectashield mounting medium. analyzed under the fluorescent microscope, by changing filters: a green color is seen for the N-terminal region of dystrophin, and a red color for the C-terminal region of dystrophin. The results are illustrated in **Fig. 2A**.

For the control of the plasma membrane and basal lamina preservation, it is also recommended to study dystrophin through a double reaction, with an antibody for a muscle membrane protein, such as β-spectrin, and α2-laminin *(41)*.

3.2. Western Blot Methodology

The most used analytic methodology is based on the SDS-Polyacrlyamide gel Electrophoresis (SDS-PAGE) of proteins, with the methodology described by Laemmli *(43)*, and Western blotting according to Towbin et al. *(44)*.

The electrophoresis of proteins is carried out in polyacrylamide gels under conditions with strong anionic detergent SDS in combination with a reducing agent and heat, which dissociates the proteins into their individual polypeptide subunits. The denatured polypeptides bind SDS—proportionally to their molecular weight—become negatively charged, and the complex migrates through polyacrylamide gels in accordance to the size of the polypeptide *(1)*.

The SDS-PAGE is carried out with a discontinuous buffer system, with different pH and ionic strength in the buffer, in the reservoirs, and in the one used in gel. When an electric current is passed between the electrodes, the applied proteins in the sample are

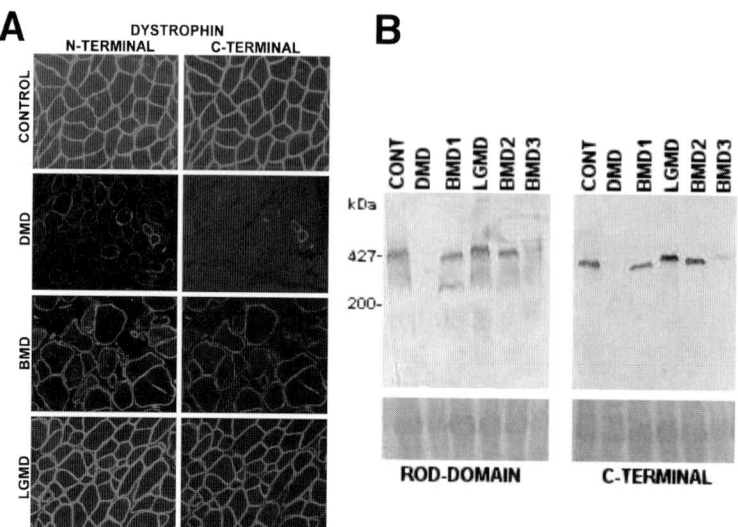

Fig. 2. Dystrophin analysis through the use of N-terminal and C-terminal antibodies. (**A**) Double imunohistochemical analysis in a normal control, and in DMD, BMD and LGMD patients. Note the revertent positive fibres in the DMD patient, and the weaker pattern of labeling in the BMD patient. (**B**) Western blot analysis in the same patients, showing no 427 kDA band in DMD, a double 400/250 band (with N-term antibody) in patient BMD1, who has a deletion of exons 45–48, normal bands in a LGMD patient, a band with a smaller size (410 kDa) in patient BMD2, and a weaker 427 kDa band in patient BMD3. (See color plate 10 appearing in the insert following p. 82)

separated according to their molecular weight (MW). The ability of discontinuous buffer systems to concentrate all the complexes in the sample into a very small volume greatly increases the resolution of SDS-PAGE *(1)*.

Subsequently, the electrophoretically separated proteins are transferred from the gel to a nitrocellulose filter, and are probed with antibodies. The bound antibody is detected by one of several secondary immunological reagents, such as anti-immunoglobulin coupled to horseradish peroxidase or alkaline phosphatase.

The technique is very sensitive and because electrophoretic separation of proteins is almost always carried out under denaturing conditions, problems of solubilization, aggregation, and co-precipitation of the target protein are minimized. On the other hand, individual immunoglobulin may preferentially recognize a particular conformation of its target epitope, and consequently, not all monoclonal antibodies [MAbs] are suitable for use as probes in Western blots, where the target proteins are denatured *(2)*.

Many different methodologies have been published, but the one used in our lab is based on the methodologies described by Zubrzycka-Gaarn et al. and Bulman et al. *(45–46)*, with some modifications adapted from Ho-Kim et al. *(47)*.

3.2.1. Preparation of the Sample

1. Small fragments of muscle tissue (~50 mg) are crashed, in a liquid nitrogen bath. The frozen muscle powder is transferred to an eppendorf tube (always maintained frozen).
2. The sample is immediately homogenized with 100–200 µL of boiling solubilization solution.

Table 1
Gel Mixture Preparation

Gel concentration	6%	10%	13%
H$_2$O	10.85 mL	8.3 mL	6.2 mL
Separating buffer (4X)	5 mL	5 mL	6 mL
Acrylamide	4 mL	6.7 mL	8.7 mL
TEMED	5 µL	5 µL	5 µL
APS 10% (see **Note 10**)	150 µL	150 µL	150 µL
Use	Dystrophin Multiplex	Calpains	Sarcoglycans

Table 2
Stacking Gel Mixture Preparation

Gel concentration	Stacking 4%
H$_2$O	3.175 mL
Stacking buffer (4X)	1.25 mL
Acrylamide	0.575 mL
TEMED	3.75 µL
APS 10%	6.25 µL

3. The sample is placed in boiling water for 2–5 min, vortexing and crashing the extract with a small rod several times during the boiling process and spun in an eppendorf microcentrifuge for 15 min at 18,000g.
4. The supernatant is separated (muscle extract) and protein amounts are measured using routine methodology for quantifying proteins.
5. The extract samples are aliquoted, pipeting 50 µg of proteins (in a volume up to 10 µL) and adding the same volume of sample buffer (2X).
6. The samples are placed in boiling water-bath for 2 min just before applying.

3.2.2. Preparation of the Gel

1. The inner surfaces of glass plates should be well cleaned with alcohol, assembled in cassettes, and placed in casting stand.
2. The gel mixture is prepared, according to the size of the protein to be studied (see **Table 1**).
3. For 2 gels: 20 mL of the resolving gel and 5 mL for the stacking gels are used.
4. The resolving gel is poured into the cassette, avoiding bubbles and leaving about 4 cm on the top for the stacking gel and combs.
5. About 2–3 mL of water across the top of the gel is pipeted gently.
6. The gel is placed in a vertical position at room temperature and allowed to set for ~1 h (see **Note 11**).
7. When ready, the water from the gel is poured, and drained with folded filter paper.
8. The stacking gel is prepared and poured in the cassettes (see **Table 2**).
9. The combs are gently inserted, avoiding bubbles.
10. The gel is placed in a vertical position at room temperature and allowed to set for ~45 min.
11. The combs are removed, and the wells are rinsed out with water.

3.2.3. Running the Gel

1. The cassettes are fit in the central core of the electrophoresis tank and the upper chamber is filled with reservoir buffer, diluted 1/10.
2. 20 µL/lane of the samples are applied to the bottom of the wells.
3. A known molecular weight marker is also applied in the first lane of the gel.
4. The cassettes/upper tank unit is placed in the main tank, topped up with reservoir buffer, and the lid is put on.
5. They are run at 70–80 V, for approx 2 h, until the samples reach the lower part of the gel.
6. When finished, the power supply is turned off, and the gel cassettes are removed and processed for staining or blotting steps.

3.2.4. Staining/Destaining the Gel

For each new muscle sample, first stain the gel with Coomassie blue, to evaluate the quality of the extract and the amount of muscle proteins, such as myosin.

1. The gel is put in a glass recipient, covered with Coomasie blue staining solution, and left shaking for 1 h at room temperature.
2. The stain is removed (and can be saved for future use).
3. The gel is briefly washed with water.
4. Destaining solution is added, soaking the gel.
5. Three to four sheets of Kleenex paper are added (*see* **Note 12**), and the glass recipient is covered with film and incubated under agitation, overnight.
6. All the color will be taken off, and the bands on the gel will be evident.

3.2.5. Western Blotting the Gel

1. The assembling of blotting is previously prepared as follows:
2. The nitrocellulose membrane is first immersed in the transfer buffer, for 1 h.
3. Each blotting "sandwich" is assembled, as follows, on the black negative side of the cassette:
   a. Soaked Scotch Brite pad.
   b. Soaked 3 sheets of Whatman filter paper.
   c. Gel.
   d. Soaked nitrocellulose filter.
   e. Soaked 3 sheets of Whatman filter paper.
   f. Soaked Scotch Brite pad.
4. The absence of air bubbles between the gel and the nitrocellulose is avoided by gently rolling with the fingers over the last filter paper layer. The sandwich is closed.
5. The cassettes are inserted into the blotting tank, being careful with the negative (black) and positive (red) terminal markers.
6. All the cube is inserted in a refrigerator (or cold-room) and connected to the power supply: 100 V - 250 mA for 1 h.
7. After 1 h, the power supply is switched off and the cassettes are opened.
8. With a pencil, the upper and bottom part of the blot are marked (this is important for calculating the MW).
9. The nitrocellulose blots are removed and dried between paper towels, in a 37°C oven, overnight. This step is very important for fixing the proteins on the membrane.

3.2.6. Pre-Staining the Blot

1. The blots can be stained with total protein reversible stain Ponceau S for evaluation of the quality of the transfer.
2. The blots are immersed in the Ponceau S solution for 15 min, with gentle agitation.

3. The Ponceau S is taken out and the blots are washed 3 times with water and dried between two sheets of filter paper.
4. The lanes are identified and the blots are scanned or photographed.

3.2.7. Reacting the Blot with Antibodies

1. The blots are placed in well-sealed plastic bags.
2. All the next steps are done with the blots inside plastic bags, at room temperature and gentle agitation.
3. The blots are first destained with TBST-washing solution, which removes the Ponceau S.
4. Unreacted binding sites on the nitrocellulose are blocked by incubation in blocking buffer solution for 1 h, at room temperature.
5. The blocking solution is removed, and the blots are washed briefly with washing buffer and incubated with the primary antibody.
6. The time of incubation depends on the quality of the antibody, varying between 1 h and 12 h.
7. The dilution of the antibodies also depends on their concentration and quality. MAbs are usually diluted 1/10–1/20, whereas PAbs 1/100–1/1000.
8. The antibody is removed from the bag (see **Note 13**), and the blots are washed 3–4 times for 15 min with TBST.
9. The blots are incubated for 1 h with the secondary antibody (anti-immunoglobulin according to the animal in which the antibody was made, anti-mouse, anti-rabbit, anti-sheep).
10. Usually, the concentration is 1/1000, for alkaline-phosphatase conjugated antibodies.
11. The antibody is removed and the blots are washed 3–4 times for 15 min with TBST.

3.2.8. Revealing the Reaction on Blots

1. The blots are briefly washed with water and incubated with the staining reaction, shaking it while the development of color occurs (1–5 min).
2. Bands of purple color will appear in the expected MW and are analyzed for presence of the protein, its size, and its abundance.

3.2.9. Analysis of the Protein

According to the analyzed sample, the protein band may be normal, abnormal in size or intensity, or absent.

3.2.9.1. ANALYSIS OF THE PROTEIN SIZE

The relative mobility of a protein is determined by measuring the distance from the top of the resolving gel to the protein band and dividing by the distance from the top to the bottom of the gel run. These measures are compared in a curve where the $\log_{10}$ of the polypeptide molecular mass is plotted against the Rf obtained running a standard serie of proteins of known size (2).

In some cases, when a small reduction on the MW of a protein should be confirmed, the patients' sample can be mixed with a normal control protein and blotted. In this case, a double band should appear.

3.2.9.2. ANALYSIS OF THE PROTEIN ABUNDANCE

The blot is scanned and submitted to densitometric analysis, using commercial densitometry packages. The amount of the pathological band must be compared to a normal control band run in the same blot.

Because pathological muscle can contain fat and connective tissue, a correction of the true muscle proteins loaded must be done, routinely using the myosin heavy chain on the Ponceau pre-stained blot.

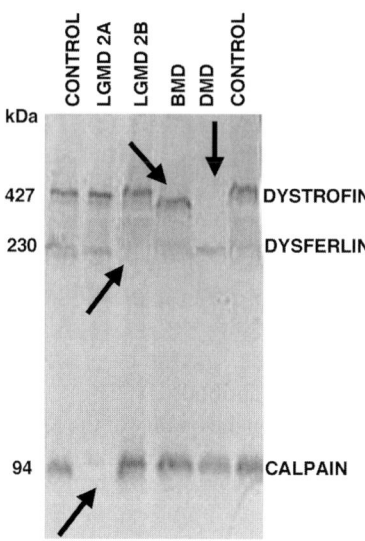

Fig. 3. Multiplex Western blot analysis for dystrophin (30 kDa antibody), dysferlin, and calpain in a normal control and in patient with LGMD2A, 2B, BMD, and DMD. The arrows are indicating the altered proteins bands. (See color plate 10 appearing in the insert following p. 82)

3.2.9.3. MULTIPLEX BLOT

The blot can be probed with a mix of several different antibodies, for proteins of different MW, and reacted together. This methodology is very efficient for extensive screening of muscle proteins deficiencies in large samples of muscle biopsies, as illustrated in **Fig. 3**.

3.3. Results

3.3.1. Xp21 Muscular Dystrophies

Dystrophin, the primary defect in Xp21 MD, is a large 427 kDa rod-shaped subsarcolemmal protein, coded by a gene on chromosome X. The amino terminus of dystrophin binds to actin, and the carboxyl terminus, which is rich in cysteine, links dystrophin to a complex of glycoproteins in the sarcolemma *(18,19,48)*. This dystrophin-glycoprotein complex (DGC) is vital for normal muscle functioning. Dystrophin deficiency causes the Duchenne/Becker muscular dystrophies *(3,4)*.

The DGC forms a bridge across the muscle membrane, between the inner cytoskeleton (dystrophin) and the basal lamina (merosin). It is accepted that the DGC stabilizes the sarcolemma and protects muscle fibers from long-term contraction-induced damage and necrosis. The DGC comprises the dystroglycan (DG), the sarcoglycan (SG), and the syntrophins/dystrobrevin subcomplexes **(Fig. 1)**. In addition to the mechanical and structural function of the dystrophin- glycoprotein complex (DGC), it has been recently shown that this complex might play a role in cellular communication *(49)*, as well as interacting with the sarcomeric network, through the binding of dystrophin to F-actin (*see* revision in **ref.** *(50)*).

About two-thrids of the DMD patients have a detectable frame-shifting deletion in the dystrophin gene, while the remaining have nonsense point mutations or small deletions or rearrangements. All these mutations lead to the deficiency of the protein dystrophin in the muscle *(5)*.

Dystrophin double IF studies have been done in more than 400 DMD patients from our laboratory. Almost all DMD patients are classified as dystrophin deficient using the C-terminal antibody. However, a variable proportion (4–30%) of revertent dystrophin positive isolated or grouped fibers are usually observed in the majority of them, mainly with the N-terminal antibody (**Fig. 2A**). This small amount of dystrophin is observed as faint bands on WB (**Fig 2B**), but with no correlation with the clinical course *(40,51)*. Counting of the number of native revertent fibers is very important, due to the possibility in the future of therapeutic trials with the replacement of dystrophin competent genes.

Patients with an intermediate clinical phenotype between DMD and BMD (outlier patients) usually show the same dystrophin deficient pattern as typical DMD.

In 100 patients affected by BMD, IF showed a positive pattern in about 90% of the cases, with a variable degree of patchiness (**Fig 1C**). However, some mildly affected patients showed a significant deficiency of dystrophin, while some severely affected patients showed high amount of the protein in the muscle *(52)* (**Fig. 2**). In addition, BMD patients with mutations of exons 45–49 often present two bands of dystrophin, using N-terminal antibodies *(53)*.

3.3.2. Sarcoglycanopathies

The four known components of the SG complex include α-SG, β-SG, γ-SG, and δ-SG. They are assembled in a complex that is inserted into the membrane. Mutations in any one of the 4 SG proteins lead to sarcoglycanopathies: LGMD2C, 2D, 2F and 2E, which are relatively common LGMD forms among the more severely affected patients *(54)*. Many different mutations have already been identified in all 4 sarcoglycan genes, including missense, splicing, nonsense, small and large gene deletions are listed on the Internet (*see* Website: http:// www.nl.dmd).

In the majority of muscle biopsies from patients with a primary sarcoglycanopathy, the primary loss or deficiency of any one of the four sarcoglycans, in particular of β- and δ-SG, leads to a secondary deficiency of the whole subcomplex (**Fig. 4**) *(7,49,55–58)*. However, exceptions may occur, such as the finding of a deficiency of γ-SG with a partial preservation of the other three SG in LGMD2C *(57)* or the partial deficiency of only α-SG with the retention of the other 3 in LGMD2A *(56,59)*. These findings also have important implications for diagnosis as they may indicate which gene should be first screened for mutations.

An important observation is the secondary reduction in dystrophin amount that may be seen in patients with primary sarcoglycan mutations, particularly in patients with primary γ-SG deficiency, leading to the suggestion that γ-SG might interact more directly with dystrophin *(57)*. Therefore, dystrophin deficiency may occur in non Becker MD, which should be taken in consideration for differential diagnosis.

3.3.3. Calpainopathy

Calpain 3, the calcium activated neutral protease 3, is the muscle-specific 94 kDa enzyme that binds to titin. As a cysteine protease, it plays a part in the disassembly of sarcomeric proteins, but it may also have a regulatory role in modulation of transcription factors. The loss of its function leads to activation of other proteases.

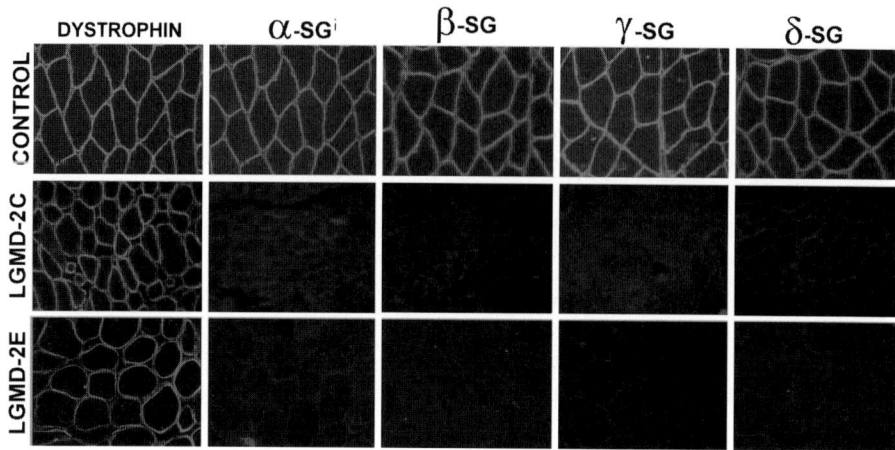

Fig. 4. IF analysis for dystrophin and the 4 sarcoglycans proteins in a control, in one LGMD2C, and one LGMD2E patients. (See color plate 10 appearing in the insert following p. 82)

Mutations in the calpain 3 gene at 15q15 may cause deficiency of the protein and LGMD2A *(34)*. LGMD2A patients present a wide range of distinct pathogenic mutations distributed along the entire length of the calpain 3 gene *(35)*. Interestingly, screening of Brazilian LGMD2A patients showed some prevalent mutations concentrated in three exons *(60)*.

The study of calpain 3 protein in muscle can only be done through WB, because the available antibodies do not react on sections. A first trial can be done through the multiplex WB analysis for dystrophin, dysferlin, and calpain (**Fig. 3**). If a reduction is suspected, a new blot specific for calpain antibodies is done (13% gel), and the presence of the three possible bands is analyzed (**Fig. 5**). The analysis of calpain on blot is not easy, since degradation of the muscle extract is frequently the cause of doubtful results. Calpain in LGMD2A patients can show a total, partial, or no deficiency at all, and no direct correlation has been observed between the amount of calpain and the severity of the phenotype. Very low levels or no expression of calpain 3 was seen in European and Brazilian patients with a clinical course varying from mild to severe *(61,62)*. LGMD2A patients with missense mutations may present faint 94kDA calpain 3 bands *(61)*, suggesting that some mutations may affect the protein function, which is not removed from muscle.

A normal amount of calpain was found in sarcoglycanopathies *(61,62)*, as well as normal SG proteins in LGMD2A *(57)*, suggesting no direct relation between calpain 3 and the sarcoglycan complex. In addition, normal calpain bands in LGMD2G patients suggest no correlation with telethonin as well *(63)*.

However, an unexpected reduction of calpain 3 was observed in LGMD2B patients, suggesting a possible association between calpain 3 and dysferlin, which requires further studies *(62,64)*.

3.3.4. Dysferlinopathies

Dysferlin, coded by a gene on 2p12-14, is an ubiquitously expressed 230 kDa molecule that is localized in the periphery of muscle fibers, linked to the sarcolemmal

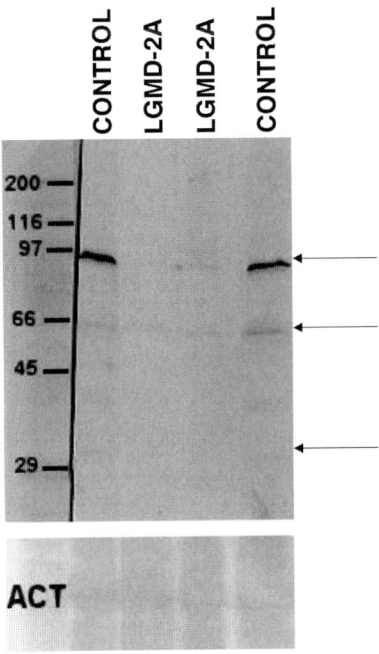

Fig. 5. Western blot analysis for calpain3, in a 13% gel, showing the three detectable calpain bands in the controls, a total deficiency in the LGMD2A patient on the left, and a partial deficiency in the LGMD2A patient on the right. (See color plate 11 appearing in the insert following p. 82)

membrane. Two distinct phenotypes are associated with mutations in this gene: Miyoshi myopathy, which is a predominantly distal muscle wasting *(37)*, and LGMD2B, with a proximal weakness *(36)*. Only a few mutations have been identified, due to the large size of the dysferlin gene (55 exons), and no apparent hot spot for mutations. Therefore, muscle protein analysis is very helpful.

Protein analyses in LGMD2B have shown a total deficiency of dysferlin, both through immunofluorescence and western blot **(Fig. 6)**. Although a partial deficiency has been reported in LGMD2B patients *(65)*, dysferlin deficiency seems to be specific to LGMD2B in our patients, and has not been seen as a secondary effect in other forms of MD *(66)*. Dysferlin is an ubiquitously expressed protein, and can be detected also in the skin and in corionic villus biopsy **(Fig. 6)**.

In DMD and sarcoglycanopathies, a normal localization and molecular weight (MW) for this protein was found, suggesting no interaction between dysferlin and the DGC.

3.3.5. Telethoninopathy

The sarcomere is the unit of skeletal and cardiac muscle contraction. In the past few years, there have been many studies focusing the role of skeletal and cardiac muscle proteins *(67,68)*. Mutations in several sarcomeric proteins such as telethonin *(33)*, myotilin *(9)*, actin *(69)*, tropomiosin 3 and 2 *(70,71)*, nebulin *(72)*, troponin T1 *(73)*, have been associated to human muscle diseases. We have recently detected one nemaline myopathy affected patient, with a deficiency of only the SH3 domain of nebulin, through Western blot analysis *(74)*.

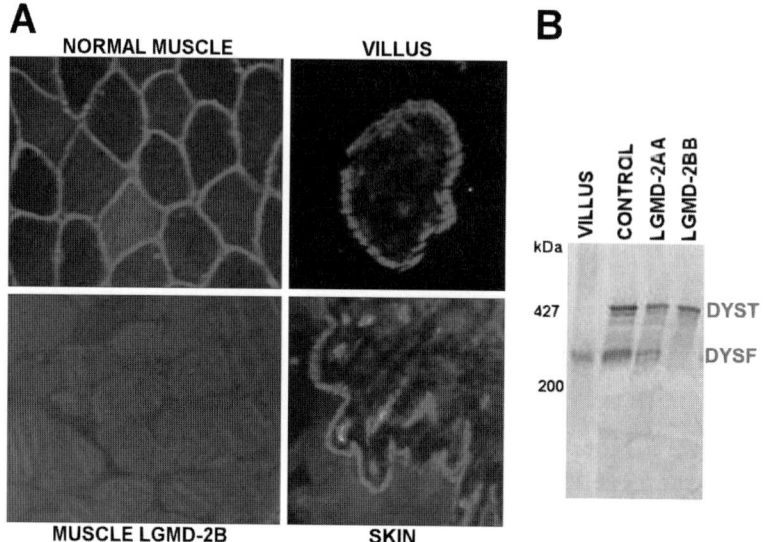

Fig. 6. **(A)** Dysferlin through IF analysis showing the normal sarcolemmal pattern in the control muscle, the positive labeling in the normal villus and in skin, the negative pattern in the muscle from one LGMD-2B affected patient. **(B)** Multiplex Western blot analysis for dysferlin (with dystrophin, at 427 kDa), showing the presence of the dysferlin 230 kDa band in the normal villus sample, in the control, and in the LGMD-2A patient, and no dysferlin band in the LGMD-2B patient. (See color plate 11 appearing in the insert following p. 82)

The role of the majority of these proteins is still unknown. However, their presence in affected patients with the aforementioned conditions suggests an essential role in the constitution of the sarcomere, since total deficiencies are probably incompatible with life. New methodologies to detect possible alterations in sarcomeric proteins have to be developed to elucidate their role.

Telethonin is a sarcomeric protein of 19 kD, present in the Z disk of the sarcomere of the striated and cardiac muscle *(38)*. Mutations in the telethonin gene at 17q cause LGMD2G *(33)*. Telethonin was found to be one of the substrates of the serine kinase domain of titin, that acts as a molecular ruler for the assembly of the sarcomere by providing spatially defined binding sites for other sarcomeric proteins. The specific function of telethonin and its interaction with the other proteins from the muscle is still unknown.

Protein analysis in 6 LGMD2G Brazilian patients, from four unrelated families, showed deficiency of telethonin in all of them **(Fig. 7)**, associated with frameshifted mutations in the LGMD2G gene. The possibility of other mutational mechanisms in this sarcomeric gene associated with the presence of the protein in the muscle cannot be ruled out yet.

Additional protein studies in these patients have shown normal expression for the proteins dystrophin, sarcoglycans, dysferlin, calpain, and titin. Immunofluorescence analysis for α-actinin-2 and myotilin showed a cross-striation pattern, suggesting that at least part of the Z-line of the sarcomere is preserved. Ultra-structural analysis confirmed the maintenance of the integrity of the sarcomeric architecture. Therefore, mutations in the telethonin gene do not seem to alter the sarcomere integrity *(63)*.

Telethonin was clearly present in the rods, in muscle fibers from patients with nemaline myopathy, confirming its localization in the Z-line of the sarcomere. The analysis of telethonin on muscle biopsies from patients with LGMD2A, LGMD2B, SGpathies, and DMD showed normal localization, suggesting that the deficiency of calpain, dysferlin, sarcoglycans, and dystrophin does not seem to alter telethonin expression *(63)*.

3.3.6. Caveolin 3 Deficiency

Caveolin3 is the protein present in the Caveolae, small invaginations in the plasma membrane that are present in most types of cells, and is probably involved in signal transduction.

Mutations in the caveolin-3 gene (CAV-3) with a negative dominant effect and reduction of the protein expression cause autosomal dominant LGMD1C muscular dystrophy *(11)*. It has been suggested that CAV-3 mutations might also cause the AR-LGMD form *(12)*. However, recent screening for mutations in the CAV-3 gene in 61 Brazilian LGMD patients and 100 normal controls has not confirmed the existence of the AR caveolin deficiency *(75)*.

3.3.7. Congenital MD with Merosin Deficiency

Laminin 2 is a constituent of the basal lamina, which links to dystroglycan and which provides structural support in the extracellular matrix. It is composed by three chains: α-2, β-1 and γ-1. Laminin α-2 deficiency due to mutations in the LAMA2 gene at 6q2 is the cause of the autosomal recessive Congenital MD *(76,77)*. Laminin α2 is totally deficient in muscle biopsies from patients with the severe typical congenital dystrophy phenotype. However partial deficiencies have been described in patients with heterogeneity in the clinical picture (**Fig. 8**) *(78,79)*. The protein is ubiquitously expressed, and may be detected in skin biopsy *(80)* as well as in chorionic villus, which is a very useful test for prenatal diagnosis *(81,82)* (**Fig. 9**).

We have studied 20 patients affected by the typical form of congenital MD, and detected a total deficiency of laminin α2, using both the 80 kDa and 300 kDa antibodies. In patients with partial deficiency, usually the 300 kDa antibody shows a more deficient pattern *(83)*. We also have recently detected a partial deficiency of only the 300 kDa α2-laminin antibody in 5 patients with the classical LGMD clinical course *(84–86)*. Screening for mutations in the LAMA2 gene will elucidate the primary or secondary etiology of these deficiencies.

3.3.8. Protein Study for Differential Diagnosis

Testing for defective protein expression is a powerful tool for deciding where to start the search for gene mutations (**Fig. 9**)

In adult MD forms, multiplex Western blot analysis for dystrophin, calpain, and dysferlin has shown to be very useful for preliminary screening of muscular dystrophy. With the exception of calpain 3, which may occur as a secondary effect of a dysferlinopathy, dysferlin and telethonin deficiencies seem to be the consequence of their respectively primary gene defect. Therefore the absence of dysferlin or telethonin on muscle biopsy strongly suggests a diagnosis of LGMD2B or LGMD2G, respectively.

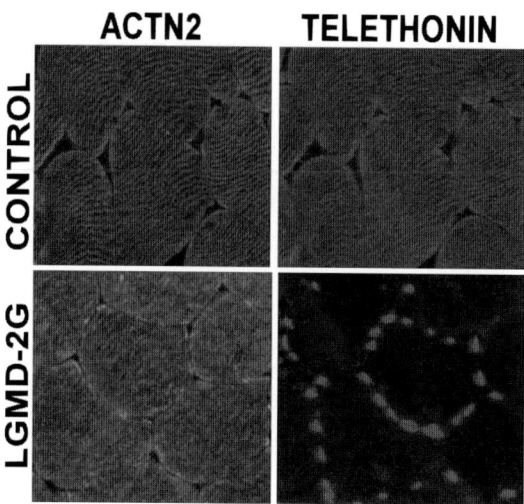

Fig. 7. Double IF analysis for α-actinin 2 (a positive marker of Z-band of the sarcomere), and telethonin in one control and one LGMD-2G patient. Note the sarcolemmal deficiency of telethonin in the patient. Some unspecific reaction are commonly seen in the nucleus, which requires further studies. (See color plate 11 appearing in the insert following p. 82)

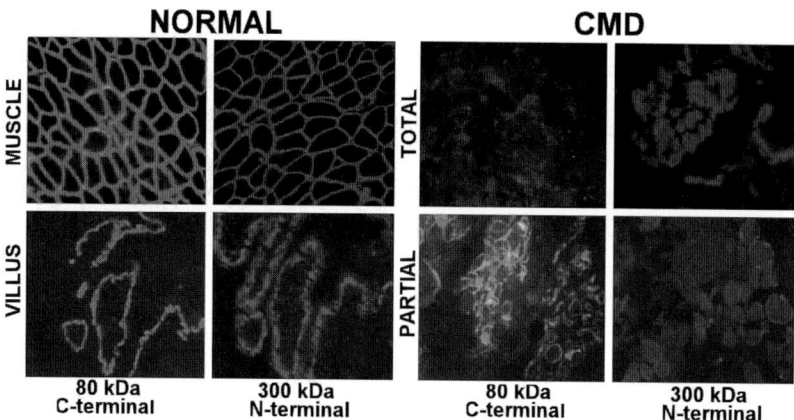

Fig. 8. IF analysis for α2-Laminin with antibodies against the 80 kDa and 300 kDa fragments showing the pattern in a normal control, the positive pattern in a normal villus, the total deficiency, through the two antibodies in one CMD severely affected patient and one patient with partial deficiency. (See color plate 11 appearing in the insert following p. 82)

If no protein or DNA alterations are found in patients with clinical diagnosis of LGMD, the possibility of spinal muscular atrophy (SMA) should be considered, due to the clinical overlap of these diseases.

Patients with suspected Xp21 dystrophy are first tested for deletions in the dystrophin gene, a less invasive test. The identification of a molecular deletion will confirm the diagnosis of DMD/BMD. If no deletion is detected (about 40% of the cases), muscle proteins are analyzed in an attempt to elucidate the possible diagnosis.

Protein Expression in Neuromuscular Diseases

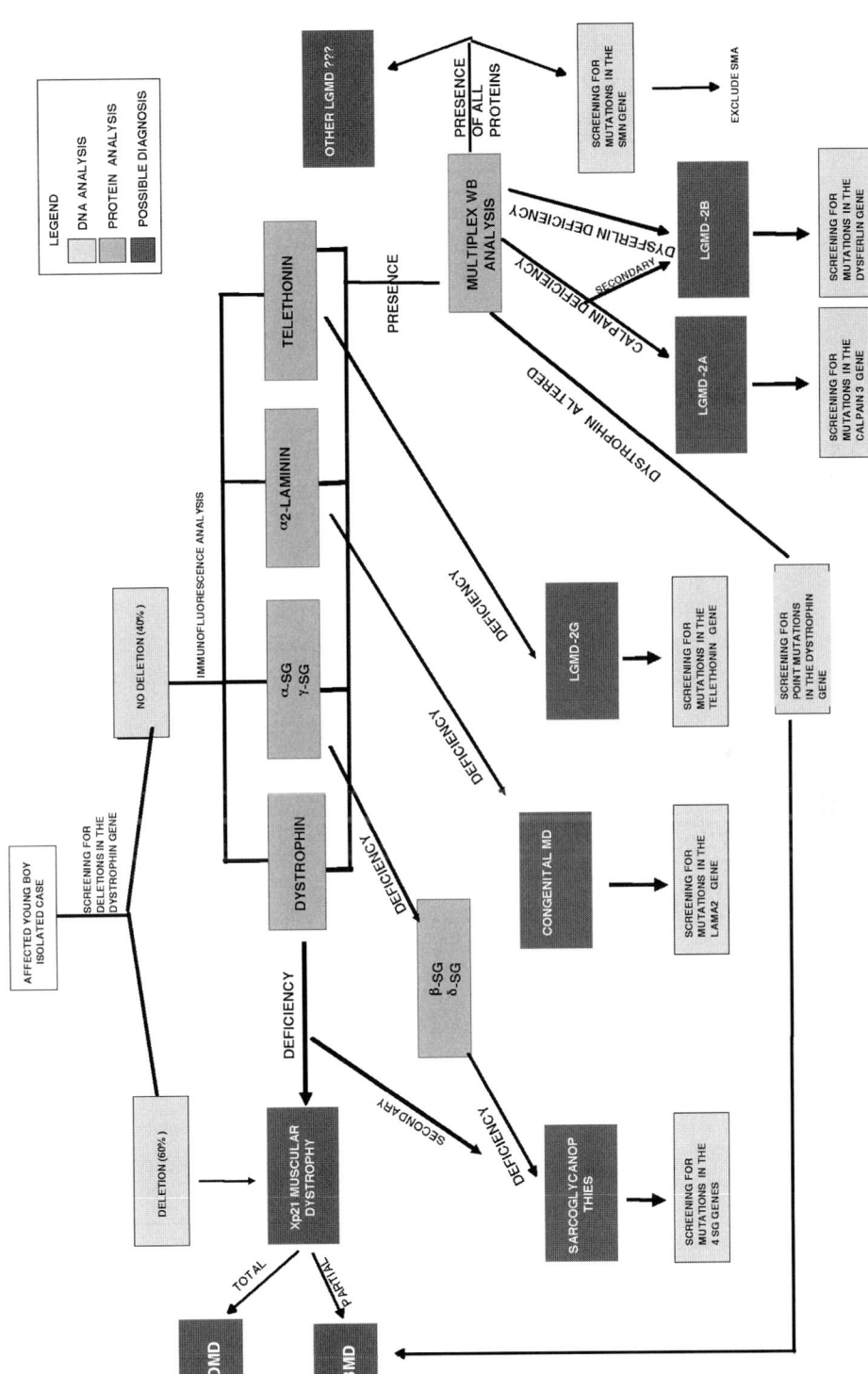

Fig. 9. Schematic representation of the procedures and methodology used for the diagnosis of NMD in our Center.

Dystrophin is the first protein to be tested, using N- and C-terminal epitopes antibodies, in a double reaction. A significant deficiency of dystrophin is suggestive of DMD or severe BMD. Complementary WB analysis will confirm the amount of present protein, and a possible prognostic.

If dystrophin is present through IF analysis, an autosomal form is suspected and complementary studies for α-SG and γ-SG in a double reaction with a N-terminal antibody for dystrophin is done. If a deficiency of any of the SG is detected, additional studies for β-SG and δ-SG are done, to confirm a possible sarcoglycanopathy. For the final diagnosis of a SGpathy, mutation screening should start with α-SG since this is the most prevalent sarcoglycanopathy. If γ-SG is predominantly absent, a γ-sarcoglycanopathy should be suspected first.

Additional IF study for α2-laminin, using the 300 kDa antibody, should be done in more severely affected patients. A total or partial α2-laminin deficiency is complemented with the study through additional antibody (80 kDa) against the N-terminal region. The study of α2-laminin in a double reaction in all patients is also very useful for the control of sarcolemmal integrity.

4. Notes

1. New fluorochromes, which are very bright and maintain the fluorescence for a longer time, have been developed by the Alexis Company.
2. Use pure SDS (for electrophoresis) from only one manufacturer. Pattern of migration of polypeptides may change quite drastically when SDS from different manufacturers is used *(1)*.
3. Acrylamide and bisacrylamide are potent neurotoxins, which are absorbed through the skin. Their effect is accumulative and skin contact must be avoided.
4. Ready made slides are commercially available (FisherScientific, PA, USA).
5. The suggested time of incubation is the minimum required, but it can be longer, according to the efficacy of the antibody.
6. Even if the dilution of the antibody is suggested, each laboratory must establish and adapt its own methodology.
7. Dilutions of primary or secondary antibodies must be done freshly and the centrifuging of the diluted antibody before use is highly recommended, which make the reaction more specific.
8. Depending on the quality of the antibody, an overnight incubation is recommended.
9. It is also necessary to test the specificity of the mixture of the two antibodies, because sometimes the double reaction may not work.
10. Polimerization will begin as soon as APS has been added. Therefore, swirl the mixture rapidly and pour the solution into the gap between the glass plates.
11. When the gel is polymerized, the two layers become again visible.
12. This absorbs the stain as it leaks from the gel.
13. Primary antibodies can be frozen and re-used several times.

Acknowledgments

The collaboration of the following persons is gratefully acknowledged: Marta Cánovas for her invaluable technical help, Dr. Eloisa S. Moreira, Dr. Rita C. M. Pavanello, Dr. Ivo Pavanello, Dr. Edmar Zanotelli, Cleber da Silva Costa, Viviane P. Muniz, Telma L. F. Gouveia, Flavia de Paula, Dr. Acary S. B. Oliveira, Alessandra

Starling, Antonia Cerqueira, Constancia Urbani, Ms. Janet Tajchman, and Lucete Cesana, for help in English corrections. Very special thanks are dedicated to Dr. Zubrzycka-Gaarn (in memorium), Dr. Peter Ray, and Dr. Ron Worton, who have opened their laboratory for our introduction to Western blot studies. We would like also to thank very much the following researchers, who kindly provided us with specific antibodies: Dr. Louise Anderson, Dr. Elizabeth McNally, Dr. Carsten Bonnemann, Dr. Louis M. Kunkel, Dr. Vincenzo Nigro, Dr. Jeff Chamberlain, Dr. Ronald Worton, Dr. Eric Hoffmann, Dr. Kevin Campbell, Dr. Eijiro Ozawa, and Dr. Georgine Faulkner. This work was supported by grants from FAPESP- CEPID , PRONEX, and CNPq.

References

1. Sambrook, J., Firtsch, E. F., and Maniatis, T. (1989) Molecular Cloning: A Laboratory Manual, 2nd ed., Detection and analysis of protein expressed from cloned genes. Cold Spring Harbor Laboratory Press, Cold Spring Harbor, NY, USA.
2. Anderson, L. V. B. (2001) Multiplex Western blot analysis of the muscular dystrophy proteins, in: *Muscular Dystrophy: Methods and Protocols, Methods in Molecular Medicine*. (Bushby, K. M. D. and Anderson,L. V. B. eds.), Humana Press, Totowa, NJ, USA, pp. 369–386.
3. Koenig, M., Monaco, A. P., and Kunkel, L. M. (1988) The complete sequence of dystrophin predicts rod-shaped cytoskeletal protein. *Cell* **53,** 219–228.
4. Hoffman, E. P., Brown, R. H., and Kunkel, L. M. (1987) Dystrophin: the protein product of the Duchenne muscular dystrophy locus. *Cell* **51,** 919–928.
5. Hoffman, E. P., Fischbeck, K. H., Brown, R. H., Johnson, M., Medori, R., Loike, J. D., et al. (1988) Characterization of dystrophin in muscle-biopsy specimens from patients with Duchenne's or Becker's muscular dystrophy. *N. Engl. J. Med.* **318,** 1363–1368.
6. Zatz, M., Vainzof, M., and Passos-bueno, M. R. (2000) Limb-girdle muscular dystrophy: one gene with different phenotypes, one phenotye with different genes. *Curr. Opin. Neurol.* **13,** 511–517.
7. Bushby, K. M. D. (1999) The limb-girdle muscular dystrophies-multiple genes, multiple mechanisms. *Hum. Mol. Genet.* **8,** 1875–1882.
8. Salmikangas, P., Mykkanen, O. M., Gronholm, M., Heiska, L., Kere, J., and Carpen, O. (1999) Myotilin, a novel sarcomeric protein with two Ig-like domains, is encoded by a candidate gene for limb-girdle muscular dystrophy. *Hum. Mol. Genet.* **8,** 1329–1336.
9. Hause, R. M. A., Horrigan, S. K., Salmikangas, P., Viles, K. D., Tim, R. W., Torian, U. M., and Anu, T. (2000) A mutation in the Myotilin gene causes limb-girdle muscular dystrophy 1A. *Hum. Mol. Genet.* **9,** 2141–2147.
10. Van der Kooi, A., van Meegen, M., Ledderhof, T. M., McNally, E. M., de Visser, M., and Bolhuis, P. A. (1997) Genetic localization of a newly recognized autosomal dominant limb-girdle muscular dystrophy with cadiac involvement (LGMD1B) to chromosome 1q11-21. *Am. J. Hum. Genet.* **60,** 891–895.
11. Minetti, C., Sotgia, F., Bruno, C., Scartezzini, P., Broda, P., Bado, M., et al. (1998) Mutations in the caveolin-3 gene cause autosomal dominant limb-girdle muscular dystrophy. *Nat. Genet.* **18,** 365–368.
12. McNally, E., Moreira, E. S., Duggan, D., Bonneman, C. ., Lisanti, M. P., Lidov, H. G. W., et al. (1998) Caveolin-3 in muscular dystrophy. *Hum. Mol. Genet.* **7,** 871–878.
13. Messina, D. I., Speer, M. C., Pericak-Vance, M. A., and McNally, E. M. (1997) Linkage of familial dilated cardiomyopathy with conduction defect and muscular dystrophy to chromosome 6q23. *Am. J. Hum. Genet.* **61,** 909–917.

14. Speer, M. C., Vance, J. M., Grubber, J. M., Graham, F. L., Stajich, J. M., Viles, K. D. et al. (1999) Identification of a new autosomal dominant limb-girdle muscular dystrophy locus on chromosome 7. *Am. J. Hum. Genet.* **64**, 556–562.
15. Feit, H., Silbergleit, A., Schneider, L. B., Gutierrez, J. A., Fitoussi, R. P., Reyes, C., et al. (1998) Vocal cord and pharyngeal weakness with autosomal dominant distal myopathy: clinical description and gene localization to 5q31. *Am. J. Hum. Genet.* **63**, 1732–1742.
16. Weiler, T., Greenberg, C. R., Zelinski, T., Nylen, E., Coghlan, G., Crumley, J., et al. (1998) A gene for autosomal recessive limb-girdle muscular dystrophy in Manitoba Hutterites maps to chromosome region 9q31-q33: evidence for another LGMD locus *Am. J. Hum. Genet.* **63**, 140–147.
17. Driss, A., Amouri, R., Ben Hamida, C., Souilem, S., Gouider-Khouja, N., Ben Hamida, M., and Hentati, F. A. (2000) A new locus for autosomal recessive limb-girdle muscular dystrophy in a large consanguineous Tunisian family maps to chromosome 19q13.3. *Neuromuscl. Disord.* **10**, 240–246.
18. Ervasti, J. M., Ohlendieck, K., Kahl, S. D., Gaver, M. G., and Campbell, K. P. (1990) Deficiency of a glycoprotein component of the dystrophin complex in dystrophic muscle. *Nature* **345**, 315–319.
19. Yoshida, M. and Ozawa, E. (1990) Glycoprotein complex anchoring dystrophin to sarcolemma. *J. Biochem.* (Tokyo) **108**, 748–752.
20. Matsumura, K., Tomé, F. M. S., Collin, H., Azibi, K., Chaouch, M., Kaplan, J. C., et al. (1992) Deficiency of the 50 kDa dystrophin-associated glycoprotein in severe childhood autosomal recessive muscular dystrophy. *Nature* **359**, 320–322.
21. Azibi, K., Bachner, L., Beckmann, J. S., Matsumura, K., Hamouda, E., Chaouch, M., et al. (1993) Severe childhoood autosomal recessive muscular dystrophy with the deficiency of the 50 kDa dystrophin-associated glycoprotein maps to chromosome 13q12. *Hum. Mol. Genet.* **2**, 1423–1428.
22. Roberds, S. L., Leturcq, F., Allamand, V., Piccolo, F., Jeanpierre, M., Anderson, R. D., et al. (1994) Missense mutations in the adhalin gene linked to autosomal recessive muscular dystrophy. *Cell* **78**, 625–633.
23. Bonnemann, C. G., Modi R., Noguchi, S., Mizuno, Y., Yoshida, M., Gussoni, E., et al. (1995) β-sarcoglycan (A3b) mutations cause autosomal recessive muscular dystrophy with loss of the sarcoglycan complex. *Nat. Genetics* **11**, 266–273.
24. Lim, L. E., Duclos, F., Broux, O., Bourg, N., Sunada, Y., Allamand, V., et al. (1995) β-sarcoglycan (43 DAG): Characterization and involvement in a recesive form of limb-girdle muscular dystrophy linked to chromosome 4q12. *Nature Genetics* **11**, 257–265.
25. Noguchi, S., McNally E. M., Ben Othmane, K., Hagiwara, Y., Mizuno, Y., Yoshida, M. H., et al. (1995) Mutations in the dystrophin-associated protein γ-sarcoglycan in chromosome 13 muscular dystrophy. *Science* **270**, 819–822.
26. McNally, E., Passos–Bueno, M. R., Bonnemann, C. G., Vainzof, M., Moreira, E. S., Lidov, H. G. W., et al. (1996) Mild and severe muscular dystrophy caused by a single γ-sarcoglycan mutation. *Am. J. Hum. Genet.* **59**, 1040–1047.
27. Passos-Bueno, M. R., Moreira, E. S., Vainzof, M., Marie, S. K., and Zatz, M. (1996) Linkage analysis in autosomal recessive limb-girdle muscular dystrophy (AR-LGMD) maps a sixth form to 5q33-34 (LGMD2F) and indicates that there is at least one more subtype of AR LGMD. *Hum. Mol. Genet.* **5**, 815–820.
28. Passos-Bueno, M. R., Vainzof, M., Moreira, E. S., and Zatz, M. (1999) The seven autosomal recessive limb-girdle muscular dystrophies (LGMD): from lgmd2a to lgmd2g. *Am. J. Med. Genet.* **82**, 392–398.

29. Nigro, V., Piluso, G., Belsito, A., Politano, Z., Puca, A. A., Papparella, S., et al (1996) Identification of a novel sarcoglycan gene at 5q33 encoding a sarcolemmal 35 kDa glycoprotein. *Hum. Mol. Genet.* **5,** 1179–1186.
30. Nigro, V., de Sa Moreira E. S., Piluso G., Vainzof, M., Belsito A., Politano, L., Puca, A. A., Passos-Bueno M. R., and Zatz, M. (1996) The 5q autosomal recessive limb-girdle muscular dystrophy, LGMD2F, is caused by a mutation in the d-sarcoglycan gene. *Nat. Genet.* **14,** 195–198.
31. Moreira, E. S., Vainzof, M., Marie, S. K., Nigro, V., and Zatz, M., and Passos-Bueno, M. R. (1998) A first missense mutation in the δ-sarcoglycan gene associated with a severe phenotype and frequency of limb-girdle muscular dystrophy type 2F (LGMD2F) among Brazilian sarcoglycanopathies . *J. Med. Genet.* **35,** 951–953.
32. Moreira, E. S., Vainzof, M., Marie, S., Sertié, A. L., Zatz, M., and Passos-Bueno, M. R. (1997) The seventh form of autosomal recessive limb-girdle muscular dystrophy is mapped to 17q11-12. *Am. J. Hum. Genet.* **61,** 151–159.
33. Moreira, E. S ., Wiltshire, T. J., Faulkner, G., Nilforoushan, A., Vainzof, M., Suzuki, O. T., et al. (2000) Limb-girdle muscular dystrophy type 2G (LGMD2G) is caused by mutations in the gene encoding the sarcomeric protein telethonin. *Nat. Genet.* **24,** 163–166.
34. Richard, I., Broux, O., Allamand, V., Fougerouse, F., Chiannilkulchai, N., Bourg, N., et al. (1995) A novel mechanism leading to muscular dystrophy: mutations in calpain 3 cause limb girdle muscular dystrophy type 2A. *Cell* **81,** 27–40.
35. Richard, I., Roudaut, C., Saenz, A., Pogue, R., Grimbergen, E. M. A., Anderson, L. V. B., et al. (1999) Calpainopathy: a survey of mutations and polymorphisms. *Am. J. Hum. Genet.* **64,** 1524–1540.
36. Bashir, R., Britton, S., Stratchan, T., Keers, S., Vafiadaki, E., Richard, I., et al. (1998) A novel mammalian gene related to the C. elegans spermatogenesis factor fer-1 is mutated in patients with limb-girdle muscular dystrophy type 2B (LGMD2B). *Nat. Genet.* **20,** 37–42.
37. Liu, J., Aoki, M., Illa, I., Chou, F. L., Oeltjen, J. C., Hosler, B. A., et al. (1998) Dysferlin, a novel skeletal muscle gene, is mutated in Miyoshi myopathy and limb girdle muscular dystrophy. *Nat. Genet.* **20,** 31–36.
38. Valle, G., Faulkner, G., Antoni, A., Pacchioni, B., Pallavicini, A., Pandolfo, D., et al. (1997) Telethonin, a novel sarcomeric protein of heart and skeletal muscle. *FEBS Lett.* **415,** 163–168.
39. Dubowitz, V. (1998) Muscle biopsy: A practical approach. 2nd ed. Bailliere Tindall, London, UK.
40. Nicholson, L. V. B., Johnson, M. A., Gardner-Medwin, G., Bhattacharya, S., and Harris, J. B. (1990) Heterogeneity of dystrophin expression in patients with Duchenne and Becker muscular dystrophy. *Acta Neuropathol.* **80,** 239–250.
41. Sewry, C. A. and Lu, Q. (2001) Protein analysis in the muscular dystrophies: immunological reagents and amplification systems, in: *Muscular Dystrophy: Methods and Protocols,* (Bushby, K. M. D., and Anderson, L. V. B., eds.), Humana Press, Totowa, NJ, USA, pp. 325–338.
42. Vainzof, M., Zubrzycka-Gaarn, E. E., Rapaport, D., Passos-Bueno, M.,R., Pavanello, R. C. M., Pavanello, I., and Zatz, M. (1991) Immunofluorescence dystrophin study in Duchenne dystrophy through the concomitant use of two antibodies directed against the carboxy-terminal and the amino-terminal region of the protein. *J. Neurol. Sci.* **101,** 141–147.
43. Laemmli, U. K. (1970) Cleavage of structural proteins during the assembly of the head of bacteriophage T4. *Nature* **227,** 680–685.
44. Towbin, H., Staehelin, T., and Gordon, J. (1979) Electrophoretic transfer of proteins from polyacrylamide gels to nitrocellulose sheets: procedure and some applications. *Proc. Natl. Acad. Sci. USA* **76,** 4350–4354.

45. Zubrzycka-Gaarn, E. E., Bulman, D. E., Karpati, G., Burghes, A. H., Belfall, B., Klamut, H. J., et al. (1988) The Duchenne muscular dystrophy gene product is localized in sarcolemma of human skeletal muscle. *Nature* **333,** 466–469.
46. Bulman, D. E., Murphy, E. G., Zubrzycka-Gaarn, E. E., Worton, R. G., and Ray, P. N. (1991) Differentiation of Duchenne and Becker muscular dystrophy phenotypes with amino- and carboxy-terminal antisera specific for dystrophin. *Am. J. Hum. Genet.* **48,** 295–304.
47. Ho-Kim, M-A., Bedard A., Vincent, M., and Rogers P. A. (1991) Dystrophin: A sensitive and reliable immunochemical assay in tissue and cell culture homogenates. *Biochem. Biophys. Res. Comm.* **181,** 1164–1172.
48. Campbell, K. P. and Kahl, S. D. (1989) Association of dystrophin and an integral membrane glycoprotein. *Nature* **338,** 259–262.
49. Hack, A. A., Groh, M. E., and McNally, E. M. (2000) Sarcoglycans in muscular dystrophy. *Microsc. Res. Tech.* **48,** 167–180.
50. Cohn, R. D. and Campbell, K.P. (2000) Molecular basis of muscular dystrophies. *Muscle Nerve* **23,** 1456–1471.
51. Vainzof, M., Pavanello, R. C. M., Pavanello, I., Passos-Bueno, M. R., Rapaport, D., and Zatz, M. (1990) Dystrophin immunostaining in muscles from patients with different types of muscular dystrophy: Brazilian study. *J. Neurol. Sci.* **98,** 221–233.
52. Vainzof, M., Passos-Bueno, M. R., Pavanello R. C. M., and Zatz, M. (1995) Is dystrophin always altered in Becker muscular dystrophy patients? *J. Neurol. Sci.* **131,** 99–104.
53. Vainzof, M., Passos-Bueno, M. R., Rapaport, D., Pavanello, R. C. M., Bulman, D. E,, and Zatz, M. (1992) Additional dystrophin fragment in Becker muscular dystrphy patients: Correlation with the pattern of DNA deletions. *Am. J. Med. Genet.* **44,** 382–384.
54. Vainzof, M., Passos-Bueno, M. R., Pavanello, R. C. M., Marie, S. K., and Zatz, M. (1999) Sarcoglycanophathy is responsible for 68% of severe autosomal recessive limb-girdle muscular dystrophy. *J. Neurol. Sci.* **164,** 44–49.
55. Bonnemann, C. G., McNallly, E. M., and Kunkel, L. M. (1996) Beyond dystrophin: current progress in the muscular dystrophies. *Curr. Op. Pediatr.* **8,** 569–582.
56. Bönnemann, C. G. (1999) Limb-girdle muscular dystrophies: an overview. *J. Child. Neurol.* **14,** 31–33.
57. Vainzof, M., Passos-Bueno, M. R., Moreira, E. S., Pavanello, R. C. M., Marie, S. K., et al. (1996) The sarcoglycan complex in the six autosomal recessive limb-girdle (AR-LGMD) muscular dystrophies. *Hum. Mol. Genet.* **5,** 1963–1969.
58. Vainzof, M., Moreira, E. S., Ferraz, G., Passos-Bueno, M. R., Marie, S. K., and Zatz, M. (1999) Further evidences for the organization of the four sarcoglycans proteins within the dystrophin-glycoprotein complex. *Eur. J. Hum. Genet.* **7,** 251–254.
59. Vainzof, M., Moreira, E. S., Canovas, M., Suzuki, O. T., Pavanello, R. C. M., Costa, C. S., et al. (2000) Partial α-sarcoglycan deficiency associated with the retention of the SG complex in a LGMD2D family. *Muscle Nerve* **23,** 984–988.
60. Paula, F., Moreira, E. S., Bernardino, A. L. F., Kai, A., Passos-Bueno, M. R., Vainzof, M., and Zatz, M. (2000) Recurrent LGMD2A (Calpainopathy) mutations in Brazilian patients. *Am. J. Hum. Genet.* **67,** 251.
61. Anderson, L. V. B., Davison, K., Moss, J. Á., Richard, I., Fardeau, M., Tome, F. M. S., et al. (1998) Characterization of monoclonal antibodies to calpain 3 and protein expression in muscle from patients with limb-girdle musclar dystrophy type 2A. *Am. J. Pathol.* **153,** 1169–1179.
62. Vainzof, M., Anderson, L. V. B., Moreira, E. S., Paula, F., Pavanello, R. C. M., Passos-Bueno, M. R., and Zatz, M. (2000) Calpain 3: Characterization of the primary defect in LGMD2A and analysis of its secondary effect in other LGMDs. *Neurology* **54,** A436.

63. Vainzof, M., Moreira, E. S., Passos-Bueno, M. R., Faulkner, G., Valle, G., Zanotelli, E., et al. (2000) the effect of telethonin deficiency in LGMD-2G and its expression in other forms of muscular dystrophies and congenital myopathies. *Am. J. Hum. Genet.* **67,** 379.
64. Anderson, L. V. B., Harrison, R., Pogue, R., Vafiadaki, E., Davison, K., Moss, J.A., et al. (2000) Secondary reduction in calpain 3 expression in patients with limb girdle muscular dystrophy type 2B and miyoshi myopathy (primary dysferlinopathies) *Neurom. Disord.* **10,** 553–559.
65. Anderson, L. V. B., Davison, K., Moss, J. Á., Young, C., Cullen, M. J., Walsh, J., et al. (1999) Dysferlin is a plasma membrane protein and is expressed early in human development. *Hum. Mol. Genet.* **8,** 855–861.
66. Vainzof, M., Anderson, L. V., McNally, E. M., Davis, D. B., Faulkner, G., Valle, G., et al. (2001) Dysferlin protein analysis in limb-girdle muscular dystrophies. *J. Mol. Neurosci.* **17,** 71–80.
67. Gregorio, C. C., Granzier, H., Sorimachi, H., and Labeit, S. (1999) Muscle assembly: a titanic achievemen? *Curr. Opin. Cell Biol.* **11,** 18–25.
68. Laing, N. G. (1999) Inherited disorders of sarcomeric proteins. *Curr. Opin. Neurolog.* **12,** 513–518.
69. Nowak, K. J., Wattanasirichaigoon, D., Goebel, N. H., Wilce, M., Pelin, K., Donner, K., et al. (1999) Mutations in the skeletal muscle alpha actin gne in patients with actin myopathy and nemaline myopathy. *Nat. Genet.* **23,** 208–212.
70. Laing, N. G., Wilton, S. D., Akkari, P. A., Dorosz, S., Boundy, K., Kneebone, C., et al. (1995) A mutation in the alpha tropomyosin gene TPM3 associated with autosomal dominant nemaline myopathy NEM1. *Nat. Genet.* **10,** 249–250.
71. Donner, K., Ollikaine, M., Pelin, K., Grönholm, M., Carpén, O., Wallgren-Pettersson, C., and Ridanpää. (2000) Mutations in the β-trpomyosin gene in rare cases of autosomal dominat nemeline myopathy. *Neuromusc. Disord.* **10,** 342–343.
72. Pelin, K., Hilpelä, P., Donner, K., Sewry, C., Akkary, P. A., Wilton, S.D., Wattanasirichaigoon, D., et al. (1999) Mutations in the nebulin gene associated with autosomal recessive nemaline myopathy. *Proc. Natl. Acad. Sci. USA* **96,** 2305–2310.
73. Johnston, J. J., Kelley, R. I., Crawford, T. O., Morton, D. H., Agarwala, R., Koch, T., et al. (2000) A novel Nemaline Myopathy in the Amish caused by a muatation in troponin T1. *Am. J. Genet.* **67,** 814–821.
74. Gurgel-Giannetti, J., Reed, U. C., Bang, M. L., Pelin, K., Donner, K., Marie, S. K. N., et al. (2001) Nebulin expression in Nemaline Myopathy. *Neuromusc. Disord,* **11,** 154–162.
75. Paula, F., Vainzof, M., McNally, E. E., Kunkel, L. M., and Zatz, M. (2001) Screening of mutations in the caveolin-3 gene in Brazilian limb-girdle muscular dystrophy patients? *Am. J. Med. Genet.* **99,** 303–307.
76. Hillaire, D., Leclerc, A., Faure, S., Topaloglu, H., Chiannilkulchai, N., Guicheney, P., et al. (1994) Localization of merosin-negative congenital muscular dystrophy to chromosome 6q2 by homozygosity mapping. *Hum. Mol. Genet.* **3,** 1657–1661.
77. Tomé, F. M. S., Evangelista, T., Leclerc, A., Sunada, Y., Manole, E., Estournet, B., et al. (1994) Congenital muscula dystrophy with merosin deficiency. *C R Acad Sci III.* **317,** 351–357.
78. Philpot, J., Sewry, C., Pennock, J., and Dubowitz, V. (1995) Clinical phenotype in congenital muscular dystrophy: correlation with expression of merosin in skeletal muscle. *Neuromusc. Disord.* **5,** 301–305.
79. Vainzof, M., Reed, U. C., Schwartzman, J. S., Pavanello, R. C. M., Passos-Bueno, M. R., and Zatz. M. (1995) Deficiency of merosin (Laminin M or α2) in congenital muscular dystrophy associated with cerebral white mater alterations. *Neuropediatrics* **26,** 293–297.

80. Sewry, C. A., D'Alessandro, M., Wilson, L. A., Sorokin, L. M., Naom, I., Bruno, S., et al. (1997) Expression of laminin chains in skin in merosin-deficient congenital muscular dystrophy. *Neuropediatrics* **28,** 217–222.
81. Naom, I., Sewry, C., D'Alessandro, M., Topaloglu, H., Ferlini, A., Wilson, L., et al. (1997) Prenatal diagnosis in merosin-deficient congenital muscular dystrophy. *Neuromuscul. Disord.* **7,** 176–179.
82. Voit, T., Fardeau, M., and Tome, F. M. (1994) Prenatal detection of merosin expression in human placenta. *Neuropediatrics* **25,** 332–333.
83. Sewry, C. A., Naom, I., D'Alessandro, M., Sorokin, L., Bruno, S., Wilson, L. A., et al. (1997) Variable clinical phenotype in merosin-deficient congenital muscular dystrophy associate with differential immunolabelling of two fragments of the laminin alpha 2 chain. *Neuromuscul. Disord.* **7,**169–75.
84. Naom, I., D'Alessandro, M., Sewry, C. A., Philpot, J., Manzur, A. Y., Dubowitz, V., and Muntoni, F. (1998) Laminin alpha 2-chain gene mutations in two siblings presenting with limb-girdle muscular dystrophy. *Neuromuscul. Disord.* **8,** 495–501.
85. Naom, I., D'alessandro, ., Sewry, C. A., Jardine, P., Ferlini, A., Moss, T., et al. (2000) Mutations in the laminin alpha2-chain gene in two children with early-onset muscular dystrophy. *Brain* **123,** 31–41.
86. Brockington, M., Sewry, C. A., Herrmann, R., Naom, I., Dearlove, A., Rhodes, M., et al. (2000) Assignment of a form of congenital muscular dystrophy with secondary merosin deficiency to chromosome 1q42. *Am. J. Hum. Genet.* **66,** 428–435.

Index

A

Angelman syndrome (AS),
   bisulfite restriction analysis,
      bisulfite treatment, 213, 214
      DNA extraction, 213
      electrophoresis and digestion
         patterns, 214, 215
      materials, 213
      polymerase chain reaction, 214
      principles, 210, 211
      troubleshooting, 215
   DNA methylation abnormalities, 210
   gene mutations, 209
   polymerase chain reaction-based
      methylation testing, 210
AS, *see* Angelman syndrome

B

Bisulfite restriction analysis (BRA),
   bisulfite treatment, 213, 214
   DNA extraction, 213
   electrophoresis and digestion
      patterns, 214, 215
   materials, 213
   polymerase chain reaction, 214
   principles, 210, 211
   troubleshooting, 215
BRA, *see* Bisulfite restriction analysis

C

CAG repeat expansion, *see* Polyglutamate; Trinucleotide repeat expansion
Calpain-3, defect analysis, 365, 366
Caveolin-3, defect analysis, 369
Charcot-Marie-Tooth disease (CMT),
   clinical features, 177, 223
   gene mutations,
      MPZ, 178
      PMP22, 177, 178, 225
   heredity, 222
   Southern blot analysis of CMT1A,
      blood culture,
         harvest, 227, 228, 230, 233, 234
         slide preparation,
            228, 20, 231, 234
      blotting, 179, 181
      denaturation and hybridization
         of blood slides, 229, 232, 235
      densitometry, 182, 183
      detection, 229, 233
      digestion of DNA, 179, 180
      DNA isolation, 178–180, 184
      gel electrophoresis, 179, 181
      hybridization, 179, 181, 192
      materials, 178–180, 184, 227–230
      principles, 225, 226
      probe,
         labeling, 228, 229, 231, 233–235
         precipitation, 229, 232, 235
         types, 179, 180, 184
      washing, 229, 232
   types, 177, 223
CMT, *see* Charcot-Marie-Tooth disease
Complementary DNA microarray,
      Emery-Dreifuss muscular
         dystrophy analysis,
adenovirus infection, 256, 258
detection, 256, 258
hybridization, 257
materials, 254–256
microarray,
   commercial microarrays, 254
   preparation, 254–257
principles, 253, 254
RNA isolation from fibroblasts, 255
scanning and data processing, 258–261
COS cell models, *see* Pelizaeus-Merzbacher disease

D

Denaturing gradient gel electrophoresis (DGGE),
  Duchenne muscular dystrophy diagnostics,
    DNA isolation from lymphocytes, 166, 167, 170, 173
    electrophoresis, 167, 170–174
    materials, 166, 167, 173
    polymerase chain reaction, 167, 170, 171
    primers, 168, 169
    sequencing, 167, 172, 174
  principles, 166
Denaturing high-performance liquid chromatography (DHPLC),
  applications, 120, 127, 128
  assay design, 119, 120, 125
  MECP2 mutation detection in Rett syndrome,
    agarose gel electrophoresis, 124
    chromatography, 124
    DNA extraction, 122
    heteroduplex formation, 124
    materials, 122
    mutation types, 120
    overview, 121, 128
    polymerase chain reaction,
      amplification reaction, 124
      primer design, 120, 127
    sequencing of amplification products, 125
  principles, 119, 120, 125
  sensitivity, 120, 125, 127
DGGE, *see* Denaturing gradient gel electrophoresis
DHPLC, *see* Denaturing high-performance liquid chromatography
DIRECT,
  advantages, 74, 75
  applications, 74
  autoradiography, 76, 78
  CAG probe preparation, 75–78
  cloning, 76, 78, 79
  genomic blot preparation, 75, 76, 79, 80
  hybridization, 75, 76, 78
  materials, 75, 76, 79, 80
  prehybridization, 75, 78
  selective hybridization principles, 73, 79
  washing of blots, 76, 78
DMD, *see* Duchenne muscular dystrophy
Drosophila melanogaster,
  brain degeneration mutants, 241
  human gene homologs, 241
  polyglutamine disease models,
    biochemical analysis, 246
    classification, 242, 243
    genetic approaches for modifiers, 244–246
    immunocytochemical characterization, 246
    progressive neuronal degeneration, 243
    protein context in toxicity, 243
    proteomics, 246, 247
    resources, 247, 248
    tissue specificity of expression, 243
    toxic conformation of proteins, 243, 244
  transgenic fly generation, 242
Duchenne muscular dystrophy (DMD),
  denaturing gradient gel electrophoresis diagnostics,
    DNA isolation from lymphocytes, 166, 167, 170, 173
    electrophoresis, 167, 170–174
    materials, 166, 167, 173
    polymerase chain reaction, 167, 170, 171
    primers, 168, 169
    principles, 166
    sequencing, 167, 172, 174
  dystrophin protein mutation analysis, 364, 365
  gene dosage determination with quantitative polymerase chain reaction,
    amplification reaction, 8, 9

carrier identification, 3
gel visualization without blotting, 10
genomic DNA isolation, 6
internal standard, 4, 5, 9–11
linearity, 4
materials, 6–8
principles, 4
quantification, 5
Southern blots, 4, 9
gene mutation types, 165
protein truncation test,
  see Protein truncation test
single-strand conformational
  polymorphism testing, 165
Dysferlin, defect analysis, 366, 367
Dystrophin, defect analysis,
  364, 365, 370, 372

E

EDMD, see Emery-Dreifuss muscular
  dystrophy
ELISA, see Enzyme-linked
  immunosorbent assay
Emery-Dreifuss muscular dystrophy
  (EDMD),
  complementary DNA microarray
    analysis,
    adenovirus infection, 256, 258
    detection, 256, 258
    hybridization, 257
    materials, 254–256
    microarray,
      commercial microarrays, 254
      preparation, 254–257
    principles, 253, 254
    RNA isolation from fibroblasts, 255
    scanning and data processing,
      258–261
  gene mutations, 253
  heredity, 253
Enzyme-linked immunosorbent assay
  (ELISA),
  FMR1 antibodies, 346, 349, 351, 353
  galactosemia polymerase chain
    reaction–enzyme-linked
    immunosorbent assay
    of mutations,

allele-specific hybridization, 115
digoxigenin enzyme-linked
  immunosorbent assay
  and interpretation, 115–118
DNA isolation, 114, 115
materials, 112–114, 116
multiplex polymerase chain
  reaction, 115
principles, 112
troubleshooting, 117, 118
prion protein, 307

F

Facioscapulohumeral muscular
  dystrophy (FSHD),
  clinical features, 153
  gene rearrangements, 153, 155
  pulsed field gel electrophoresis
    diagnostics,
    apparatus, 159
    DNA isolation in agarose blocks,
      157, 159, 160, 162
    electrophoresis, 157, 160, 162
    hybridization, 158, 161, 162
    lymphocyte,
      agarose block embedding,
        157, 159
      isolation, 157, 159
    materials, 157–159
    principles, 155
    restriction digestion reactions, 157,
      160, 162
    Southern blot analysis, 158, 161
FISH, see Fluorescence *in situ*
  hybridization
Fluorescence *in situ* hybridization
  (FISH),
  advantages, 219
  cell cycle considerations, 221, 222
  hereditary neuropathy with liability
    to pressure palsies and CMT1A
    diagnostics,
    blood culture,
      harvest, 227, 228, 230, 233, 234
      slide preparation, 228, 20,
        231, 234

denaturation and hybridization
    of blood slides, 229, 232, 235
detection, 229, 233
materials, 227–230
principles, 225, 226
probe,
    labeling, 228, 229, 231, 233–235
    precipitation, 229, 232, 235
    washing, 229, 232
Pelizaeus-Merzbacher disease
    diagnostics,
    blood culture,
        harvest, 227, 228, 230, 233, 234
        slide preparation, 228, 20,
            231, 234
    denaturation and hybridization
        of blood slides, 229, 232, 235
    detection, 229, 233
    materials, 227–230
    principles, 227
    probe,
        labeling, 228, 229, 231, 233–235
        precipitation, 229, 232, 235
        washing, 229, 232
principles, 219, 221
resolution, 219, 221
FMR1,
    AGG repeats, 31, 32
    CGG repeat expansion, 29, 30
    function, 345
    hypermethylation, 29
    monoclonal antibody production,
        bleeds, 346, 349, 350, 353
        enzyme-linked immunosorbent
            assay, 346, 349, 351, 353
        fusion, 346, 349–353
        immunization, 346, 349–351, 353
        immunoprecipitation, 349, 350,
            352, 353
        materials, 349, 350
        single-cell cloning, 347, 350,
            352, 353
    mutation types, 31, 38
    polymerase chain reaction testing
        of premutation alleles,

amplification and gel
    electrophoresis, 37
principles, 30, 37
solutions, 34
premutation allele phenotypes
    and transmission risks, 30
sequence homology between species,
    345, 346
Southern blot testing of premutation
    alleles,
    complexity of patterns, 30, 31, 37
    hybridization and probing, 35, 36
    principles, 30, 37
    probes, 34, 35, 37, 38
    solutions, 32–34
Fragile X syndrome, *see* FMR1
FSHD, *see* Facioscapulohumeral
    muscular dystrophy

G

Galactosemia,
    GALT mutations, 111, 112
    polymerase chain reaction–enzyme-
        linked immunosorbent assay
        of mutations,
        allele-specific hybridization, 115
        digoxigenin enzyme-linked
            immunosorbent assay and
            interpretation, 115–118
        DNA isolation, 114, 115
        materials, 112–114, 116
        multiplex polymerase chain
            reaction, 115
        principles, 112
        troubleshooting, 117, 118
    screening, 111
GALT, *see* Galactosemia
GeneScan, *see* Huntington's disease

H

Hamartin, *see* Tuberous sclerosis
    complex
HD, *see* Huntington's disease
Hereditary neuropathy with liability
    to pressure palsies (HNPP),
    clinical features, 223, 225

gene mutations, 225
heredity, 222
Southern blot analysis,
   blood culture,
      harvest, 227, 228, 230, 233, 234
      slide preparation, 228, 230, 231, 234
   denaturation and hybridization of blood slides, 229, 232, 235
   detection, 229, 233
   materials, 227–230
   principles, 225, 226
   probe,
      labeling, 228, 229, 231, 233–235
      precipitation, 229, 232, 235
   washing, 229, 232
Hexosaminidase A, see Tay-Sachs disease
High-performance liquid chromatography, see Denaturing high-performance liquid chromatography
HNPP, see Hereditary neuropathy with liability to pressure palsies
Huntington's disease (HD),
   CAG repeat expansion, 101
   clinical features, 101
   fluorescence polymerase chain reaction,
      amplification reactions, 103–105
      DNA extraction, 102–104
      gel electrophoresis of amplification products, 103–105
      GeneScan analysis, 104–107
      materials, 103
   genetic testing indications, 101
   huntington,
      polyglutamate repeats, 277
      transfection in cells,
         materials, 277, 278
         plasmid preparation, 279, 281
         rationale, 277
         selection, 280, 281
         stable transfection, 279, 281
         subcellular localization, 281, 283
         transient transfection, 279, 281
         Western blot analysis, 280–282
   mouse models, 277

I

Immunoprecipitation, FMR1, 349, 350, 352, 353
Imprinting,
   definition, 209
   diseases, see Angelman syndrome; Prader-Willi syndrome,
Ion channels
   gene polymorphism detection, see Single-strand conformational polymorphism
   heterologous expression systems,
      calcium phosphate transfection, 288, 291–293
      electrophysiology studies, 285, 286
      materials, 287, 288
      overview, 285, 286
      Xenopus oocytes,
         microinjection, 288, 290–292
         procurement, 287–289, 292
         transcription, in vitro, 287, 288, 290, 293

L

Laminin a-2, defect analysis, 369
Leber hereditary optic neuropathy (LHON),
   clinical features, 199, 200
   diagnostics,
      DNA isolation, 201, 203
      gel electrophoresis, 201, 203
      materials, 201, 202
      polymerase chain reaction, 201, 203
      restriction digestion, 201, 203
   genetic counseling, 204
   mitochondrial DNA mutations, 200–204
LGMD, see Limb-girdle muscular dystrophy
LHON, see Leber hereditary optic neuropathy

Limb-girdle muscular dystrophy (LGMD),
  calpain-3 defect analysis, 365, 366
  caveolin-3 defect analysis, 369
  clinical features, 355
  differential diagnosis using protein studies, 369, 370, 372
  dysferlin defect analysis, 366, 367
  dystrophin defect analysis, 364, 365, 370, 372
  gene mutations, 355, 356
  immunohistochemistry,
    antibodies, 356, 357, 372
    double immunofluorescence analysis, 359, 372
    primary antibody incubation, 358, 359, 372
    solutions and buffers, 357, 358, 372
    tissue samples, 357, 358, 372
  laminin a-2 defect analysis, 369
  sarcoglycanopathy protein defect analysis, 365
  telethonin defect analysis, 367–369
  types, 355
  Western blot analysis of proteins,
    analysis, 363
    antibodies, 356, 357, 372
    antibody staining, 363
    apparatus, 357
    blotting, 362
    detection, 363
    electrophoresis, 362
    gel casting, 361
    multiplex blot, 364
    pre-staining of blot, 362, 363
    principles, 359, 360
    sample preparation, 360, 361
    solutions and buffers, 357, 358, 372
    staining and destaining, 362

M

MALDI-MS, *see* Matrix-assisted laser desorption/ionization mass spectrometry

Mass spectrometry, *see* Matrix-assisted laser desorption/ionization mass spectrometry
Matrix-assisted laser desorption/ionization mass spectrometry (MALDI-MS),
  DNA analysis applications, 93, 99
  instrumentation, 93, 94, 98
  principles, 91
  time-of-flight mass spectrometry principles, 91, 92
  trinucleotide repeat analysis,
    advantages, 92, 93
    DNA purification, 93, 95
    genomic DNA isolation, 93, 95
    materials, 93, 94
    matrix, 92
    polymerase chain reaction, 93, 95
    sample preparation, 95, 96
    spectra acquisition, 96, 97
MECP2 mutation detection, *see* Denaturing high-performance liquid chromatography
Microsatellite expansion cloning, *see* DIRECT; Repeat analysis, pooled isolation, and detection cloning; Repeat expansion detection
Mitochondrial DNA,
  diseases, *see also* Leber hereditary optic neuropathy,
    deletions and rearrangements, 191, 192
    point mutations, 186, 187
  genetics overview, 199
  genome features, 185
  maternal heredity, 185, 199
  mitotic segregation, 186, 199
  mutation analysis,
    materials, 187, 190
    muscle DNA isolation, 187, 188, 195
    polymerase chain reaction-restriction fragment length polymorphism analysis of point mutations, 188, 189, 195
    sequencing and interpretation of point mutations, 189, 191
    Southern blot analysis of deletions, 193–195

Index

Muscular dystrophy, *see* Duchenne muscular dystrophy; Emery-Dreifuss muscular dystrophy; Facioscapulohumeral muscular dystrophy; Limb-girdle muscular dystrophy

N

Neurofibromatosis type 1 (NF1),
  gene, *see* Neurofibromin
  incidence, 315
Neurofibromin,
  mutations,
    detection techniques, 317
    types, 316, 317
  protein truncation test,
    gel electrophoresis, 322–325
    in vitro transcription and translation, 321, 323, 324
    materials, 321, 322
    overview, 320
    polymerase chain reaction, 321–324
    reverse transcription, 321–324
    RNA extraction, 321–324
  structure, 315, 316
  tumor suppression, 316
NF1, *see* Neurofibromatosis type 1

P

Parkin,
  gene structure, 13
  mutations in Parkinson's disease, 13
  semiquantitative multiplex polymerase chain reaction dosage assay,
    amplification reaction, 16, 19
    calculations, 22–24
    exponential phase verification for amplification, 17
    gel electrophoresis, 16, 17, 19
    interpretation, 17, 19, 20
    materials, 14–19
    primers, 15
    principles, 13, 14
PCR, *see* Polymerase chain reaction
Pelizaeus-Merzbacher disease (PMD),
  clinical presentation, 226, 263
  COS-7 cell transfection studies of PLP proteins,
    agarose gel electrophoresis, 265, 268, 271
    COS cell culture, 266, 269, 271, 272
    immunocytochemistry, 267, 270, 271, 273
    materials, 265–267, 270, 271
    plasmid purification, 265, 269, 271
    plasmid restriction digestion, 265, 267, 268, 271
    rationale, 263, 264
    site-directed mutagenesis, 266, 269, 271
    subcloning, 265, 268, 269, 271
    transfection, 266, 267, 269, 271, 272
    transformed bacteria culture, 265, 268, 271
  mouse models, 264
  PLP1
    gene dosage imbalance, 226, 227
    protein isoforms, 263
  Southern blot analysis,
    blood culture,
      harvest, 227, 228, 230, 233, 234
      slide preparation, 228, 230, 231, 234
    denaturation and hybridization of blood slides, 229, 232, 235
    detection, 229, 233
    materials, 227–230
    principles, 227
    probe,
      labeling, 228, 229, 231, 233–235
      precipitation, 229, 232, 235
    washing, 229, 232
  subtypes, 226
PFGE, *see* Pulsed field gel electrophoresis
PLP1 gene, *see* Pelizaeus-Merzbacher disease
PMD, *see* Pelizaeus-Merzbacher disease
PMP22, *see* Charcot-Marie-Tooth disease

Polyglutamate, *see also* Trinucleotide repeat expansion,
  aggregation assay,
    aggregate preparation, 298
    biotinylated peptide preparation, 297–299
    extension assay, 297–300
    materials, 297, 299
    overview, 295–297
    solubilization of peptides, 297, 298, 300
    standard curve preparation, 298
  disease overview, 73, 83, 295
  Drosophila disease models, *see Drosophila melanogaster*
  pathology, 295
  size range in disease, 83
  Western blot using antibody 1C2
    affinity of antibody, 84
    applications, 83, 84
    blotting, 86–88
    disease discovery, 84, 85
    electrophoresis, 85, 87
    gel casting, 85–88
    immunodetection, 86–88
    materials, 85, 86
    protein extraction, 85, 86, 88
Polymerase chain reaction (PCR),
  bisulfite restriction analysis, *see* Bisulfite restriction analysis
  denaturing gradient gel electrophoresis, *see* Denaturing gradient gel electrophoresis
  denaturing high-performance liquid chromatography, *see* Denaturing high-performance liquid chromatography
  fluorescence polymerase chain reaction, *see* Huntington's disease; Tay-Sachs disease
  fragile X syndrome diagnostics, *see* FMR1
  galactosemia diagnostics, *see* Galactosemia
  gene dosage determination, *see* Duchenne muscular dystrophy; Parkin; Spinal muscular atrophy
  Leber hereditary optic neuropathy diagnostics, 201, 203
  mitochondrial DNA point mutations, PCR-restriction fragment length polymorphism analysis, 188, 189, 195
  single-strand conformational polymorphism analysis, *see* Single-strand conformational polymorphism
  spinocerebellar ataxia assays, *see* Spinocerebellar ataxia
Prader-Willi syndrome (PWS),
  bisulfite restriction analysis,
    bisulfite treatment, 213, 214
    DNA extraction, 213
    electrophoresis and digestion patterns, 214, 215
    materials, 213
    polymerase chain reaction, 214
    principles, 210, 211
    troubleshooting, 215
  DNA methylation abnormalities, 210
  gene mutations, 209
  polymerase chain reaction-based methylation testing, 210
Prion disease,
  pathogenesis, 305
  prion protein characterization,
    biosafety considerations, 307
    enzyme-linked immunosorbent assay, 307
    immunohistochemistry,
      diagnosis of disease, 307
      materials, 308, 309
      staining, 311, 312
    purification,
      centrifugation, 310, 311
      homogenate preparation, 310
      materials, 308
      overview, 306, 307
    subtypes of PRP^{Sc}, 305, 306

Western blot,
  diagnosis of disease, 306
  electrophoresis, blotting,
    and detection, 309
  materials, 308, 311
  sample preparation, 309, 311
  types, 305
Protein truncation test (PTT),
  advantages and limitations, 317, 319, 320
  applications, 315
  neurofibromin analysis,
    gel electrophoresis, 322–325
    in vitro transcription and
      translation, 321, 323, 324
    materials, 321, 322
    overview, 320
    polymerase chain reaction, 321–324
    reverse transcription, 321–324
    RNA extraction, 321–324
  principles, 315, 316
  RNA-based analysis, 317
  tuberous sclerosis complex proteins,
    agarose gel electrophoresis, 335, 338, 342
    autoradiography, 335, 339
    B lymphocyte transformation, 334, 336, 341
    cloning, 335, 339, 340, 343
    complementary DNA synthesis, 334, 337, 342
    genomic DNA extraction, 334, 336, 337
    in vitro transcription and
      translation, 335, 338, 343
    materials, 334–336
    overview, 330, 332, 333
    polyacrylamide gel
      electrophoresis, 335, 339, 343
    polymerase chain reaction, 334, 335, 338, 342
    primers, 331
    RNA extraction, 334, 337, 341, 342
    sequencing, 336, 341, 343
    T lymphocyte culture, 334, 336, 341

PTT, *see* Protein truncation test
Pulsed field gel electrophoresis (PFGE),
  facioscapulohumeral muscular
    dystrophy diagnostics,
  apparatus, 159
  DNA isolation in agarose blocks, 157, 159, 160, 162
  electrophoresis, 157, 160, 162
  hybridization, 158, 161, 162
  lymphocyte,
    agarose block embedding, 157, 159
    isolation, 157, 159
  materials, 157–159
  principles, 155
  restriction digestion reactions, 157, 160, 162
  Southern blot analysis, 158, 161
PWS, *see* Prader-Willi syndrome

R

RAPID cloning, *see* Repeat analysis,
  pooled isolation, and detection
  cloning
RED, *see* Repeat expansion detection
Repeat analysis, pooled isolation, and
  detection (RAPID) cloning,
  cultures, 65, 71
  library amplification, 66
  ligation reaction, 65
  materials, 62, 63, 70
  phagemid excision from lambda Zap
    II vector, 66, 67
  phage,
    packaging, 65
    titering, 66
  postcloning enrichment, 67, 69, 70
  principles, 61, 62
  repeat expansion detection,
    optimized one-dimensional
      analysis,
      gel electrophoresis and blotting, 64
      hybridization of CAG
        nucleotides, 64
      reactions, 63, 64
    two-dimensional analysis, 64, 65, 70

Repeat expansion detection (RED),
    cloning strategy,
        cloning of enriched fragments, 57
        enrichment of long trinucleoide
            repeats, 56, 57, 59
        hybridization and plaque selection, 58
        materials, 54, 55
        overview, 52, 53
        phagemid DNA preparation, 58, 59
        plaque lifts, 57, 58
        plate library, 57
        sequencing, 59
        titering, 57
    electrophoresis and blotting, 55
    hybridization and autoradiography
        of blots, 55
    interpretation, 55, 56
    materials, 53, 54, 59
    principles, 51, 52, 75
    reaction mixture, 55
    repeat analysis pooled isolation and
        detection cloning, see Repeat
        analysis, pooled isolation, and
        detection cloning
Rett syndrome (RTT),
    clinical features, 120
    genotype-phenotype correlation,
        120, 121
    MECP2 mutation detection, see
        Denaturing high-performance
        liquid chromatography
RTT, see Rett syndrome

S

Sarcoglycanopathy, protein defect
    analysis, 365
SCA, see Spinocerebellar ataxia
Single-strand conformational
        polymorphism (SSCP),
    denaturing high-performance liquid
        chromatography, see Denaturing
        high-performance liquid
        chromatography
    Duchenne muscular dystrophy
        testing, 165

ion channel gene mutation detection,
    diseases, 143, 144
    gel electrophoresis, 145, 147, 148
    materials, 145, 146
    polymerase chain reaction,
        optimization, 145, 146–148
        primer design, 145, 146, 148
        radiolabeling of products,
            145, 147, 148
    sequencing,
        gel sequencing, 146, 148–150
        template preparation,
            146, 147–149
    principles, 143, 144
    sensitivity, 143
SMA, see Spinal muscular atrophy
Southern blot,
    Charcot-Marie-Tooth disease
        diagnostics,
        blotting, 179, 181
        densitometry, 182, 183
        digestion of DNA, 179, 180
        DNA isolation, 178–180, 184
        gel electrophoresis, 179, 181
        hybridization, 179, 181, 192
        materials, 178–180, 184
        probes, 179, 180, 184
    FMR1 allele testing,
        complexity of patterns, 30, 31, 37
        hybridization and probing, 35, 36
        principles, 30, 37
        probes, 34, 35, 37, 38
        solutions, 32–34
    gene dosage determination,
        Duchenne muscular dystrophy, 4, 9
        spinal muscular atrophy, 9
    mitochondrial DNA deletion analysis,
        193–195
    pulsed field gel electrophoresis
        analysis for
        facioscapulohumeral muscular
        dystrophy, 158, 161
    repeat expansion detection,
        see Repeat expansion detection
    spinocerebellar ataxia, see
        Spinocerebellar ataxia

Spinal muscular atrophy (SMA),
  clinical features, 5
  gene dosage determination with
      competitive polymerase chain
      reaction,
    amplification reaction, 9, 10
    controls, 10
    genomic DNA isolation, 6
    internal standards, 6, 10, 11
    materials, 6–8
    principles, 6
    Southern blot, 9
  survival motor neuron gene deletion,
      5, 6
Spinocerebellar ataxia (SCA),
  anticipation, 41
  CAG repeat expansion, 41
  polymerase chain reaction/Southern
      blot assay for extreme
      expansions in SCA2 and SCA7
    agarose gel electrophoresis,
        45, 46, 48
    blotting, 46, 48, 49
    DNA extraction, 43, 47, 48
    hybridization, 46, 49
    importance of family history and
        results from standard assays,
        43, 47
    interpretation, 47
    labeled 1kb ladder preparation,
        45, 48
    materials, 42, 43, 47
    overview, 41, 42
    polymerase chain reaction, 44, 45, 48
    probe preparation, 46, 48
    signal detection, 47
    washing of blots, 47
SSCP, see Single-strand conformational
    polymorphism

T

Tay-Sachs disease,
  allele-specific amplification
      fluorescent polymerase chain
      reaction assay,
    amplification reactions,
        135, 136, 139, 140
    gel electrophoresis on fluorescence
        detection apparatus, 136, 137
    genomic DNA isolation, 135, 138
    interpretation, 137, 139, 140
    materials, 133–135, 138, 139
    primers, 134
    principles, 132, 133
    quality control, 137
  biochemical screening, 131
  epidemiology, 131
  hexosaminidase A gene mutations,
      131, 132
Telethonin, defect analysis, 367–369
Trinucleotide repeat expansion, see also
      Huntington's disease; FMR1;
      Polyglutamate; Repeat expansion
      detection; Spinocerebellar ataxia,
  cloning of microsatellite expansions,
      see DIRECT; Repeat analysis
      pooled isolation and detection
      cloning; Repeat expansion
      detection
  disease overview, 73, 83
  genomic abundance, 51
  mass spectrometry assay, see Matrix-
      assisted laser desorption/
      ionization mass spectrometry
  motifs in human disease, 51
  number of repeats, 73
TSC, see Tuberous sclerosis complex
Tuberin, see Tuberous sclerosis
    complex
Tuberous sclerosis complex (TSC),
  clinical features, 329
  gene mutations and functions, 329, 330
  protein truncation test analysis of
      tuberin and hamartin,
    agarose gel electrophoresis,
        335, 338, 342
    autoradiography, 335, 339
    B lymphocyte transformation,
        334, 336, 341
    cloning, 335, 339, 340, 343
    complementary DNA synthesis,
        334, 337, 342

genomic DNA extraction, 334, 336, 337
in vitro transcription and translation, 335, 338, 343
materials, 334–336
overview, 330, 332, 333
polyacrylamide gel electrophoresis, 335, 339, 343
polymerase chain reaction, 334, 335, 338, 342
primers, 331
RNA extraction, 334, 337, 341, 342
sequencing, 336, 341, 343
T lymphocyte culture, 334, 336, 341

W

Western blot,
huntington expression in transfected cells, 280–282
limb-girdle muscular dystrophy proteins,
analysis, 363
antibodies, 356, 357, 372
antibody staining, 363
apparatus, 357
blotting, 362
detection, 363
electrophoresis, 362
gel casting, 361
multiplex blot, 364
pre-staining of blot, 362, 363
principles, 359, 360
sample preparation, 360, 361
solutions and buffers, 357, 358, 372
staining and destaining of gel, 362
polyglutamate detection using antibody 1C2
affinity of antibody, 84
applications, 83, 84
blotting, 86–88
disease discovery, 84, 85
electrophoresis, 85, 87
gel casting, 85–88
immunodetection, 86–88
materials, 85, 86
protein extraction, 85, 86, 88
prion protein,
diagnosis of disease, 306
electrophoresis, blotting, and detection, 309
materials, 308, 311
sample preparation, 309, 311

X

Xenopus oocyte, ion channel expression,
microinjection, 288, 290–292
oocyte procurement, 287–289, 292
transcription, in vitro, 287, 288, 290, 293